2nd CCPS China Conference on Process Safety 2014

Qingdao, China
28-29 August 2014

ISBN: 978-1-5108-1970-2

Printed from e-media with permission by:

Curran Associates, Inc.
57 Morehouse Lane
Red Hook, NY 12571

Printed by Curran Associates, Inc. (2016)

For permission requests, please contact AIChE
at the address below.

AIChE
120 Wall Street, FL 23
New York, NY 10005-4020

Phone: (800) 242-4363
Fax: (203) 775-5177

www.aiche.org

Additional copies of this publication are available from:

Curran Associates, Inc.
57 Morehouse Lane
Red Hook, NY 12571 USA
Phone: 845-758-0400
Fax: 845-758-2633
Email: curran@proceedings.com
Web: www.proceedings.com

TABLE OF CONTENTS

How to Effectively Manage PSM in a China Chemical Plant - Dupont Learning and Experience Sharing

Wu, M.,

How to Effectively Manage PSM in a China Chemical Plant ? DuPont Learning and Experience Sharing

By Mike Wu, Senior Consultant, DuPont Sustainable Solutions

As the world is making rapid progress in science and industry, PSM has proved to be the only holistic and effective management system to ensure the safe operation of modern and sophisticated manufacturing facility. PSM is developed as a result of many serious incidents around the world during the period of 1965 ? 1980. Thru the incident investigation and analysis, it was found that multiple means of protection are required to safe-guard the integrity of an operating facility.

As one the original developer and implementation of the PSM concept, DuPont is well experienced in all aspects of PSM utilization in industry. Such experience can be quite beneficial to Chinese chemical sites who are trying to standardize management process using proven scientific approaches.

According to DuPont method, PSM shall be one integrated management system. Only toughing one or several pain points cannot effectively solve systematic issues. There are several stages for effective implementation of PSM

a. Build and nurture Safety Culture

b. Provide leadership?s commitment and felt-leadership

c. Establish a comprehensive Process Safety and Risk Management Plan and System

d. Employee capability development

e. Achieve operation excellence thru operating discipline

Detailed description of each of the five stages will be covered in the full article. There will be discussion of current situation and condition in China. Suggestions and recommendations will be provided for each stage.

Pipeline Transportation of Hazardous Materials – an Updated Quantitative Risk Assessment Methodology

Colin Hickey
Adeyemi Oke
DNV GL
200 Great Dover Street, London, SE1 4YB
Colin.Hickey@dnvgl.com

Keywords: QRA; pipeline; transportation risk analysis and assessment; Safeti; Phast

Abstract

Pipeline transportation of hazardous materials is increasing globally. Major accidents involving pipelines continue to occur. A wide range of guidance is available from various sources for robust quantitative risk assessment (QRA) of mobile and pipeline transportation of hazardous materials. Derived from the aforementioned sources, this paper describes new hazardous pipeline risk assessment capabilities implemented in the Safeti QRA software package. The methodology includes automatic loss of containment case generation based on pipeline construction, topography and operating parameters. The geometrical nature of pipelines and their associated risk footprint relating to people, surrounding assets and the environment are often long and relatively narrow. Creation of risk contours around such geometries requires use of high aspect ratio grids. The calculation algorithm required for high aspect ratio hazard and risk zones such as those presented by pipelines is a challenge to traditional risk contour based risk assessment techniques for process plants. The new methodology introduces a method for adaptive grid solution for long pipelines for optimal computation and result accuracy. The new adopted risk analysis method also introduces a time-varying fire simulation to account for the time-varying flow phenomenon typically experienced with pipeline breaches. In summary this paper demonstrates how these QRA methodology enhancements provide analysts with the tools necessary to perform rigorous assessment and management of pipeline risks in a consistent and efficient manner using validated models.

1. Introduction

Safeti was developed in the 1980s for quantitative risk analysis of major accident hazards associated with process facilities. Safeti has since been continuously developed and has been adopted by a global user base in a range of applications through the process industry including oil and gas, pharmaceuticals, insurance, chemicals and petrochemicals. Positive change is evident in the process industry – process safety is becoming common parlance amongst process engineers, while hazard and risk analysis are becoming part of the standard engineering skill-set. Hazard and Risk analysis has moved from a legislation driven, land use planning requirement to a standard practice in risk based design and risk based operation. This change is coupled with an increase in transportation of hazardous materials and declining tolerance of risk to the public, property and the environment. It is evident that a standard, best practice method is required for consistent analysis of transportation risks.

Accidents highlighting the need for thorough analysis, communication and management of pipeline risks include:

- Qingdao, China, 2013: Oil pipeline leak and explosion, 62 fatalities, 136 hospitalized. (Wikipedia - 2013 Qingdao pipeline explosion, 2014)
- Dalian, China, 2010: Oil release to sea from port for 90km, covering 946km^2. Fatalities and injuries occurred, number not reported. Extent of environmental damage also not reported. (Wikipedia - 2010 Xingang Port oil spill, 2013)
- San Bruno, California, natural gas pipeline explosion, 8 fatalities. (Wikipedia - 2010 San Bruno pipeline explosion, 2014)
- Ghislenghien, Belgium 2004: 24 fatalities, 120+ injuries. (French Ministry of Sustainable Development, 2009)

The scope of transportation risk analysis (TRA) can move through three dimensions, as advised by CPPS (CCPS, 1995) and shown in

Figure 1. You can choose to cover a limited, generalized or detailed set of failure cases. You can perform qualitative, semi-quantitative or quantitative analysis. Finally you can perform consequence, frequency or risk analysis. Primarily it is not feasible to be at the deepest extent of this cube, and a pragmatic process is required whereby transport hazards across the enterprise are screened and increased depth and focus given to areas identified as being of significant hazard.

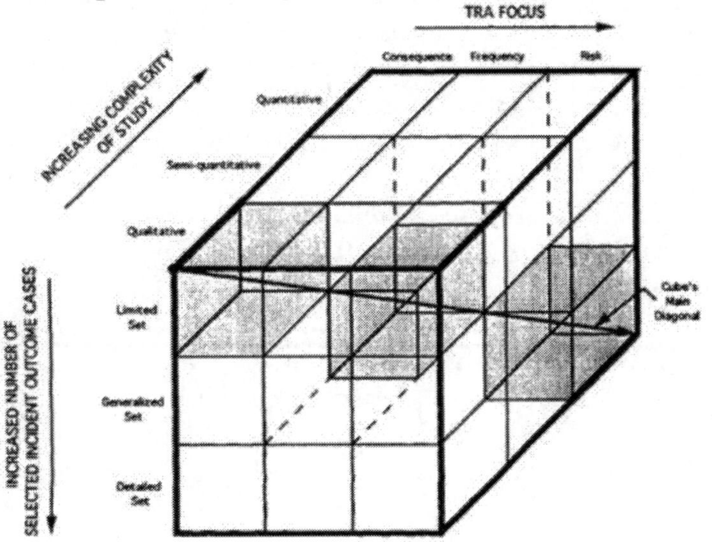

Figure 1. CCPS TRA study cube

This paper describes enhancements to the Safeti software, used to accurately model consequences arising from loss of containment events from pipelines and their associated risks. The paper then describes how Safeti supports the CCPS TRA methodology. This new approach will be available in Safeti 7.2.

2. Modelling enhancements

Whether performing hazard analysis of a single loss of containment scenario or a holistic risk analysis across an entire enterprise involving several sites interconnected by different transport modes, accurate modelling of the loss of containment scenario is fundamental. For this reason, Safeti has been reviewed and areas of improvement identified to more accurately simulate hazardous outcomes resulting from breaches from pipelines.

The sequence of events that occurs when there is a release from a pipeline is: discharge, pool formation (if liquid present and rainout occurs), and vapour cloud dispersion. At each of these stages people may be exposed to toxicity if the material is toxic. If the material can act as an asphyxiant (such as CO_2, Nitrogen) then hazard from displacement of oxygen will occur. If the material is flammable, upon breach immediate ignition may occur resulting in a fireball which transitions to a jet fire, in a two period phenomenon, as observed by Lutostansky (Lutostansky, Modeling of Underground Hydrogen Pipelines, 2012). If the released flammable material experiences delayed ignition a flash fire or explosion may occur. If a liquid pool has formed, a pool fire may occur. Each of these current capabilities in Safeti have been assessed for their suitability to pipeline modelling and the following improvements have been identified and implemented

2.1 Discharge source term

The work by Lutostansky (Lutostansky, Modeling of Underground Hydrogen Pipelines, 2012) presented validation of Phast and Safeti's models against gas (Hydrogen) pipeline rupture experiments. The discharge rate from Phast and Safeti's pipeline release model was found to give good results. For buried pipelines a momentum modification factor of 0.25 should be used and for non-buried pipelines, no modification factor need be applied.

Phast and Safeti have pipeline release models for gas (Webber & Witlox, Gaspipe Theory, 2010) and two phase (Webber & Witlox, Pipebreak Theory, 2010) containment. No model is currently available for sub-cooled liquids.

Best practice guidance for definition of pipeline release source terms have been taken from OGP report 434-7, Consequence Modelling (OGP, 2010 - 434-7). This guidance recommends size category and orientation of releases.

Recommendations from OGP 434-7 have been combined with experimental findings and valve closure logic. From this it is possible to establish a set of standard failures to be used at a single location on a pipeline, as shown in Table 1. These would be accompanied by cases in which upstream and downstream valves close, and where only the upstream valve succeeds in closing and finally where only the downstream valve succeeds in closing. Releases are governed by:

- Hole size
- Elevation (below or above ground)
- Release direction
- Valve behaviour

Safeti contains inputs for controlling each of the above factors.

Hole Size Considerations

Hole sizes are an input to Safeti as they are pipeline specific. Information from organisations such as EGIG (EGIG, 2011) can help to define typical hole sizes.

The definition provided by EGIG on the size of the breaches for their data is:

- Pinhole crack: diameter of defect equal to or less than 2 cm
- Hole: diameter of defect more than 2 cm and equal to or less than the diameter of the pipe
- Rupture: diameter of defect more than the pipe diameter

In Safeti un-isolated small and medium leaks are represented by steady state orifice model calculations and not the time-varying long pipeline (gas-pipe and two-phase pipe-break) models. Isolated small, isolated medium and all full-bore rupture releases are calculated using the time-varying pipeline release models. This application of steady state and time-varying approaches balances the need for accurate modelling of the disturbance zone within the pipeline (for which the un-isolated small and medium cases have less need) and the computation time of time-varying discharge calculations.

Elevation Considerations

When pipelines are buried, upon release there is interaction with the over-burden. When the over-burden is removed a crater remains. Safeti takes account of interaction with the over-burden and crater by offering a momentum adjustment factor. This reduces the velocity of the release whilst maintaining the mass flowrate. Lutostansky found that 0.25 aligns with experimental findings for a hydrogen pipeline for discharge rate and consequent jet fire behavior (Lutostansky, Modeling of Underground Hydrogen Pipelines, 2012). A release angle of +45° is also applied to horizontal releases from buried pipeline to account for interaction with the crater.

Release Direction Considerations

As discussed by OGP 434-7, releases can be vertical, horizontal and downward. Some special considerations can be applied to some circumstances such as small releases from buried pipelines, which are unlikely to have sufficient energy to displace the over-burden. They can therefore be modelled as vertical with reduced velocity to represent permeation through the substrate. An example of an approach is given in Table 1 for the buried small horizontal case.

Valve Behaviour

Valves will be present upstream and downstream of breach locations. Emergency shut-down (ESD), non-return, excess flow and manual valves may be present on the pipeline. Safeti can take account of ESD, non-return and excess flow valves. The inputs to control valve behaviour in Safeti are the valve location, valve type and for closure valves the closure time. When a breach occurs the control system is designed to isolate the breached section to minimise outflow of the hazardous material. A change of operating conditions would be detected and this could instigate automatic closure of ESD valves. The distance from the breach to detectors and isolation valves will impact the response time. The closure time is not provided automatically by Safeti as this is scenario specific. A useful reference for closure times can be found in the UK HSE contract research report 206/1999 (UK HSE, 1999). For excess flow valves the input of mass flow set point is required by Safeti. Reverse flow from pipeline downstream of the breach will be limited by non-return valves in Safeti.

The variation on accidental release source terms is governed by the event tree shown in Figure 2. In this way Safeti can provide support for decisions about effectiveness on investment and location of pipeline control systems.

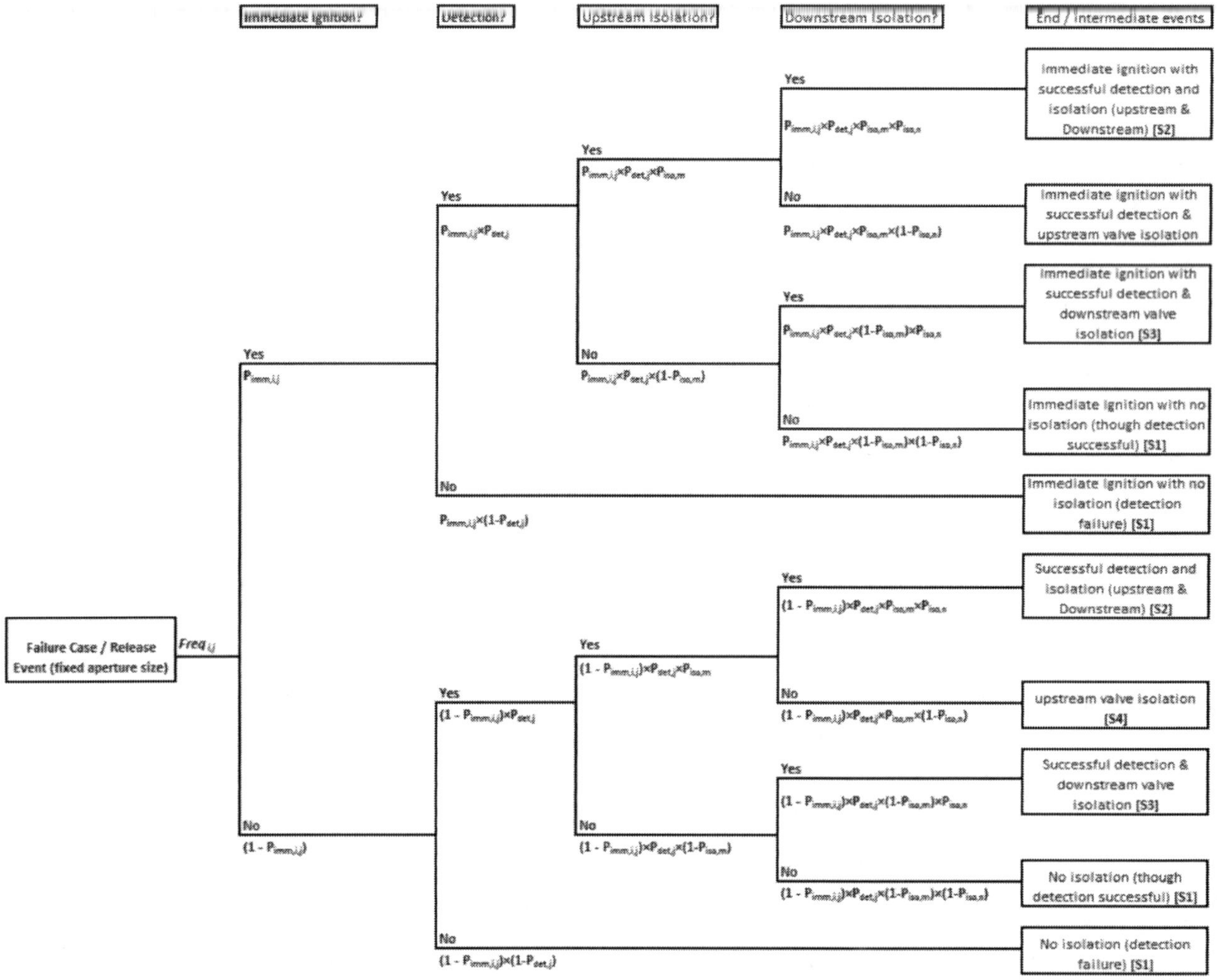

Figure 2. Event tree of valve success/failure for immediate and delayed ignition

Table 1. Suggested failure cases at pipeline breach location for above/below ground

	Size	Horizontal	Vertical	Downward
Buried	Small	1.Calculate as normal 2. Remodel to match flowrate by adjusting hole size for: 0.1barg for <10bar and 1bar for >10bar 3.Vertical	Same as small, buried, horizontal	Same as small, buried, horizontal
	Medium	+45° release direction, 0.25 momentum	Vertical	Vertical @ 5m/s
	Rupture	+45° release direction, 0.25 momentum	Vertical	Vertical @ 5m/s
Above ground	Small	Horizontal	Vertical	Downward
	Medium	Horizontal	Vertical	Downward
	Rupture	Horizontal	Vertical	Downward

2.2 Failure Frequency

Using information from the European Gas Pipeline Incident Group (EGIG, 2011) it is possible to derive the failure frequency of pipeline breaches per length of pipeline. As shown in Table 2. Due to the useful format in which the EGIG data is reported, with information on cause of breach and size of breach (column and row in Table 2 respectively) it is also possible to determine the relative probability and therefore the frequency of the various breach sizes. The inputs to Safeti are failure rate per distance. Sections of pipeline can be assigned specific failure rates per distance. In this way it is possible to account for local factors such as burial depth, pipe wall thickness. Breach categories in Safeti are assigned relative probabilities. Referring to data shown in Table 2 one might assign small, medium and rupture breaches with relative probabilities of 49%, 36.5% and 14.5% respectively.

Table 2. Frequency of breach type, EGIG 2011

	External interference	Corrosion	Construction Defect or Material Failure	Hot tap made by error	Ground movement	Total Frequency	Relative probability
	(/1000km.yr)						
Pinhole crack	0.045	0.054	0.041	0.012	0.006	0.158	0.490
Hole	0.09	0.002	0.013	0.005	0.008	0.118	0.365
Rupture	0.033	0	0.004	0	0.01	0.047	0.145
						Total = 0.323	

Using the information from EGIG and the class of "secondary failure frequencies" (EGIG, 2011 p.25) it is possible to investigate the effect of variation of pipeline design on risk. For example, using the relationship between cover depth and failure frequency it is possible to perform sensitivity analysis in a QRA to see the risk reduction achieved by increasing the burial depth of a pipeline. The same process could be applied to see the effect of wall thickness on risk. In this way, Safeti can take the inputs in the best-practice format discussed above, but does not offer default values, as these may be misleading for any given pipeline configuration.

2.3 Automatic failure case generation

With the above methods defining the source term, how it behaves after release and its likelihood, it is possible to define rules for automatic creation of failure cases. Continuously variable events are too computationally onerous to perform. Therefore the pipeline is broken into segments, and mid-point representative release cases are used per segment. The choice to use representative release cases is to gain efficiency at an acceptable cost of accuracy.

The pipeline route should be divided into sections along which the pipeline conditions are "approximately constant". To achieve this, the following rules are applied:

- Sections upstream and downstream of valves that could be closed (including Excess Flow Valves but excluding manual valves) following a release should be modelled separately.
- The absolute pressure should change by no more than 20% over the length of a section.
- Section lengths should be at least 500 m and no more than 20% of the total pipeline length.
- Releases from the section should be modelled using conditions (pressure, temperature) at the mid-point of each section.

Using the above rules, it is possible for a user to draw a pipeline route, define process conditions, insert valves and then click to automatically create the array of failure cases.

2.4 Results to Grid

Due to the traditional risk analyses' location specific individual risk and societal impact calculations, it has been necessary to perform the calculations of risk to an area using a cellular grid method. This requirement comes at some cost to computation and requires some new logic for optimal calculation of transport risks. The two issues related to optimization of transport risk on grids are that of failure event spacing and efficient grid use.

2.4.1 Failure event spacing

In the existing mobile transportation risk model within Safeti 6.7 the user is required to define event spacing along the route. If the spacing between events is larger than the release's hazard zones the risk summation will contain gaps between the events. If the spacing between events is much smaller than the hazard zones then extraneous calculations are being performed by Safeti's risk summation model. While the latter problem only wastes time, the former problem can lead to incorrect results. For this reason the pipeline route model in Safeti will determine the optimum failure spacing automatically. The algorithm will look at a distance to lethality from flammable and toxic hazards. Different failure event spacing is applied to different size breaches modelled along the pipeline in a manner which optimizes results resolution and computation time.

2.4.2 Grid area optimization

Safeti, by default, uses a 200 x 200 cell grid for calculation of risks. This approach is ideal for process plants. Pipelines and mobile transportation routes will, by their nature, have high length to width ratio. This results in sub-optimal usage of grid cells when risk summation calculations are being performed. A (multi-)linear transport risk grid method has been added to Safeti. This approach will support adaptive (with finer grid resolution close to the pipeline), multi-layered, non-rectangular (radial) irregular grids. This results in higher resolution results in the near field around the pipeline and follows the route of the pipeline over its length. It is envisaged that this enhancement will add further broad benefits in the future in Safeti for other cases (including non-transport) such as flammable and toxic materials where near-field and far-field effects must be accounted for in the same release and would cause problems for the non-adaptive grid based method.

2.5 Fireballs and Jet Fires

In earlier versions of Phast and Safeti, the impulsive fireball / jet fire generated following immediate ignition of a pipeline release may be represented by a short duration fireball and a time-averaged jet fire. The jet fire contribution to this approach has a limitation in that the jet fire is represented by the time-averaged mass flowrate over 20 seconds (or the release duration, whichever is shorter). This under-predicts the maximum hazard zone size and may under-predict the radiation dose at various locations. The approach has therefore been improved by introduction of a time-varying fire model.

Experimental/field evidence (Cracknell & Carsley, 1997), (Hirst, 1986) suggests that upon immediate ignition of a flammable release from a pipeline a buoyant fireball is formed shortly after ignition followed by a quasi-steady state jet fire. The new time-varying fire model (TVFM) developed for Phast and Safeti makes use of the existing efficient, flame, radiation and numerical integration calculations. Radiation effects are calculated for multiple mass flowrates during the time-varying release; this leads to different flames and associated radiation intensities, while radiation dose, probit and lethality at various observer locations over a given duration of interest are determined using established numerical integration methods. Figure 3 shows a transect of radiation intensity versus upwind and downwind distance from the release. The transect is shown for several times of interest as the flame develops. The time-dependent dose at various locations of interest along the transect are determined via numerical integration as illustrated in Figure 4. Figure 4 also shows the predicted radiation dose along the transect for the scenario illustrated in Figure 3 using the "20 second" time-averaged steady-state fire analysis (with and without an initial fireball) implemented in earlier versions of Safeti as compared against the time-varying fire and radiation analysis (with an initial fireball) introduced within Safeti 7.2. Note how in Figure 3 that the radiation intensity starts at a peak level and is centered around 0m downwind distance. This is due to the flame being at maximum mass flowrate with high momentum modelled starting as a fireball centered at the point of release. As time progresses and the mass flowrate to the fire decreases, the flame decays and the resulting radiation field decreases in size for a given level of radiation intensity. Figure 3 also shows that at later times as the flame decays the radiation versus time profiles move downwind, indicative of the flame losing vertical momentum and therefore being tilted by the wind. This leads to a very dynamic picture for the received dose at locations upwind and downwind from the release point. In all the results are significantly more accurate for flammable pipeline releases.

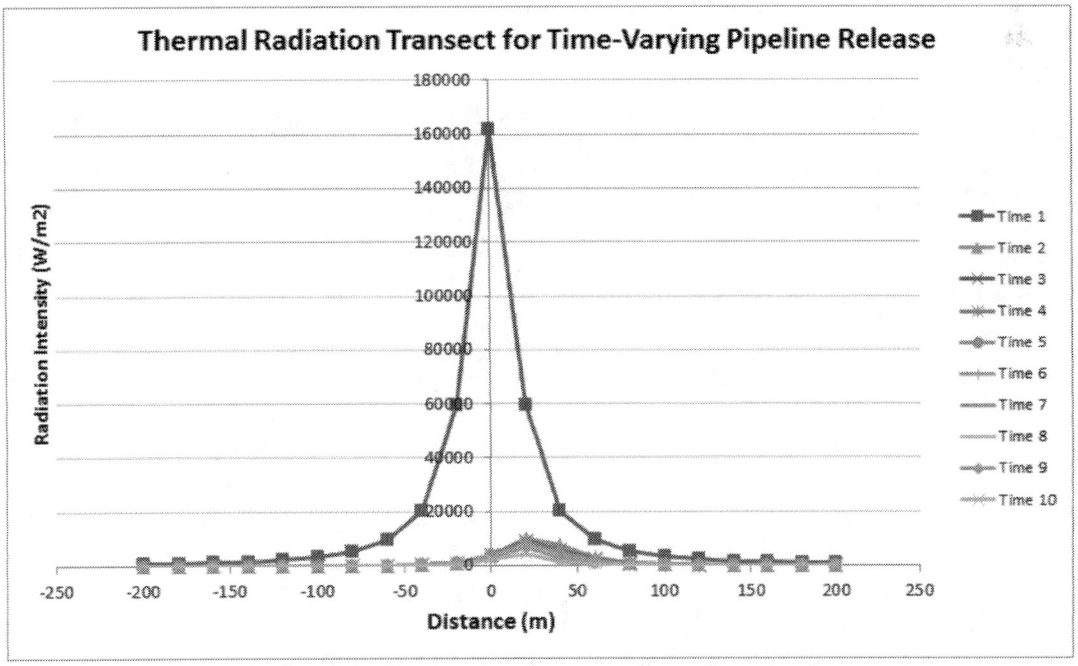

Figure 3. Example time varying fire results

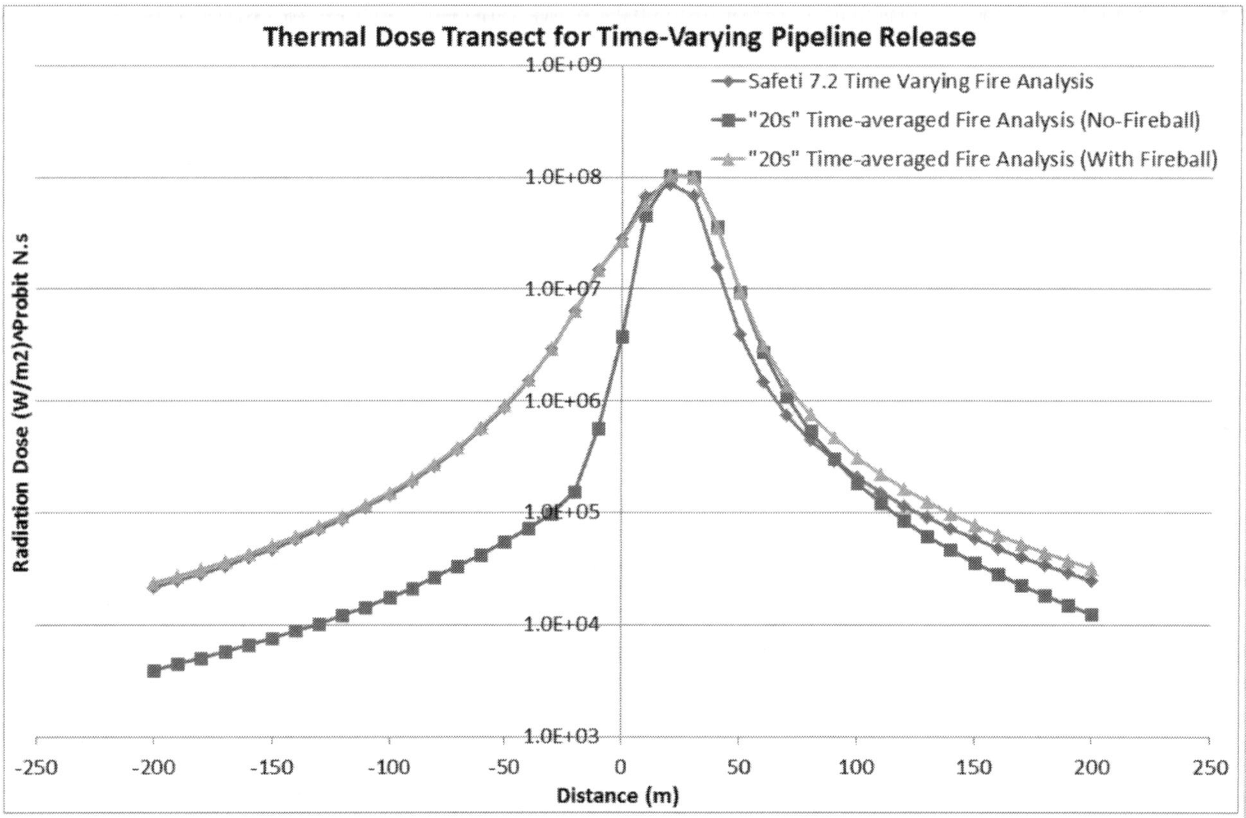

Figure 4. Example thermal dose transect for time-varying pipeline release scenario illustrated in Figure 3: comparison of Safeti 7.2 time varying fire and radiation analysis results against the "20 second" time-averaged fire analysis (with and without an initial fireball) as implemented in earlier versions of Safeti.

3. Risk Assessment Methodology

The purpose of the pipeline hazard and risk assessment methodology is to understand the nature of a hazard from a phenomenological perspective and to include the event at the correct level of priority amongst the many other risks that are managed by an enterprise. As discussed above, detailed, risk analysis of all transportation risks is unlikely to be possible in a single project. Note that the underlined terms here represent the axes of the CCPS TRA study cube shown in Figure 1. CCPS' transport safety, security and risk management guidance offers a systematic approach to managing enterprise transport risks (CCPS, 2008). The process begins with mapping out all plants, transport routes, their hazardous materials and modes of transport. Figure 5 illustrates how transportation of a range of feed and product materials can be organised for qualitative risk screening on a large geographical and operational scale. This process recommends that one considers the conditions of roads, the performance of 3rd party service providers and other practical factors which may influence the likelihood and consequence of a hazardous event. This screening process will highlight segments which require more detailed semi-quantitative and quantitative analysis. For segments of this network which are supported by pipeline one can move onto the next step with the new Safeti pipeline risk model.

Segmented Routes (Chemicals/Modes of Concern)

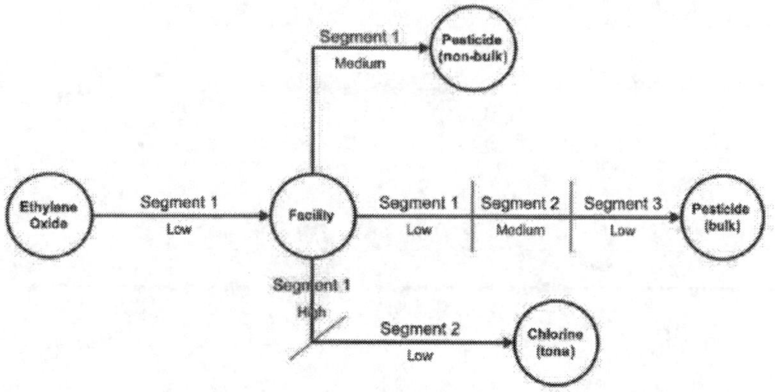

Figure 5. CCPS network of transport operations with chemicals and modes of concern

The large geographical area covered by pipelines can lead to large amounts of input information being required. It is impractical to provide detailed information on population and delayed ignition sources for the entire pipeline length. This infers that we need to perform a relatively coarse pipeline QRA for the overall pipeline route as a first step. One can do this by using coarse blocks of population and delayed ignition along the route. The variable nature of the pipeline process and release cases (governed by process pressure, elevation, proximity to valves and hole sizes) result in various failure cases interacting with the population and ignition. The results from a coarse analysis can be considered to be semi-quantitative. However, the failure cases which have been automatically generated are full, detailed simulations of the hazard phenomenon – this is a large step forward for the analyst, freeing up their time to where it is best spent: reducing and managing the identified risks. It also helps the analyst move on to the next step, focusing the analysis on areas of concern which have been identified.

Detailed information is required around the pipeline when there are factors which may increase the risk. These may be:

- Significantly vulnerable population
- High population density
- Increased likelihood of ignition
- Large hazard zones
- Areas with potential for escalation (e.g. near process plant)

Within the broad study of the entire pipeline, one can zoom in on areas of concern and add information on population, ignition and any other local factors. This will then provide more detailed results around these locations. One can then identify risk prevention and mitigation options. For example we might choose re-routing, or culverting at vulnerable locations such as road or rail crossings or human activity. One may choose to increase wall thickness, introduce more proximate detection and emergency shutdown measures or work with construction and access exclusion zones around the pipeline corridor. One can very efficiently make changes to the Safeti study reflecting the suggested prevention measures and perform repeated analyses to see what benefits these return in terms of risk reduction.

In summary, to conduct the process described by the CCPS TRA study cube as shown in Figure 1 an approach is to first perform screening of a large subject set, and to then drill down with more detail on the areas of concern which have been revealed. The process supports efficient analysis of broad subject matter with the potential for more detailed analysis by adding details to the same study. Such a study may appear as shown in Figure 6, where an entire route has been analysed and then a local, vulnerable location has been investigated in more detail.

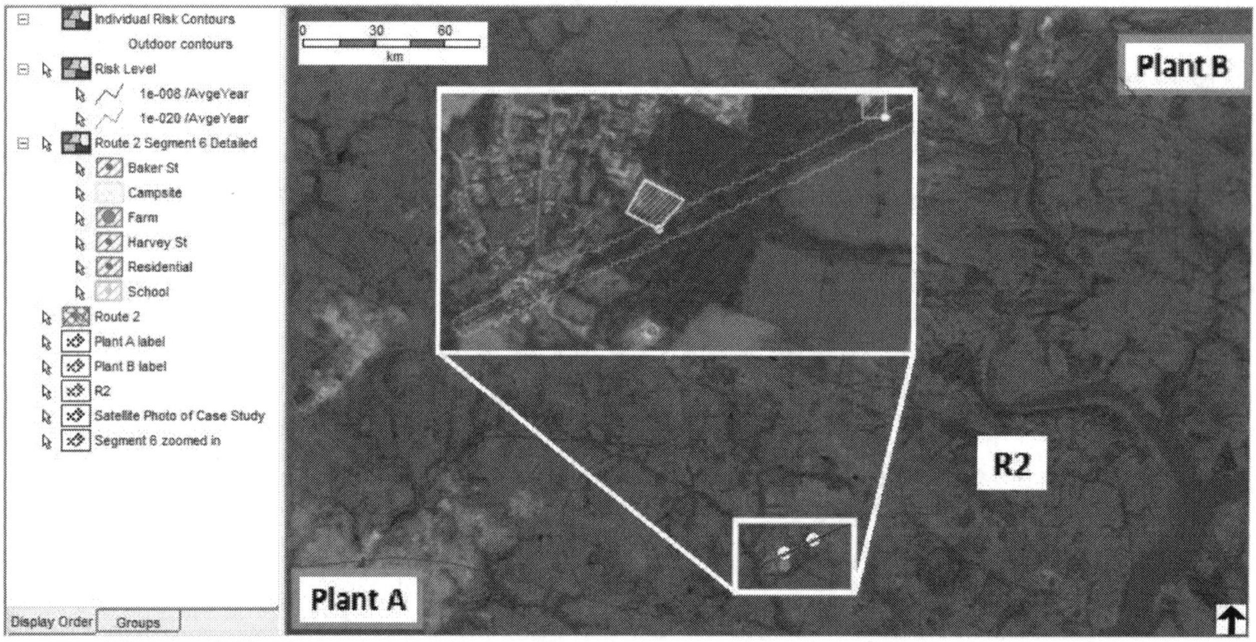

Figure 6. Entire route with detailed segment shown in zoomed area, in Safeti

4. Conclusion

Safeti has long had a model for transport risk. This model was suitable for mobile transport units but required manual work to apply to pipelines. Areas have been identified to improve Safeti for risk analysis for transportation of hazardous materials through pipelines. Improvements include automatic creation of failure cases for small, medium and large releases above and below ground. The failure cases are generated for the entire pipeline length taking account of pressure change within the pipeline and location of shut-down valves. Safety systems are taken account of by inclusion of a new configurable event tree that allows improved modelling of safety-system performance by accounting for the impact of different isolation actions, detection times and probabilities and shutdown probabilities for upstream and downstream isolation. Time-varying mass flow and flame behavior are simulated by a new model giving rise to time-varying radiation, dose and lethality. The new methodology includes improved support for event frequency according to the data format available from EGIG and the OGP methodologies.

In all this takes more accurate account of accidental breaches and presents a more realistic assessment of major accident hazard risks stemming from transportation of hazardous materials through pipelines and the improvements introduced to Safeti help better support the CCPS TRA methodology.

The improvements described in this paper allow analysts to more accurately and efficiently identify key risk drivers and undertake detailed benefit analysis of potential remedial measures for risk based design / management and/or decision making.

5. References

[1] CCPS. (1995). Guidelines for Chemical Transportation Risk Analysis. New York: AIChE.
CCPS. (2008). *Guidelines for Chemical Transportation Safety, Security and Risk Management.* Hoboken: Wiley.
[2] Cracknell, R. F., & Carsley, A. J. (1997). Cloud fires – A methodology for hazard consequence modelling. *ICHEME Symposium Series, No. 141*, (pp. 139 -150).
[3] EGIG. (2011). *8th Report of the European Gas Pipeline Incident Data Group.* Groningen: European Gas Pipeline Incident Data Group.
[4] French Ministry of Sustainable Development. (2009, 9). *Rupture and ignition of a gas pipeline July 30, 2004, Ghislenghien, Belgium, No. 27681.* Retrieved 5 1, 2013, from www.aria.developpement-durable.gouv.fr: http://www.aria.developpement-durable.gouv.fr/ressources/fd_27681_ghislengheinv_jfm_anglais.pdf
[5] Hirst, W. J. (1986). Combustion of Large Scale Releases of Pressurized Liquid Propane. *Proceedings 3rd Symposium on Heavy Gas Risk Assessment.* Dordrecht, Netherlands: Reidel.
[6] Lutostansky, E. (2012). Modeling of Underground Hydrogen Pipelines. *Global Congress on Process Safety.* AIChE CCPS.
[7] Lutostansky, E., Shork, J., Ludwig, K., Creitz, L., & Jung, S. (2013). Release Scenario Assumptions for Modeling Risk From Underground Gaseous Pipelines. *Global Congress on Process Safety.* AIChE CCPS.
[8] OGP. (2010 - 434-7). *Consequence Modelling Report - 434-7.* London: : International Association of Oil & Gas Producers.
[9] UK HSE. (1999). *Assessing the risk from gasoline pipelines in the United Kingdom based on a review of historical experience.* Norwich: UK HSE.
[10] UK HSE. (1999). *Risks from gasoline pipelines in the United Kingdom crr 206/1999.* Norwich: Crown.
[11] Webber, D., & Witlox, H. (2010). *Gaspipe Theory.* London: DNV.
[12] Webber, D., & Witlox, H. (2010). *Pipebreak Theory.* London: DNV.
[13] Wikipedia - 2010 San Bruno pipeline explosion. (2014, 07 30). *2010 San Bruno pipeline explosion.* Retrieved from http://en.wikipedia.org/: http://en.wikipedia.org/wiki/2010_San_Bruno_pipeline_explosion
[14] Wikipedia - 2010 Xingang Port oil spill. (2013, 5 1). *2010 Xingang Port oil spill.* Retrieved 5 1, 2013, from Wikipedia/Xingang_port_oil_spill: http://en.wikipedia.org/wiki/Xingang_Port_oil_spill
[15] Wikipedia - 2013 Qingdao pipeline explosion. (2014, 07 23). *2013 Qingdao oil pipeline explosion.* Retrieved from http://en.wikipedia.org/: http://en.wikipedia.org/wiki/2013_Qingdao_oil_pipeline_explosion

Natural Gas Pipeline Integrity Management and Accident Preventive Measures Research

Zhou, Y., Nanjing University of Technology
Wang, S., Nanjing University of Technology

Natural gas pipeline integrity management and accident preventive measures research

Abstract The world's gas pipeline safety status is not optimistic currently,miserable accidents brought huge economic losses.Pipeline service are divided into three stages.The Failure rate is higher in the first stage,it is due to design flaw and the quality problem;the failure rate is low and stable in the second stage,it is always due to wrong operation and third party damage;in the third stage ,the failure rate increasing with time,because of the corrosion, aging,fatigue and etc. In the daily safety management,the gas pipeline can be evaluated on the basis of this feature. Through comprehensive analysis, optimize the original pipeline integrity management model and draw a flow chart.To some extent,strengthen the evaluation and analysis level.Integrity management is a process that through monitoring, testing and inspecting to achieve equipment integrity information and evaluate the risk of equipment and applicability, enterprises can timely take corresponding measures to ensure the integrity of the equipment on the physical and functional.The modified system mainly includes:data collection(daily records ;raw data), data management,risk assessment,integrity evaluation(ECDA;NDTtechnology and other detection technology), repair and maintenance(repair plan;regular check),performance evaluation.Each procedure includes several contents.The natural gas pipeline running environments are complex, generally can be divided into four forms:the design and construction defects, corrosion, third party damage and natural factors.Through consult the literature ,analyze the four primary factors roundly.In the end,put forward five suggestions. The five suggestions mainly includes technology and management.Suggestions are roughly as follows.In the aspect of technology,use analysis software(ANSYS) to analysis the residual strength of the pipeline,compared with the theoretical analysis of the results closer to reality. The welding technology and aseptic technique.Embedded devices based on the technology of IOT.In the aspect of management, carry out the relevant laws and regulations, formulate emergency plans and so on.

Key Words:gas pipeline;third party damage;integrity management;IOT

An Oil and Gas Pipeline Accident Model Based on Failure of Generalized Management

Jiang, S., College of Urban Construction and Safety Engineering, Nanjing Technology University
Zhang, L., College of Urban Construction and Safety Engineering, Nanjing Technology University
Tao, G.,
Wang, Z.,

An Oil and Gas Pipeline Accident Model Based on Failure of

Generalized Management

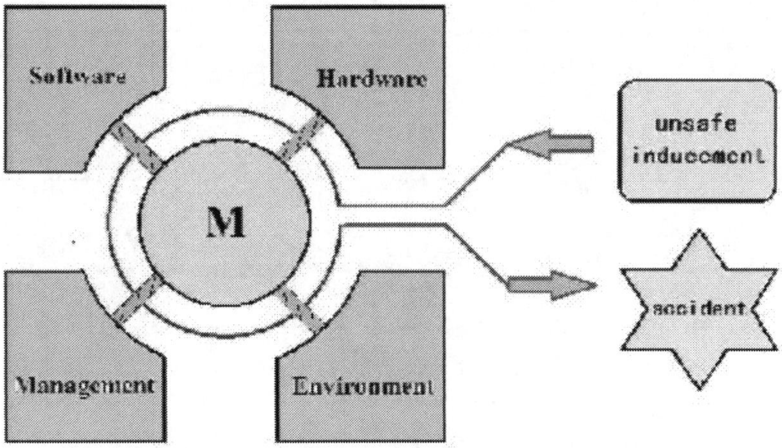

Abstract: With the arrival of China's oil and gas pipeline network era, the security of oil and gas pipelines confront great challenges. Scientific pipeline safety management methods and concepts are being urgently needed. Oil and gas pipeline accidents are usually caused by a series of generalized management failure factors. In this paper all the oil and gas pipeline accidents can be explained by SHEM model which composed of hardware, software, environment and management. Generalized management is the core of model which formed M-H, M-S, M-M and M-E interfaces with other parts. When closely cooperated, these interfaces can build a solid protective barrier, or they will be penetrated by unsafe inducement because of management failure. Then oil and gas pipeline accidents will occur. The SHEM model is shown in Fig.1 below.

Fig.1 SHEM Model

The SHEM model of oil and gas pipeline accidents contains four interfaces, namely management-hardware, management-software, management-management and management-environment. The connotation of each interface is interpreted below. Management-hardware interface means the management of physical factors interface, such as the pipeline itself, ancillary facilities and equipments. Management-Software interface means the management of the main non-physical interface, such as legal regulations and management files. Management-Management interface means the interaction between each management subject, such as collaborative management between different departments. Management-Environmental interface means the management of internal and external environment, such as temperature and pressure in pipelines or social environment along the pipeline. In order to prevent management failure evolving into accidents, the control system of management failure is established which contains four control interfaces and twenty control points. The detailed introduction is shown in Tab.1 below.

1

Tab.1 Control System of Management Failure

Interface Control Point

H1 pipe (design flaws, security sign)

M-H M-S M-M

M-E

Internal environment

External environment

H2 physical equipments in maintenance (excavator hammerâ?¦) H3 ancillary facilities

S1 laws, regulations and standards to protect pipelines

S2 rules and regulations prescribed by pipeline company

S3 emergency plans and procedures

S4 equipment archives and work handbook

S5 risk assessment management

S6 inspection technology

S7 anti-corrosion technology

M1 participants of pipeline construction

(Pipeline companies, design units, municipal units, government regulators) M2 departments in pipeline companies

E1 internal corrosion

E2 internal parameters (temperature, humidity, noise, pressureâ?¦) E3 internal medium

---

E4 pipeline company

E5 natural environment

(corrosion ,tide, Landslide and debris flow) E6 social environment along the pipeline

(destruction of residents along, human theft, vehicle extrusion) E7 political and economic environment

---

E8 corporate culture system

According to SHEM model of pipeline accidents and control system of management failureï¼?oil and gas pipeline accident model based on failure generalized management is established. The cases resulting in accidents are various such as case1, case2, case3, etc. In summary, the similar point is all the fours protective barriers are penetrated because the failure holes exist. In certain conditions unsafe inducements havenâ??t been blocked effectively and lead to accidents. Oil and gas pipeline accidents model is shown in Fig.2 below.

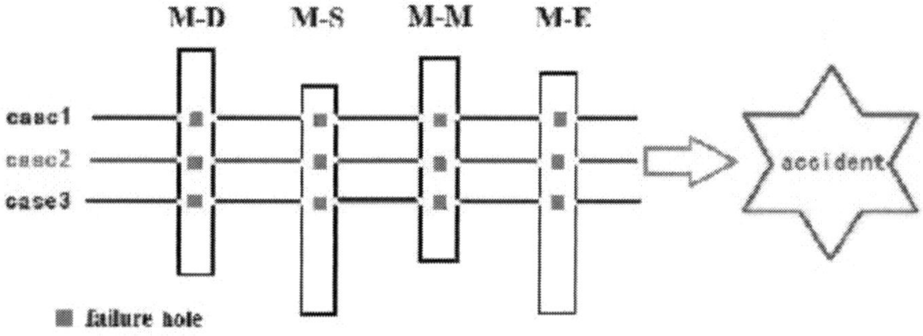

Fig.2 Oil and Gas Pipeline Accidents Model

2

The Qingdao oil pipeline explosion accident is discussed thoroughly by the oil and gas pipeline accident model mentioned above. First failure holes are abstracted from the process of accident scenario reproduced. Because the importance is different, control points are classified into core points, important points and normal points. Then according to the principle of giving priority to core control point, the preventive precautions are proposed to control the risk of pipelines.

The oil and gas pipeline accident model is put forward from the view of the generalized management failure and accidents are largely due to this. The factors affecting pipeline safety are summarized and classified comprehensively aiming at finding effective ways to improve the safety of pipeline in essence. In theory, the more failure holes it exists, the more accidents occur easily and more serious the consequences will be. So our target is to reduce failure holes and take measures to make up for the failure. The model can be used to cognize the process of the accident and investigate the cause of accidents. Once failure holes are found, many problems will be readily solved. At the same time the model provides methods for accident prevention and construction of safety management system for oil and gas pipeline.

The Idea of In-service Oil (gas) Pipeline Segment Management Based On the 3D Technology

Abstract: Long distance oil (gas) pipeline has become an important transportation facility for energy between countries. The transmission medium leakage accident has caused great damage and adverse effect to society. Corrosion is the main reason to the leakage of pipeline. However, the nondestructive testing technology, the fault detection technology and the security evaluation technologies for long pipeline have certain limitations. Building a scene graph of pipeline based on 3D technology by using modern digital technology, clearly marking all the villages, towns, factories, roads, bridges, rivers, geology and buried pipeline and overhead pipeline situation within a certain width range by this route, is not only convenient for daily safety management, but also can help choose appropriate nondestructive testing technology, fault detection technology and safety evaluation technology for pipeline, and can provide support for emergency rescue. The application of 3D technology provides a new idea for safety management of oil (gas) pipeline.
Keywords: 3D technology; oil pipeline; segment; management; idea

近年来，随着西气东输二线、俄气南供、陕京复线、忠武线、兰成渝成品油管线、西南成品油管线、甬沪宁成品油管线、冀鲁宁管线、仪长管线等重大工程的实施，我国油气管道干线联网的雏形已经形成。早期兴建的管线已经运行到寿命的中后期，不少管线已因腐蚀多处发生过泄漏[1]。去年发生在青岛的“11•22”中石化东黄输油管道泄漏爆炸特别重大事故，事故的直接原因是“输油管道与排水暗渠交汇处管道腐蚀减薄、管道破裂、原油泄漏，流入排水暗渠及反冲到路面；原油泄漏后，现场处置人员采用液压破碎锤在暗渠盖板上打孔破碎，产生撞击火花，引发暗渠内油气爆炸”。事故的直接原因并不复杂，可人们对该事故间接原因复杂性的思考却是深远的。本文正是要试图解决这些深层次原因，以保障长输管道的长久安全，保障沿途人民的生命财产安全和环境安全。

1. 现阶段，在役长输管道在安全管控方面的不足

1.1 长输管道的无损检测技术方面

对在役长输管线的无损检测，不能仅仅局限于对管体本身，还应对管道敷设的大气环境、土壤环境、阴极保护效果、管道防腐层及管道内外壁、管道壁厚、材质等方面进行全面检测，才能最大限度地保障长达几十千米、几百千米甚至几千千米的长输管道不出现泄漏故障。

用于在役长输管道无损检测的技术有多种，如磁场衰耗法检测管道防腐层绝缘电阻、电磁法检测评价管道阴极保护效果、电位电阻法检测评价管道环境腐蚀性、电位法检测土壤中散杂电流干扰腐蚀、磁场衰耗异常法检测管道内外壁腐蚀、漏磁法检测管道内壁腐蚀、瞬变电磁法检测管壁厚度、磁场涡流法检测管材质量等技术。每一种检测技术虽然各有其独特的优势，但它们的缺陷也非常突出，有的使用方法非常繁琐、有的适用条件非常苛刻，有的需要开挖管道。另外，地下管线有时在水下，有时在山石中；有的地方埋得深一些，有的地方埋得浅一点；有时是单管埋设，有时是复管或多管布置；有时在其它管线之下，

有时又在其它管线之上；绝大部分是直管，还会有弯头或膨胀节等等。长输埋地管道设置的复杂性，致使单一的无损检测技术都无法对长输管道作出全面准确的无损检测。

1.2 故障诊断技术方面

目前，用于在役长输管道的故障诊断技术有电位差法诊断管道防腐层破损点、电涡流法诊断长输管道内外壁缺陷、超声波与C扫描复合诊断管道缺陷、电磁干扰地区管道隐患故障点诊断、远距离电磁遥测管道故障点、磁力断层摄影（MTM）诊断管体腐蚀点、负压波法诊断管道泄漏点、红外热像法诊断管道泄漏点等，但没有一种技术能够对所有的管道缺陷进行准确诊断，有的受故障位置、故障大小影响；有的受土壤性质影响；有的受检测设备的操作空间所限；有的受地下长输管道的纵横交错影响；有的是检测设备本身的精度有缺陷；有的是检测技术运用的费用太昂贵，超出了许多企业的承受能力。

1.3 安全环保评估技术方面

油（气）长输管道输送的都是易燃易爆且有毒的危险化学品，一旦泄露不仅会发生火灾爆炸事故，造成附近的人员伤亡、财产损失，还会导致环境污染，破坏当地的生态环境，污染地下水资源等，非常有必要对输油（气）管线全线进行全面的安全环保风险分析。

油（气）长输管道的评估技术，不应是定性的分析，而应是定量的或既定性又定量的分析技术。进行安全环保评估分析，不仅需要有可行的评估技术，最主要的是进行定量评估所需的数据库。美国运输部DOT、加拿大的国家能源委员会NEB、欧洲输气管道事故数据组织EGIG、英国陆上管道供应商协会UKOPA、俄罗斯GOSGORTEHNADZOR标准、澳大利亚管道工业协会APIA，都进行了大量的管道事故分析与研究工作，建立了各自的事故管道数据库。然而这些数据库之间存在着很大的区别，不仅在于原始统计数据的范围和详细程度不同，还在于对原始数据的分析方法不同。这使得应用这些数据进行安全、环保评估时所得出的结论会相距较大[2]。

随着人们对预防各种重大灾害认识水平的不断提高，一种新的评估方法——脆弱性分析逐步被人们所熟知。脆弱性分析是近几年来迅速兴起的一种与扰动敏感性、系统易损性、后果承受力及灾害应对力密切关联的安全评价方法，它与一般风险分析只关注风险的致灾因子不同，还关注致灾环境，最主要的是在脆弱性分析中增加了人们对灾害承载体承载能力的关注。虽然脆弱性分析技术本身已经被世界各国的专家学者所接受，目前同样由于没有完善的数据库而无法应用其对油（气）长输管道进行安全、环保的定性评估。

1.4 管道泄漏应急方面

通过长管输送油（气）的单位，一般都是具有一定规模的大型企业，有包括管道输送应急预案在内的预案体系，并配备有自己的救援队伍和救援物资。但由于长输管道的特殊性，按预案编制导则编制的预案并不能穷尽长管输送过程中所有异常事件的应对措施；异常事件发生后，指挥调度中心不能及时、准确地判断出泄漏位置与泄漏现状；应急队伍和应急物资不能在较短的时间内到达异常事件的现场以实施有效地救援。

1.5 日常巡检力量方面

尽管通过长管输送油（气）的单位都有专门的管理部门负责长输管道的安全，但其日常的巡检也只能对管道在地上的部分进行一般性地检查，如管道损伤及变形情况、管道防腐层和绝热层破损情况、管道附件和安全装置缺失情况、管道标志桩和标志牌及电法保护系统的完好情况等，对于长输管道埋地部分故障的检查主要靠巡线员的经验。另一方面，对于在山区、沿河流、穿越沟壑布置的长达几十千米、几百千米甚至几千千米的长输管道，无法保障巡线员的巡检质量。

1.6 管道安全管理机构需掌握的风险分析资料方面

从事长管输送油（气）的单位虽设有专门的安全管理机构，配置了与管理内容相适应的人力和装备，逐级建立了安全管理网络，健全了各级安全生产责任制，并编制了包括管道工艺流程图、操作工艺指标、开停车操作程序、异常情况处理措施及汇报程序、防冻防凝安全要求、清管操作程序等内容的管道安全技术操作规程。但长输管道沿途所经过的城镇规划情况、道路与管道的交叉情况、各地地质的酸碱沙石情况、是否处于地震带、道路改造对管道的影响、管道周边化工厂特别是新建化工厂所排废气废水对管道及防腐层潜在的破坏作用、锚固墩松动及人为的破坏情况等，管道安全管理机构很难及时获取，也无法在安全技术操作规程中如实描述。另一方面，由于人员变动，也很难保证管道安全管理机构的所有成员及巡线员对管道安全技术操作规程及沿途的变化完全掌握。这些情况均是管道日常安全管理、进行安全环保评估所必需的内容，如不能及时、准确、全面地掌握，也许会对长输管道的安全造成灾难性的破坏。

2. 制定全方位的长输管道管控措施

2.1 引入三维全景技术

三维全景技术也称为全景摄影或虚拟实景，是基于静态图像的虚拟现实技术。把相机环 360°拍摄的一组照片拼接成一个全景图像，用一个专用的播放软件在互联网上显示，让使用者能用鼠标控制环视的方向，可左可右、可近可远观看物体或场景。

2.2 长输油（气）管道分段管控思路

长输油（气）管道距离长、沿途周边情况复杂、无损检测与泄漏故障诊断技术都有一定的局限性、长管输送单位的应急力量不能保证及时应对油（气）输送过程中的各种突发事件。鉴于此，我们可借数字技术的成果，建立输油（气）长管的三维全景图，将沿途一定宽度范围内所有村庄、工厂、城镇、公路、桥梁、河流、地质及管道埋地与架空情况标注清楚，并注明各种境况的详细信息，从而使长管的每一段及每段周围的所有境况均可时时、随时被管理者、检测人员、评估人员获得，便于需要时对长管全程或某段做出迅速、准确且全面地分析判断。既方便于管道日常的安全管理，又有助于选择恰当的管道无损检测、故障诊断及安全环保评价技术的建立与完善，还可为应急救援提供支持。

表 1 输油（气）长管的三维全景图需标注的内容

序号	须标注内容	详细情况描述
1	敷设时间	自某年某月某日起开始破土动工，于某年某月某日通过竣工验收
2	设计	设计单位名称、资质、营业范围、法人代表、设计负责人
3	监理	监理单位名称、资质、监理范围、法人代表、项目监理负责人
4	供应商	管材供应商名称及营业执照、法兰供应商名称及营业执照、防腐料供应商名称及营业执照

5	总承包商	总承包商名称、资质、营业范围、法人代表
6	各种审批手续	安全、环保、土地等各种审批手续（备案证明）
7	是否符合现行法规规定	国家历年颁布的相关法规、标准及法规、标准颁布后的整改记录
8	管道管理制度	与现行法规、标准的符合性，与本单位实际的符合性
9	管道安全技术操作规程	要在风险分析的基础上编制、完善各种意外事件的应急处置措施、有关于操作规程的培训记录、岗位人员每人一本操作规程、操作规程保持最新版本、操作规程变更应有专家评定结论
10	应急预案	建立完善的长输管道应急预案体系，包括综合预案、专项预案、现场处置方案；预案应有备案证明
11	总体试车方案	总体试车方案要具体、详细，各项试车方案要有责任人
12	巡线员的素质	熟悉管道安全技术操作规程、熟悉应急预案、违章记录、培训记录
13	管体	管体材质、管内径、壁厚、焊接要求、设计压力
14	防腐层	材质及型号、设计厚度、防腐施工队伍名称及法人代表
15	输送过的油（气）	输送过的油（气）生产厂家、含硫分析记录、含水分析记录等
16	高速、国道、省级公路	与道路交叉、并行情况；在道路下方穿过时的受力分析报告；从道路下方穿过时是否用套管或碎石予以保护
17	城镇	标明居住区及商业中心、公园等人员密集场所距长输管道的距离；标明学校、医院、影剧院、体育场（馆）等公共设施距长输管道的距离
18	乡村	标明乡村距长输管道的距离
19	化工厂	标明化工厂距长输管道的距离；化工厂简介、工艺特点（属于典型危险工艺的要叙述详细一点）、危化品的危险特性、危化品的储存量、是否属于重大危险源、有腐蚀品的要将腐蚀品的危险特性、主要理化数据标注清楚
20	车站、码头、通信干线、通信枢纽、铁路、道路干线、水路干线	标明车站、码头、通信干线、通信枢纽、铁路、道路交通干线、水路交通干线距长输管道的距离
21	军事禁区、军事管理区	标明军事禁区、军事管理区距长输管道的距离
22	河流	标明河流距长输管道的距离、标明管道横穿河流的长度、表明河流的历史最高水位、表明河流底部是否有地热存在
23	重要电力设施	标明重要电力设施距长输管道的距离
24	长输管道各段地质	标明长输管道沿途地址存在盐、碱、碎石、沙土、穿过隧道的位置；标明历史上出现山体滑坡位置和记录、标明属于地震带的位置
25	与其它管线交叉、并行情况	标明各段与其它管线交叉、并行时，其它管线的输送介质、压力及与长输管道的交叉或平行距离
26	管线故障、更换、维修记录	标明管线故障、更换、维修段落及原因
27	其它变更对长输管道的影响	标明长输管道所经道路、桥梁、城镇、工厂、乡村变更记录，分析变更对长输管道安全性的影响
28	防腐保护措施及运行记录	标明防腐保护措施，运行中的防腐维护记录
29	故障诊断	标明故障诊断技术与诊断、处置记录
30	中间监测点	保证中间监测点完好有效的措施
31	弯头位置、位号	表明长输管道所有弯头、弯曲管线的详细内容，并标注位号
32	油（气）日常分析记录	标明油（气）日常分析记录

33	无损检测技术与检测记录	标明无损检测技术与检测记录、标明检测出缺陷的具体位置、对检测出的缺陷进行详细描述
34	加药保护记录	标注保护长输管线药剂的名称、型号、特性
35	管道评估	标明对长输管道的评估时间、评估单位、评估结论及建议
36	日常巡检记录	标明日常巡检记录
37	本单位应急队伍、物资	标明本单位应急队伍、物资的详细资料
38	其他应急力量	标明沿途其他应急救援力量（院校、消防队等）分布位置、特长、能力、应急救援协议
39	沿途当地政府机构	与沿途各地安监、环保、职业卫生、环保部门及化工企业的应急通讯联络、应急救援协议
40	锚固墩	标明锚固墩的位置、锚固墩曾经发生过松动的时间及原因
41	埋地深度	标明长输管道设计埋地深度、标明长管大于或小于设计埋地深度的位置及实际埋地深度
42	落差	标明长输管道穿过隧道、桥梁、沟壑、丘陵前后的高度落差
43	清管	标明清管记录、清管过程中发现的问题
44	防冻防凝记录	标明详细的防冻防凝方案
45	阀门、法兰设置	标明各段所有阀门、法兰的位置、标明曾经发生故障的阀门、法兰
46	汛期	标明各地汛期最大水位及汛期曾经造成山体滑坡、泥石流、河流溃堤的记录
47	管道事故统计	进行国内外长输外管道事故统计，标明事故原因、后果、处置措施
48	对重大危险段的措施计划	对可能发生重大危险的段落，标明风险分析过程、结论、应急处置措施
49	其它	

注：对于有较多内容的记录，为避免三维情景图过于复杂，可采用链接的方式进行处理

由于长输管线距离长且不同阶段的境况不同，对于无损检测、日常巡检特别是发生泄漏故障时的应急需要，宜对长输管道实施分段管理，与不同段落附近的应急队伍、应急技术支持力量、故障诊断及无损检测机构签订合作协议，并将本单位制作的“输油（气）长管的三维全景图”发给他们，以保证签订协议的单位能有效履行协议规定的相关义务。

3. 展望

1、近年来，由于腐蚀、城市规划、道路修建、修渠筑坝等因素，使已经敷设的油（气）埋地长输管道受到不同程度的影响甚至损坏，给沿途城市、村庄人们的安全带来很大的威胁，危化品长输管道的安全已引起国家的高度重视。目前，高等院校和科研机构正在就长输管道的无损检测技术、故障诊断技术、安全环保评价技术、应急救援技术及相关标准、法规组织专门的机构进行研究，新的研究成果将会促进长输管道的安全管理。

2、每个通过埋地长管道输送油（气）企业，都有能力独立或聘请专业技术人员完成管道三维图情景的设计，实现埋地长输管道在室内对全景进行透视性的“观察”，有利于对长输管道全程的风险分析、重点管控，并可在室内对各段的安全、环保、应急管理实施遥控指挥，从而使对长输管道的管控更加全面、准确、及时。

4. 参考文献

[1] 戴厚明.地下管道电磁无损检测与隐患故障诊断[M].北京：中国石化出版社，2013.
[2] 黄贤滨，刘小辉，谢守明，叶成龙. 长输管道事故故据库的对比分析[J].安全、健康和环境，2012，12(11):48-51.

2nd CCPS CHINA CONFERENCE ON PROCESS SAFETY™

Paper Title: The New Corrosion Control Technology of Oilfield Equipment

Presenter/Principal Author: Wang Shuaijun
Affiliation: Department of Environmental Engineering of China University of Petroleum
Address: No. 66, Changjiang West Road, Huangdao District, Qingdao , China , 266580
Email address: 1055439500@qq.com

Other Author(s): Jiang Huichao
Affiliation(s): Department of Environmental Engineering of China University of Petroleum
Address (es): No. 66, Changjiang West Road, Huangdao District, Qingdao , China , 266580
Email address (es): 1052797504@qq.com

Paper Title: The New Corrosion Control Technology of Oilfield Equipment

Presenter/Principal Author: Wang Shuaijun
Affiliation: Department of Environmental Engineering of China University of Petroleum
Address: No. 66, Changjiang West Road, Huangdao District, Qingdao , China , 266580
Email address: 1055439500@qq.com

Other Author(s): Jiang Huichao, Zhao Chaocheng
Affiliation(s): Department of Environmental Engineering of China University of Petroleum

Keywords: Equipment corrosion, New technology, Comprehensive control, Corrosion control

Abstract

With the application of enhanced oil recovery, equipment corrosion problems become increasingly acute. Equipment corrosion affects the normal operation of the device and brings about a series of related problems. In solving these problems, a new technology on oilfield equipment corrosion control was proposed, the comprehensive control technology, combining oilfield equipment corrosion characteristics and traditional methods. This technology will effectively reduce the loss caused by corrosion. Nanophase anti-corrosion coating, biological corrosion probes and cathodic protection technology are currently being used to retard corrosion in some developed countries and proved to be feasible with significant efficiency.

1. Introduction

Equipment corrosion is responsible for an enormous loss to the national economy [1], amounting to as much as 2200 billion dollars all over the world. Particularly serious losses due to corrosion occur in the oilfield equipment. With the application of enhanced oil recovery, the total salinity of oilfield produced water increased, corrosive, scaling ions raised, dissolved oxygen, aggressive carbon dioxide, sulfide and other corrosive media exist and SRB, TGB [2] bacteria multiply, resulting in equipment corrosion, fouling, erosion. Equipment corrosion affects the normal operation of the device and brings about a series of related problems: economic benefit, safety production and environmental protection etc. In this regard, lots of researchers have been devoted to the exploration of anti-corrosion control technology.

2. Traditional corrosion control method

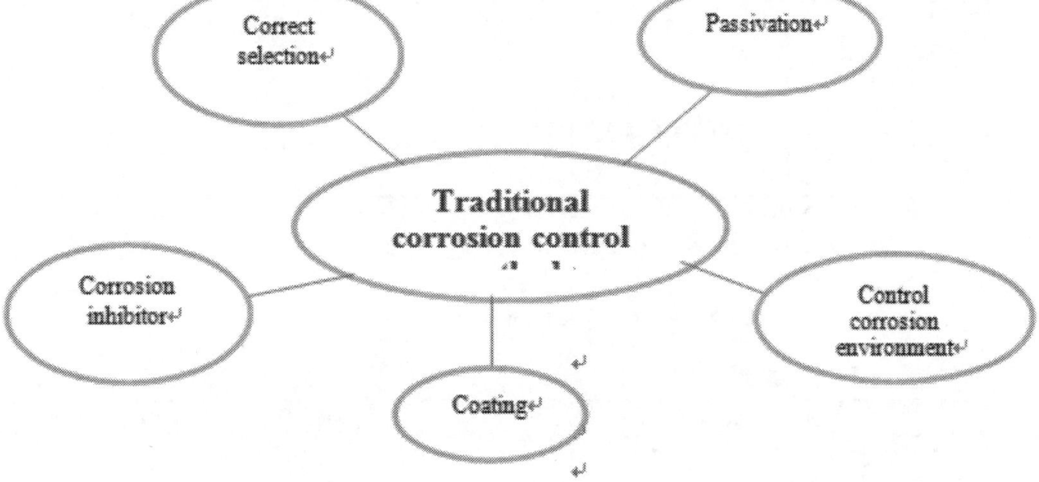

Fig.1Traditional corrosion control method

The main traditional control methods [3]: ① Correct selection. There is big difference between corrosion rate and corrosion spontaneity, when using different material in different environment. Hence, choosing the material that the corrosion rate is slow and corrosion spontaneity chance is low to meet requirement. ② Passivation. The passive film of metal surface make diffusion residence larger, corrosion basically ceased and a number of metals can apply this method. ③ Corrosion inhibitor. The corrosion rate was greatly reduced when the corrosion inhibitor was added in the solution. ④ Coating. Coating is the most widely used means of corrosion, it is usually prepared from a synthetic resin, vegetable oil, rubber and solvent slurry, covering the metal surface, forming a thin porous membrane after dried. Although the metal and corrosive media can't be completely isolated, the diffusion resistance and solution resistance of media through the porous was greatly increased, the corrosion current was decreased. ⑤ Control corrosion environment. Eliminate environmental factors directly or indirectly caused by corrosion and corrosion will stop.

However, traditional anti-corrosion methods just focus on correct selection and try to solve the problem when the equipment is in trouble, it turned out to be a poor effect.

3. Oilfield equipment corrosion characteristics

As an increasing amount of produced water, the corrosion phenomenon seems more and more serious. The characteristics of oilfield equipment corrosion [4] is various :① Produced water with high salinity, strong conductivity and even electrochemical corrosion. ② High content of bacteria, serious microbiological corrosion. ③ The corrosion and hydrogen embrittlement caused by dissolved gas(Oxygen, carbon dioxide, hydrogen sulfide, etc.). LI Wenfei [5] report that the effect of H_2S on CO_2 corrosion varies with the H_2S concentration. H_2S can either accelerate the corrosion by acting as a promoter of anodic dissolution through sulfide adsorption and affecting the pH, or it can decrease the CO_2 corrosion rate by the formation of a protective sulfide scale. ④ High content of scaling ions, serious corrosion.

In solving the defects of traditional control methods, new technology on oilfield equipment corrosion control was proposed, comprehensive control technology, combining oilfield equipment corrosion characteristics and traditional methods. This technology will reduce the loss of control by corrosion effectively.

4. New corrosion control technology

New corrosion control technology also called comprehensive control technology; it turns corrosion control from passive to active and retards corrosion in a comprehensive avenue. It includes design, manufacture, storage installation, operation, maintenance and scientific research, management, education, benefit and other measures.

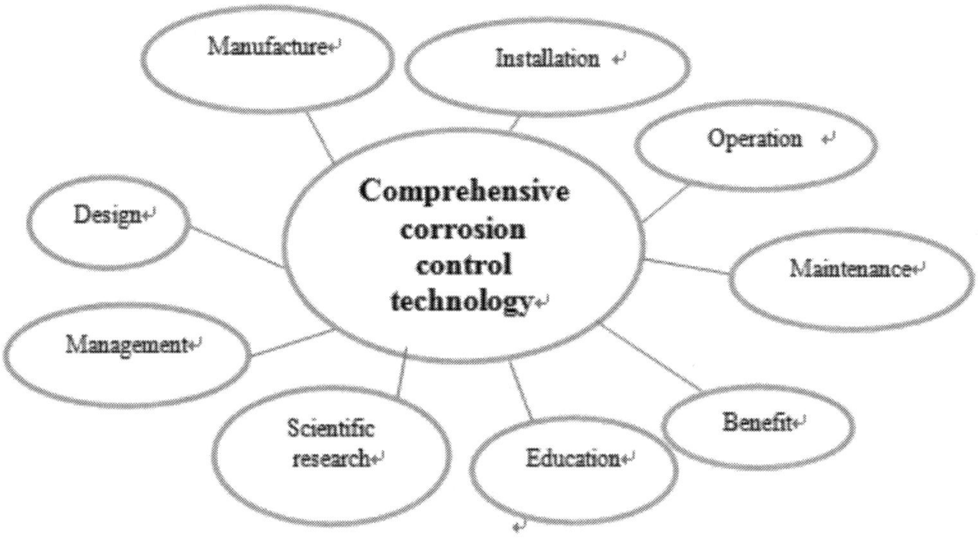

Fig.2 Comprehensive corrosion control technology

As the fig.2 shows, Comprehensive corrosion control technology can take control of corrosion throughout the life circle, thus, it may more effectively than traditional control method in corrosion control aspects.

4.1 Corrosion control during design

During oilfield corrosion design, both static conditions and dynamic conditions should be taken into consideration. Structural design should be reasonable, otherwise it may cause accumulation of corrosive media, local overheating and local stress concentration in the structure, so as to result in pitting, crevice corrosion, contact corrosion or stress corrosion

For uniform corrosion, the design of certain corrosion allowance is necessary. However, as regard to local corrosion, the correspondence measures are critical in designing. For example, for the sake of the convenience of the construction and maintenance, structure should be as simple as possible; in order to reduce stress corrosion tendency of structural parts, arc transition should be used to take place of sharp corners.

4.2 Corrosion control during manufacturer

4.2.1 Heat treatment

Without effective measures, large amount of corrosion not only occur in the course of using, but also during manufacturing process, even leave a serious potential risk to the device. The residual stress in the work piece can be reduced during by stamping molding process in the annealed condition. In the high-temperature heat treatment process, the metal is prone to high temperature oxidation, vacuum heat treatment, protecting the heat treatment atmosphere or protective coating can be used to eliminate stress after the final heat treatment.

4.2.2 Casting

Minimize pores and voids in casting process, in the case of corrosion by gathering material.

4.2.3 Connection process

The crevice corrosion caused by riveting and stress concentration of expansion should be avoided.

4.2.4 Surface coverage process

Quality of the metal surface has a great influence on the corrosion resistance, the smoother of the surface, the higher corrosion resistance to the environment, so it is an effective mean to make a treatment to the surface of device or article for the corrosion control. For the purpose of good effect of anti-corrosion, strong corrosion resistance materials are required to cover the surface of the metal, isolating the corrosive media and metal. Oxide skin, rust, burns and other surface defects on the device surface should be completely removed, getting good anti-corrosion effect. Protective coatings can be divided two categories: metallic and non-metallic coating. Alloy steel with low concentration of aluminum, silicon, chromium shows excellent resistance to high temperature oxidation and sulfuration, among which the effect of aluminum and silicon is better than chromium.

4.3 Corrosion control during installation

Before installation, anti-corrosion measures should be taken into consideration, such as whether there is danger to metal liner, whether the parts have been corrode, etc. Reasonable measures should be taken to reduce the installation stresses while the device is installed. Using tools and instruments reasonably and the surface of device can avoid being scratched. If required, the site should be sealed during installation process; for the permanent installation, corrosive inhibitor and anti-corrosion primer should be used. To prevent stress corrosion cracking, account must be taken of controlling the chloride ion content when doing water pressure test. After test, the water should be drained and dried promptly in case of rust.

4.4 Corrosion control during operation

To impede corrosion by sudden change of various parameters, open parking process should strictly enforce. In general, the corrosive rate increases with the raising of temperature and pressure of system media. The use of corrosion inhibitors or electrochemical protection equipment should strictly follow the rules, if the amount of corrosion inhibitor added improper, not only can't play a protective role, sometimes it will accelerate corrosion. If the device leakage occurs, maintenance should be conducted in time to prevent equipment corrosion exacerbated.

4.5 Corrosion control during maintenance

Invest the technical condition of equipment, find out the reason of corrosion, establish files and make reasonable corrosion maintenance program during maintenance process. In the selection of repair material and the substrate material, galvanic corrosion, large cathode corrosion and small anode corrosion should be prevented. Lining layer repair of substrate surface should avoid being not cleaned up.

Maintenance records should be keeping well, as the proof of inspection. Posterior maintenance control, including analyze and repair for corrosion problems encountered in the process, observe whether have a link with the design, manufacture, storage, transportation, installation and operation and how to take actions to control erosion, is pretty critical

4.6 Scientific researches

The direction of research should be combined with comprehensive corrosion controls, deal with various corrosion studies, further find the mechanism of corrosion and provide basic data for the production of anti-corrosion. Develop corrosion resistant materials, which can be applied under different circumstance, to minimize corrosion of equipment. Providing reliable data to the designing, manufacturing, installing, operating and maintaining process is one of a major works.

4.7 Management

Establish the necessary organization and rules and regulations, formulate codes and standards of anti-corrosion designing, unify inspection criterion and construction procedure, etc., exchange and collaborate corrosion technology regularly, plan corrosion database, project corrosion protection work scientifically and reasonably, anti-corrosion technology and equipment should also be commercialized and standardized.

4.8 Education

Just as solving any other questions, education and training are vital. All levels of people should receive education, so as to be aware of the importance of the hazards and anti-corrosion of corrosion. Conduct elaborated training on corrosion and protection to technical personnel, management personnel, foremen and operators, including the aspects of design, manufacture, installation, operation, maintenance etc.

4.9 Benefit

To obtain the optimum benefit, all sorts of programs are carried out, which is an important indicator to measure the effect of anti-corrosion technology. Economic assessment, feasibility analysis and studies are vital to corrosion control programs and to determine best solution.

5. New corrosion control technology in developed countries

Although the corrosion control technologies have been studied by China researchers, there is still gap with the developed countries. Nanophase anti-corrosion coating, biological corrosion probes and cathodic protection technology are currently being used to retard corrosion in some developed countries and proved to be with significant efficiency.

5.1 Nanophase anti-corrosion coating

Italian researchers apply nanoparticles into anti-corrosion coating technique, [6] obtaining a high level effect of corrosion protection. The coating can be used for steel and concrete structure, to enhance the ability of the substrate isolate from oxygen, moisture and other chemical agents, to reduce the permeability and to improve its coating cracking performance, and also to enhance its elastic modulus and strength index. Technical focus on the development of innovative nano-composite material, in particular including thermosetting and thermoplastics nanoclay, so as to realize the innovation of antiseptic and fire retardant coating and polymer.

5.2 Biological corrosion probes

Because of oilfield produced water contains a large number of SBR bacteria, which will take a serious hazard to the oilfield equipment. The bio-corrosion probe technology was worked out by Canada scientists, which can monitor the activities of SBR online and corrosion rate at the same time. It is helpful for monitoring the material of microbiologically influenced corrosion and thus optimizing the process environment to prevent the occurrence of microbiological corrosion, it can also be treated as a mean of evaluations of the effect of fungicides. The probe consists of a biological enzyme electrode, a working electrode, a counter electrode and a reference electrode etc., among which biological enzyme electrode is the most unique one with the function of detecting sulfide oxides. It can detect sulfide ions in solution and not affected by other ions and substances. When the bio-corrosion probe find the activity of sulfate-reducing bacteria and produce sulfide ions and then will send a signal to the collaborative electrode, thereby producing an electric current that can be detected.

5.3 Cathodic protection technology

Cathodic protection technology [7] is the connection of direct current power of anode and the protected object, the positive and difficult to soluble auxiliary anode, which provides protection to current. In the past of years, lots of researchers have been devoted to the application of cathodic protection technology on oilfield equipment. The main applications are as the following:

(1) The cathodic protection of buried steel pipeline.

(2) Regional cathodic protection.

(3) The bottom outside wall of large steel tank.

(4) The inner wall of the container.

(5) The steel structure and subsea pipeline of marine oil and gas field in the development environment

(6) Cathodic protection on oil well casing

Cathodic protection technologies are related to the application [4] of software designing and scheme optimization, power supply equipment development and the remote control technology. Aiming at the serious corrosion of oilfield storage tank, research is currently operated on the tank cathodic protection method, material testing technology.

6. Conclusions and prospect

The new corrosion control technology, comprehensive control technology, is one of the most effective and economic technologies, and also the most widespread technology in oilfield equipment. It was proposed combining oilfield equipment corrosion characteristics and traditional methods, including design, manufacture, storage installation, operation, maintenance and scientific research, management, education, benefit and other measures. Despite the excellent effect have been achieved in the application of the new corrosion control technology, there is still urgently need of coordination and efforts should be paid on the operation. With the common efforts, our further on corrosion control aspect will be brighter and brighter.

References

[1] V. G. D'yakovand A. V. Shreider, CORROSION CONTROL AND SELECTION FOR REFINERY EQUIPMENT OF MATERIALS [M], No.12, pp.27-36, December, 1976.

[2] Wang YuJiang. Application of corrosion control technology in Shengli Oilfield Surface Engineering [J]. CORROSION SCIENCE AND PROTECTION TECHNOLOGY, 2012, 24(5): 433-435.

[3] Huang WenCheng, Lu JianXiang. Discussion of general corrosion control technology equipment [J]. Equipment management and maintenance, 2013,27-29.

[4] Wang ZhiGang. Application of corrosion control technology in Shengli Oilfield Surface Engineering [C].Third Corrosion and Protection Technology Symposium Shandong Petroleum Institute. Committee Forum, 2002,47-52.

[5] LiWenfei, Zhou Yanjun, Xue Yan. Corrosion Behavior about Tubing Steel in Environment with High H_2S and CO_2 Content [J]. Journal of Wuhan University of Technology-Mater, 2013,1038-1043.

[6] Wei Renhua. Innovation in Corrosion control technologies [J]. Coatings technology and abstracts, 2012， 35-37.

[7] Du Ronggui, Huang Ruoshuang, etc. Status and Research Progress of cathodic protection of reinforced concrete structures technology [J]. Material protection, 2003,36(4):11-14.

Near- miss Management and its Application in Long- distance Pipeline

XING Jin-duo[1,2], ZHAO Dong-feng[1,2], HAN Feng-lei[1,2]

(1 The Environment and Safety Technology Center of China University of Petroleum, Qingdao, Shandong, 266580, China;

2 The Center for Chemical Process Safety China Section, Qingdao, Shandong, 266580, China)

Key words: near-miss; management; risk matrix; long-distance pipeline; third -party damage

Abstract

Near- miss is the suspension state of the accident evolution. The research on near- miss can effectively prevent and control the approaching accidents. Confronted with the situations of neglect, blocked report and enormous quantity, the coping measures including the duty-exemption reporting, anonymous reporting and near- miss database are put forward. Subsequently, the risk matrix and report management process are given for long- distance pipeline near- misses. The study has shown that the near- miss management is an indispensable part of the risk management system. Enhancement of the near– miss management is conductive to lower the accident risk level. Long-distance pipeline near- miss prevention methods are listed, which can improve the of risk assessment efficiency and emergency response capabilities of the pipeline.

0 Introduction

With the sharp increase in oil and gas consumption long-distance pipeline has been rapid developing as the main way of oil and gas transportation. According to the report our oil (gas) pipeline has reached a total mileage of 106,000 kilometers. It is predicted that long-distance pipeline will reached 150,000 kilometers in 2015 [1].Long-distance pipeline transport medium are flammable, explosive. Besides transport media and transport outer environment can lead to the pipeline corrosion. Pipeline rupture, leak, shutdown, etc. may not only cause huge property losses and casualties, but also lead to environmental pollution [2]. Thus it is very important to make sure safety operation for long-distance pipeline.

There are many signs of no damage things or minor damage things occurred before every major incident, but these events are not caused the relevant staff put highly valued, so that the unsafe condition of things, human interaction unsafe behavior continues, eventually evolved into the accident. We can put much importance to the investigation of the minor damage on long-distance pipeline incident before the accident including statistics, surveys, analysis. It can help to improve the level of production employees to identify risk factors for the job to run, to prevent accidents, while improving emergency response capabilities. Management leader can improve safety management based on the analysis of the management of minor incidents or near-miss events.

Near-miss event study to explore the risk factors that exist in the production process, in line with "prevention first" production safety guidelines for the prevention and control of accidents has an important role.

1 The abstract of Near-miss

Near-miss event is the cause or circumstance could cause an accident, the result did not produce disease, injury, damage or other loss events; accident means to cause death, illness, injury, damage or other unexpected loss [3]. Near-miss has the similar mechanism with the accident. [4]. It will happen under the interaction of things' unsafe state and human beings' unsafe actions. Generalized near-miss events concept including human unsafe behavior, unsafe conditions and can cause serious physical harm to small events, but the damage did not occur that may occur [5]。 The famous Ice Theory has indicated that accidents are caused by danger and does not meet the conditions of existence, the event evolved into an accident. Heinrich investigated 5000 similar incidents and pointed that in 330, including 300 from non-injury accidents, 29 cases of minor accidents and1cases of serious injuries. That is the famous Heinrich law [6], as can be seen in Figure 1.

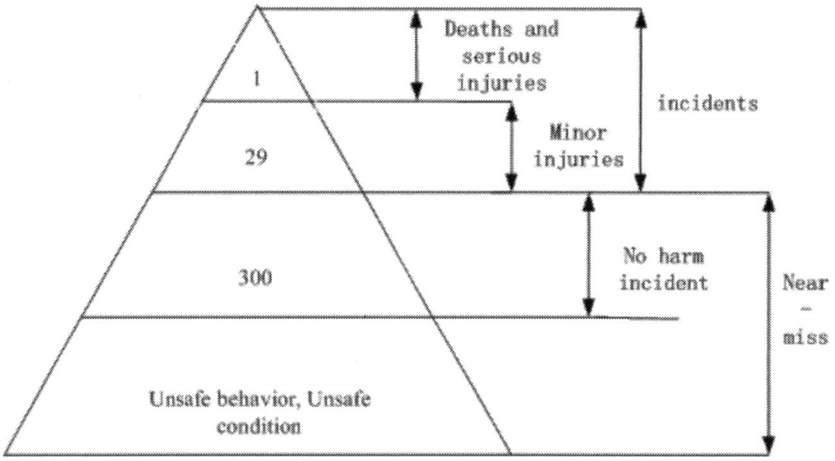

Fig.1 Heinrich incident law

Figure 1 shows that, under normal circumstances, events are less likely to cause serious consequences, there are a lot of minor injuries or attempted incidents; attempted incident did not cause serious consequences, but the process takes place as its causes and continues to evolve, that will greatly increase the probability of a serious accident.

Study the mechanism of near-miss events and analysis and manage near-miss. As a result can improve enterprise safety management and employee security awareness, effective prevention and control of accidents and promoting safety performance improvement[7]. From the enterprise management level analysis, event study attempted to make security management that did not at the expense of serious consequences to find loopholes. Automation and control systems in the enterprise and security can take advantage of attempted incident report, improve the system database, modify process safety management level[8]. In the aspect of employees' consciousness, we can turn incident learning mode to accident prevention mode. The safety consciousness are changed from passive acceptance to active demand [9].At last safety awareness level is improved.

The purpose of collecting near-miss reports is to grasp the development of small mistakes in the mechanism of failure or near-miss incidents or accidents, near misses combined effect of a single factor or factors, to within a statistical range of reliable qualitative analysis [10]. The near-miss reports learning can help employees be alert to danger especially when the actual injury rate and the accident rate has been below the required management level.

2 Near-miss management analysis

Near-miss management has become an important part of safety management in abroad, for example Seveso that has suggested to manage Near-miss reports in our daily operations [11].The American James R Phiminster has put forward seven stages frame of near-miss management.

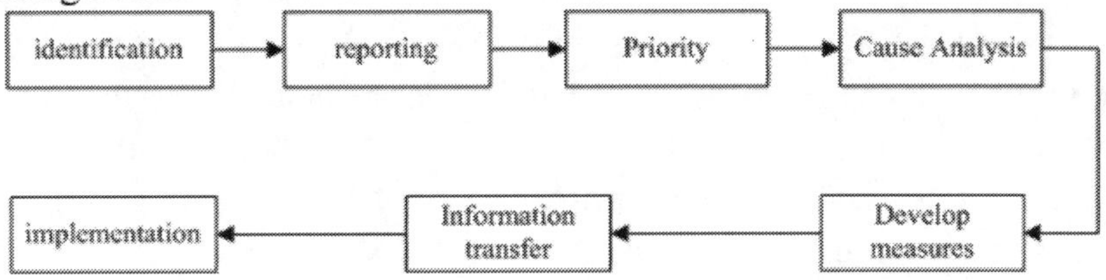

Fig.2 James R Phiminster near-miss seven stages frame

Near-miss event management include near-miss event information collection, analysis, archiving, study, involving many aspects of corporate personnel, including managers, operations staff, security personnel, etc. Thus to develop rational management analysis process is necessary.

2.1 Near-miss management contents

Based on Near-miss seven stages frame from James R Phiminster we divided near-miss management into three stages including information collection, events analysis, control measures.

To play the maximize value of the near-miss event, you must attempt to identify the events, including the understanding of the serious consequences that may result, conditions [12]. The stuff has to know the concept of Near-miss and believe the event has a big influence on safety production.

In order to have a better identification for near-miss the enterprise can designate the appropriate level of Near-miss events according to their level of risk acceptable. The level was designate from two aspects the likelihood and severity. It is benefit for determining reporting and tracking governance sector [5]. For instance high risk near-miss are carried out by the business-led, sector organizations to develop cause analysis and corrective action tracking, low-level risk near-miss can be attempted rectification by the workshop or team. The method reduce near-miss management burden for the enterprise to some extent.

The basic contents of the attempted incident investigation section includes description of an event scenario, and indirect cause analysis, hazard prevention and control measures to be taken, as well as the investigation process temporarily stranded problems can not be solved directly by hand tracking records to confirm the implementation of corrective measures.

In addition, to fully explore the role of incident reports attempted training personnel in education, so that employees understand the mechanism incidents and accidents of evolution.

2.2 Near-miss event management bottlenecks and solutions

(1) As can be seen from near-miss event definition that near-miss did not cause serious casualties, property damage, or equipment damage, easily government, corporate management and employee ignored. [6]. This is the most important reason for the low level of near-miss management.

(2) Near-miss incidents reported blocked. Attempted accident site work itself lead to unsafe behavior, because the accountability system or the poor cognitive ability the near-miss event can easily be concealed.

(3) To define the near-miss event is difficult because there is no uniform standard, and the security personnel and the level of cognitive ability to accept different, so in a lot of unexpected events, the delineation of near-miss events blurred.

(4) By accident pyramid shows that relatively high frequency of attempted incidents occurred in the accident itself, attempted to collect event information, as well as subsequent analysis and event control, management of large workload report.

In response to these problems, companies need to strengthen safety education and training, improve staff security awareness, attempted to fully understand the importance of the event for the prevention of accidents and incidents reported attempted to carry out the process of education and learning, so that workers can be positive, proactive, detailed, accurate and comprehensive attempted event reporting analysis, internal advocate "attempted Disclaimer event reporting system" or attempted incidents anonymous reporting system [13], Enterprises can take advantage of the enthusiasm of the staff involved in the attempted event management, continuous depth investigation and analysis. There is no uniform standard for the attempted division events, the preparation of an acceptable level of risk matrix attempted events according to their own safety. In addition, the establishment of specialized databases attempted Event Management Event Management attempted to reduce the load. Scientific design database structures, including case upload, fix, maintain, display and other features to learn, to achieve the attempted retrieval, learning automation, improve accident prevention and control levels.

3 Near-miss management for long-distance pipeline

3.1 The level of long-distance pipeline near-miss management

Pipeline attempted incident classification is based on the likelihood and severity of incidents into account two aspects. Possibility is based on the frequency of such incidents, near misses from the consequences of general security, economic and environmental point of view [14]。 Pipeline attempted event severity according to the severity of injury and the transmission medium leakage, ignition delay determination. Pipeline attempted event risk matrix shown in Table 1

Table 1 Near-miss risk matrix

Occurrence possibility	3	3	6	9
	2	2	4	6
	1	1	2	3
Low	Middle	1	2	3
high	Grading standards	The severity of the consequences		

The possibility of Near-miss event is based on the likelihood of an event occurring frequency. That can take many factors into account including actual operating life of the pipeline, pipeline material, crossing geographical environment, the transmission medium to determine. The extent and consequences of injury severity index include losing time that the event caused, the amount of medium leakage, economic losses and environmental pollution.

For low risk level events can be archived record attempt, without taking immediate steps to rectify; moderate level of risk for attempted incident investigation records required, and to take measures at the right time correction; attempted a high risk event to be a priority investigation and analysis, and to take immediate measures control and prevention of similar incidents evolve.

3.2 Near-miss management flow for Long-distance pipeline

Pipeline operating conditions are complicated, complex factors, including terrain conditions, third party damage, design factors, such as corrosion, mainly by the valve chamber in the region, station management control and maintenance. Attempted long-distance pipeline incident, the general line first by patrol officers found that collect information reported to the administration section of pipeline safety, the forensic analysis to determine the level of risk by risk grade level, determine the approval of the record department and tracking corrective measures department, and finally attempted event information sharing and learning resources. Specific management process are shown in Figure 3。

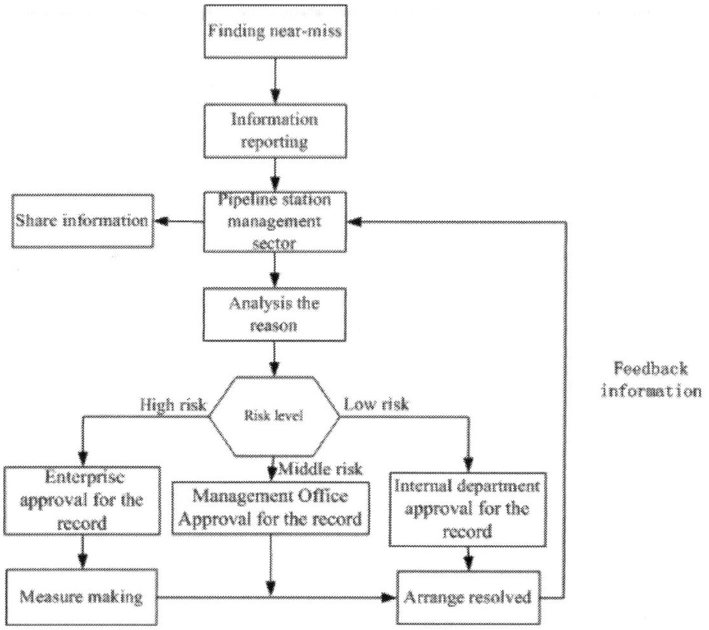

Fig3 Near-miss management flow for Long-distance pipeline

Long-distance pipeline near-miss analysis attempted to know the cause of the event, the need to fully identify the root cause of the incident, such as the rules unreasonable, weak awareness of security personnel, management issues, design issues, etc. [11]. From pipeline operations management department, investigate the event step by step to prevent near-miss event evolved into an accident.

Near-miss event is impossible to completely eliminate, how fast and effective response measures after the incident, so as to minimize the damage is another important theme of the near-miss event management. Repairing work is very important because of the security of long-distance pipeline transportation of flammable and explosive media, pipeline leaks after the attempted pipe run departments to event analysis, preparation of specialized maintenance procedures for similar incidents in the emergency repair work on the pipeline repair standards and technologies, such as with oil flare fire technology, pipe plugging, pipeline ruptured tube replacement measures. Meanwhile, a comprehensive investigation of the oil and gas that are prone to leak, fire, explosion should be carried out. And carry out emergency exercise in targeted key areas.

In our long-distance pipeline operations management process, with large-scale urban construction and improvement of public facilities construction, and due to poor planning and inadequate communication between enterprises, tying up some areas by pipeline, pipeline corrosion aging, does not meet the requirements of cross-crossing, event security issues resulting from the lack of frequent. Suggestions are given from both the prevention and control [15]including:

（1）According to years of pipeline operations, pipeline corrosion inspection should be carried out on a regular basis and take timely measures to safeguard against corrosion region. For some cases operation enterprise should prepare contingency plans.

（2）Third party damage accidents in recent years is frequent. We should establish and improve the urban underground construction approval system involving oil and gas pipelines in the town of ground excavation and underground construction strict examination and approval. Construction should be prohibited without approval and in the case of unknown, blind underground construction.

（3）Pipeline safe is a widespread problem thus reforming efforts is necessary. For some high-risk areas, particularly in business linkage area, good coordination efforts related to the rectification should be made timely.

4 Conclusions

（1）Reinforcing near-miss event management can facilitate targeted to carry out safety supervision and inspection. And it is benefit to develop safety standards, so that we can turn accident management from passive to active control in accident prevention concept, to ensure safe and efficient operation.

（2）Developed a unified long-distance pipeline near-miss grading standards. The near-miss event will be attempted in accordance with the size and cause of the incident loss of classified management. It can help improve enterprise security management efficiency.

（3）The special event management processes on long-distance pipeline are determined Near-miss event reporting transparency, accuracy of data, analysis is the basis to achieve full production of operational safety. Near-miss event for the long-distance pipeline may establish specialized databases for collecting near-miss event information, share near-miss event management experience. It is a strong increase long-distance pipeline safety management.

5 References

[1] State Council. "the 12th Five-Year Plan" Comprehensive Transportation System Planning[R].Beijing：2012

[2] ZHANG Qi-bo, JIA Ying, YAN Xiao-jing. Hazard Analysis of Long-distance Oil and Gas Pipelines[J]. China Safety Science Journal, 2008,(07): 134-138

[3] General Administration of Quality Supervision , Inspection and Quarantine of the People's Republic of China. GB/T28001-2001 Occupation health safety management system[S]. 2001

[4] Linda W, van der Schaaf T. Accident versus near miss causation a critical review of the literature, an empirical test in the UK railway domain, and their implications for other sectors[J]. Journal of Hazardous Materials, 2004, 111: 105-110.

[5] WANG Guang-yu Analysis on definition and management process of Near-Miss[J]. Journal of Safety Science and Technology, 2012, 8(06): 197-200.

[6] WU Qiong, XU Kai-li. Safety Management Science[M]. China Coal Industry Publishing House, 2002.

[7] HE Ling-cheng, LI Ji-zu. Coal Mine Near-MIss Study Based on BBS[J]. Journal of Taiyuan University of Technology, 2014, 45(03): 394-397.

[8] Oktem U G, Seider W D, Soroush M. Improve process safety with near miss analysis[J]. Cep, 2013: 20-27.

[9] Gnoni M G, Andriulo S, Maggio G, et al. "Lean occupational" safety: An application for a Near-miss Management System design[J]. Safety Science, 2013, 53: 96-104.

[10] Van Der Schaaf T W, Lucas D A, Hale A R. Near miss reporting as a safety tool[M]. Butterworth-Heinemann, 1991.

[11] Jones S, Kirchsteiger C, Bjerke W. The importance of near miss reporting to further improve safety performance[J]. Journal of Loss Prevention in the Process Industries, 1999, 12(1): 59-67.

[12] Phimister J R, Oktem U. Near - Miss Incident Management in the Chemical Process Industry[J]. Risk Analysis, 2003, 23(3): 445-459.

[13] WANG Bi-qing, DANG Chun-cai, GUAN You-li. Research on Management of Attempted Events in Coal Mines[J]. Technology and Innovation Management, 2013, 34(01): 40-42.

[14] DU Man, ZHAO Dong-feng, MENG Yi-fei. Long Distance Natural Gsa Pipeline Leakage Accident Consequence Assessment Method and Application[J]. Oil &Gas Storage and Transportation, 2012, 31(5): 340-344.

[15] ZHANG Sheng-zhu, WU Zong-zhi. Risk based optimization method for long-distance oil and gas transmission pipeline route[J]. Acta Petrolei Sinica, 2014, 35(3): 570-576.

Analysis of Environmental Risk Assessment and Emergency Monitoring in the Hazardous Chemicals Pipeline Accidents

YANG Wei, LIU Fang

(College of Chemical Engineering, China University of Petroleum, Qingdao 266580, China)

Abstract: The frequency of hazardous chemicals pipeline accidents is increasing every year, this has become a world-wide environment issue. In order to deal with hazardous chemicals pipeline accidents, the characters and causes of environmental pollution in them are approached, as well as the concepts, preliminary identification methods and emergency measures of environmental risk assessment, environmental emergency monitoring and environmental emergency response, which are significant for dealing with them. The importance and urgency are indicated of the environmental risk assessment, environmental emergency monitoring and environmental emergency response for hazardous chemicals pipeline accidents. Meanwhile for the existing problems, some constructive ideas are proposed.

Keywords: Hazardous chemicals, pipeline accidents, environmental risk assessment, environmental emergency monitoring, environmental emergency response

2nd CCPS
China Conference
On Process Safety

Safety Management of Hazardous Chemicals Diagnostic Pipeline

Qingdao, China
August 28 - 29, 2014

WenHui Han

China Petroleum Dushanzi Petrochemical Company
Safety and Quality Department of Environmental Protection

Email:hanwenhui@petrochina.com.cn
Tel: 13579152017

CCPS CENTER FOR CHEMICAL PROCESS SAFETY

Thinking - challenge of Accident

On March 23, 2005, an explosion at The third largest oil r efinery, BP, which is located in the United States caused serious consequences over 170 people were injured and 15 people were killed; In July 28, 2005 evening another explosion occurred in this refinery, blazing fires burning a full nine hours;

And on July 28, in the UK north sea of the BP has under gone a serious deepwater oilfield fires.

CCPS CENTER FOR CHEMICAL PROCESS SAFETY

Thoughts and Questions

What is wrong with our safety management ?
"And why BP which was considered as a very good company in safety management in the outside world happened major accidents continuously?

The challenge for the Domestic Process Safety

On July 28, 2006, the reaction kettle occurred explosion and caused 22 people killed in the Fu Yan Chemical company, Sheyang country, Yancheng city in Jiangsu province.

On August 26, 2008 Guangxi Vinylon Group Co.,Ltd. occured chemical plant exploded and 21 people were killed.

June 3, 2013, in Changchun City, Jilin Province Baoyuanfeng Poultry Industry Co. , Ltd occurred great fire explosion accident as the liquid nitrogen leaking and caused 121 died.

For the rate of the oil and gas pipeline accident is far more than in other countries, such as in the November 22, 2013 morning, Sinopec Qingdao east yellow explosion accident because of the oil pipeline leaking occurred 62 people killed.

The challenge for the Domestic Process Safety

Dalian petrochemical happened eight explosion accidents in four years

Thoughts and Questions

The establishment， implementation， measurement and ascension of management system …

- 'A fall into the pit, a gain in your wit' Be learn and progress based on 'experience' =pain
- 'Others fall into the pit, a gain in your wit' Be learn and progress based on 'meditation'=wise
- 'Need not to fall into the pit, a gain in your wit' Through continuous learning, identification, evaluation and continuous improvement=the most scientific

Safety supervision and management of the three department（2010）No.186

- Strengthening process safety management to improve the level of intrinsic safety （eight articles）
  - Strengthening **the safety management of constructio n projects**。*Design and construction of eugenics desig n and can give a good care*。
  - Carry out risk analysis process actively 。Chemical enterprises should in accordance with the (AQ/T3034-2010) ， Strengthening the safety management in chemical process in an all-round way.
  - Ensure the integrity of equipment and facilities
  - Improve the level of process automation control and security instrument vigorously。
  - Strengthening the **management of change**。
  - ……

CCPS CENTER FOR CHEMICAL PROCESS SAFETY

Safety supervision and management of the three department（2012）No.87

- Carry out the special action of "illegal rule violations" in the dangerous chemicals area in depth.
- Accelerate enterprise automation control system reform work that related with 'two focus and one major 'SIS、DCS、FCS
- Strengthen the storage device design and safety management of dangerous chemicals production. In service/new
- Carry out special projects of dangerous chemical pipelines through the public areas .
- Promote relocation of the dangerous chemicals enterprise in urban densely populated area.
- Push the key work and advance the level of safety management of dangerous chemicals. PSM/Standardization
- Improve employees and professional quality of access conditions .（Full-time safety management personnel must not be hold other department's work）
- Time: January 2013 to June 2015

2010 ,Guidelines for process safety management of chemical enterprise,(AQT3034)

CCPS CENTER FOR CHEMICAL PROCESS SAFETY

Summary of Evaluation

- 211 items on-site problem have been identified and evaluated by application of observation and communication, site survey, checklists, HAZOP, JHA, risk matrix and other methods;
- Include: entry and ability of managers, commitment and demonstration, division of institutions and responsibilities, perfection of system, improvement of training and competence ensurance of the implementation and execution, mechanism of inspection, supervision and incentives, summarization, implemention and improvement, etc. To understand the entire management process, to understand the company's current main strengths, weaknesses, threats and opportunities, to measure the status of its management system, which made more applicable improvement programs and strategies..

Discovery of Evaluation

- 2.1 Strengths:
- 2.2 Weaknesses:
- 2.3 Threats:
- 2.4 Opportunities:
- 2.5 Strategy:

Discovery of Evaluation

- 2.3 Threats：
- 2.3.1 No risk assessment management requirements and gui delines.
- 2.3.2 No valid acceptance in the whole project construction.
- 2.3.3Data transfer and training of equipment manufacturer ;A ccidents would happen constantly because of the contractor' s management, subcontract and lack of qualifications.
- 2.3.4Fire-fighting equipment has not been accepted, at the s ame time, modification rates of scene has repeatedly found i n site are very low, the problems of rectification have not solv ed from the system, mechanism, system and preventive .

Discovery of Evaluation

- 2.3.5 Important protective layer， for example，
- Safety interlock is realized by DCS control system without an ind ependent set. So it does not conform to the specifications and 'tw o focus and a significant' requirements made by Safety supervisi on Bureau ; Relief valve before hand valve was without lock man agement and failed to verify regularly; Boiler room has a big ris k. It used LPG with interlocking of electromagnetic valve of the scene and combustible gas detection, but the solenoid valve has b een damaged and have not been replaced and combustible gas de tector has not for testing and maintenance, etc.
- 2.3.6 The accident and event collection platform and learning sy stem and Learning effect validation program have not been estab lished;
- 2.3.7 Because of lack of management and technical personnel, w orkers began to appear loss, resignation, etc.

Discovery of Evaluation

- 2.5 Strategy：
- 2.5.1 Pay close attention to short slab
- 2.5.2 Change idea
- 2.5.3 Carry out felt leadership
- 2.5.4 To further strengthen the training
- 2.5.5 Promoting the construction of process safety management system and continuous improvement persistently

The monkeys pick up corn“A system, conscientiously, is culture。”

CCPS CENTER FOR CHEMICAL PROCESS SAFETY

Conclusion

- Values and driving force of leaders are the core of leadership；
- Powerful supervision from safety supervision departments at various levels is support; Fully considering the integration principle in the application；
- Secondly, in order to realize the good operation by insist on doing compulsory exercise only, no optional exercise consistently of each work and each operation；
- Each staff in department and unit always follow PDCA when carry out any work, and achieve continuous improvement；
- Normalized， regard various evaluation methods of PHA process hazard analysis as the gripper , and discover the problems, hidden dangers and risks in the process of production constantly , and according to this line of thinking to analyze and solve and off problems 。
- In this way, the concept of security is rooted in daily work and the seemingly invisible security be relected through every concrete work, that is the best explanation for the security. Reach the safety broad road eventually。

Thank you very much！

CCPS CENTER FOR CHEMICAL PROCESS SAFETY

2nd CCPS CHINA CONFERENCE ON PROCESS SAFETY™

The analysis about corrosion and corrosion control technologies of pipelines

Yan Xi
Liu Fang
Gao Ya
Zhang Li

China University of Petroleum
Qingdao economic area
Yanxi19911024@163.com
liufangfw@163.com
395071235@qq.com
940947932@qq.com

Prepared for Presentation at
2nd CCPS China Conference on Process Safety
Qingdao,Shandong
August 28-29,2014

UNPUBLISHED

The analysis about corrosion and corrosion control technologies of pipelines

Yan Xi
Liu Fang
Gao Ya
Zhang Li

China University of Petroleum
Qingdao economic area
Yanxi19911024@163.com
liufangfw@163.com
395071235@qq.com
940947932@qq.com

Keywords: Pipeline, corrosion control, prediction methods, inspection techniques

Abstract

Pipeline, as a reliable and economical infrastructure, plays a very important role in many industries like petrol, chemical and nuclear industries. Corrosion is the major indication of pipeline failure. Both internal and external corrosions are very effective in damaging pipelines. Thus, pipeline corrosion becomes a severe danger of the running pipeline system nowadays, which can lead to transported matter escapes, conduits ruptures and reduced lifetimes and all that. The disastrous consequences brought by these accidents are severe, such as economic loss and casualty as well as environment pollution. Therefore, controlling of pipeline corrosion is very vital. To control pipeline corrosion successfully and prolong pipeline life effectively, this article separately introduces corrosion phenomena and corrosion mechanism in detail. After this, the prediction methods were presented to reduce the loss caused by corrosion. Besides, the inspection techniques can analyze the reasons of the corrosion and also be applied to estimate and evaluate the corrosion control measures outcome, which make great contribution to future corrosion control. This paper proposes effective countermeasures concerning the corrosion control based on the corrosion mechanism, introduces corrosion control techniques both internally and externally.

1. Introduction

Pipeline transportation, as an economic and efficient mean of transport, has great advantage in the delivery of bulk items such as gas, liquid and slurry. Long-distance pipelines are universally used for the transportation of crude oil, natural gas and oil products and so on. However, the long-distance pipelines have to pass through desert, swamp or river, which offer pipelines a corrosive environment. Besides, the components of the transfer medium can be very complicated, which can lead to internal corrosion easily. Researches indicate that corrosion is the most important damaging mechanism of pipelines.[1] Pipeline corrosion can not only cause the transported matter escape but also lead to conduits rupture, which bring huge economic loss and environmental pollution at the same time. More importantly, these accidents put a direct threat on human health. For example, the corrosion of natural gas pipelines may results in an explosion. Considering the disastrous consequences of pipeline corrosion, it's significant and urgent for us to take pipeline corrosion under control. Internal and external corrosions are very effective on the security and integrity of pipelines for a long time, thus pipelines always need continuous inspection.[2] Besides, it's urgent and vital that effective countermeasures concerning the corrosion control be proposed.

2. Main Reasons

2.1 reasons of external pipeline corrosion

Complicated and changeable outer environments may lead to pipeline failures by corrosion. Soil, atmosphere and ocean are all able to create disasters like pipeline corrosion.

2.1.1 soil corrosion

Soil is the mixture of solids, liquids and gases. Gases and liquids filled in between the soil colloids can transfer electron. The soil ionic conductivity causes pipelines corrode easily in soils. Due to the complexity of soil, the corrosion mechanism in soil is more sophisticated than in other media. The corrosion of pipelines in soils is a major problem for managers and owners of water, sewerage and oil and gas distribution systems.

The main types of pipeline corrosion that can occur in soils are chemical etching and electrical corrosion. (a)The uneven distribution of oxygen may lead to pipeline corrosion. Dissolved oxygen in soils is partially remained in the water, while some remained in the gap between soils. The soil oxygen content is associated with the soil moisture and structure. Thus, once a buried pipeline go through soils that have varied properties, on the surface of which may form corrosion-caused galvanic cells. (b) Stray current corrosion. The portion of the current leaving the original specified conductor is called stray current. It mainly comes from the leakage of electric trains, DC welding, underground railroads and electrolyzers. Stray current existed in soils will accelerate the process of electrical corrosion. (c) Microbiologically influenced corrosion (MIC). MIC is responsible for around 20–30% of total corrosion costs.[3] Some microorganisms can induce or accelerate the reactions of the electrochemical corrosion process, this kind of material degradation is called MIC. The main microorganisms are sulfate reducing bacteria, sulphur-oxidising bacteria, methanogens. For example, the richness of sulfate in soils favors the growth of sulfate reducing bacteria, which accelerate the reaction of sulfate reduction. Produced-H_2S strengthen the acidity of the soils, which accelerate the corrosion rate.

2.1.2 atmospheric corrosion

Atmospheric corrosion is a common type of steel corrosion, which can be divided into industrial atmosphere, marine atmosphere, country atmosphere, suburb atmosphere according to the surrounding environment. The most necessary factors for atmospheric corrosion are water and oxygen. And water vapor content is the factor that decides the rate of pipeline corrosion. Thus, more close to the coast, more severe the corrosion is. In addition to the surrounding humidity and oxygen content, sulphur dioxide and chloride are the two dominant constituents that influence the corrosion rates of pipelines exposed to atmosphere. For example, in coastal atmosphere, the cathodic corrosion process is that FeOOH and dissolved oxygen reduces. In the anodic process, the chloride ions will attack the steel substrate directly after they penetrate through the rust layer. In industrial atmosphere, with the existence of SO_2 and O_2, iron can be corroded into $FeSO_4$. After that, $FeSO_4$ can be oxidized to rust, during the process of which, SO_4^{2-} can be freed. The freed SO_4^{2-} will corrode more iron. This cyclic process can produce H_2SO_4 and accelerate the rusting process.[4][5]

2.1.2 ocean corrosion

Pipelines exposed to the seawater environment are at high risk of corrosion, particularly after many years in service. The influencing factors of ocean corrosion process are as follows: salt content, dissolved oxygen, temperature, PH value and microorganisms. For example, once the temperature fluctuates, the solubility of SO3, SO2 and CO2 may increase, thus reduce the PH value and create an acidic environment to corrode pipelines. Moreover, H2S dissolved in ocean may be a source of hydrogen ions that can react with iron to form sulphide compounds, which will generate sulphur and iron oxides after its reaction with oxygen. Then sulphur may further combine with rust to form corrosion products, flakes of whom may detach from the pipeline. Hence, exposed fresh steel surface will accelerate corrosion process.[6] In an aerobic environment, the microbiologically-induced corrosion (MIC) dominates the corrosion processes in ocean instead of the usual corrosion [7]. Additionally, deep sea water with various nutrients and low oxygen levels provides an ideal anaerobic condition for bacteria such as sulphate-reducing bacteria(SRB) to proliferate which also may accelerate pipelines corrosion.

2.2 reasons of internal pipeline corrosion

The conveying medium is directly contacted with the inner wall of the pipeline. Corrosive impurities existed in the conveying medium are carbon dioxide, hydrogen sulfide, dissolved oxygen, chloride ion, sulfate reducing bacteria, high salinity and so on. Considering the impact of temperature, pressure, velocity and alternating stress, the interaction of these impurities may increase the risk of pipeline corrosion. According to the statistics [8], pipeline accidents caused by internal corrosion are more frequent than accidents caused by external corrosion. And the main reasons for pipeline internal corrosion are electrochemical corrosion and chemical corrosion.

In electrochemical corrosion process, there are isolated anodes and cathodes on the surface of the pipeline, the currents between which cause the corrosion of the pipeline. While in chemical corrosion process, there is no current exist. In the process of internal pipeline corrosion, both electrochemical and chemical reactions are vital.

2.2.1 acid corrosion

Under the condition where CO_2 is present may enhance corrosion of pipelines. According to the statistics [9], the rates of general corrosion and localized corrosion of carbon steel, sometimes as pitting corrosion can be accelerated. When CO_2 relative humidity is less than 60%, it won't corrode steel, but CO_2 can associate with water to form carbonic acid, the process of which may decrease PH value. Thus, the pipelines are at higher risk of corrosion. The pipeline surface reactions are as follows:

$$Fe + CO_2 + 2H_2O \rightarrow Fe(CO_3)_2 + 4H^+$$
$$Fe + H_2CO_3 \rightarrow FeCO + 2H^+$$

Furthermore, it has been acknowledged [10] that the corrosion rate increases as CO_2 solubility increases with the increasing temperature and CO_2 partial pressure.

2.2.2 hydrogen sulfide corrosion

Hydrogen sulfide corrosion can occur really fast, it can cause a great loss of metal in a very short period of time. H2S associate with water can form hydrosulfuric acid, which can cause pitting corrosion and cracking of the pipeline inner wall. And insoluble iron sulfide will form too in this process.

$$H_2S \rightarrow H^+ + HS^-$$

$$HS^- \rightarrow H^+ + S^{2-}$$

anodic reaction $$Fe \rightarrow Fe^{2+} + 2e^-$$

$$H^+ + 2e \rightarrow H_2$$

cathodic reaction $$O_2 + H_2O + 2e^- \rightarrow 2OH^-$$

$$Fe^{2+} + s^{2-} \rightarrow FeS$$

$$Fe + O_2 + H_2O \rightarrow Fe(OH)_2$$

$$4Fe(OH)_2 + O_2 + 2H_2O \rightarrow 4Fe(OH)_3$$

With lower solubility in water, amorphous ferric hydroxide will precipitate on the surface of the pipeline, which can form corrosion products that have loose construction. The loose construction makes the corrosion products easier to complex with hydrocyanic acid to generate new products which can accelerate the corrosion rate.

2.2.3 oxygen corrosion

The principle of oxygen corrosion of internal pipeline corrosion is similar with oxygen corrosion in soil. For the high oxygen concentration on the surface and the low oxygen concentration in deeper places of the pipeline, the steel may have different electrode potentials on the same surface. The electric potential difference between cathode with high oxygen concentration and anode with low oxygen concentration tends to make electric flows which cause pipeline pitting corrosion. The products of pitting corrosion covered on the metal surface loosely and discontinuously, which strengthen the uneven distribution of oxygen surrounding the products. For the products cutting off the oxygen transmission process, anoxic anode area can be formed blow the pitting corrosion products. Therefore, there will be an electric potential difference between this area and cathodic area surrounding the products that hastens the pitting corrosion process.

2.2.4 microbiologically induced corrosion

Microorganisms can act as inducers or accelerators of the reactions of the electrochemical corrosion process. Just as previously introduced, anaerobic condition is ideal for sulphate-reducing bacteria(SRB) to proliferate which has capacity to reduce sulphite and ferric iron to produce hydrogen sulphide. Hydrogen sulphide can attack metals surfaces directly or associate with iron to produce ferrous sulfide resulting in anaerobic pipeline corrosion and pit formation. In addition to SRB, there are other microorganisms that also play important roles in pipeline corrosion, such as iron-reducing bacteria and so on.

2.2.5 stress corrosion cracking

Stress corrosion cracking in medium transmission pipelines is also an important issue, for the leakage and failure of the pipelines can always be a potential threat. The corrosive environment in the pipeline promotes the initiation of cracking. During operation, the cracks formed before can grow from primary sizes to critical sizes which can lead to leakage, particularly in thin wall pipelines or unexpected sudden failure especially at elevated temperature. And this process is called stress corrosion cracking. Due to the chemical inhomogeneity surrounding the weld, it is easier for micro batteries to form, which accelerate the corrosion rate of this area. The pipeline suffered from stress corrosion cracking can still appear shiny and bright on the surface while the inside already filled with microscopic cracks. This is why it is hard to detect this kind of corrosion before pipeline failure.

3. Corrosion Rate Prediction Methods

Many efforts have been devoted to predicting pipelines remaining life to ensure the reliability of pipelines. Methods used to predict corrosion rate of pipelines are grey theory, statistics analysis, neural networks and reliability function etc., in which, neural networks was applied mostly. [11]

3.1 neural networks

BP neural network was applied widely in corrosion field.[12] Prediction of pipelines corrosion rate based on BP neural networks including learning modeling and outside forecast. During this process, a sample set will firstly be established on the basis of measurements of all the corrosion factors. After this, BP Neural Network will predict the unknown pipelines corrosion rates by automatically learning and acquiring knowledge. This way can predict the pipeline service life accurately and conveniently without considering various impacts on the pipeline that different factors have. However, the structure is complex. And according to Yan [12], the prediction results of BP neural network are not as good as RBF neural network model' s when predicting steel corrosion. The RBF neural network is a novel feed-forward artificial neural network, which has a simple structure and much excellent performance such as best approximation, global optimum, convergence rate and so on. He's [11] investigation shows that the precision of ANFIS model is better than that of RBF model on the prediction of steel pipelines corrosion rate in soil. ANFIS was based on the firsrt-order Sugeno fuzzy model. Rule base, database and reasoning mechanism are the three conceptual components of a fuzzy inference system (FIS) basic structure.

3.2 reliability function analysis

The probability distribution of pipeline corrosion failure can be achieved after collecting the information about the sizes of corrosion defects and the development law of corrosion, which can be used to build a probability distribution model of pipeline corrosion rate. The relationship between failure probability and time then can be calculated. Give the pipeline remaining lifetime a risk assessment then get its reliability.

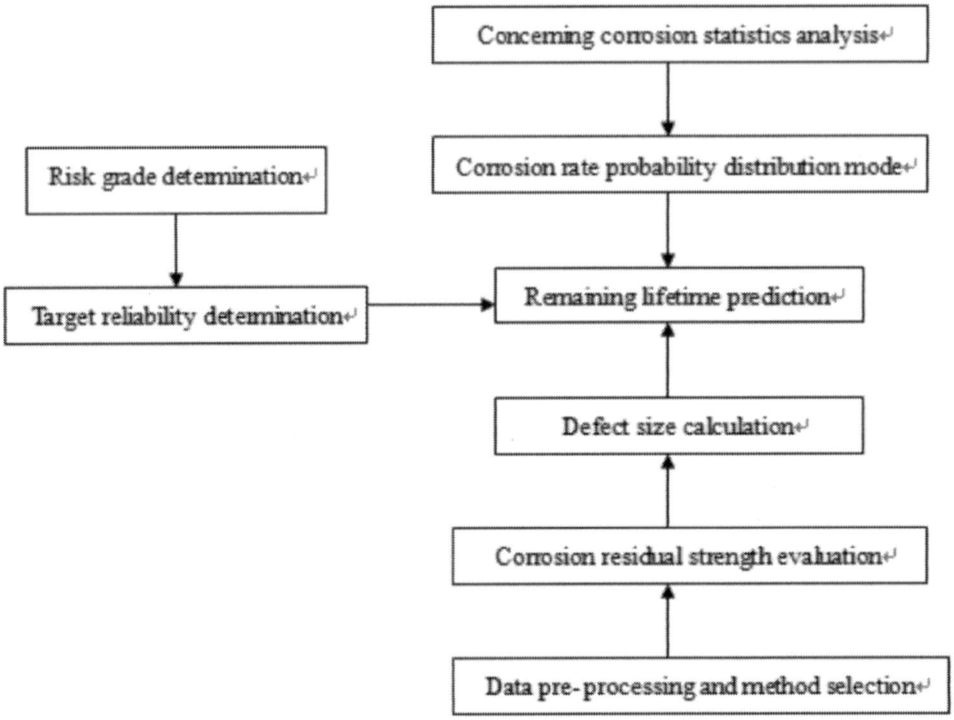

3.3 statistics analysis

Establish the probability distribution model of defects by Webb probability density distribution function according to the statistics of defect characteristics, which can describe corresponding pipeline corrosion failure characteristics of different stages to reflect the stage of corrosion failure or the severity of it. [13] This method on the prediction of pipeline corrosion rate is more scientific for the data is on the basement of both measurement and statistics, which provide a reliable basis for pipeline remaining lifetime assessment.

3.4 grey theory

In the 1980s, Chinese scholar Professor Deng Julong initiated the grey system theory which is a scientific and systematic theory. [14] Currently, it has been widely used in the prediction of pipeline corrosion rate. The theory committed to solve uncertainty problems without empirical date. According to the grey theory, the limited and seemingly unrelated data need to go through generation processing to get generated data to establish prediction model, which can reveal the potential rules of the system. Using dynamic memory characteristics of corrosion rate to establish grey differential equation model so that the trend of pipeline corrosion can be predicted.

4. Corrosion Control Techniques

4.1 external corrosion control techniques

4.1.1 anti-corrosive layer

Researches indicate that surface pipelines external corrosion is one of the most important reasons for making leakage. [1] Thus, all the pipelines should be protected with anticorrosion coating. Anti-corrosive layer is a primary mean of mitigating pipeline corrosion, which can protect pipeline from corrosion medium touching to fundamentally prevent electrochemical corrosion. The materials of insulating layer must be corrosion resistant, water-proof and chemical properties stable. Additionally, to ensure the anti-corrosive layer a higher efficiency, gapless insulting layer must be guaranteed. Current coatings used on pipeline corrosion control are FBE (fusion bonded epoxy), coal tar, asphalt enamels and 3LPE (3-layer poly ethylene).

4.1.2 catholic protection

As coatings cannot stop pipeline corrosion completely, a supplementary system is needful to provide protection for pipelines. This system is called catholic protection, which can provide extra currents to stop or decrease the corrosion rate of pipelines. There are two cathodic protection systems: sacrificial (or galvanic) anode system and impressed-current system. (1) Sacrificial anode system. Connect a metal or an alloy anode with lower electric potential to the protected pipeline for forming a new corrosion cell. Thus, the new formed anode will corrode and discharge an electric current to the pipeline. As long as the residual current is equal to or lesser than the cathodic protection current cathodic, the residual current can be mitigated. Therefore, the pipeline which becomes a cathode can be protected. The most commonly used anode materials are magnesium based alloys, aluminum based alloys and zinc based alloys. (2) Impressed-current system. In this system, the cathode of an external current source can be attached to the pipeline. Thus, the current flows from the current source auxiliary anode to the pipeline's external surface for the electric potential difference between the anode and the pipe to achieve cathodic polarization so that corrosion can mitigate or stop. The anode of the external current source can be graphite, high silicon cast iron, cast iron, carbon steel, magnetic iron oxide etcetera.

4.1.3 connecting line

As previously introduced, stray currents coming from the leakage of electric trains, DC welding, underground railroads and electrolyzers can accelerate the process of pipe corrosion. Connecting line is the proper way to take stray currents under control, thereby stop this kind of corrosion.

4.2 internal corrosion control techniques

4.2.1 anti corrosion coating

Just as external anti-corrosive layer can mitigating pipeline corrosion for protecting pipeline from corrosion medium touching, this technology can also be used in internal corrosion control. FBE and cement mortar etcetera are proper anti corrosion coating material for sewage pipelines and water injection pipelines.

4.2.2 corrosion inhibitors

The polar groups of these corrosion inhibitors' molecules can absorb on metal surfaces by physical or chemical action. The purpose of the absorbed inhibitors is transforming the charge distribution and interfacial properties of the surface metal to stabilize the surface energy states. Besides, the exposed non-polar groups can form a layer of hydrophobic and protective film to stop the transfer of corrosion reaction related electric charge or substance. Dosing inhibitors is an effective anti-corrosion measure. According to Liu [15], under the optimal conditions (the dosage of inhibitor HZ-2 is 350mg/L), the corrosion inhibitor rate reached 89.78%. Only slight corrosion was observed on the surface.

4.2.3 natural gas desulphurization and dehydration

Dried sour natural gas does not corrode pipelines under room temperature. Hence, when gas desulphurization can not eliminate pipelines corrosion completely, the most reasonable method is gas dehydration. Control the dew point temperature below the minimum operating temperature to prevent electrochemical corrosion and SSC.

4.2.4 sterilization

In addition to countermeasures like inhibitors and anti corrosion layer, agents such as fungicides and deoxidizers should be adopted to avoid pipeline corrosion. In areas where sulphate-reducing bacteria exist abundantly, fungicides and UV disinfection should be adopted at the same time.

5. Pipelines Corrosion Inspection

To obtain corrosion information of pipelines, pipe inspection is an effective way. Modern domestic and overseas inspection methods are introduced to improve the competence of pipeline inspection and evaluation.

5.1 external inspection methods

The key point of external inspection methods is to inspect the effectiveness of coating and cathodic protection. Use non-digging and non-destructive detection technologies to inspect the corrosion state of pipelines.

5.1.1 DCVG

The principle of DCVG is to detect the coating failures by inspecting the voltage gradient change of pipelines. When a DC signal is applied to the pipe, there will be a voltage gradient between pipe breakage spot and soil. For the current density increases at the place near the damaged spot, the voltage gradient increases correspondingly. And voltage gradient is generally proportional to the exposed area. This method can determine the corrosion state of the defect spot by the current flow direction. Currents flow from pipe to soil indicating the part near the breakage spot is under the state of corrosion.

5.1.2 PCM

PCM can be used to detect the breakage spot and the dimension of it, ensure the pipe distribution and inspect the validity of the cathodic protection system. Just as the principle of DCVG, when a AC signal is applied to the pipe, there will be a voltage gradient between damaged pipe and soil. Since the coating resistance and the distributed capacity exist, the strength of the current signal show a exponential decay trend as the pipe length increases. If the coating has a defective point, the currents would flow from the point to soil, therefore, the strength of which would decrease suddenly. The insulation capacity of the anti-corrosion layer can be evaluated by the decrement of the currents.

5.1.3 Pearson

Pearson is a method based on the principle of DCVG. AC signal applied to the pipe can produce currents, which cause a voltage gradient between the pipe defecting point and the soil. Then the detector would acquire the potential difference to convert it to amplified audio signal so that the location of the breakage point can be found.

5.1.4 CIPS

The results of CIPS can estimate the validity of cathodic protection system. The detector is made up of three parts: high-sensitive millivoltmeter, Cu/CuSO4 half cell probe and tail wheel. During measuring, the interrupter is in series with the cathodic protection power output line. Therefore, when the interrupter disconnects or connects with the output line in a certain period, the collector can automatically record the protection potential and the nature potential. CIPS combined with DCVG can not only evaluate the state of pipe cathodic protection system comprehensively but also detect the coating leakage point precisely.

Table 1. The Comparison of External Inspection Methods

Items	Application	FunctionColumn Title
DCVG	few	detect coating damage

PCM	moderate	detect pipe distribution and coating insulation capacity
Pearson	widespread	locate the breakage point
CIPS	widespread	estimate cathodic protection validity; detect coating failure area
DCVG&CIPS	research phase	estimate cathodic protection validity; locate the breakage point

5.2 internal inspection methods

External inspection technologies cannot achieve a comprehensive outcome for its indirection. While internal inspection methods can sincerely detect not only the dimension and basic condition of the pipeline but also the corrosion situation of it.

Nowadays, there is a new method for inspecting pipelines, called pigging. A pig is a device which travels freely through a pipeline to inspect the condition of the pipe and location of any problem. After this, it provides certain date about pipe integrity so that potential problems can be mitigated before it is created. [16] Magnetic flux leakage and ultrasonic tool are the key technologies of internal pipe inspection.

5.2.1 magnetic flux leakage tool

MFL pig induces a strong magnetic flux into the pipeline wall. The distribution of magnetic lines of force is uniform when the pipe is intact. But a flaw in the pipe wall can cause flux leakage which can be detected by the sensors. It is found that there is deformation of the magnetic lines of force.

5.2.2 ultrasonic tool

UT uses ultrasonic technology to detect the flaw in the pipeline. In this method, a transducer transmits signals to the pipe wall at first. Then the time taken to receive the signals both from the internal wall and the external wall can be used to determine the thickness of the pipe wall. But this technology requires the pipe wall to be clean. Besides, it works well with heavy-wall pipe, but not with thin-wall pipe. So it does not use widely as the MFL tool. [17]

6. Conclusion

Pipelines transport many dangerous items such as natural gas and crude oil. Thus, pipeline safety is a very significant matter. As corrosion is the most important damaging mechanism of pipelines, it has to be prevented and controlled. Thus, the forming mechanism and prediction methods of corrosion must be learned at first. Then, anti-corrosion layer, cathodic protection system and other technologies are used for corrosion control. But, sometimes corrosion occurs, even by using these countermeasures. Therefore, pipeline inspection is also a very necessary part for corrosion control. Then plans come out for corrosion management in pipelines systems.

7. References

[1] MKarami, "Review of Corrosion Role in Gas Pipeline and Some Methods for Preventing It," *Journal of Pressure Vessel Technology*, Volume 134, NO. 5, 2012

[2] Choi, et.al.. "Development of Limit Load Solutions for Corroded Gas Pipelines," *Pressure Vessels Piping*, Volume 80, 2003

[3] Corinne Whitby, "Applied Microbiology and Molecular Biology in Oilfield Systems," ISBN978-90-481-9251-9

[4] J.H.Dong, "Rusting Evolution of Mn–Cu Alloying Steel in a Simulated Coastal Environment," *Corros. Sci. Prot. Technol*, Volume 22, 2010

[5] Wenjuan Chen, et.al.. "Effect of Sulphur Dioxide on the Corrosion of a Low Alloy Steel in Simulated Coastal Industrial Atmosphere," *Corros. Sci*, Volume 83, 2014

[6] Yikun Wang,et.al.. "Ultimate Strength Analysis of Aged Steel-plated Structures Exposed to Marine Corrosion Damage : A review," *Corros. Sci*, 2014

[7] D.J.Howarth, "Its Impact on Modern Shipping with Particular Reference to Crude Oil Tankers, in: Lloyd's Register Technology Days," http://www.lr.org/Image/proceedings2011tcm155-206831.pdf (accessed August, 2013)

[8] Bullard S J, et.al.. "Soil Corrosion Monitoring near A Pipeline under CP. Florida: Natural Gas Institute," *Natural Gas Technologies*, 2005

[9] S.Nesic, "Key Issues Related to Modeling of Internal Corrosion of Oil and Gas Pipelines — A Review," *Corros. Sci*, Volume 49, 2007

[10] L.Zhang, "Corrosion and Stress Corrosion Cracking Behavior of X70 Pipeline Steel in a CO2-Containing Solution," *Journal of Materials Engineering and Performance*, Volume 18, NO.3, 2009

[11] San He, "Application of RBF Neural Network and ANFIS on the Prediction of Corrosion Rate of Pipeline Steel in Soil," *Electrical Engineering*, Volume 124, 2012

[12] Wang, S., et.al.. "Application of MATLAB Neural Network in Soil Corrosion Evaluation of Pipeline," *OGST Journal*, Volume 28, 2009

[13] Haiqiu Wang, et.al.. "Corrosion Failure Probability Statistics and Prediction Models of Oil Pipelines," *Oil-Gasfield Surface Engineering*, Volume 26, 2007

[14] Xuemei Shen, Zhengnan Lu, "The Application of Grey Theory Model in the Predication of Jiangsu Province's Electric Power Demand," *Science Direct*, Volume 7, 2014

[15] YuchengLiu, et.al.. "Influence of Produced Water with High Salinity and Corrosion Inhibitors on the Corrosion of Water Injection Pipe in Tuha Oil Field," *Engineering Failure Analysis*, Volume 45, 2014

[16] Kishawy,H.A., Gabbar,H.A., "Review of Pipeline Integrity Management Practices," Int.J. *Pressure Vessels Piping*, 2010

[17] Sadeghi Meresht,E., et.al.. "Failure Analysis of Stress Corrosion Cracking Occurred in a Gas Transmission Steel Pipeline," *Eng. Failure Anal.*, Volume 18, 2011

Seismic performance analysis of pipeline with different corrosion defects

Guan Youhai, Wang Yazhou, Jia Juanjuan, Su Ning

1 Associate Professor, Collage of Pipeline and Civil Engineering, China University of Petroleum,66# Yangtze River Road West, Qingdao Economic-Technological Development Area, Shandong, China,266580; Tel:13864220186;Fax:0532-86981138□Email:guanyouhai@upc.edu.cn;

2 Graduate Students, Collage of Pipeline and Civil Engineering, China University of Petroleum,66# Yangtze River Road West, Qingdao Economic-Technological Development Area, Shandong, China,266580; Tel:13553061563;Fax:0532-86981138□Email:dongewyz@gmail.com;

3 Graduate Students, Collage of Pipeline and Civil Engineering, China University of Petroleum,66# Yangtze River Road West,Qingdao Economic-Technological Development Area, Shandong, China,266580; Tel:15264252651;Fax:0532-86981138□Email:845985540@qq.com;

4 Graduate Students, Collage of Pipeline and Civil Engineering, China University of Petroleum,66# Yangtze River Road West,Qingdao Economic-Technological Development Area, Shandong, China,266580; Tel:13687689424;Fax:0532-86981138□Email:147405236@qq.com;

Abstract: To compare the seismic performance analysis of pipeline with different corrosion defects, the finite element software ADINA was used to build pipeline under earthquake and liquid-solid coupling model of internal fluid. By comparing different etching depths (selecting corrosion depth of 30%, 50%, 70%, 90% of the intact pipeline) with the peak strain and stress value in different corrosion positions (internal corrosion and external corrosion) and strain and stress at a particular point, we can analyze effects of corrosion depth and internal and external corrosion on the seismic performance of the pipeline and have a significance of guidance to the pipeline maintenance in practice as well. It shows that external corrosion has greater effects on the seismic performance of pipeline than internal corrosion; with the increase of corrosion depth, the growing speed of pipe stress is increasing; either internal or external corrosion, the pipeline is in danger of breaking under earthquake when pipeline corrosion depth is more than 50%.

A Brief Review on Hazard Material Management and Incident Response for Eastern Asia Countries

Jao-Jia Horng
Department of Safety, Health, and Environmental Engineering, Yunlin University of Science and Technology (YunTech)
horngjj@gmail.com

Kuang-Yu Liao
Liao Yuan-Chun
Graduate School of Engineering Science and Technology, Yunlin University of Science and Technology (YunTech) Emergency Response Information Center
liaoky@gmail.com
steveycliao@gmail.com

Keywords: Chemical; Hazardous Material (HAZMAT); Regulation; Incident response; Development trend

Abstract

Chemicals have brought us prosper civilization, successful industrialization, high living standard, many convenient products and, hence, many chemical products. Thus, those useful hazardous materials (HAZMAT) were also threatening our daily life through their dangerous properties during production processes and abundant applications. Since the Bhopal disaster in 1984, there were many stunting events by the leaks, fires, and explosions of dangerous chemicals and even caused international pollution incidents. Those events often attracted attentions from general public as well as international societies and organizations. Many developed countries had their eyes on the safety and fates of chemicals, and they developed regulations of management and responsibility, encouraged research and development on green chemistry, and prepared toward emergency incident response with scientific knowledge and applications. In addition, many international norms were discussed and implemented, including GHS (Global Harmonized System), NIOSH(National Institute of Occupational Safety and Health), REACH (Regulation for Registration, Evaluation, Authorization and Restriction of Chemicals), SAICM (UN Strategic Approach to International Chemical Management), WHO (World Health Organization) and WSSD (World Summit on Sustainable Development). This paper briefly reviewed the management and response regulation and measures developing in Japan, South Korea, Taiwan and China as a comparison of development trends for the past decades for dangerous chemicals.

1. Introduction

Chemicals have brought us prosper civilization, successful industrialization, high living standard, many convenient products and, hence, many chemical products. Thus, those useful dangerous materials (HAZMAT) were also threatening our daily life through their dangerous properties during production processes and abundant applications. Since the Bhopal disaster in 1984, there were many stunting events by the leaks, fires, and explosions of dangerous chemicals and even caused international pollution incidents. Those events often attracted attentions from general public as well as international societies and organizations. In 1989, ABS resin plant explosion incident in Korea caused 13 death more than 20 injured and 3 billion won loss [1]. On March 20, 1995, the Tokyo subway was attacked by terrorist using sarin killing 13 people, injuring 50 and causing temporary vision problems for 1,000s [2]. In 2001, there was a terrible run-away explosion left several reaction and storage tanks at Fu-Kao chemical plant in Taiwan causing 1 death, 112 injured people and huge financial losses [3]. In 2005, China Jilin benzene plant exploded causing 5 kills, 60 injured and river pollution extending to Soviet. This incident has become a international environmental pollution event because of the large amounts of wastewater from fire-fighting and chemicals infused into the river toward ocean [4].

With increasing HAZMAT incidents of leaks, fires, explosions around the world in recent years, the management and response to HAZMAT incidents have caught attentions in many developed and developing countries. A large number of manpower and resources were invested, many specialized agencies were set up, and related laws and regulations were established to form scientifically-based chemical incident response systems [5]. Since 1992 World Summit discussed the safety management of hazardous chemicals [6], many international norms were formed to demonstrate national and international-wide management commitment on chemical use, including GHS (Global Harmonized System) [7], NIOSH (National Institute of Occupational Safety and Health) [8], REACH (Regulation for Registration, Evaluation, Authorization and Restriction of Chemicals) [9], UN's SAICM (Strategic Approach to International Chemical Management), WHO (World Health Organization) and WSSD (World Summit on Sustainable Development) [10].

In order to establish good management and response system, we must understand the development trends firstly. Therefore, this paper reviewed the status of management and response to HAZMAT incidents in four East Asian Countries, including China, Japan, Korea and Taiwan.

2. Developments of HAZMAT management in China, Japan, Korea, and Taiwan

2.1 China

The regulation of safety management regarding dangerous chemicals was established in 1987 to enhance the safety supervision, process safety, preserve the environment and public health, and it was replaced by the regulation on the safety administration of dangerous chemicals in 2002. There were more than 10 departments were involved in those management efforts. Then, it was revised again on December, 2011 as the main regulation of Decree No. 591 [11] and followed by several Measures on management hazardous chemicals administrated by various departments.

On 19 Jan 2010, the Ministry of Environmental Protection (MEP) of China released the revised version (Order No. 7) of the Provisions on Environmental Administration of New Chemical Substances. The new regulation replaced the old regulation issued in 2003 and came into force on 15 Oct 2010. This regulation is similar to EU REACH and is also known as "China REACH" and the inventory was updated in 2013. The catalog of hazardous chemical was revised in 2013 to form China GHS with several national standards as GBs since 2006 [12]. The import/export of toxic chemical was regulated by the Provision early in 1994 and revise in 2009 and 2011. Their chemical inventory showed that there are listed on the Inventory of Existing Chemical Substances Produced or Imported in China (IECSC) for 45,612 chemicals (2013) existed in China currently and 3270 were listed in "the List of Dangerous Chemicals" and 335 as the toxic chemicals.

Under the regulation, all companies need to submit new chemical substance notification to the Chemical Registration Centre (CRC) of the Ministry of Environmental Protection (MEP) for the new chemicals of certain annual tonnage. Thus, those chemicals other than those 45,612 substances currently listed on IECSC.

China's chemical legislation also covers new chemical substance, existing substance, toxic chemicals, hazardous chemicals, restricted or prohibited chemicals, food additives, cosmetics and its ingredients, pesticides and dangerous goods, etc. With most of those regulations have been revised by Chinese government since 2011, the determination of government to overhaul the chemical management is obvious.

2.2 Japan

With early industrialization, Japanese Chemical Substances Control Law (CSCL) was initially implemented on 16 April 1974, and the latest amended law was published on 20 May, 2009 and effective on 1 April ,2011. This CSCL is commonly referred to as Japanese REACH Regulation [13,14]. The three government bodies responsible for the implementation of CSCL are the Ministry of Economy, Trade and Industry (METI), Labor and Welfare (MHLW), and the Ministry of the Environment (MOE). Although Japan CSCL covers general industrial chemical products only, it include both new chemicals and existing chemicals. Thus, food or feed additives, pharmaceuticals, cosmetics, agricultural chemicals and fertilizers are subject to different laws and acts. And, general industrial chemical products might also be subject to the requirements of the Industrial Safety and Health Law (ISHL), the Poisonous and Deleterious Substances Control Act, the Air Pollution Control Law, and the Water Pollution Control Act.

Japan CSCL regulates the following categories of chemicals (Fig. 1) [15]: new chemical substances; general chemicals; priority assessment chemical substances; monitoring chemical substances (~38 substances); Class II specified chemical substances (23 substances); and Class I specified chemical substances (16 substances). It followed the chart below. New rules and regulations laid down the main hazard from the past into a risk-based one, from April 2011 it start a new law Concerning the Examination modify the new chemical species distinguish then gradually mining risk assessment classification. From now on Japanese regulations (CSCL) managed a total about 20,000 kinds of chemicals, it's classified as:

- Class I specified chemicals: Persistent, highly bio-accumulative, a total of 30 kinds.
- Class II specified chemicals: The risk of long-term toxicity to humans or animals and plants in the physical environment of human life, a total of 23 kinds.
- Monitoring of chemical substances: Monitoring chemicals are persistent and bio-accumulative high, the long-term toxicity to humans or animals, prey high trophic level unclear substance, a total of 37 kinds.
- Priority Assessment Chemical Substances: Long-term toxicity of substances to humans or animals and plants is not yet clear, to be confirmed, a total of 212 kinds.
- The new release of chemical substances: Not classified as Class I specified chemical substances, monitoring a substance type II or type III monitoring chemical substances of chemical substances, a total of 6778 kinds.

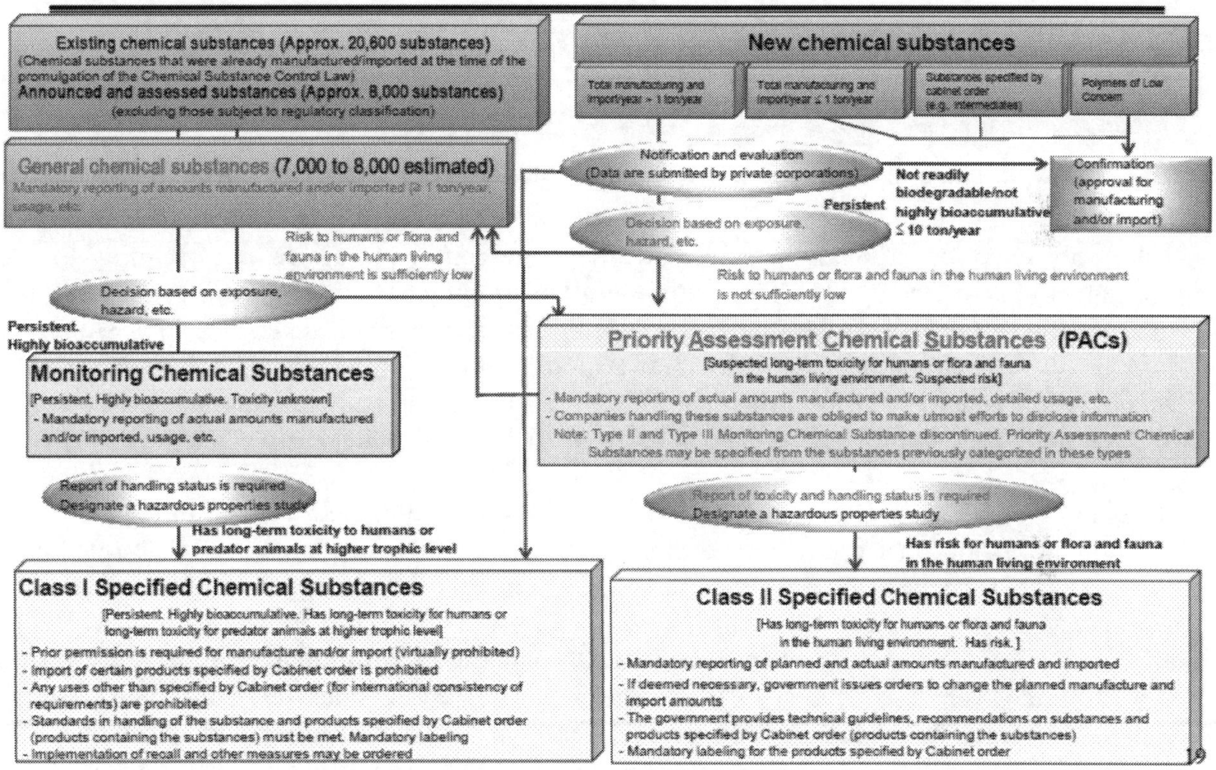

Fig. 1. Regulation Structure of Japan's CSCL

Japan's chemical managed also extended to several fronts.

- Mandatory Reporting Requirements: Manufacturers and importers of the chemical substances below will be required to report their quantities, shipment and usage (use category) to the government annually between 1 April and 30 June starting from 2011:
- Some substances are exempt from mandatory reporting, for example, intermediates (excluding imported intermediates), low risk substances (water, etc), substances for R&D, substances with tonnage band less than 1 ton per year, and PLC.
- Information Delivery in the Supply Chain: When Monitoring Chemical Substances or Priority Assessment Chemical substances are transferred between legal entities, the entity transferring the substance is obliged to make utmost efforts to inform the recipient that

the substance transferred is specified as Monitoring Chemical Substance or Priority Assessment Chemical substances. This is very similar to the communication obligations related to substances of very high concern (SVHC) under REACH.

- GHS Implementation in Japan: The main law introducing GHS to Japan is the Industrial Safety and Health Law (ISHL). ISHL requires GHS classification data on labeling in MSDS.
- Article 57 of ISHL regulated 99 designated chemicals and SDSs for 640 designated chemicals.
- Besides ISHL, the law concerning reporting, etc. of the release to the environment of specific chemical substances and promoting improvement in their management (PRTR law) has also required MSDS for over 400 designated chemicals.

Since Japan's government already revised and enforced those regulations since 2011, her chemical management was relatively advanced among the countries in East Asia.

2.3 Korea

The main regulation in was Toxic Chemicals Control Act (TCCA) and it began to formulation from 1963-1990 to prevent the occurrence of Hazard poisoning. In the second phase of 1990-1996, all chemicals began a systematic management. Due to Korea joining the OECD (Organization for Economic Co-operation and Development) [16] after 1996. From 1996 to 2005, many test was introduced like the U.S. Environmental Protection Agency's TRI (Toxic Release Inventory) [17], Laboratory GLP Principles (Good Laboratory Practice) and Risk Assessment [18]. Since 2006, Korea began to implement the new management of chemical hazards by prohibiting and restricting the use of chemicals by the health risk assessment.

In 2011, Korea began practice REACH (Fig. 2) and established evaluation system aimed to provide an institutional framework of preventive chemical management. Updated in June 2013, the Act on the Registration and Evaluation of Chemicals (known as Korea REACH) passed the National Assembly in Korea on April 30, 2013 and will come into force on Jan 1, 2015 (showed as Fig. 3). This Act is to protect public health and the environment through these provisions: Registration of chemical Substances; Screening of hazardous chemical substances; Hazard and risk assessment of products containing chemical substances and hazardous substances; Sharing information of chemical substance. The Ministry of Environment (MoE) was given the responsibility for the registration and evaluation of chemical substance under this Act.

The current TCCA which came into force on 2 Feb 1991 will be divided into Korea REACH and Chemicals Control Act (CCA) on Jan 1, 2015. KOREA REACH focuses on registration and evaluation of substance while CCA focuses on the control of hazardous substance and response to chemical accidents. Korea REACH does not apply to many products as the follows.

- Radioactive substances of the Atomic Energy Act;
- Pharmaceuticals and non-pharmaceutical drugs of the Pharmaceutical Affairs Act; Narcotics of the Act on the Control of Narcotics;
- Cosmetics and materials of the Cosmetics Act;
- Ingredients and agrochemicals of the Agrochemicals Control Act; Fertilizers of the Fertilizer Control Act;

- Food, food additives, tools, containers and package of the Food Sanitization Act;
- Livestock feeds of the Act on Control of Livestock and Fish Feeds;
- Ammunitions of the Act on the Control of Firearms, Swords, Explosives; Military supplies of the Act on the Management of Military Supplies and of the Defense Acquisition Program Act (except for commercial goods under Article 3 of the Act on the Management of Military Supplies;
- Health functional food of the Health Functional Food Act; Medical devices of the Medical Device Act.

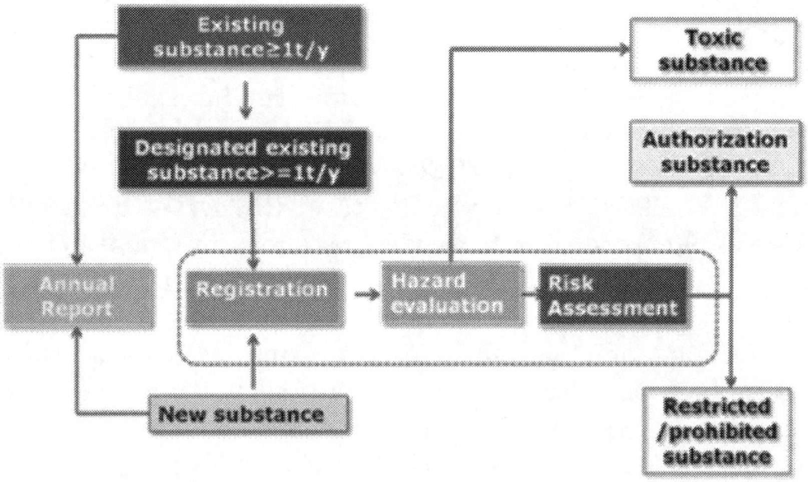

Fig. 2. Summarization of the Korea REACH

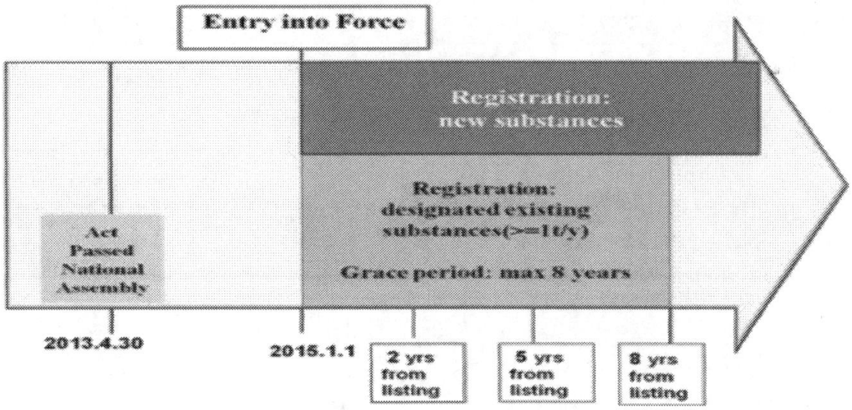

Fig. 3. Implementation Schedule of Korea REACH

Manufacturers and importers of the chemical substances will be required to report their quantities and uses to MoE annually on new chemical substances and existing substances manufactured, imported or sold more than 1 ton per annum. At present, Korea regulation managed the total of about 42,000 kinds of chemicals, Classified as, the follows [19, 20]:

- Toxic Chemicals: harmful to human health or the environment, a total of 688 kinds.
- Observation chemicals: Being likely to be harmful to human health or the environment, a total of 107 kinds.
- Restrict or banned chemicals: Severely harmful to human health or the environment, a total of 92 kinds of restrictions (such: PCBs, etc.) / disable Total 104 kinds (such as

carbon tetrachloride, etc.).

Accident prevention chemicals: acutely toxic and explosive, High acute toxicity, explosive hazard, etc. and thus presents a high risk at an accident, a total of 69 kinds (such as phosgene, chlorine, etc.).

2.4 Taiwan

Taiwan's Toxic Chemical Substances Control Act (TCSCA) was issued in Nov 1986 and has gone through 5 revisions with the latest amendment done in 2007. It is the main legislation for industrial chemicals in Taiwan and its primary focus is on the control of toxic chemical substances, requiring business operators to apply for handling permits for designated toxic chemical substances from Environmental Protection Administration (EPA). The newly revised Toxic Chemical Substances Control Act (TCSCA) in Taiwan was officially promulgated on 11 December 2013. This revised TCSCA took the spirit of REACH which requires enterprises to register new substances 90 days prior to production or importation and register designated existing substances manufactured or imported above a certain quantity. The Act also strengthens the management of its Class 4 toxic chemical substances. Enterprises are required to declare relevant toxicological information and obtain approval from competent authorities for Class 4 toxic chemical substances prior to handling. The provisions pertaining to chemical registrations are expected to come into force on 11 Dec 2014 while other provisions will come into force at the date of promulgation. Table 1 below showed the definitions of 302 chemicals categorized to Class 1, 2, 3 & 4 as they regulated under TCSCA.

Table 1. Definitions of 302 Chemicals Categorized to Class 1, 2, 3 & 4 under TCSCA

<table>
<tr><th>Category</th><th>Class 1</th><th>Class 2</th><th>Class 3</th><th>Class 4</th></tr>
<tr><td>Hazard</td><td>Persistence or hard to break down in environment and could pollute the environment and endanger human health due to bio-accumulation, bio-concentration, or bio-transformation.</td><td>may cause cancer, infertility, birth defects, genetic mutations, or chronic diseases.</td><td>chemicals exposure will immediately endangered human health and environment</td><td>Risk to endanger human health and the environment.</td></tr>
<tr><td>Management</td><td colspan="3">1.need permit on manufacture, import or sell toxic substances more than the large scale handling standard
2.require registration(use, store and dispose toxic chemical substance more than the large scale handling standard) – much like reporting;
3.require approval(manufacture, import, sell, use, store or dispose toxic chemical substance less than the large scale handling standard)</td><td>Need submit and approve Toxicological information of toxic chemical substances.</td></tr>
<tr><td>Label and SDS</td><td>yes</td><td>yes</td><td>yes</td><td>Yes</td></tr>
</table>

The main reason for this amendment is to develop chemical registration scheme in Taiwan. After many East Asian countries (Korea, China, Japan) have implemented REACH-like chemical legislation and required risk assessment of chemical substances, Taiwan may become a test ground for new substances if not having this new amendment. In addition, an event of the food plasticizer contamination incident in 2011 made government realize that no sufficient power under TCSCA (thus, request for toxicology data or requiring license) for controlling the environmental pollution or endangering human health. In addition to the establishment of a chemical registration scheme, the revised TCSCA also strengthens the management of Class 4 toxic chemical substance.

Taiwan's Ministry of Labor completed chemical substances login and listed 79,000 chemical substances with 94% chemical imported. The operation more than one tone per year included 31,600 chemicals. The management authority of hazardous chemicals or substances in Taiwan scattered among government agencies and departments, including the Ministry of Labor, Council of Agriculture, Ministry of Economic Affairs, Environmental Protection Agency, Ministry of Transportation, and Ministry of the Interior fire department, etc. There were briefed as follow.

- Ministry of Labor on Occupational Safety and Health Facilities for the explosive, fire resistance, oxidation resistance, flammable liquids and other dangerous items and organic solvents, specific chemical substances;
- Council of Agriculture on pesticide management;
- Ministry of Economic Affairs on explosives and petroleum products.;
- Ministry of Transportation on dangerous goods transportation safety;
- Fire Department of Ministry of Interior onpublic hazardous materials and flammable high pressure gas.

Among those East Asian countries, the classification of chemicals could be listed as the follows.

Table 2. Comparison of Chemicals Classification in China, Japan, Korea and Taiwan

Classification	China	Japan	Korea	Taiwan
Bioaccumulation, bioconcentration, biotransformation	335 highly toxic chemicals and 3270 in confidential section	Class III Monitoring Chemical Substances	Toxic Chemicals	Class I: Persistenc e
Cause cancer, impaired fertility, teratogenic		Class II designated	Restriction / prohibition Chemicals	Class II
Immediate harm human health or biological life		Class I specified chemical	Accident Prevention Chemical	Class III
may pollute the environment or endanger human health		Priority Assessment Chemical	Observation Chemicals	Class IV

3. Development of HAZMAT response in China, Japan, Korea and Taiwan

It is known that chemical leaking or fire may cause life and property losses as well as environmental consequences. Hence, the prevention measures and emergency responses are essential to decrease the impacts of those events. In responding to those needs, most industrialized countries set up their own chemical emergency information center (CERCs) and poison control center (PCCs). Those centers provide 24-hour emergency call system to provide emergency response information. While PCCs provide emergency medical advice by medical professionals during toxic emergency, CERCs provide chemical emergency information to emergency responders of fires, environmental spills and vapor dispersion. Sometimes, CERCs could dispatch professionals to assist at the site in high-level emergency services. The operations of CERCs were supported from initiatives of private sectors (CHEMTREC of American Chemistry Council and ICE of European chemical industrial association (CEFIC) or by government (CANUTEC of Canada by Transport Canada). Those organizations were shown in the Fig. 4.

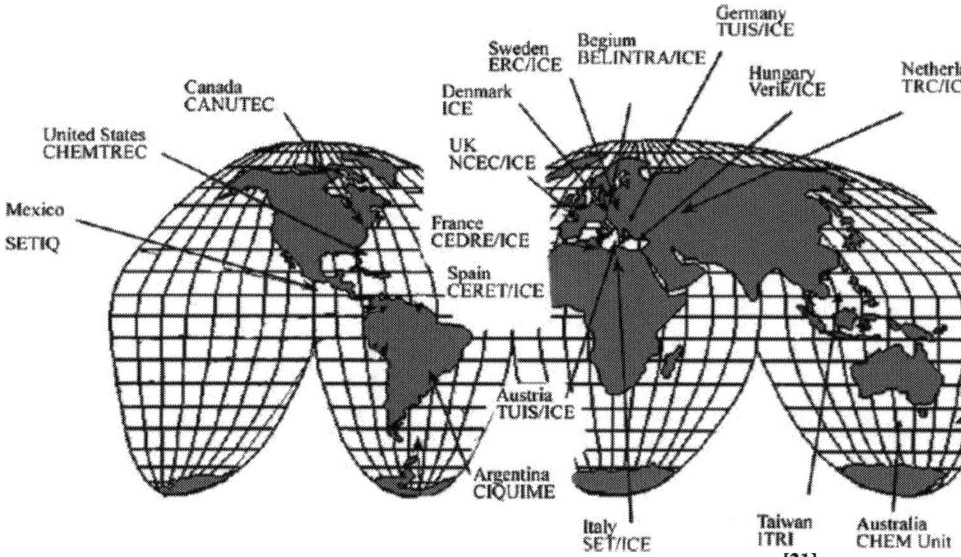

Fig. 4 The Distribution of CERCs Globally [21]

3.1 China

China's response to environmental pollution incidents developed relatively late but with rapid development as incidents and public attentions both raising quickly. National Environmental Protection Administration set up the emergency measures investigation team of maritime pollution On April,1984, and it further established a emergency program for marine pollution in 1988 [22-23]. With more and more development of chemical related industries in recent years, it has been important for responding to chemical accidents actively [22]. In 1994, the Ministry of Chemical Industry (MCI) formulated and promulgated the "Measures for the administration of chemical accident emergency rescue". In 1996, the Ministry of Chemical Industry (MCI) and the State Economic and Trade Commission jointly issued "The notice of the formation "chemical accident emergency rescue system" and established a national command center of chemical accident emergency rescue and formed eight centers for responding to chemical accident by region. The command center has the responsibility for the command task of significant and large chemical accident emergency rescue. The office has the responsibility for communication task of significant and large chemical accident emergency rescue, and guided operation for rescue center. The response centers have the responsibility for emergency response work of significant and serious chemical accident site and then carry out emergency response training and consulting services

Beside the above organizations, there are two agencies which provide integrated emergency management. They are the National Civil Defense (CD) and the Emergency Management Office (EMO). The Civil Defense covers aerial defense, CBRN, HazMat and other common accidents in cities. The Emergency Management Office covers emergency planning, natural disasters, technological accidents, public sanitation issues, social security concerns, and recovery and reconstruction activities. The CD has their own resource and tactical team to response, while the EMO has more coordination role between the Ministry of Civil Affairs (MCA), the State Administration of Work Safety (SAWS), the Ministry of Public Security (MPS), the Ministry of Health (MOH) and other related agencies, such as Minitry of Environment. The Chinese central government established the EMO in December 2005, which is a milestone of the modern emergency management system. China was starting to build up all kinds of emergency plans based on this framework. By March 2009, 51 national level emergency plans have been developed for the country. Although China did not have one centralized powerful agency like FEMA or DHS to handle emergencies yet, the establishment of EMO showing the trend of centralization is positive. Moving forward, China will need more professional emergency management staff. Thus, they will need more education program from local universities.

Following its 5,000 year-long history, China will continue to fight with natural disasters, technological accidents and more terrorist attacks and criminal activities. The Chinese government has implemented many efforts to establish a modern emergency management system. These include new laws and local regulations. The policy changes to support these efforts came quickly. They include short term reactions as well as long term consequences… Understanding this tough situation, the Chinese government is pushing top-down to enhance the emergency management system by establish a new emergency management office, equip it with new technology and tools, provide more emergency management training/education, and communicate with other countries to share information and experience [24].

3.2 Japan

The framework of responding to dangerous goods incidents laid upon the Fire Department of the Japanese Ministry of Internal Affairs. In the event of the large-scale events, such as chemical weapons terrorist attacks, the Fire Disaster Response Headquarters took responsible for intelligence, planning and response measures. to extend the response capabilities, the program included the support units as follows: mutual aid agreements, prefectures support, Emergency Fire Assistance Team, Self-Defense Forces (SDF), and medical institutions [25].

Being one of the most disaster prone countries in the world Japan has developed sophisticated and all-embracing disaster management system. With high frequency and unprecedented levels of recurring natural disasters challenging disaster response system of Japan, the existing response system has been form relatively short time frame with sophisticated mechanism in order to mobilize forces and resources to respond in a comprehensive manner. It will establishment of ad hoc headquarters during large-scale disasters at all levels to serve for quick mobilization of forces, better coordination of response activities and disaster information sharing. Based on 3-layered from national, prefecture to municipal government system, rapid development of the country during this period enabled it to make considerable investments on DMS and integrate latest technological achievements [26].

Rather than being managed by one central body, disaster management system of Japan is decentralized and growing in recent years. One prominent characteristic features of this decentralization system enables more government agencies involving in disaster management with various extent and responsibilities to foster and to develop capability regional for enhancement of overall system. Thus, 3-layered hierarchy of the system fosters comprehensive supervision and management. It also allows tackling the disaster and/or accident on relevant level with relevant resources depending on its scale and implications. Activities at national, prefectural and municipal are taken in concerted and coordinated manner and supervised by the higher level for engaging efficacy at the high frequency of disasters. In addition, the integration of disaster management measures by specific bodies allows to handle specific disaster and management issues more professionally [26].

Japan Poison Information Center (JPIC) was set up by Japan Emergency Medicine Association of health care workers to provide 24 hours online aid to emergency recommendations service in July 1986. JPIC function to identify the chemical hazards to proper treatment for chemical incidents to terrorist attacks, even dispatch experts and staff to field sites.

Once the incident notification was issued to the Disasters Prevent and Response Headquarters, the response started with dispatch of cooperative units—fire department and others. The fire command post will be responsible for directing action. Other units included the follows.

- Police authorities (local police);
- City, town and village (the municipalities Disaster Prevention Unit);
- Medical institutions;
- Prefectures (Fire and Disaster Prevention Division);
- Ministry of Internal Affairs Fire Department;
- Japanese Poison Information Center (JPIC);
- SDF.

3.3 Korea

The Ministry of Environment was responsible for terrorist event and also plays a role as a support institute for the Central Disaster Response Division (Ministry of Public Administration and Security) in case of chemical spill/HAZMAT accident. The Ministry established and provided chemical data, properties, response plan, counter- measures, monitoring and testing, and performs impact assessment after the accident. It also implemented training or educations responding to chemical accident or terrorism for personnel of fire fighters and police, etc.

Furthermore, it maintained the Chemical Accident Response Information System (CARIS) which provided information of chemical properties, treating facility, dispersion prediction, control measures, resident evacuation know-how for initial responser in order to effectively assist and promptly respond to HAZMAT accident.

In case of severe chemical or HAZMAT accident, the Regional Disaster Safety Response Division headed by a governor or mayor to conduct site responses and to direct actions. The head of regional environmental office would set up headquarter to assist the responses of the local government. As the event escalating, the comprehensive responses at the governmental level would initiated including, the Central Disaster Response Division at the Ministry of Public Administration and Security and Central Accident Resolution Division at the Ministry of Environment [21].

The Inje University studied the chemical incidents in Korea and concluded that the functions of CERC, such as MSDS and response information, were essential to the HAZMAT responders. The PCCs in Korea were set up in several attempts since 1998 and the Korea National Response Information Center was set up as a prototype at InJe University in 2001 by the Ministry of Environment and officially 24 hour operation on 2004 by the supports of the Korea Responsible Care (KRCC) [21].

3.4 Taiwan

Since 1988, the chemical management and HAZMAT response were in-charged by the Taiwan Environmental Protection Administration (TEPA) and the Fire Department. The promotion of mutual aids was supported by the enterprises and formed On-Scene Prevention Coordinator Team (OSPCT) in 1992. As the HAZMAT incidents transformed to the responding to toxic chemicals in 2000, the TEPA set up new era to work with universities to initiate on-scene response aids. As a runaway reaction incident in an acrylic plant in Hsinchu, northern Taiwan, resulted in 1 fatality and 109 injuries and raised great awareness of the proper response of toxic chemical incidents in 2001. The explosion and fire also resulted in significant fire, smoke, and chemical and gas spillage and was aid by ITRI on-scene actions of technical advices, coordination for locating resources, monitoring. TEPA appreciated the importance of the on-scene response and decided in late 2001 to setup level-two response centres in northern, central, and southern Taiwan for a more prompt and direct response of toxic chemicals incidents. ITRI, National Yunlin University of Science and Technology (NYUST), and National Kaohsiung First University of Science and Technology (NKFUST) were chosen for the setup response centres in northern, central, and southern Taiwan, respectively. ITRI was chosen for its proven experience in emergency response while NYUST and NKFUST were chosen for their combined expertise in environmental and safety engineering. In five years' time from 2002 to 2006, more than five hundred on-scene response services were provided. The incidents ranged from a leak of a toxic gas cylinder to a large-scale fire in a high-tech facility. The effective response service has greatly reduced the potential impact from these incidents.

In 2007 the growing number of incidents required TEPA proposed a Plan to Strengthen Toxic Chemical Safety Management and Incident Response to boarder the timely response works. This involves commissioning the establishment of information and monitoring centres and seven Emergency Response Teams (ERTs) geographically with increased staffs and equipment to reduce the response load and timing [27]. Among the seven teams, six teams are located in the west, while one team is located in the east, reflecting the fact that most industrial parks and small industries are alsolocated in the west. The Chung Yuan Christian University operated three ERTs in the northern and eastern, National Yunlin University of Science and Technology operated two ERTs in the central, an National Kaohsiung First University of Science and Technology operated two ERTs in the southern. And, there is one Information Centre and a Control Centre both operated by Industrial Technology Research Institute as shown the Fig. 5.

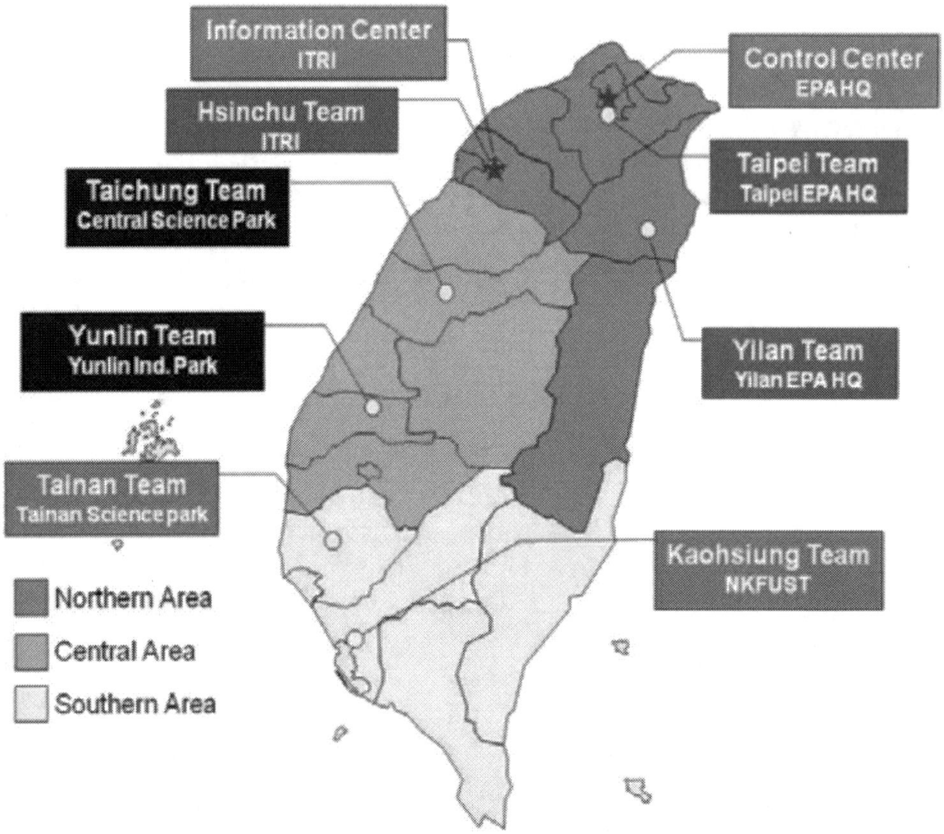

Fig. 5. Geographic Location of the Seven ERTs [27]

The seven ERTs provide round-the-clock on-scene emergency response aid to a toxic chemical incident within one hour after the notice of incident. The aid includes technical advice, coordination of the response, PPE for the responders, kits for stopping leaks or spills, advice on clean up, on-scene air pollution monitoring with PID, FID, GC-MS and FT-IR as well as surface water, soil, solid waste sampling and analysis, aand incident investigation. The ERTs also assisted local EPA bureaus with non-emergency information service, response training, large-scale drills for the response of toxic chemicals, and inspecting and auditing the operating facilities.

In order to effective and prompt service notification for the ERTs, the Article 24 of the Toxic Chemical Substances Control Act [28] has regulated that the operating facilities must report the incident to the local EPA office within one hour after occurrence to avoid the delay in response. In the cases when the incident was not caused directly by the listed toxic chemicals, the reporting may be delayed. The best practice found is to have a direct link between the Information Centre and the local fire brigades where most incidents, regardless of the causes, were reported. Upon receiving incident messages from the fire brigade, the Information Centre immediately check and confirm for the chemicals involved and provide relevant information to the ERTs and the fire brigade. On-scene emergency response of the ERTs is activated, depending on the cause of the incident and possible chemicals involved.

Although the establishment of the ERTs helped greatly the control of the toxic incidents, the number of responded incidents reached a record high in 2007, among which more than 20 % were transportation incidents as shown in Fig. 6. The transportation incidents which consisted mainly tank truck incidents on the road not only require significant resources and efforts to respond but also put nearby communities into risk. Regulatory efforts were initiated in 2008 and revised regulations were issued for transportation of toxic chemicals. Hence, the incidents were reduced.

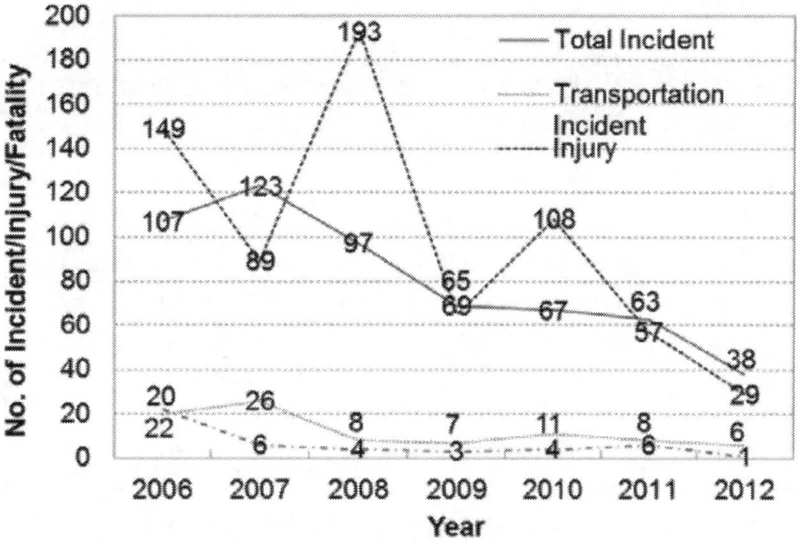

Fig. 6. Incidents Responding by ERTs from 2006 to August 2012 [27]

It worthy mentioned that Taiwan set up the ERIC and ERTs to assist from the central to local and even toward business response to HAZMAT events. Following the mutual aids and regional/national ERTs, the essential functions included the response information, consultation, material/equipment routing, expert/specialist on-site help and environmental monitoring. This type of organization might be worthy to invest as the local enterprises mostly small or medium companies as well as transportation incidents that could not develop their HAZMAT response effectively and cost-effectively.

3.5 comparison

By studying the information regarding to HAZMAT response organization for China, Japan, Korea and Taiwan, several features were similar with different development time-frame. The PCCs and CERCs were all function well except no CERCs in Japan. It might be because its decentralization and good integration of response capability with 3-levels. Except Korea, others all have set up offices in the central government to lead the possible HAZMAT events/disasters. No environmental agencies was specifically mentioned in Japan's HAZMAT response system.

China, Korea and Taiwan all have strong involvement of Ministry of Environment which probably related to their chemical regulation on management. However, the involvement of the Fire Departments was unavoidable since many HAZMAT events started with emergency calls. Among them, Taiwan uniquely set up ERIC and 7 ERTs to assist the EPA and local EPB on-site monitoring and expert assistant. They functioned as mutual aids to local enterprise, mostly small or middle sized manufactures. The mutual aids from large companies were also included into the whole response system to further strengthen the capacity of HAZMAT response.

Table 3. Comparison for PCCs, CERCs. Leading government agencies and the involvement of environmental agencies for China, Japan, Korea and Taiwan

	China	Japan	Korea	Taiwan
PCCs	Yes, 2003	Yes, 1986	Yes, 1998	Yes, 1985
CERCs	Yes, 2012 (NRCC?)	?	Yes, 2004?	Yes, ITRI 1996
Leading Gov. Agency	EMO?	Cabinet office	Mini. of Env.	Cabinet, EPA, NFA
Invol. EPA	Yes,	?	Yes	Yes,

4. Conclusion

In conclusion, the rapid industrialization brought about the abundant use of chemicals and the proper and careful management is essential. Since 1992 World Summit called on the safety management of hazardous chemicals, many international norms, regulations and conventions all encourage national and international-wide commitment on managing chemicals, including GHS, REACH, and UN's SAICM, in order to achieve the sustainable goals in 2020. Those efforts also promoted risk-based regulation restructure on chemical management. In the East Asia, industrialized Japan started early in new chemical regulation following REACH by Ministry of Labor and Economic Affair. China and Korea followed the trends to develop new chemical management system as China and KOREA REACH initiated by the Ministry of Environment. Taiwan also followed the pack in December 2013. Although some differences among those, such as the chemical categorization or management details, it seemed that those REACH-like regulations will be all put into enforcement by 2015.

For proper chemical management, the REACH-like regulations will not complete eliminate the exposure risks for hazardous chemicals. Thus, proper response systems need to be put in place to mitigate, reduce and stop the HAZMAT accidents or disasters. Assessing the current status of HAZMAT response mechanism, China, Japan, Korea, and Taiwan all placed their chemical response closely with the Fire Departments and establish experts by PCCS and CERCs with the environmental agencies to perform the protection toward public safety and environmental protection properly. It worthy mentioned that Taiwan set up the ERIC and ERTs to assist from the central to local and even toward business response to HAZMAT events. Following the mutual aids and regional/national ERTs, the essential functions included the response information, consultation, material/equipment routing, expert/specialist on-site help and environmental monitoring. This type of organization might be worthy to invest as the local enterprises mostly small or medium companies that could not develop their HAZMAT response effectively and cost-effectively.

5. References (examples given to show format)

[1] "Tokyo subway sarin attack," http://en.wikipedia.org/wiki/Tokyo_subway_sarin_attack (accessed Jun. 29, 2014).

[2] "South Korea's happiness ABS resin plant fire and explosion," http://www.chemicalsafety.org.cn/detail.php?oneid=10083 (accessed Jun. 29, 2014).

[3] Shan, K., Chen, F.L., Hu, G.H., "Fu-Kao analysis of the chemical accident review," ITRI Center for Environmental Safety, 2001.

[4] "Jilin Petrochemical Company double benzene plant explosion case," May 2007, http://www.cpc.com.tw/big5/files/%E6%A1%88%E4%BE%8B1%E5%90%89%E6%9E%97%E6%A1%88.pdf (accessed Jun. 29, 2014).

[5] Meyer, Eugene, "Chemistry of Hazardous Materials," Pages 353-370, 2004.

[6] "UNITAR-supported SAICM Enabling Activities and Guidance for eveloping SAICM Implementation Plans," SAICM Seminar, Tokyo, Japan, 2011.

[7] "Globally Harmonized System of Classification and Labelling of Chemicals GHS," First edition, Fourth revised edition, United Nations, New York and Geneva, 2011.

[8] "NIOSH Pocket Guide to Chemical Hazards." http://www.cdc.gov/niosh/npg/ (accessed Jun. 29, 2014).

[9] "REACH - Registration, Evaluation, Authorisation and Restriction of Chemicals" http://ec.europa.eu/enterprise/sectors/chemicals/reach/index_en.htm (accessed Jun. 29, 2014).

[10] "Strategic Approach to International Chemicals Management," http://www.unep.ch/saicm/index.php?menuid=2&pageid=256 (accessed Jun. 29, 2014).

[11] People's Republic of China, People's Republic of China State Council decree," Dangerous Chemicals Regulations," 2011.

[12] "China GHS and dangerous chemicals regulations Profile," http://www.52rd.com/Blog/Detail_RD.Blog_angel840901_27018.html (accessed Jun. 29, 2014).

[13] "Japanese Chemical Substances Control Law (CSCL)," April 1974, and the latest amended law May 2009, http://www.ron.gr.jp/law/law/dokugeki.htm (accessed Jun. 29, 2014).

[14] Chan, C.H., "Firefighter's perception and training needs on special type of fires in Taiwan," Thesis of Master's Degree in Chinese, Graduate School of Engineering Science and Technology, YunTech, Taiwan, ROC, 2013.

[15] "Enforcement of hazardous chemical substances management," http://eng.me.go.kr/eng/web/index.do?menuId=163 (accessed Jun. 29, 2014).

[16] "Organization for Economic Co-operation and Development," http://www.oecd.org/ (accessed Jun. 29, 2014).

[17] "Toxic Release Inventory," http://www2.epa.gov/toxics-release-inventory-tri-program (accessed Jun. 29, 2014).

[18] "Good Laboratory Practice," http://www.accessdata.fda.gov/scripts/cdrh/cfdocs/cfcfr/CFRsearch.cfm?CFRPart=58 (accessed Jun. 29, 2014]).

[19] "Ministry of Environment Korea," http://eng.me.go.kr/eng/web/index.do?menuId=70&findDepth=1 (accessed Jun. 29, 2014).

[20] "Enforcement of hazardous chemical substances management," http://eng.me.go.kr/eng/web/index.do?menuId=163 (accessed Jun. 29, 2014).

[21] Chung, S.T. Kim, K. I., "Case studies of chemical incidents and emergency information service in Korea," *Journal of Loss Prevention in the Process Industries,* Volume 22, Pages 361–366, 2009.

[22] Hsieh, Y.T., "Explore Chemicals Disasters through News Media and Questionnaire," Thesis of Master's Degree in Chinese, Graduate School of Engineering Science and Technology, YunTech, Taiwan, ROC, 2012.

[23] Guo, Z.R., Zhang, J.M., Li, W.X., "Sudden Environmental Pollution Accident Prevention and Emergency Response," *China Environmental Science Press*, 2006.

[24] Victor Bai, "Comparative Emergency Management: Understanding Disaster Policies, Organizations, and Initiatives from Around the World- Emergency Management in China," 2013.

[25] "Unit rescue operations manual fire hazard of chemical and biological weapons of terror attacks," Japanese Fire Department emergency rescue lesson, March 2005.

[26] Nazrov, E., "Emergency Response management in Japan. Final Research report," ASIAN Disaster Reduction Center, FY2011A Program, 2011.
[27] Chena, S.L., Lua, P.T., Lee, H.L., Lua, C.H., Yuana, S.Y., Tsaib, H.Y., Chen, J. R., "Risk Management and Regulatory Control of Toxic Chemicals in Taiwan," *Chemical Engineering*, Volume 31, 2013.
[28] "Toxic Chemical Substances Control Act," EPA (Environmental Protection Administration, Taiwan), http://law.epa.gov.tw/en/laws/788537580.html (accessed Aug. 28, 2012).

Reliability Analysis of Urban Gas Transmission and Distribution System Based on FMEA and Correlation Operator

Li, S.,

In order to improve the safety management of urban gas transmission and distribution system, failure mode and effects analysis (FMEA) was used to construct the reliability analysis system of the pipeline network. To solve the problem of subjectivity and uncertainty of the multi-expert decision making, the author introduces the correlation operator into the calculation of the risk priority number (RPN). Using FMEA along with weight analysis and expert investigation approach, FMEA evaluation table is given including five failure modes, risk priority numbers, failure causes and effects, as well as corrective actions. The results show that correlation operator can directly process the linguistic terms and quantify the priority of the risks.

Quantitative Fire Risk Assessment of Gas Pipeline Leakage
He, J.,

Quantitative fire risk assessment of gas pipeline leakage

HE Jing, College of urban construction and safety engineering, Nanjing university of technology, Nanjing 210000, C hina;
Abstract: O nce gas pipeline breaks, compressed gas rapidly expand. The electrostatic discharge spark energy can ignite the gas, causing a large range of gas combustion and explosion, and leading to catastrophic losses. Due to the characteristics of natural gas are flammable and explosive, there exist explosion, fire risk, casualties and damage in transport, storage and using process. In addition, if leakage continues long time, it can cause long-term harm because of continuous gas supplying. The factors contributing to gas pipeline leakage most are third party damage, corrosion, design and incorrect operation.
In this paper, the accident consequences of gas pipeline leakage and quantitative risk assessment method of gas pipeline leakage consequence are analyzed. Gas pipeline leakage patterns mainly include pipeline completely rupture leakage, big hole leakage and the middle hole leakage. Disaster patterns mainly include flash fire and vapor cloud explosion. In order to achieve quantitative fire risk assessment of gas pipeline leakage, flash fire model, vapor cloud explosion model and thermal radiation damage criterion are analyzed. F irstly, the gas leakage quantity is estimated. Then, the incident thermal radiation intensity, damage range and damage area are got by using flash fire model, vapor cloud explosion model and the thermal radiation damage criterion respectively. Combined with an example of gas pipeline, damage ranges of different disaster patterns are estimated according to different gas pipeline leakage pattern. At the same time the damage degree is assessed.

ï□¬ Under the same leakage pattern, the damage ranges varies with different disaster patterns. When the flash fire caused by pipeline completely rupture leakage occurs, people will 100% die in 1min and 1% die in 10s within 244m range. In

contrastï¼?when the vapor cloud explosion caused by pipeline completely rupture leakage occurs, people will 100% die in 1min and 1% die in 10s within 60m range. When the range extend to 107m, people will 100% die in 1min and seriously injured in 10s. When the range extend to 176m, people will 1% die in 1min and first-degree burn within 10s.

ï□¬ Under the same disaster patterns, the damage ranges varies with different leakage patterns. When the flash fire caused by pipeline completely rupture leakage occurs, people will 100% die in 1min and 1% die in 10s within 244m range. In

contrastï¼?when the flash fire caused by the middle hole leakage occurs, people will 100% die in 1min and 1% die in 10s within 90m range.
The quantitative risk assessment method for hazard evaluation of natural gas pipeline accident can be applied in estimating the gas leakage consequences. Through comparing the accident consequences of gas pipeline leakage, this paper provides some reliable scientific references for gas pipeline management departments of operation, construction and others.
Keywords: gas pipeline; leakage; flash fire; vapor cloud explosion;
quantitative risk assessment

AHP-HAZOP 半定量分析方法在天然气长输管道系统风险分析中的应用

陈晓玮[1]，赵东风[2]，孟亦飞[1]

（1. 中国石油大学（华东）机电工程学院， 山东青岛 266580；2. 中国石油大学（华东）化学工程学院，山东青岛 266580）

摘要： 天然气长输管道系统的特点是压力高、易燃、连续运行，由于管道腐蚀、材料缺陷、人为操作失误等原因，在长期运营过程中可能发生泄漏、破裂等，引发事故，造成严重的人员伤亡和财产损失。对天然气长输管道系统存在的风险利用 HAZOP 方法进行辨识，提出合理的建议措施，在 HAZOP 分析的基础上引入层次分析法，建立工艺参数的评价指标体系，实现风险分析的半定量化，并以天然气长输管道系统末站为例说明 AHP-HAZOP 半定量分析方法的应用。

关键词： 层次分析法；HAZOP；半定量

On Application of AHP-HAZOP Semi-quantitative Method in Risk Analysis of Long-distance Natural Gas Transmission Pipeline

Chen Xiao-wei[1] , Zhao Dong-feng[2] , Meng Yi-fei[1]

(1 College of Mechanical and Electronic Engineering, China University of Petroleum(East China), Qingdao 266580, Shandong, China;2 College of Chemical Engineering, China University of Petroleum(East China), Qingdao 266580, Shandong, China)

Abstract: Characteristics of long-distance natural gas transmission pipeline system are high pressure, flammability and continuous working. During Long-term continuous operation, leak, fire and explosion may happen resulting in accidents with serious casualties and property losses due to the pipeline corrosion, material defects, human error, etc. Apply HAZOP to identify the risk of long-distance natural gas transmission pipeline system, coming up with reasonable suggestions. On the basis of HAZOP analysis introduce the analytic hierarchy process to establish evaluation index system of process parameters and analyze the risk semi quantitatively. Take the terminal station of long-distance natural gas transmission pipeline system as an example to illustrate this approach.

Key words: AHP; HAZOP; Semi quantitative

引言

随着天然气资源不断被开发，世界各地区天然气管道总长迅速增加。我国天然气管道建设始于上世纪 60 年代，截至 2010 年底我国已有天然气管道总长为 4 万公里，并且根据国家能源局发布的《天然气发展“十二五”规划》，“十二五”期间我国建设管道总长度将超过 4 万公里[1]。

由于天然气的管道输送是在连续密闭的系统中进行的，任何一处出现问题都将影响整个系统的运行，同时天然气管道具有压力高、易燃、连续运行等特点，再加上材料缺陷、管道腐蚀、人为操作失误等原因，天然气长输管道在长期运营过程中可能发生泄漏、破裂等，引发事故，造成严重的人员伤亡和财产损失[2]。因此，对天然气管道系统进行风险分析显得尤为重要，有助于减小天然气管道事故严重程度，确保管道安全正常运行。

我国天然气长输管道系统风险评价技术研究起步较晚，风险评价技术基本上还处于理论研究阶段，定性方法采用较多，如故障类型和影响分析（FMEA），危险性和可操作性研究（HAZOP）。对天然气长输管道系统进行风险分析时，利用 HAZOP 分析方法定性描述工艺参数偏离对系统的影响，并在 HAZOP 分析的基础上引入层次分析法（AHP），针对复杂节点建立相应的指标体系，计算工艺参数相对于系统的权重，得到各个参数发生偏离对系统影响的程度，实现天然气长输管道系统安全评价的半定量化。

1 HAZOP 分析方法

HAZOP 分析方法是英国帝国化学工业公司（ICI）为解决除草剂制造过程中的危害，于 1960 年代发展起来的一套以引导词为主体的危害分析方法，用来检查设计的安全及危害的因果来源。

HAZOP 分析是由各专业人员组成的分析小组，通过对系统工艺过程和操作详细的进行检查，以确定过程的偏差是否导致不希望的后果发生，HAZOP 分析将列出引起偏差的原因、产生的后果、以及针对这些偏差及后果已使用的安全保护措施，当分析组确信对这些偏差的保护措施不当时，将提出相应的改进措施。

HAZOP 分析的基本理念和程序如图 1 所示。

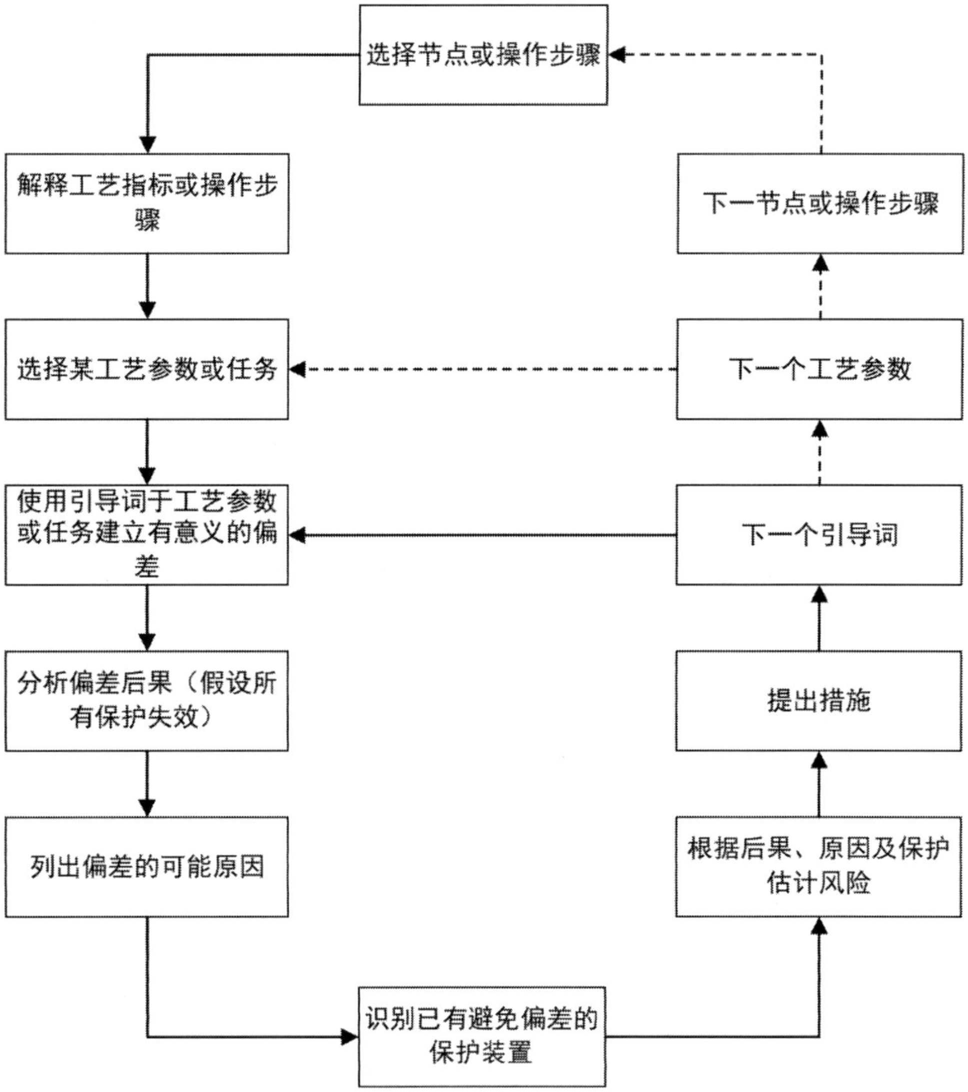

图 1　HAZOP 分析基本理念及程序图

Figure 1　Basic Concept and Procedure Chart of HAZOP Analysis

2 层次分析法（AHP）

2.1 方法简介

层次分析法 (Analytic Hierarchy Process)，是定性与定量结合的多目标决策方法。上世纪 70 年代初由匹兹堡大学教授萨提(Satty) 首先提出，是确定权重有效的方法。层次分析法的原理是把复杂系统中的多种因素通过划分为相互关联的有序层次结构并条理化，根据对一定客观现实的主观判断结构（两两比较判断）建立判断矩阵，将每一层中的元素两两比较的重要性进行定量描述。而后，利用数学方法计算反映每一层次元素的相对重要性次序的权值，通过所有层次之间的总排序计算所有元素的相对权重并进行排序[3]。

2.2 **分析步骤**

采用层次分析法对系统进行分析的一般步骤如下：

（1）构造层次分析模型。将要分析的对象层次化，形成一个多层次的分析结构模型，从高层次到低层次分别包括目标层、规则层、方案层等，从而将系统分析转变为确定最低层（方案层）相对于最高层（目标层）的综合相对权重，分析模型层次结构如图 2。

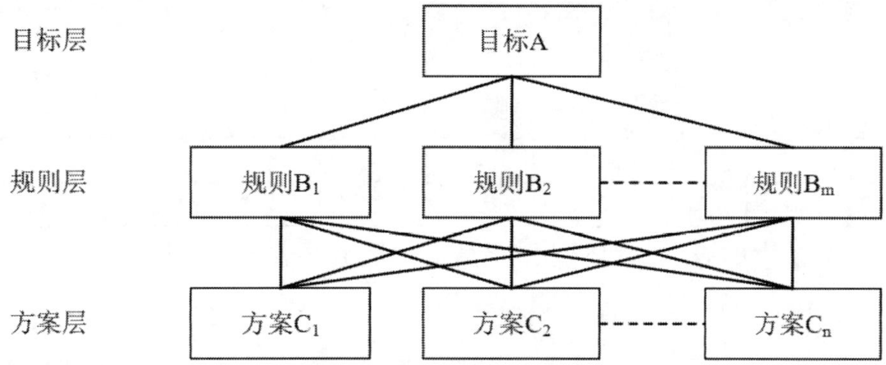

图 2　层次结构图

Figure 2　Hierarchical Structure Chart

（2）构造判断矩阵。分析下层每个因素相对于上一层某个因素的相对权重，对下层因素重要性进行两两比较，并引入数量标度方法，通常使用 1~9 比例标度法，见表 1，使比较定量化，得到两两比较的判断矩阵，如表 2，计算判断矩阵的最大特征值和对应的特征向量。

表 1　标度法

重要性比较	取值	重要性比较	取值
i 与 j 同等重要	1	i 与 j 同等重要	1
i 比 j 稍微重要	3	i 比 j 稍稍不重要	$\frac{1}{3}$
i 比 j 明显重要	5	i 比 j 明显不重要	$\frac{1}{5}$
i 比 j 强烈重要	7	i 比 j 强烈不重要	$\frac{1}{7}$

i 比 j 极端重要	9	i 比 j 极端不重要	$\frac{1}{9}$
2，4，6，8，1/2，1/4，1/6，1/8为介于两相邻重要性比较之间的值			

表 2　判断矩阵

规则B_1	方案C_1	方案C_2	…	方案C_n
方案C_1	a_{11}	a_{12}	…	a_{1n}
方案C_2	a_{21}	a_{22}	…	a_{2n}
…	…	…	…	…
方案C_n	a_{n1}	a_{n2}	…	a_{nn}

（3）一致性检验（适度原则）。对于构建的每一个判断矩阵求得的最大特征值均需要进行一致性检验，通常认为随机一致性比率$C.R.=\frac{CI}{RI}<0.1$时判断矩阵具有满意的一致性，一致性指标$CI=\frac{\lambda_{max}-n}{n-1}$，$\lambda_{max}$为判断矩阵最大特征值，n为判断矩阵的阶数，平均随机一致性指标RI的取值见表 3。

表 3　随机一致性指标 RI

n	1	2	3	4	5	6	7	8	9
RI	0	0	0.58	0.9	1.12	1.24	1.32	1.41	1.45

（4）层次单排序。判断矩阵的最大特征值通过一致性检验后，其对应的特征向量即为得到下层每个因素相对于上一层某因素的相对权重系数。

（5）层次总排序。利用层次单排序得到的权重值进行加权综合计算底层方案层每个因素对于顶层目标的相对权重。

3 层次分析法在天然气长输管道系统末站 HAZOP 分析中的应用举例

末站的主要功能是接收上游分输站来气。站内主要有清管器接收装置、过滤、计量、调压、加热等功能。

以某天然气管道工程的末站为例，该站的工艺流程为：上游站场来气，经过滤分离器除去可能带入的粉尘、杂质后，共分 3 路，第 1 路在此经过计量、调压后分 2 路，1 路进入门站，1 路为工业用户直供预留口；第 2 路为预留给 CNG 汽车加气标准站用气；第 3 路为工业用户直供预留口。站内设有一套自用气调压计量撬系统，用于站内职工生活和锅炉用气，如图 3。

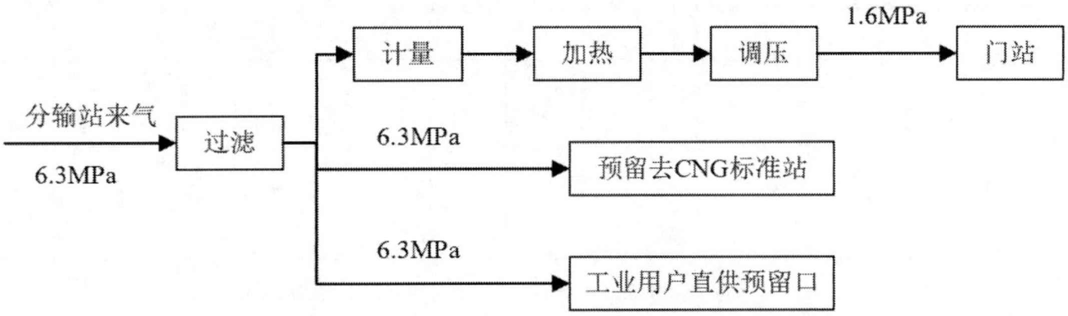

图 3 末站工艺流程图

Figure 3 P&FD of the Terminal Station

HAZOP 分析中将该末站流程直接划为一个分析节点，分析了流量、压力、温度、检维修、泄压、引燃、腐蚀/侵蚀等参数，从节点范围内找出导致流程参数发生偏离的原因，分析偏离的后果和已有的安全措施，并根据风险等级和已有安全措施的情况提出合理可行的建议措施，最终形成 HAZOP 分析表，本文仅考虑流量、压力和温度 3 个参数（表 4），分析这 3 个参数发生偏离对末站流程的影响程度。

表 4 末站 HAZOP 分析表

参数	引导词	原因	后果	已有保护措施	建议措施
流量	过低/无	1、过滤器 GF2402 和 2401 堵塞。	1、站场进气流量过低，影响下游用户用气。	1、过滤器前后设置压差显示及高报警 PIA2402 和 PIA2401。	
压力	过低/无	1、末站或上下游管线发生泄漏或火灾。	1、造成大量的天然气泄漏，扩大事故影响。	1、末站进出口均设置远程电动阀 2201 和 2501。	14、建议末站进出站电动阀门 2201 和 2501 接入 ESD 系统。
	过高	1、调压控制系统 PIC2401 或 PIC2402 故障。	1、调压失效，可能引起下游管路和设备超压。	1、调压系统采用串联设置独立的安全截断阀、监控调压阀和工作调压阀的方式; 2、调压后的管路上设置安全阀 2507。	15、鉴于目前末站调压后的压力等级分界划分在调压阀之后，若调压失效可能会造成站场内阀门、管线超压破裂的事故，建议设计院考虑在调压后管路上增设压力泄放系统。
温度	过低	1、热水加热器热水中断。	1、造成节流后的气体温度过	1、出站气体温度远传显示 TI2501。	16、建议末站出站气体温度显示 TI2501 增加低温

			低，可能会影响气体输送。		报警。
		2、环境温度过低。	过滤器底部的排污管线盲端若积水，易发生冻堵和冻裂。		17、建议设计院考虑在过滤器 GF2402/2401 和旋风分离器 XF2301/2302 底部排污管线盲端增设伴热。

经过 HAZOP 分析后，对流程参数发生偏离后可能导致的风险有了大致的了解，但各个参数发生偏离对流程影响的程度到底有多大，在确保末站安全运行方面所占的比重是多少，这些都无法通过 HAZOP 分析来得到。因此，在 HAZOP 分析的基础上引入层次分析法，以定性和定量相结合的方法分析得出参数的重要性，从而为现场人员有针对性地对运行参数进行监控提供参考依据。

（1）构建判断矩阵。以末站安全运行为目标，流量、压力和温度 3 个参数为指标，构造针对末站的判断矩阵，如下表 5。

表 5　判断矩阵

	流量	压力	温度
流量	1	1/5	1/3
压力	5	1	4
温度	3	1/4	1

求得最大特征值为 3.086949，特征向量为 $(0.103847,0.66507,0.231082)^T$。

（2）一致性检验。$C.R.=\frac{(3.086949-3)/(3-1)}{0.58}\approx 0.074956<0.1$，判断矩阵具有良好的一致性，取值合理。

通过上述计算可以得出，判断矩阵的特征向量可以用来衡量 3 个指标参数对末站安全运行的重要性，即压力＞温度＞流量。

4 结论

该末站参数之间的相对关系是根据实际情况和专家经验得到的，主观性比较强。因此，构造判断矩阵时，要多征求现场操作人员和专家的意见，使判断矩阵的取值尽可能地反映实际工况。

判断矩阵取值的正确与否是可以通过一致性进行检验的，在随机一致性比率大于 0.10 的情况下要对判断矩阵进行修正，所以利用层次分析法计算参数在确保末站安全运行方面的权重是合理可行的。判断矩阵的特征向量值直接反映了 3

个参数在确保末站安全运行上的权重大小，层次分析法是从另一个角度对HAZOP 分析结果进行了验证和补充。

采用 HAZOP 方法与层次分析法相结合的方法，可以实现天然气长输管道系统末站安全评价的半定量化，具有较好的实用性。通过层次分析，可以将复杂的分析对象进行分解，筛选出对系统安全运行最为重要的指标。

参考文献

[1] 戚爱华. “十二五”我国油气管道运输发展趋势分析[J]. 综合运输,2011,04:15-18.

[2]刘聪. 油气管道安全管理问题及加强措施[J]. 科技风,2013,12:136.

[3]叶义成, 柯丽华, 黄德育编著. 系统综合评价技术及其应用 [M]. 冶金工业出版社, 2006.

2nd CCPS CHINA CONFERENCE ON PROCESS SAFETY™

Oil Transportation Pipeline Leakage Emergency and Environmental Restoration

Zhang Li
China University of Petroleum
No. 66 , Changjiang West Road , Huangdao District , Qingdao , China
940947932@qq.com

Yan Xi
Liu Fang
Gao Ya
China University of Petroleum
No. 66 , Changjiang West Road , Huangdao District , Qingdao , China
Yanxi19911024@163.com
liufangfw@163.com
395071235@qq.com

UNPUBLISHED

Oil Transportation Pipeline Leakage Emergency and Environmental Restoration

Zhang Li
China University of Petroleum

Yan Xi
Liu Fang
Gao Ya
China University of Petroleum
No. 66 , Changjiang West Road , Huangdao District , Qingdao , China
Yanxi19911024@163.com
liufangfw@163.com
395071235@qq.com

Keywords: oil; pipeline; pollution; environmental remediation

Abstract

With the development of economic construction and social development, dangerous chemicals used in the production and life are increasingly common. The production, storage and the use of dangerous chemicals are inseparable from transport links. Pipeline transportation has become the fifth mode of transport because of its low transport cost, none pollution, and less wear and tear, etc. Pipeline transportation, which transports fluid such as gas and liquid by pipeline, is developed with the growth of the production of oil and gas. And it is responsible for 70% of the crude oil and 99% of the gas transportation. Recent years, many factors such as corrosion, sudden natural disaster and accidents caused by human, made oil and gas pipeline emergency happen frequently. This paper is about to introducing several methods of emergency treatment and environmental remediation caused by the oil pipeline leaks.

1. Introduction

Pipeline transport is developed along with the increase of oil and gas production and transportation technology. Because of the advantages of the high environmental benefit, no harmful substances emissions, small transportation quanlity, less floor space, no firmland occupition, small energy consumption, more safe and reliable, low cost and small wastage, pipeline transport has now become the main mode of transport for oil, gas, coal particle transport on the land. Moreover, pipeline transport proportion of offshore oil and gas transport is also growing. While, due to the material aging, corrosion, condensation in the tube, natural disasters and other factors as wrong operations, major accidents such as tube explosion, pipe-brokens and leaks are on the rise. On May 15, 2014, in Los Angeles, an oil pipeline burst, about 10,000 gallons of crude oil spread to the streets of Los Angeles. According to the British "Daily Mail" reported on May 15, the Los Angeles City Fire Department confirmed that the oil pipeline rupture resulting in tens of thousands of gallons of oil spill into the city streets, covering more than 800 meters range, and in some places the oil spill up to knee-deep, which may leads to environmental disaster. On June 30, 2014, CNPC oil pipeline in Dalian occurred continuous spill, causing oil leakage flow into the municipal sewage pipe network and flames appeared in the Sewer exit, and the film was found at sea. Oil leak not only causes economic losses and natural environment pollution, more importantly, it could be directly threat people's lives and property,and even more serious fatalities. Therefore, emergency and environmental restoration issues after oil pipeline leak has become a priority.

2. Offshore oil spill treatment measures

Tanker accidents, offshore oil pipeline leak, leaking and sewage on navigation of the ship is the main method of oil pollution of the oceans. Tanker accidents is the worst and also payed most concerned among them. But with offshore oil and gas field exploration, oil spills due to the blowout and pipeline leaks has significantly increased, which leads to the environmental and ecological risks can not be ignored. With the continuation of the operating time, the submarine pipeline damage incidents has increased every year. Ecological risk caused by undersea pipeline leak may include: structural imbalances in marine ecology, landscape ecology decline in the value and coastal human health risks[1].

2.1 Offshore oil emergency mechanism

In order to reduce the harm caused by the oil spill, firstly, we must improve oil spill emergency response capacity and improve emergency response mechanisms; secondly to develop the oil pollution prevention techniques. To improve oil spill emergency response capabilities, the first thing is to predict, including aspects of determining oil spill point, oil spillage, oil spill drift paths; secondly the need to strengthen the ability to configure and allocate oil emergency equipment[1,2] . Emergency Decision is based on diffusion and drift in the oil spill forecast , environmentally sensitive areas map, oil spill follow-up report, and information from the oil spill monitoring system reporting database, etc., and then the command staff make further assesment to oil spill for further assessment, and make emergency response Measures program.

Our system of environmental emergencies contingency plans relatively late start and compared with foreign countries there still exists a huge gap. We can refer to the overseas emergency system, make improvements in specific aspects of the legislation and plans for the development and execution[3,4]. One of the aims to develop emergency decision is to reduce oil spill pollution on the shoreline. However, due to equipment or options, not all endangered shoreline can get protection and governance. Problems encountered in the process of decision-making is that a decision can't meet all the requirements of regional or standard. To solve this problem, Liu et al point out that computing and evaluation model system can be used to guide decision making[1,2,5].

2.2 Marine oil pollution treatment technology

Sea oil pollution prevention techniques can be divided into physical, chemical and biological methods and combinations of them.

2.21 physical method

Physical method is simple and effective measures to recovery and processing of oil leakage and this kind of technology is relatively mature. By physical methods and mechanical devices the oil on the water and shoreline can be eliminated. Common physical methods include oil booms, mechanical, adsorption method[6-9].

When the oil leaking into the water, the first using oil booms to surround and prevent its spread, and then processed and recycled. Oil booms is with hysteresis, wave resistance and other properties. Oil booms is commonly used in relatively calm waters, like port terminals. Mechanical method includes clean-up vessels and mechanical recovery equipments. Recovery equipment has many categories, such as disc, weir, vacuum, etc. According to different site environments and weather conditions, different devices can be used. Adsorption method recycles waste oil by recycling oil sbsorbment material which is hydrophobic and oleophilic and oil can be adsorbed on the surface of them. Adsorption materials include polymeric materials, inorganic materials and fibers of the porous.

Physical treatment methods continue to develop in the direction toward the light and efficient. Brojv and others designed the surface of oil skimmers and oil strippers to exactly match a series of triangular structure, which makes the oil recovery efficiency improved by three times[10]. MouraFCC produced a hydrophobic porous material taking CH4 gas as the reaction gas and it has a high adsorption effect for various types of oils[11]. Besides the common methods, and the study attempts to use the vacuum pumping and laser machine to processing,or add settlement materials for precipitation and other methods of governance, but also because of high cost or not easy to operate has not been widely applied[12-14].

2.22. chemical method

Chemical method is mainly used after physical treatment or when physical mothed can't be used. Chemical methods that is used most are chemical reagent and combustion. Chemical reagent method is to spray chemicals to leak oil on the sea that can make the dispersed state of oil changes and be cleared. According to the differend dispersion state of the oil, gelling agent, dispersing agent and demulsifier can be used. Different reagents and its role as follows:

① **Dispersants**. Disperse oil particles into small droplets. Making it easy to mix with water in favor of degradation. But it can only handle low concentrations of oil and it is necessary to consider its own toxicity when used..

② **Oil thickener.** Condense oil into a viscous that is easy to remove by mechanical means.

③ **Set of agent.** Increase the surface tension of the oil, increase in oil film thickness, and then removed by physical means. The law requires regular medication, and a larger amount.

④ **settlement agent**. Settlement the oil to the bottom of f water, but that would endanger the benthic water.

Essence of chemical reagent treatment method is the transfer of pollutants, and chemicals used may cause new pollution. Some chemicals even have a certain chemical toxicity, and therefore agents should be used cautiously. Currently, sea oil processing chemicals is developing in the direction toward non-toxic, highly efficient, low cost, biodegradable. Combustion method is applied to the oil spill in situ processing, generally requires fire-resistant booms to prevent the spread of oil. Combustion method doesn't need complex devices, and low processing costs, but exists impact on marine life and the environment, so this treatment object is usually a massive oil spill or the high seas quite far from the coast. what is worth noting is, in addition to having an adverse effect on the ecological environment, the combustion process is also a waste of energy. And incomplete combustion will release smoke, resulting in a large number of aromatic compounds, which still contaminates water and the atmosphere.

2.23 biological method

Biological method is through the use of oil as a microbial metabolism of nutrients to degrade, so as to achieve the purpose of the removal of oil. Microorganisms can completely degrade petroleum hydrocarbons that make them non-toxic, so offshore oil leak biological treatment technology attractes widespread attention[15]. According to recent documents, there are 79 genera of bacteria, 9 genera of cyanobacteria, 103 genera of fungi, 14 genera of algae can degrade or converte hydrocarbons[16].Compared with the chemical, physical methods, the advantages of biological methods is that the impact on people and the environment is small, and repair costs 30% to 50% of traditional physical chemical repair costs[17], especially for a thin oil layer when a mechanical device can not be cleared and is limited to the use of chemical agents, the superiority of biological treatment is more significant. The disadvantage is that once the massive oil spill or a relatively thick, due to the insufficient supply of nutrients and oxygen, bacterial growth is inhibited thus affect the biological effects of oil spill treatment[18,19]. Measures taken in response to these circumstances, one is to add dispersing agent, increase capacity through emulsifying , improve the availability of petroleum hydrocarbons microorganism[20]. Second, adding N, P nutrients to increase the degradation rate of degrading bacteria. In present study, nitrogen nutrition mostly use urea and nitrate, phosphorus nutrition mostly dodecyl phosphoric acid and phosphate. These methods also apply for coastline oil pollution, in addition, measures like manual skimming, washing, spraying dispersants or cleaning agents can be taken in accordance with its specific characteristics.Washing usually used for oil removal on wharf rock reefs and other top. By using pressurized hot water or cold water to flush out the oil, the temperature of hot water used is 60℃. Washed with a rinse characteristics, but rinse with hot water will cause harm to plants and animals on the coast, and damage to fragile ecosystems. Besides, washing is also limited by water sources and supply of contaminated areas channels[2]. Dispersants can be used to deal with oil pollution on rocky coast, the dispersants sprayed directly in the oil on rock reef and structure surface, so dispersants contacting with the oil for certain time, and then using natural seawater rinse or brush oil with the help of tide , dispersants can also be used artificially to scrub the rock or reef to accelerate the dispersion process of oil[21]. Surface Cleaner is the use of cleaning mechanism to remove the oil on sand rocks and other solid surfaces.

3. Oil spill treatment measures on land

3.1 Hazards soil of contamination

Onshore oil pipeline leaks can contaminate soil, water and damage ecosystems, even cause fire or explosion. Oil-contaminated soil is an important environmental pollution caused by land-based oil pollution, and an important environmental problem oil production and transportation enterprises need to solve. After petroleum contaminants into the soil, due to its special physical and chemical properties and difficulty to remove and the characteristics of the residual long time, causing a series of hazards to soil and the environment, adversely affected ecology, crops and human health of the contaminated areas. Damage mainly in the following four areas:

① **Change soil.** Petroleum substances into the soil, can change the composition and structure of soil organic matter, causing carbon-nitrogen ratio and phosphorus ratio soil organic matter changes[22].

② **Adversely affect the growth and development of crops.** Mainly as follows: low germination rate of emergence, delayed growth period, late maturity, setting rate decreased, decreased resistance to pests and diseases. In addition, the oil is rich in reactive groups that can be combined with inorganic nitrogen and phosphorus and limit nitrification and dephosphorylation, so that the available nitrogen and the phosphorus contentin soil reducing , affect the absorption of crops. Oil also adheres to the surface of the plant roots to form mucous membranes, hinder roots breathing and absorbing, causing root rot, effecting crop root growth, and even cause death crop.

③ **Cause harm to humans.** Oil is a mixture of various components, and the toxicity of various petroleum fractions (component) is not the same[23], among which polycyclic aromatic hydrocarbons in oil (PAHS) has a greater toxicity . Polycyclic aromatic hydrocarbons carcinogenic, mutagenic and teratogenic effects, etc. PAHs are carcinogenic, mutagenic and teratogenic effects, etc, while the low-boiling fuel oil and lubricants can cause the body's anesthesia, asphyxia, chemical pneumonitis, and dermatitis[24]. Crop and food have absorb residual effect to petroleum pollutants. These toxic substances can indirectly affect human health through the food chain. In addition, the composition of pollutants that is not easy to be absorbed by soil could infiltrate into underground with ground rainfall, contaminate shallow groundwater environment and affect the quality of drinking water, and ultimately harm to human health.

④ **Bring huge economic losses to the country.** Petroleum contaminated soil not only bring great harm to the environment and ecosystem, but also cause irreparable damage to the national economy, companies and businesses. The losses includes losses to agriculture, the cause of damages to human body health, pollution treatment facilities running costs, governance project investment cost.

3.2 Treatment techniques for oil contaminated soil

In recent years, countries all over the world paid attention to the management technology of contaminated soil, the European and American countries successively invested a lot of manpower material resources for restoration of contaminated soil and governance. Soil technology has become a new hotspot in the research of environmental protection engineering science and technology. According to different pollution condition,a series of petroleum contaminated soil remediation technologies has formed. according to the dominant repair adopted by the existing soil pollutionduring the processing, repair technology can be simply divided into three categories, namely, physical repair method, chemical method and biological repair methods[25].

3.21 physical and chemical methods

Physical methods include landfill excavation, air stripping, electrolysis, washing method, isolation, soil replacement method, and so on, mainly belongs to heat treatment. Most of the organic pollutants in the soil can be purified by these methods, but the same the soil structure and composition is damaged. Furthermore, the fuel used and the equipment is expensive and therefore it's difficult to implement. Chemical methods include oxidant oxidation, photochemical oxidation, thermal decomposition, extraction and chemical shed method, mainly for the chemical leaching and washing, which can also get a good degreasing effect. Traditional physical and chemical treatment often requires the construction of a fixed treatment facilities, chemical dosing, also need to deal with temperature, pressure, power supply and other conditions in the running process, needing high processing costs, while also produce secondary pollution. Soil Bioremediation is a techniques use of microbial degradation of petroleum contaminated soil to achieve ecosystem restoration, the treatment costs about traditional physical and chemical methods of 1/3 to 1/5. The technology gets widespread attention and rapid development due to its low processing cost, no secondary pollution, mild processing conditions and other good technical characteristics,and has become the hotspot of domestic and foreign environmental control technology, expected to become a very promising environment alternative technologies pollution control.

3.22 bioremediation techniques

Microbial remediation of petroleum contaminated soil can generally be divided into two categories: in-situ repair and ectopic repair.

In situ bioremediation technology is the technology based on the soil without damaging the original structure, put in nitrogen, phosphorus and other nutrients to the contaminated soil, thereby increasing the metabolic activity of indigenous microorganisms or efficient microbial strains inoculated, improve the efficiency of the microbial degradation of petroleum hydrocarbons. In situ bioremediation technologies include biological culture, bioaugmentation and biological ventilation method.

Biological culture refers to the direct use of indigenous soil microbes, through regular delivery to petroleum contaminated soil nutrients and oxygen to stimulate the indigenous microbial activity to meet their growth and reproduction needs, so as to achieve the purpose of the degradation of pollutants[26].

Refers to the basis of the original microorganisms on petroleum contaminated soil to put exogenous bacteria, while providing N, P, S, K, Ca, Mg, Fe, Mn and other microbial growth necessary nutrients.

Is through the discharge of carbon dioxide in the soil and supplemental oxygen, to enhance physiological activity of microorganisms, increase the rate of degradation of contaminants in the soil. Biological ventilation method commonly used in mild contaminated soil bioremediation caused by underground tanks leak, this method has been applied to the Air Force base in addressing aviation oil contaminated soil remediation[27].

Ectopic bioremediation technology refers to the transfer of petroleum-contaminated soil and concentrated biodegradable treatment. This method applies to severe oil pollution and a smaller area of soil remediation. Ectopic bioremediation technology has its unique advantages: on the one hand, it can limit the spread and migration of the pollutants by the beginning of contamination and reduce the scope of pollution in the soil; and on the other, the reaction controller or bioreactor can be designed and installed to promote the rate of biodegradation. . Ectopic microbial remediation technologies including bio-accumulation, bio-composting and bioreactor.

Biological accumulation method is an improvement of the degradation of soil tillage techniques, usually petroleum-contaminated soil piled into strips, middle left "ridge", pipeline covered with piled soil providing degradation of water and perforated collector pipes featured for leachate collection. This systom also has air tube and air pump to stabilize the oxygen supply. Li Peijun[28] and his patners use bio-accumulation method for different types of petroleum-contaminated soil restoration process, after running for nearly 60d, the total removal rate of oil hydrocarbons reached 38.37% ~ 56.74%. The advantage of this method ist less secondary pollution and higher processing efficiency.

Add to petroleum contaminated soil twigs, straw, manure, sawdust and other modifiers, using mechanical or pressure systems to provide oxygen, while adding lime to adjust the pH, provide basic energy for the growth of microbial and degradation of petroleum substances, after 30 ~ 120d most contaminants can be degraded. This method is suitable for high-volatile, high concentrations of petroleum contaminated soil remediation[29].

Bioreactor also called bio-slurry method, adjusting petroleum contaminated soil into a slurry with water, and then loaded into the reactor, adding microorganisms and nutrients and other substances, and then stirred vigorously under oxygen-rich conditions. After treatment compliance, soil dehydration, finally transported back in place[25]. This method is suitable for handling petroleum-contaminated topsoil.

Another scholar proposed treatment of contaminated soil using plant extraction. That's planting crops on contaminated soil, so that the transfer of pollution to the crop fruit and straw, and then do follow-up treatment. While application of this method is rare in oil-contaminated soil remediation.

Despite so many advantages, there are limits of bioremediation as follows: First of all, not all the pollutants into the environment can be degraded by microorganisms, difficult biodegradability, infusibility, and also combined with soil or humus in soil offen makes bioremediation can't happen. Secondly, bioremediation needs a detailed and expensive specific study about the situation and the presence of pollutants in place. For example, in some of the low permeability of the soil may not use bioremediation technology because such soil or water injection wells in such soil will be blocked due to excessive bacterial growth. Thirdly, specific microbial degradation of specific type of chemical. Slight change in the state of the compound could not be destroyed in the same microbial enzyme. The fourth is the microbial activity affected by temperature and other environmental conditions. Fifth, in some cases, bioremediation of pollutants can not be removed completely, because when the pollutant concentration is too low to maintain a certain number of bacteria, the residual contaminants will remain in the soil. These limitations will be continuously improved with the deepening of research progress and people's constantly on the special environmental microbiology research.

4. Conclusion

At present oil pollution emergency response mechanism is mainly used in maritime accidents and land oil spill response system remains to be further improved. Bioremediation technology becomes the most promising treatment technology for its superior characteristics to the traditional physical and chemical techniques. However, as a new promising treatment technology, biological treatment technology has yet to be further improvement and development. Biological treatment technology combined with physical and chemical methods is the main way of pollution treatment. To overcome the limitations of the prior art and research new treatment technology will be the next major research directions.

5. Refereces

[1] Zeng Jiangning, Xu group, Shou Lu, Liao Yibo, Chen Zhen whole, Zheng Ping , "Ecological Risks of Undersea Oil Pipeline Spill and Countermeasures [J]," *Ocean Development and Management,* 2007,03: 120-123.

[2] Shenhong Chen, Wang Jian line, Cheng Yu Tao, Wei sent, Zheng Jianwei, Su Yufeng, "High Orientation Offshore Oil Spill Accident Hazards and The Emergency Treatment [J]," *Environmental Engineering*, 2011,06: 110-114 + 85.

[3] Huang Hui. "Theory of environmental emergencies contingency plans for the system [D]." Shandong University, 2012.

[4] Wang Hui.," Comparative Study —— Sino-US Maritime Oil Spill Emergency Response Mechanism to The Gulf of Mexico Oil Spill andThe Dalian Bay Oil Spill as An Example [J]." *China University of Political Science and Law*, 2011,03: 84-91 + 159 .

[5] LiuX, WirtzKW. "Decisionmaking of Oil Spill Contingencyoptions with Fuzzy Comprehensive Evaluation ［J］." *Water Resources Management*, 2007, 21(4) : 663-676.

[6] Pu Wen Hong, Zhou Li Xin, Yang Fan, et al. "Marine Oil Spill Prevention Technology Research Progress [J]." *Marine Science*, 2005,29 (6): 73-76.

[7] Zhang Weinuo. "Offshore Oil Leak Clean-up Techniques [J]." *Wash Technology*, 2003 (9): 31-34.

[8] BanerjeeSS，Joshi MV，JayaramRV. "Treatment of Oil Spill by Sorption Technique Using Fatty Acid Grafted Sawdust [J]," *Chemosphere*，2006，64(6) : 1026-1031.

[9] Wen Kui, Zheng Yuanyang. "Oil Spill Control Technology and Equipment [J]." *Petroleum Processing and Petrochemicals*, 1994,26 (8): 44-49.

[10] BrojeV, Keller AA. "Improvedmechanical Oil Spill Recoveryusing Anoptimized Geometry for The Skimmer Surface[J].", *Environ Sci Technol*，2006，40(24) : 7914-7918.

[11] MouraFCC，LagoRM. "Catalytic Growth of Carbon Nanotubes and Nanofibers on Vermiculite to Produce Floatable Hydrophobic

"Nanosponges"for Oil Spill Remediation [J] "*Applied Catalysis Benvironmental,* 2009, 90(3-4) : 436-440.
[12] Zhang Weinuo. "Offshore Oil Leak Clean-up Techniques [J]. " *Wash Technology*, 2003 (9): 31-34.
[13] GhannamM T, Chaalal O. "Oil Spill Clean up Using Vacuum Technique [J]. " *Fuel*, 2003, 82(7) : 789-797
[14] Liu Guangjiang, Geng Hung, History Lite. "Clear Water Spill Research Methods [J]." *China Water Transport*, 2010 (2): 32-33.
[15] YangSZ, JinHJ, Wei Z, et al. "Bioremediation of Oil Spills in Cold Environments: Areview [J]," *Pedosphere*, 2009, 19(3) :371-381.
[16] Head IM, Jones DM, Roling WFM. "Marine Microorganisms Make A Meal of Oil[J]." *Nature Reviews Microbiology*, 2006, 4(3):173-182.
[17] Hicks BN, Caplan JA. "Bioremediation: Anaturalsolution[J]." Pollution Engineering, 1993, 25(2):30-33
[18] Chao, Du Min, Zhang Zhenghui, "Petroleum and Petrochemical Enterprises Massive Oil Spill Site Emergency Response Techniques [J]," *Security*, 2013,05: 17-19.
[19] Pei Yuqi, Chu Shengli, Du Yin, Yang Fang, "Technical Analysis of The Status Quo Disposal of Oil Pollution [J] ," *Oil and Gas Fields of Environmental Protection*, 2011,01: 49-52 + 62.
[20] Wang Yang, Qiu-yu Wang. "The Application of Biosurfactant in Bioremediation of Oil Pollution [J]." *China Agriculture Bulletin*, 2009, 25 (24) : 455-458.
[21] KoranKM, VenosaAD, Luedeker CC, et al. "Development and Testing of Anewprotocol for Evaluating The Effectiveness of Oil Spill Surfacewashingagents [J] ," Marine PollutionBulletin, 2009, 58(12) : 1903-1908
[22] Xia Lijiang, WenXiaoLe. "Living Garbage Filling Area Change and Its Pollution Condition of The Characteristics of Che Surrounding Soil [J]." *Journal of Soil and The Environment,* 2001, 01:17-19.
[23] Ma Zhihua. "How Much Oil on The Marine Environment Pollution [J]." *Forest and Mankind,* 2002, 12:8-9.
[24] Zhu Menglong. "Refined Oil and Harmful to Health Protection [J]." *Journal of Oil Pool and Gas Stations,* 1999 01:40-39 + 5.
[25] Bao-liang zhang. "Oil field Petroleum Contaminated Soil and in situ Bioremediation Research [D]." Daqing Petroleum Institute, 2007.
[26] Jiang-feng Guo, Sun Jinhe. "Contaminated soil bioremediation methods [J]." *Journal of Environmental Science*, 1995, 3 (5) : 62-68.
[27] Bao-ming Li. "Microbial Remediation of Petroleum Contaminated Soil Research [D]." Beijing: science press, 2001309-368.
[28] Pei-jun Li, Shu-hai Guo, Sun Tie-bu h, etc. "The Different Types of Crude Oil Contaminated Soil Bioremediation Technology Research [J]." *Journal of applied ecology,* 2002, 13 (11) : 1455-1458.
[29] Hong-jun Luo, Xu-yuan Wang, Zhao Qian, etc. "The Different Types of Crude Oil Contaminated Soil Bioremediation Technology Research [J]." *Journal of Sichuan Environment*, 2007, 26 (3) : 13.

高硫原油加工过程中减薄腐蚀类型及控制技术研究

才向磊[1],赵东风[2],李石[2],王文东[2]

(1.中国石油大学(华东)机电工程学院　山东青岛 266580；2.中国石油大学(华东)化学工程学院　山东青岛 266580)

摘要：由于我国原油品质向含酸、高硫、重质等劣质化方向发展，导致原油加工过程中硫腐蚀问题严重。本文对常见减薄腐蚀类型如盐酸（HCl）腐蚀、湿硫化氢（H2S+H2O）腐蚀、酸性水腐蚀、高温硫腐蚀、高温硫+环烷酸腐蚀、高温 H2/H2S 腐蚀、高温氧化腐蚀等的腐蚀环境、机理、影响因素和影响装置几方面进行分析，从而通过注入缓蚀剂、提高材料等级、加强原料及中间产物脱硫、提高材料防腐能力等措施减缓或消除设备腐蚀，提高设备的可靠性，从而确保设备的长周期连续运行。

关键词：高硫原油；减薄腐蚀；机理；影响因素；预防措施

Study on Thinning corrosion type and control technology in High-sulfur Crude Oil Processing

Cai Xiang-lei[1], Zhao Dong-feng[2],Li shi[2],Wang Wen-dong[2]

（1.College of Mechanical and Electronic Engineering, China University of Petroleum (East China), Shandong Qingdao 266580,China;

2.College of chemical Engineering, China University of Petroleum (East China), Shandong Qingdao 266580,China）

Abstract：The quality of crude oil of our country tends to be acid-included, high-sulfur and heavy quality,leading to severe sulfur corrosion problems in crude oil processing.Corrosion environment, mechanism, influence factors and influence equipment in common thinning corrosion type,such as hydrochloric acid (HCl) corrosion, wet hydrogen sulfide (H_2S + H_2O) corrosion, acidic water corrosion, high temperature sulfur corrosion, high temperature sulfur + naphthenic acid corrosion, high temperature H_2 / H_2S corrosion are analyzed such as hydrochloric acid (HCl) corrosion, wet hydrogen sulfide (H_2S + H_2O) corrosion, acidic water corrosion, high temperature sulfur corrosion, high temperature sulfur + naphthenic acid corrosion, high temperature H_2 / H_2S corrosion, high temperature oxidation corrosion and corrosion.By injecting corrosion inhibitor, improving the level of material, strengthening raw material and intermediate desulfurization, improving the ability of anti-corrosion materials and other measures equipment corrosion is reduced or eliminated, and the reliability of equipment is improved, to ensure that the equipment run continuously for a long period of time.

Key words: high sulphur crude oil,Thinning corrosion,The mechanism,Influencing factors, Preventive measures

1 前言

由于我国原油品质向含酸、高硫、重质等劣质化方向发展,随之而来的炼制含硫原油带来的问题日益严重,特别是由于硫腐蚀而引发的各种炼油、储运设备泄漏事故，火灾事故频繁发生，使企业造成巨大的经济损失，职工生命安全受到严重威胁[1-2]。例如 1998 年 6 月 22 日, 茂名石化焦化装置加热炉进料泵出口管线上的大小头由于硫腐蚀减薄开裂, 又错用材质, 造成严重的人身伤亡事故[3]。类似事故近几年频繁发生，因此，如何解决由于硫含量增高带来的安全问题显得尤为重要。

2 减薄腐蚀类型分析

主要的均匀或局部减薄腐蚀类型包括: 盐酸（HCl）腐蚀、湿硫化氢（H_2S+H_2O）腐蚀、酸性水腐蚀、高温硫腐蚀、高温硫+环烷酸腐蚀、高温H_2/H_2S腐蚀、高温氧化腐蚀。

2.1 盐酸（HCl）腐蚀

a.腐蚀环境

炼油厂的盐酸腐蚀通常是在常压塔、减压塔、分馏塔或汽提塔顶部物流中的含水和氯化氢的蒸汽凝结导致的露点腐蚀。在炼油厂，盐酸腐蚀通常又称为$HCl+H_2S+H_2O$腐蚀，其冷凝液中的第一滴水酸性非常高（低pH），会导致高的腐蚀速率。

b. 破坏机理

采出原油中都含有不同数量的盐和水，除少量结晶盐外，绝大多数盐都溶解在水中形成盐水，虽然电脱盐能将大部分无机盐脱除，但仍有剩余的无机盐，主要为氯化钠、氯化镁和氯化钙，这些氯化物在蒸馏塔内高温下部分发生水解生成 HCl。主要反应方程式：

$$MgCl_2+2H_2O \rightarrow Mg(OH)_2+2HCl$$

$$CaCl_2+2H_2O \rightarrow Ca(OH)_2+2HCl$$

通常氯化钠在蒸馏装置不发生水解，但在原油中含有环烷酸或某些金属元素时，氯化钠可以再 300℃以下水解，生成氯化氢：

$$NaCl+H_2O \rightarrow NaOH+HCl$$

氯化氢在原油加工过程中随轻油气从塔顶挥发线馏出达到露点温度时有水冷凝，吸收氯化氢形成强酸性腐蚀环境：

$$2HCl+Fe \rightarrow FeCl_2+H_2$$

当有H_2S存在时，还发生下述腐蚀：

$$H_2S+Fe \rightarrow FeS+H_2$$

$$2HCl+FeS \rightarrow FeCl_2+H_2S$$

但冷凝系统的腐蚀主要石油盐酸引起的，尤其是气液两相转变的露点部位，随着冷凝过程进行，冷凝水量不断增加，盐酸溶液被稀释，腐蚀相对缓和。

c.影响因素

影响盐酸腐蚀的因素主要包括盐酸浓度、温度和合金组成。

（1）盐酸浓度：腐蚀严重度随着盐酸浓度增大。在换热器和管道中的氯化铵盐沉积物容易吸收物料和注水中的水，在沉积物下形成氯化氢水溶液。干燥的氯化氢气体通常不具有腐蚀性，但在与水形成盐酸时具有很强腐蚀性。

（2）温度：腐蚀严重度随着温度升高而增高。

（3）合金组成：碳钢和低合金钢在的 pH 值低于 4.5 左右的任何浓度盐酸中腐蚀严重。对任何浓度或温度的盐酸，300 和 400 系列不锈钢都不具有有效的抗腐蚀性。合金 400、钛和某些其它镍基合金对稀盐酸具有良好的抗腐蚀性。氧化剂(氧气、铁离子和铜离子)存在会增大合金 400 和 B-2 合金的腐蚀速率。钛在氧化性条件中表现出很好的抗腐蚀性，但在干盐酸环境中会迅速失效。

d. 影响装置

炼厂中所有常见材料均能遭受盐酸腐蚀。受盐酸腐蚀影响的炼油装置主要包括原油蒸馏、加氢处理和催化重整装置。蒸馏装置主要有常压塔顶冷凝冷却系统、减压塔抽空器和塔顶冷却设备。加氢装置主要有反应流出物系统换热器、硫化氢汽提塔顶部。催化重整装置为反应流出物管线、再生系统、稳定塔、脱丁烷塔和进料/预热换热器等部位。

2.2 湿硫化氢（H_2S+H_2O）腐蚀

a.腐蚀环境

湿硫化氢（H_2S+H_2O）腐蚀通常是指在碳钢或低合金钢在pH4.5 到7.0 之间含H2S的酸性水溶液中发生的腐蚀。

b.破坏机理

在低温下，硫化氢溶解在水中与铁发生发应：

$$H_2S+Fe \rightarrow FeS+H_2$$

H_2S+H_2O腐蚀主要表现为均匀减薄腐蚀，也会发生局部腐蚀或局部垢下侵蚀，尤其是在有氧存在条件下。

c.影响因素

湿硫化氢腐蚀的影响因素包括 H2S含量、温度、pH值、流速和氧含量，在一定的压力下，酸性水中的H_2S浓度随温度升高而降低；在pH高于4.5 时，会形成一个厚的多孔的硫化铁膜，会限制腐蚀速度。空气或氧化剂的存在会增强腐蚀，并产生点状腐蚀或沉积物下腐蚀。

d.影响装置

湿硫化氢腐蚀发生在高 H2S 和低 NH3 含量的催化裂化和焦化装置分馏塔顶及冷凝冷却系统、富气压缩系统、吸收稳定系统的设备和管线；以及加氢装置硫化氢汽提塔顶、分馏塔顶系统等部位。

2.3 酸性水腐蚀

a.腐蚀环境

酸性水腐蚀是指含H_2S和NH_3的水溶液造成的腐蚀，也通常称为硫氢化铵（NH_4HS）腐蚀。酸性水腐蚀对碳钢表现为均匀减薄腐蚀，当NH_4HS浓度大于 2wt%时，在方向变化或湍流区可能存在相当高的局部腐蚀；当NH_4HS浓度较高及低流速工况，换热器或空冷器可能由于结垢导致堵塞和功能下降。NH_4HS会迅速腐蚀海军黄铜管和其它铜合金。

b.破坏机理

原油中的硫、氮杂质在加氢条件下H_2S和NH_3，二者相遇生成硫氢化氨：

$$RSR+2H_2 \rightarrow H_2S+2RH$$

$$C_6H_5N+5H_2 \rightarrow NH_3+C_6H_{12}$$

$$H_2S+NH_3 \rightarrow NH_4HS$$

在 66℃左右的温度下，NH4HS 会从气相中析出，并导致积垢，发生堵塞并造成垢下腐蚀。

c.影响因素

酸性水腐蚀的影响因素包括 NH_4HS浓度、流速/局部紊流、pH值、温度及 合金组成。

（1）NH_4HS浓度：腐蚀随着NH_4HS浓度增大而增加。浓度低于 2wt%时，溶液一般不具有腐蚀性；浓度超过 2wt%时，溶度越高腐蚀性越强。

（2）流速：腐蚀随着NH_4HS浓速度加快而增加。

（3）温度：在低于66℃左右的温度下，NH_4HS会从气相中析出，并导致积垢和堵塞，发生垢下腐蚀，除非用洗涤水冲洗掉。同时注入洗涤水中的氧和铁可导致腐蚀增加和积垢增多。

（4）氰化物存在破坏保护性的硫化物膜，导致腐蚀加剧。

d.影响装置

受酸性水腐蚀影响的装置主要包括加氢装置、催化裂化装置、焦化装置、酸性水汽提装置等。加氢装置反应流出物冷凝冷却系统、冷高压分离器、冷低压分离器、硫化氢汽提塔顶及冷凝冷却系统、酸性水管线等部位。催化裂化装置分馏塔顶冷凝冷却系统、富气压缩及冷却系统设备及管线等部位。焦化装置分馏塔顶冷凝冷却系统、富气压缩及冷却系统设备及管线等部位。

2.4 高温硫腐蚀

a.腐蚀环境

高温硫腐蚀通常是指 220℃以上的重油部位硫、硫化氢和硫醇等活性硫化物与碳钢或低合金钢反应造成的腐蚀。高温硫腐蚀通常表现为均匀腐蚀，在高温硫腐蚀环境中，碳钢和低合金钢形成了硫化物腐蚀产物，保护范围与环境因素有关。当温度和/或硫含量足够高时，该腐蚀产物会对材料的保护逐渐削减，使得腐蚀加速。

b.破坏机理

原油中的硫存在形式很多，分为活性硫和非活性硫，高温硫腐蚀主要是活性硫化物如硫化氢、硫醇和单质硫的腐蚀，这些成分都能直接与金属发生化学反应：

$$H_2S+Fe \rightarrow FeS+H_2$$

$$RCH_2CH_2SH+Fe \rightarrow FeS+RCH{=}CH_2+H_2$$

$$S+Fe \rightarrow FeS$$

硫化氢在高温下会发生分解：

$$H_2S \rightarrow S+H_2$$

分解出来的元素硫比硫化氢活性更强，腐蚀更加剧烈。

此外还有非活性硫化物如硫醚、二硫醚、环硫醚、噻吩等，他们虽然不能直接和金属作用，但在高温下能够分解成元素硫、硫化氢等活性硫腐蚀金属。

c.影响因素

影响高温硫腐蚀的主要因素为合金组成、温度和腐蚀性硫化物浓度。

（1）硫含量：总硫含量是由许多不同的含硫化合物组成的。硫腐蚀主要由高温下硫化物热分解产生的和其它活性硫引起，因此，仅根据总硫含量预测腐蚀速率会产生误导。

（2）温度：铁基底合金的硫腐蚀开始温度高于 260℃，随着温度升高腐蚀 加剧。

（3）合金组成：一般而言，铁和镍基合金的抗硫腐蚀能力是由材料的铬含量决定的。增加铬含量显著提高抗硫化能力。含

5%、7%和 9%铬的合金在这些环境中足以保证材料的耐腐蚀性能，不锈钢如 12%Cr（410、410S、405SS）和 304SS 型可以用在较高的含硫量和温度的情况。

d.影响装置

高温硫腐蚀发生在高温环境下任何处理含硫物料的设备和管线中。典型的高温硫腐蚀环境存在于蒸馏装置常压塔和减压塔中下部、加热炉管、高温换热器及高温管线；催化裂化装置分馏塔中下部、高温换热器及高温管线；延迟焦化装置主分馏塔中下部、焦化炉管、高温换热器及高温管线等部位。

2.5 高温硫+环烷酸腐蚀

a.腐蚀环境

高温硫+环烷酸腐蚀一般发生在204-400℃温度范围内，物料中硫化物和环烷酸腐蚀介质共同与碳钢或合金钢反应造成的腐蚀。

b.破坏机理

在高温下，原油中活性硫化物很容易和铁发生反应：

$$H_2S+Fe \rightarrow FeS+H_2$$

$$S+Fe \rightarrow FeS$$

因生成的FeS不溶于油，可覆盖在钢铁表面形成保护膜，能够阻止金属的继 续腐蚀，但当有环烷酸存在时，环烷酸能直接与金属作用生成环烷酸铁，也可以与FeS反应生成环烷酸铁和H_2S，从而破坏FeS保护膜，生成的H_2S与金属表面继续反应，腐蚀金属，其反应为：

$$2RCOOH+Fe \rightarrow Fe(OOCR)_2+H_2$$

$$2RCOOH+FeS \rightarrow Fe(OOCR)_2+H_2S$$

$$H_2S+Fe \rightarrow FeS+H_2$$

通常环烷酸含量低时，腐蚀表现为减薄和/或点蚀，而环烷酸高时，表现为局部腐蚀、点蚀、或高流速区由流动导致的沟槽状腐蚀。

c.影响因素

高温硫+环烷酸腐蚀是环烷酸含量、温度、硫含量、流速和合金成分综合作用的结果。

（1）硫含量：硫促进硫化亚铁的生成，有缓蚀作用，环烷酸可以去除金属表面的硫化亚铁保护膜；当硫含量极低，甚至酸值很低时，由于起保护性的硫化物没有快速形成，环烷酸腐蚀也许更严重。

（2）温度：高温硫化物腐蚀发生在约220℃以上的温度下。环烷酸腐蚀一般发生在 204-400℃温度范围内，400℃以上时，环烷酸发生分解或变成汽相。

（3）流速流态：流速提高加速去除保护性硫化物而使腐蚀性增强，这种影响在速度很高的液-汽混合相系统中最明显。

（4）酸值：原料中的环烷酸通常用—中和指数‖或—总酸值‖（TAN）表示。组成环烷酸系的各种酸具有不同的腐蚀性。评估单个馏分的 TAN 比整个原油的 TAN 更能确定环烷酸腐蚀敏感性。

（5）合金成分：含钼至少 2.5%Mo 的奥氏体不锈钢才能具有最佳的抗环烷酸腐蚀能力。

d.影响装置

受高温硫+环烷酸腐蚀影响的炼油装置主要包括原油蒸馏、加氢处理和焦化装置。蒸馏装置易发生在加热炉炉管、常压和减压转油线、以及常减压塔内构件、填料和高酸物流凝结或高速液滴冲击的部位；高温硫+环烷酸腐蚀最容易发生在管道中高流速、湍流、流向改变的部位，如泵的内构件、阀门、弯头、三通、减压器、焊缝和热偶套管。加工高酸原料的焦化装置主要为原料进焦化炉前换热器和管线等部位。加工高酸原料的加氢装置在氢注入点之前的换热器和管线等部位易受腐蚀。

2.6 高温 H_2/H_2S 腐蚀

a.腐蚀环境

高温H_2/H_2S腐蚀发生在204℃以上，在H_2存在条件下，能显著增加高温硫腐蚀的严重程度。

b.破坏机理

$$Fe+H_2S \rightarrow FeS+H_2$$

高温H_2/H_2S腐蚀通常表现为均匀减薄腐蚀，并伴有硫化铁皮的形成，硫化铁皮大约为损失金属体积的5 倍，并可能呈现多层结构。

c.影响因素

主要影响因素包括温度、有无H_2存在、H_2S浓度及合金成分。腐蚀速度与温度、H_2S

的浓度成正比；大量氢存在会比无氢条件下腐蚀速率高；Cr含量的增加提高了耐蚀性，但如果Cr含量提高至7-9%，耐蚀性能提高不明显。

d.影响装置

高温H_2/H_2S腐蚀通常发生在加氢装置，例如加氢精制装置和加氢裂化装置。在氢注入点下游的设备和管线的腐蚀速度明显增加。

2.7 高温氧化腐蚀

a.腐蚀环境

高温氧化腐蚀发生在高温条件下，金属材料和气氛中的氧发生反应，在金属表面形成氧化物膜。

b.破坏机理

金属材料在高温氧化腐蚀中表现为金属的损失，损伤外观外表面会被覆盖一层氧化皮，不锈钢和镍合金一般都有一层很薄的暗色氧化皮。高温氧化的典型过程是，当金属达到高温氧化起始温度时开始氧化，在金属表面形成一层致密的金属氧化膜阻止金属发生进一步氧化。

c.影响因素

金属温度和合金组成是影响高温氧化作用的主要因素。

（1）温度：碳钢的在温度高于 538℃时氧化加剧，腐蚀随温度升高而增加。300 系列不锈钢在温度超过 816℃发生高温氧化腐蚀。

（2）铬含量：碳钢和其它合金的抗氧化能力是由材料中的铬含量决定的。随着金属中的铬浓度增加，氧化物膜保护性增强。

d.影响装置

高温氧化腐蚀通常发生在温度高于538℃、含氧（20%)燃烧环境中。如蒸馏装置加热炉、FCC 再生器内件、焦化炉、重整炉、加氢炉等都在高温氧化腐蚀环境下。

3 防腐措施

3.1 加强原料及中间产物脱硫

目前的脱硫工艺可以有效脱除原料中的硫化物，精脱可以脱除 99％以上的硫，这可以从根本上防止硫腐蚀，同时也可以减少工艺过程中的H_2S含量，减少中毒危险，减少FeS的产生，更能减少其自燃危险[4]。当然在脱硫工艺系统仍存在硫腐蚀的问题。原料进塔之前采用以电脱盐脱水为核心的“一脱三注”工艺，及时对脱盐后的原油含盐量、含水量及初馏塔、常压塔、减压塔顶排水的pH值、Cl^-、Fe含量进行分析并严格控制，会减少原料对塔设备的腐蚀。有效的脱硫措施可以在很大程度上缓解原料硫含量提高对于现有炼油装置的安全挑战，不仅可以减少装置的腐蚀，更可以从根本上解决高硫的危害。

原油脱盐脱水对炼油设备和炼油厂都具有重要意义。首先，脱盐脱水可以脱除原油中的氯化盐类，防止盐类水解产生盐酸，并防止硫化氢在盐酸对设备的腐蚀中的催化作用，因此，脱盐脱水可以减少腐蚀介质，减轻设备腐蚀[5]。

3.2 装置材质选取及升级

对于高温硫腐蚀防护，最主要的措施就是提高管线材质，根据《加工高硫原油重点装置主要设备设计选材导则》（SH/T 3096—1999）的要求，改用抗硫化氢腐蚀的钢材。

3.3 采取适当表面处理等方法提高材料防腐能力

表面处理。对于部分换热器的腐蚀状况，可以采用渗铝技术加以处理，渗铝技术是一种材料表面活性处理的方法. 它的机理是渗铝钢在高温(1000 ℃)下氧化生成致密$A1_20_3$·FeO尖晶石结构的化合物，这种结构堆垛密度大，无空位无空缺，其他元素不易扩散进去[5]。同时这种结构阻止铁原子与其他腐蚀性介质接触，不能产生化学反应，就形成不了化学腐蚀。在电化学防腐方面，铝的电位低于铁，铝首先受蚀，对铁起到了保护的作用[6]。另外，对低温硫腐蚀严重的富气管线等对管线进行防腐涂层等表面处理可以在很大程度上减缓腐蚀，如有机硅耐蚀涂料，表面喷铝技术等。

3.4 采用注入缓蚀剂等方法减缓腐蚀

在低温硫腐蚀严重的部位采用注入缓蚀剂的方法也可以在很大程度上减轻低温硫腐蚀的产生，因此，在如焦化装置分馏塔顶部馏出线，吸收稳定前的空冷器入口等部位注入缓蚀剂可以提高塔顶工艺防腐级别。

3.5 其他

在硫腐蚀的过程中，温度流速等过程参数对腐蚀速率具有严重影响，因此，为防止和减轻装置的腐蚀，就应在设计阶段调整工艺参数，尽量减少波动，并使设备结构尽量完善，避免热应力，要使介质均匀流动和分配，减少涡流或局部过热，防止局部应力腐蚀泄露[6]。

4 总结

对于原有炼油装置改炼高含硫原油来说，原料油硫含量的提高增加了对炼油厂安全运行的挑战。虽然无法通过本质安全化的手段来提高装置的安全性，但是可以从工艺到装置到操作规程再到管理制度的全面提高，来确保企业的安全无事故运行。

参考文献

[1] 马金秋.典型炼油装置改炼高含硫原油危险因素分析及安全对策研究[D]:中国石油大学,2010.

[2] 马金秋,赵东风,酒江波,等.改炼高硫原油典型炼油装置危险性分析及对策研究[J].安全与环境工程,2010(4):55-59.

[3] 黄丽萍.含硫原油加工中的硫化物腐蚀与防护技术探析[J].金山油化纤,2002(1):45-50.

[4] 林彩虹,潘军.加工高硫原油设备的腐蚀与防护[J].石化技术与应用,2004(6):444-447+404.

[5] 高向东.加工高硫原油装置的腐蚀与安全生产(Ⅰ)[J].石油化工腐蚀与防护,2002(4):1-7.

[6] 穆冬莉.加工含硫原油设备的腐蚀机理及防护技术[J].石油化工设计,2007(2):63-64+16.

2nd CCPS CHINA CONFERENCE ON PROCESS SAFETY™

Study on environmental risk assessment of pipeline transporting dangerous chemicals and environmental management

Gao Ya
Liu Fang
Yan Xi
Zhang Li

China University of Petroleum
Qingdao economic area
395071235@qq.com
liufangfw@163.com
Yanxi19911024@163.com
940947932@qq.com

Prepared for Presentation at
2nd CCPS China Conference on Process Safety
Qingdao, Shandong
August 28-29, 2014

UNPUBLISHED

Study on environmental risk assessment of pipeline transporting dangerous chemicals and environmental management

Gao Ya
Liu Fang
Yan Xi
Zhang Li

China University of Petroleum
Qingdao economic area
395071235@qq.com
liufangfw@163.com
Yanxi19911024@163.com
940947932@qq.com

Keywords: pipeline transportation ; dangerous chemicals ; fault tree ; transportation route.

Abstract

In this paper, for the pipeline transportation of dangerous chemicals sudden pollution events, by reading literature and surveys we get that hazardous chemicals piped environmental risks of chemicals is related to the chemical transportation pipeline, the dangerous goods in pipeline and the laying pipelines. On the basis of traditional fault tree analysis we add the impact of the leak into the ground hazard degree, to find the smallest fault tree cut sets and structural importance, then we find out the greatest impact factor analysis is the pipeline density; and combining the pipeline transportation of dangerous chemicals the path optimization model with some impact factors find out above, we attempt to find lowest combination come to environmental chemicals piped of environmental risk in order to provide the basis for the pipeline transportation of hazardous chemicals management.

1. Introduction

Hazardous chemicals are in need of particular attention from production processes to store and use because of their danger. These links are only to be achieved by means of transportation. The preferred long-distance transport of hazardous chemicals is piped. If the pipeline spill in the course of transportation, it will not only cause damage to property but lead to fire, explosion, toxic, corrosive and a series of serious accidents, endangering public safety and property of the people and causing environmental pollution. Therefore, the establishment of route optimization based on risk analysis for the protection of personnel security around the pipeline, and the safe operation of long-distance pipelines is significant.

2. Hazardous chemicals piped environmental risk assessment based on fault tree

2.1 Pipeline risk assessment methods

Based on relevant research pipeline risk assessment techniques can be divided into three categories: the qualitative risk assessment, semi-quantitative risk assessment and quantitative risk assessment.[1]Many scholars and pipeline companies have established different risk assessment methods such as expert survey method, AHP, fuzzy comprehensive evaluation method, neural networks, fault tree analysis, which are commonly used.[2]

This paper studies on the risk of pipeline transportation of hazardous chemicals, and it's a low probability of high-risk events. Therefore, we select the fault tree analysis combining the quantitative and qualitative analysis to find out the greatest impact factor.

2.2 Hazardous Chemicals pipeline accident case analysis and risk factors

Case 1: New Mexico natural gas pipeline explosion [3, 4]

Accident overview: Near Carlsbad, New Mexico, a large natural gas pipeline exploded at 5:26 on August 19, 2000. The accident leaded 12 people died. The pipeline belongs to EPNG company was founded in 1950, the wall thickness of 8.5mm, diameter of 762mm, X52. When the accident occurred, the pipeline runs under the maximum allowable operating state pressure of 80%.

Cause of the accident: The bottom of the pipe in a serious internal corrosion, then internal corrosion caused severe wall thinning is the direct cause of the accident. Pipeline corrosion is caused by joint action of corrosion of internal components such as microorganisms, chlorides, H2O, O2 and H2S.

Case 2: Washington State gasoline pipeline explosion [3,5]

Accident overview: At 15:00 on June 10, 1999, petrol pipeline (Φ406mm) located in Bellingham, Washington, USA fractured. Gasoline leaks about 89.7×10^4L, forming a diffusion length of about 850m near a creek. 30min after the pipeline leak, the gas flow into the creek was ignited and exploded, killed three people and injured eight people, and a residential and a water treatment plant had been severely damaged.

Cause of the accident: The construction of mining led to the pipeline has more than 30 cracks, which formed a rip. Hydraulic analysis of the pipeline, the pressure in the fracture site is about 9.88MPa, while its maximum allowable operating pressure is 9.93MPa. It is obviously because of external damage, the bearing capacity of pipe down, and eventually leaded accidents happen.

Case 3: Qingdao "1122" explosion

Accident overview: At 3:00 on November 22, 2013, a pipeline located in Huangdao district ruptured, causing oil pollution about 1000 square meters. Some of the oil pipeline along the rainwater pipeline ran into Jiaozhou Bay, the sea area of about 3000 square feet over oil meters. Day 10:30, exploding near the leak occurred while deflagration occurred in the estuary of oil pollution on the sea. Accident has killed 62 people, injured 136, and the direct economic loss is 750 million yuan.

Cause of the accident: Pipeline corrosion thinning and pipeline rupture caused the leakage of crude oil. Then crude oil ran into municipal drainage culverts. The municipal drainage culverts formed confined spaces for oil and gas, when they meet spark then exploded. Finally, pipeline and drainage culverts interchange, and the oil ran into drainage culvert and back to the road.

To investigate the reason after a large number of dangerous chemicals transportation accidents, we found that in addition to man-made destruction, management factors such as supervision, lack of detection, there are some objective factors. We do not discuss the human management factors in this paper, only study on the objective factors controlled. The main objective reasons for pipeline transportation of dangerous chemicals accidents is pipeline corrosion, leakage of dangerous goods their risk, pipeline density and other factors. This paper summarizes that the hazardous chemicals piped environmental risks of chemicals is related to the chemical transportation pipeline, the dangerous goods in pipeline and the laying pipelines.

2.3 Risk factors associated with improved fault tree analysis method

2.3.1 Establishing improved fault tree

Risk is usually defined as a comprehensive description of the probability of occurrence and the extent of the loss of a specific hazardous event. The mathematical expression is R=F(Pf, D), wherein Pf represents the probability of occurrence and D represents the degree of loss[6]. The degree of loss include the casualties, property damage, damage to the environment and the impact on production.

The traditional fault tree analysis can only come to the probability of a hazardous event occurs, but ignoring dangerous consequences severity and magnitude of the loss. Therefore we analysis method based on improved in the fault tree of the original, adding the concept of hazard analysis.[7] Furthermore the biggest factor in the analysis is to consider not only affect the cause of the accident, also taking into account the degree of damage. So the qualitative analysis, will be more comprehensive.

After comprehensive consideration, able to establish the risk of hazardous chemical accidents fault tree as shown:

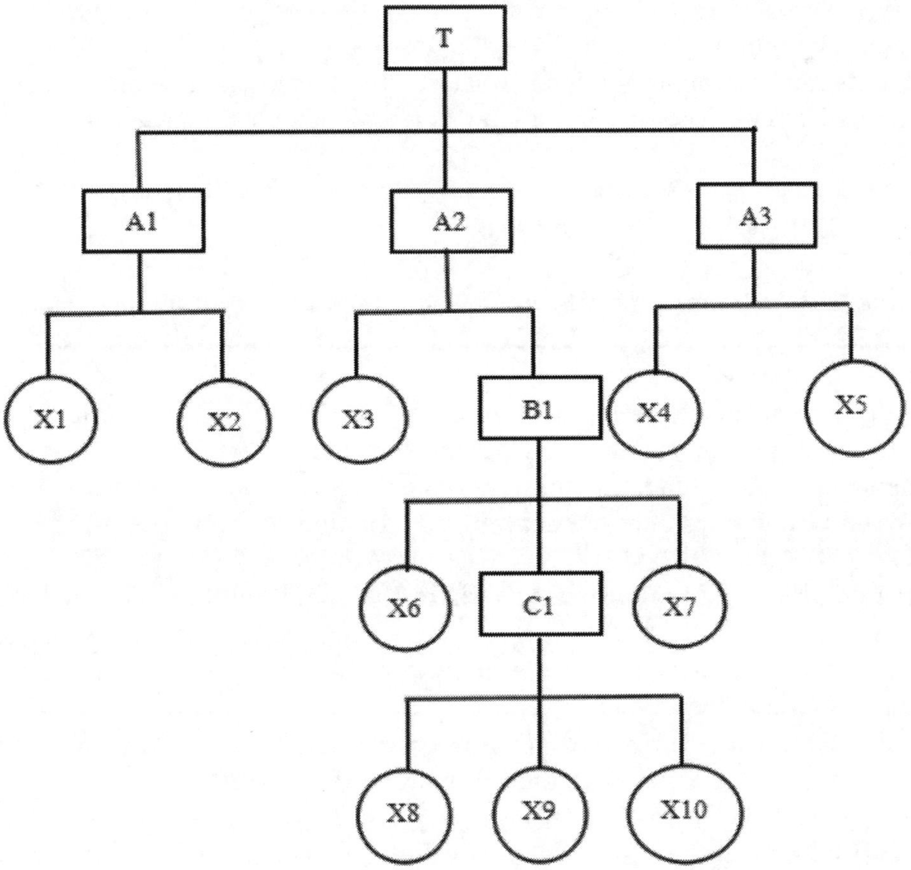

Figure 1 A fault tree hazardous chemical accident risk

Table 1. Pipeline transportation of hazardous chemical accidents tree basic event table

risk factors	name	Risk factors	name
T	pipeline transportation of hazardous chemicals risk	X3	hazardous chemicals flow factor
A1	pipeline factors	X4	pipeline density
A2	hazardous chemicals factors	X5	pipeline route
A3	pipe laying factor	X6	radioactivity
B1	hazardous chemicals nature factors	X7	toxicity
C1	dangerous chemical nature factors	X8	flammability
X1	pipeline age factor	X9	explosive
X2	piping material factors	X10	causticity

2.3.1 Establishing improved fault tree

The minimum cut sets method including determinant method, Boolean algebra method. The following is the process of simplification of Boolean algebra. After calculation, the structure function of fault tree:

$$T = A1 + A2 + A3 = X1X2 + B1X3 + X4X10 = X1X2 + (C1 + X6 + X7) X3 + X4X10$$
$$= X1X2 + X3 (X8X9X10 + X6 + X7) + X4X10$$
$$= X1X2 + X3X8X9X10 + X3X6 + X3X7 + X4X10$$

A total of nine of the fault tree fundamentals, get five minimal cut sets, which are{ X1,X2 } , { X3,X8,X9,X10 } , { X3,X6 } ,{ X3,X7 } , { X4,X10 }.

2.3.3 Analysis of structural importance

According to the minimum cut sets resulting fault tree, the following principles apply to determine the degree of importance of each basic structural sequence of events:

(1) all events, the minimal cut sets of fundamental importance biggest event structure.

(2) in the event of the same basic minimum cut centralized structure of the important factors are equal.

(3) if two basic events occurs in minimum cut sets share the same number, then in different minimum cut sets the number of basic events are the same, that means their structure are an important factor equal; Fewer occurrences, the structure important factor is small; On the contrary, the structure is important coefficient.

According to the above structure important factor approximation determine method, we derived each basic event structure importance in the following order:

$$I (3) > I (1) = I (2) = I (4) = I (5) > I (10) > I (6) = I (7) > I (8) = I (9)$$

From the above analysis we can draw significant coefficients of each basic event to arrive at its impact on the size of the fault, and then find out the weak links. In our analysis, hazardous chemicals flow in pipeline transportation is the maximum hazardous chemicals risk structure important factor, which means the most important condition for hazardous chemical pipeline transportation. However, reducing the flow to production has adverse effect, it is controllable but we do not want to control, in this paper we do not discuss this factor. Pipeline factor and pipe laying factor share approximate structure important factor for pipeline transportation of hazardous chemicals. But for the risk function of hazardous chemicals pipeline transportation, when share the same Pf, then the greater impact is D. Statistics for the event, we assign D for different value, the important structure important factor has the larger parameters. Structural importance of each basic events after the assignment results, the sequence as follows:

$$I (4) > I (5) > I (1) = I (2) > I (8) = I (9) > I (10) > I (7) = I (5)$$

Thus, pipeline density became the most important factor controllable. In actual production and life, we need to focus on design layout, to reduce the risk of the occurrence of chemical incidents. Here we have wo of the most influential factors pipeline density, pipeline route optimization discussion of pipeline transportation of dangerous chemicals.

3. Hazardous chemicals factors and pipeline route optimization model combines

For the transport of dangerous chemicals path problem, many scholars have carried out relevant research work, the research results are mainly divided into two categories: (1) selection of the optimal path based on the optimization model. Erhan Erkut etc,[8-14]are objective to minimize the risk of dangerous chemicals to build a optimization transportation network and pick up Italy Labbe region as a background empirical study and find it; Wu Zong-zhi etc,[15-16] according to the road transport of dangerous goods progress risk analysis, systematic study of the selected line path optimization, they proposed general procedure of dangerous goods road transport path optimization, building a optimization line selection evaluation system and decision-making framework. (2) the optimal path required based on evaluation model. Shen Xiayan etc, [17]based on fuzzy comprehensive evaluation model, on the basis of select build path evaluation system to achieve the optimization of the transport of dangerous choice.

In this paper, we selected the route optimization model from the paper dangerous chemicals transportation route optimization choice model.[18] On that paper an evaluation index system on the optimal selection of dangerous chemicals transportation routes was constructed by introducing the multi-objective lattice order decision making theory into the routing scheme evaluation and comparison, and then lattice ordering on the alternative schemes was realized by calculating the comprehensiveness and operability on index selection and analysis on the present data, the selection index system on transportation route of dangerous goods could be constructed from six aspects, including impact on the protection of regional water source(p1), impact on the protection of places of interests in the regions along the route(p2), impact on the protection of regional ecological environment(p3), impact on regional population scale(p4), impact on regional factories and mines scale(p5)and comprehensive transport efficiency of transportation route(p6). In addition, we add the pipeline and its surrounding pipeline laid density, we call it p7. The entropy method was applied in model system weight calculation to achieve the objectivity of the weight determination.

In the model simulation platform, we build a pipeline transportation of dangerous chemicals test scenarios. Pipeline transport virulent chlorine from production center S to processing center E. There are three transport path is chosen, the path towards was shown in Figure 2.

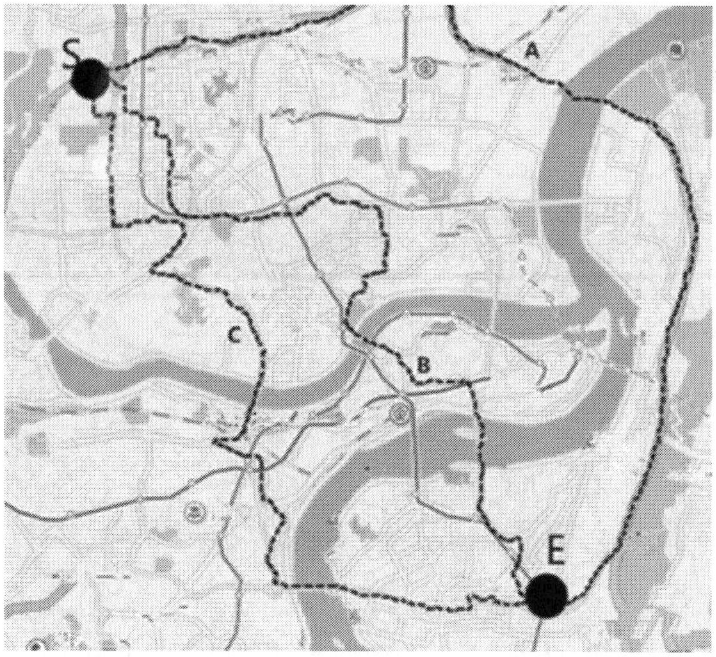

Figure 2 Transportation of dangerous chemicals path program

We evaluate from p1, p2, p3, p4, p5, p6, p7, seven aspects using the method in the model to be processed each index. Its index values calculated results are shown in Table 2.

Table 2. Pipeline transportation of dangerous chemicals path evaluation indicators program

Layout scheme	p1	p2	p3	p4	p5	p6	p7
A	0.20	0.20	0.15	0.20	0.40	0.15	0.36
B	0.32	0.50	0.56	0.63	0.10	0.35	0.91
C	0.76	0.30	0.30	0.48	0.15	0.45	0.78

We use the referred entropy method by the model to calculate the weight coefficient of each type ω, we can get that ω1=0.160,ω2=0.143,ω3=0.153,ω4=0.151,ω5=0.135,ω6=0.146,ω7=0.111. The above data were put into the given model euclidean distance between positive and negative ideal solution formula, L=0.146934. Similarly, we can obtain the euclidean distance between positive and negative ideal solution of each type. Let positive coefficient q 0.5 into the model, calculating integrated difference value of the program. The calculation results were shown in Table 3.

Table 3. Programs and euclidean distance between positive and negative ideal solution difference values

	program A		program B		program C	
	G1+	G1-	G2+	G2-	G3+	G3-
Euclidean distance	0.062561	0.039895	0.118037	0.16064	0.16986	0.143665
Comprehensive difference value	0.321791		0.875262		0.984712	

Table 4. Programs and euclidean distance between positive and negative ideal solution difference values(without considering p7)

	program A		program B		program C	
	G1+	G1-	G2+	G2-	G3+	G3-
Euclidean distance	0.060972	0.035485	0.105818	0.134494	0.166292	0.127297
Comprehensive difference value	0.328233		0.817754		0.999047	

Based on the comprehensive difference value, we sorted the path to the pipeline transporting dangerous chemicals, the result is: program C > program B > program A. That means we should select program C to be the pipeline transporting dangerous chemicals considering the pipeline density which we have analysis are the the most influential factors, No matter we consider p7 or not.

4. Conclusion

In this paper, from the improved fault tree analysis, we find out the greatest impact factor analysis is the pipeline density; and combining the pipeline transportation of dangerous chemicals the path optimization model with some impact factors. Although the pipeline density in the transportation of hazardous chemicals is the most important controllable factor, for many factors hazardous chemicals pipeline route, the effect is not significant. Through the model, we can also see that, for environmental management in the path planning problem we should consider comprehensively.

5. References

[1] Wu Zong-zhi. "Internal and External Security (Risk) Assessment Method Research and Development [J]." *Ordnance Safety Technology*,1999, 2: 37-40.

[2] Meng Xian-lin, Zhou ding, Huang Jun-li. "Practice and Development of Environmental Risk Assessment[J]." *Sichuan Environment*,2001, 20(3):1-4.

[3] Shuai Jian. "U.S. Oil and Gas Pipeline Accidents and Its Implications [J]." *Gas Storage and Transportation*,2010, 29(11): 806-809.

[4] "National Transportation Safety Board." *Pipeline Accident Report*[R/OL]. PAR-03-01, 2003[2010-03-15]. *http://www.ntsb.gov/Publictn/publictn.hrin*

[5] "National Transportation Safety Board."*Pipeline Accident Report*[R/OL]. PAR-02-02, 2002[2010-03-15]. *http://www.ntsb.gov/Publictn/publictn.htm.*

2002[2010-03-15]. *http://www.ntsb.gov/Publictn/publictn.htm.*

[6] Lv Yi-wen. "Transportation of Dangerous Chemicals Risk Analysis." *Chemical Engineering and Equipment*,2008,(09).

[7] Liu Jing. "Headwater Dangerous Chemicals Transportation Risk Assessment and Environmental Management."2012.

[8] Erhan Erkut , Vedat Verter. "Modeling of Transport Risk for Hazardous Materials" [J]. *Operations Research*, 1998, 46(5) :625-642.

[9] Erhan Erkut , Armanm Ingolfsson. "Transport Risk Models for Hazardous Material: Revisited" [J]. *Operations Research*, Letters, 2005, 33:81-89.

[10] Erhan Erkut, Vedat Verter. "A Framework Hazardous Materials Transport Risk Assessment" [J]. *Risk Analysis*, 1995, 15(5):589-600.

[11] Erhan Erkut, A. Ingolfsson. "Catastrophe Avoidance Models for Hazardous Materials Rout Planning" [J]. *Transportation Science*, 2000, 34(2): 165-179.

[12] Erhan Erkut, Osman Alp. "Designing a Road Network for Hazardous Materials Shipments" [J]. *Computer & Operations Research*, 2007,34(5): 1389-1405.

[13] B.Y.Kara, Erhan Erkut, V.Verter. "Accurate Calculation of Hazardous Materials Transport Risks" [J]. *Operations Research Letters*, 2003, 31(4) :285-292.

[14] Ethan Erkut, Fatma Gzara. "Solving the Hazmat Transport Network Design Problem" [J]. *Computer & Operations Research*, 2008, 35: 2234-2247.

[15] Wu Zong-zhi, Ren Chang-xing. "Progress of Risk Assessment and Optical Routing for Hazardous Materials Transportation by Road" [J]. *Journal of Safety and Environment*, 2007, 7(3): 120-131.

[16] Ren Chang-xing, Wu Zong-zhi. "Probe into Methods of Optimal Road Transportation Routing for Hazardous Materials" [J]. *China Safety Science Journal*, 2006, 16(6): 129-134.

[17] Shen Xiao-yan, Liu Hao-xue, Zhou Xun. "Preliminary Investigation for Route Optimization Research about Transportation of Dangerous Goods" [J]. *Journal of Safety Science and Technology*, 2006, 2(1) : 76-79.

[18] Li Jin. "Select Model of Dangerous Chemicals Transportation Route Optimization"[J]. *Journal of Safety Science and Technology*,2011,10(10).

西方发达国家非政府标准机构运行机制研究

刘国超[1] 孟亦飞[2] 赵东风[3]

（1. 中国石油大学（华东）机电工程学院，山东省 青岛市 266580；2. 中国石油大学（华东）化学工程学院，山东省 青岛市 266580; 3. 中国石油大学（华东）化学工程学院，山东省 青岛市 266580;）

摘要： 我国目前的标准机构是由政府主导，而像美、英等西方发达国家标准机构由市场主导、政府引导，分为政府与非政府成分。为加快我国标准机构的市场化进程，弥补我国相对不足，有必要对西方发达国家标准机构进行研究，重点关注其中的非政府成分。通过探究非政府成分在市场化体制下的运行机制，了解了这些民间机构与政府的关系、主要从事的业务与提供长期经营的经费来源等。最终，结合我国现状，提出一些有效的建议与措施，从而借鉴其先进经验，丰富标准机构成分，建立更加完善的标准系统。

关键词： 非政府标准机构；运行机制；标准体系；市场化

Investigation on the Operating Mechanism of Non-governmental Standards Organizations in Western Developed Countries

LIU Guo-chao[1] MENG Yi-fei[2] ZHAO Dong-feng[3]

(1. College of Electrical and Mechanical Engineering, China University of Petroleum (East China), Qingdao in Shandong Province 266580; 2. College of Electrical and Mechanical Engineering, China University of Petroleum (East China), Qingdao in Shandong Province 266580; 3. College of Electrical and Mechanical Engineering, China University of Petroleum (East China), Qingdao in Shandong Province 266580;)

Abstract: At present, Chinese standards organizations are led mainly by the government. While, the standards organizations in some western developed countries such as America and Britain are dominated by the market and composed by governmental and non-governmental parts. In order to accelerate the marketization and make up the deficiency, it's necessary to investigate the standards organizations in western developed countries especially the non-governmental part. That through exploring the operating mechanism got a whole understanding including the relationship between the non-governmental organizations and the government, main business and sources of income. Finally, proposing some suggestions taken into

Chinese current condition makes us learn from advanced experience and enrich the composition leading a result of establishing a better standard system.
Key words: non-governmental standards organizations; industrial safety; operating mechanism; standard system; marketization

一、 引言

目前，我国标准体系市场化转型尚不够成熟，其是由政府主导的自上而下的行政化体制，标准化机构从属于政府。而西方如美、英等国家的标准体系是由市场主导、政府引导的市场化体制，标准机构由政府和非政府（民间）机构组成。相比而言，我国在标准合理性与适用性方面还有待提高，标准相对老化，管理模式较为落后，有必要借鉴西方先进模式的经验。

西方国家的非政府标准机构包含两种类型：一是其特有的非政府性质的全国标准化机构，主要负责协调、管理民间标准体系。一类是民间标准制定机构，主要负责标准制定任务。西方标准体系中的非政府部分是我国与之相比最大的差异点所在，本文以此为切入点，研究其在市场经济下的运行机制，并根据我国国情，对我国目前现状提出一些建议措施，从而有利于构建我国民间标准机构成分，推进标准体制市场化转型。

二 国外非政府性质的全国标准化机构

2.1 美国国家标准学会（ANSI）

2.1.1 与政府之间关系

ANSI 是非营利性质的民间标准化团体，受政府的委托发布和管理美国国家标准，与政府只是合作关系，不受政府监管，具有较高独立性。标准制定过程中，企业是主体，政府只是作为一个利益方，反映政府意愿。

2.1.2 业务范围

ANSI 致力于协调、管理民间标准体系；将企业标准或行业标准上升为国家标准；对国家标准制定机构（SDOs）提供资格认证；作为国际标准组织秘书处五大所在地之一（其余四个：德、英、法、日），代表美国积极参加国际标准化活动。起到了联邦政府和民间标准化体系之间的桥梁作用。ANSI 本身很少制订标准，制定工作主要由 SDOs 负责。ANSI 制定了《国家标准制定程序的基本要求》作为依据审批美国国家标准。ANSI 标准编制方式如下表所示。

表 2.1-1 ANSI 标准编制方式

序号	内容
1	由 ANSI 委托有关单位起草，邀请专家或专业团体投票，将结果报 ANSI 设立的标准评审会审议批准
2	某项标准找不到合适的团体制定时，由 ANSI 授权建立一个特别技术委员会，各方代表拟订标准草案，全体委员投票表决，最后由标准评审会审核批准
3	从各专业学会、协会标准中，将较成熟的，对于全国普遍具有重要意义，经 ANSI 技术委员会审核后，提升为国家标准。目前，采用这种方式的 ANSI 标准数量最多

2.1.3 经费来源

ANSI 经费大部分来源于标准资料收入：拥有国家标准的版权和出版发行权。其次是会员会费：其拥有工业学会、协会等团体会员约 200 个，政府、企业会员约 1400 个。另外还包括合格评定项目、国际标准项目以及培训等。2002 年 ANSI 的营业收入情况可详见参考文献 1。

2.2 英国标准协会（BSI）

2.2.1 与政府之间关系

BSI 是世界上第一家国家标准化机构，是英国政府承认并支持的非营利性民间团体。英国政府与 BSI 本着平等、独立和服务于公共利益的原则，以皇家宪章和签订备忘录的形式确定了与 BSI 的关系。皇家宪章规定了 BSI 的独立法人地位、业务范围、董事会和会员组成等。备忘录中承认 BSI 的国家标准机构地位，规定了 BSI 的责任、权利。

英国贸工部标准与技术法规司是制定标准、测试和认证政策的政府主管部门，但其仅负责政策层面的管理。具体的标准制定、管理职能由 BSI 实施。

2.2.2 业务范围

英国标准协会的主要任务是编制和销售专用的、全国性和国际性标准和支持信息，制定和贯彻英国标准BS；开展产品质量合格认证和安全认证。管理规范第二方和第三方管理体系评估和认证，接受外国认证委托、按国外标准进行认证并颁发外国的认证证书和标志[2]。积极参与国际标准活动；支持标准实施和业务最佳实践方面的培训服务等。

2.2.3 经费来源

BSI 经费主要来源于三个方面：一是标准和其他出版物版权销售费用，目前 BSI 有 5 万多个出版物、书籍和标准。二是会费，目前 BSI 有捐款会员 2 万多个，委员会会员 2 万个。三是认证工作收入，作为全球最大认证机构之一，拥有全球最广泛的测试和认证能力。另外，还包含培训、咨询等资金来源。

2.3 德国标准化协会（DIN）

2.3.1 与政府之间关系

DIN 是私有化的非营利协会，政府与 DIN 签署协议，将双方合作关系予以法律明确。政府承认 DIN 是国家级标准化权威机构，但不属于政府。双方构成平等合作关系，一种合作方式是政府委托 DIN 帮助制定法律法规的实施细则，或者为满足某些法律法规要求须参考的标准；二是 DIN 向政府建议制定某些标准来实现政府法规纲领。DIN 标准不具法律效力，但德国法庭对影响深刻且广泛承认的 DIN 标准予以特殊的承认，使其成为德国产品标准的唯一权威。DIN 得到了政府有力的财政扶持，而 DIN 也承认政府对其标准化工作具有监督职能。

2.3.2 业务范围

DIN 的核心业务是与各利益方密切合作，制定协商一致标准。其次管理着产品标准并负责德国与地区及国际标准化组织间的协调事务。DIN 有两种产品认证，即“DIN 标志”和“DIN 检验和监督标志”，另外有联邦德国质量体系认证。承担企事业质量体系评定和认可工作。

2.3.3 经费来源

DIN每年的成本预算为 6000 多万欧元，其经费来源如下表所示[3]：

表 2.3-1 DIN 经费来源

具体来源	百分比
自筹资金：主要来自下属企业的经营利润，包括出版社、软件公司、IT 服务公司认证机构、培训和咨询机构，最大的利润来自出版发行各种标准文本	70%
政府项目资金	12%
工业项目资金	13%
会费：1934 名会员（2002 年），分别自工业界、各州、各工会组织、学术机构、消者协会、环保组织、专业协会	5%

2.4 法国标准化协会（AFNOR）

2.4.1 与政府之间关系

法国标准化机构形成了官助民办、政府监督的机制。AFNOR是一个由国家资助的，得到政府承认的民间团体。法国标准化专署是政府部门，内设标准化专员。专员具有批准AFNOR组织章程和办事规则，任命会长的权利。AFNOR最高领导为理事会，由非营利性团体的34名成员组成[4]。

2.4.2 业务范围

AFNOR在专员指导监督下，协调全国性标准化活动；向全国各专业标准化局传达、落实政府指令，协助他们制定标准草案、审查草案，承担标准审批工作；协调各标准化组织的活动并担任他们与政府的联络人；代表法国参加国际标准化活动；散发标准，宣传标准化；在没有行业标准化局管辖的领域，组织技术委员会，进行标准草案的制定工作。AF标准40%由自身制定，60%由与AFNOR签有协议的SDO制定。另外，AFNOR与各利益方组成产品质量认证委员会，负责NF标志产品认证工作。

2.4.3 经费来源

经费来源主要靠销售出版物，提供技术服务、会费（注册会员6000）和国家资助，国家经费占总额的三分之一[4]。

三 国外非政府性质的民间标准制定机构

3.1 美国材料与试验协会（ASTM）

ASTM是世界上最大的制定自愿性标准的非盈利组织，专注于标准制定工作。
标准制定工作中，ASTM只负责提供会议安排、投票系统、投票处理原则等行政管理工作。
标准的发起、起草、修订、投票、废止等都由会员负责。

ASTM有着一套先进的运行机制和理念原则。重视会员发展，深信并且事实也证明行业背景深厚的技术专家是了解市场动向的最佳人选，会员的广泛性是确保标准高质量的关键要素。标准工作高度市场化，ASTM技术委员会和分技术委员会的数量是动态变化的，其存在与否完全根据市场需要，由投票决定。ASTM建立了完整的会员投票、信息公开等规范程序，任何标准都要经过工作组、分技术委员会、技术委员会三个层次的投票[5]。另外，ASTM标准制定没有任何时限。ASTM确信只要规范机制，加上具有广泛代表性的会员，必然可以保证标准的及时制定。事实上如果市场需求很急切，各利益方代表很快会相互妥协、达成一致。

出售 ASTM 标准和其他出版物是年收入的主要来源，占总收入的 60%以上。此外还有会员费、培训费用、管理费用、标准相关项目费用。

3.2 美国石油学会（API）

API 是全世界范围内最早、最成功的制定标准的商会之一。API 标准是国际公认的石油机械设备行业标准，具有很高的声誉，已成为走向国家市场的通行证。API 也是美国商业部和贸易委员会承认的石油机械认证机构。包括 API 系列产品会标认证咨询和 API Q1 质量体系认证咨询。

API每年出版 30 多万册出版物，包含标准及其他一些出版物，是主要的资金来源。认证费用是资金的重要组成部分，包括申请费用、现成审核费用、每年监督费用及咨询和代理费用[6]。另外还包括API会员会费和相关培训费用。

四 启示与借鉴

通过对国外非政府标准机构的简要介绍，下面对其组建机制、业务组成、运行方式和资金来源进行一下总结分析，明晰我国非政府标准机构建设方向和细节。

1、美、英等国家政府一般通过授权、委托民间机构，成为国家标准化机构，双方构成合作关系，但合作方式略有不同。按照美、英、德、法的顺序，全国标准化机构独立性逐渐降低，政府监管控制逐渐增强，政府经费逐渐增多。为完成转型，我国也应授权或组建国家标准化机构，并通过法律或签订协议明确其地位。可按美、英、德、法的顺序，逐步实现过渡。

开始阶段，可借鉴法国官助民办、政府监督，政府仍有较强控制作用。继而像德国政府一样，对 DIN 有监督权，但无控制权。最后实现市场最大化，政府只是作为 ANSI 标准的一个利益方，表达政府意愿。当然，不能照搬西方国家标准体制，其主要提供方向性作用。我国国家标准化机构具有哪些具体权利，政府监管控制程度如何，还需结合国情和实践效果不断优化。

2、美、英、德、法国家标准化机构业务总结如下表所示。

表 4.1　西方发达国家标准化机构业务范围

业务类别	详细内容
标准制定	BSI、DIN 直接负责标准制定，由自身标准委员会完成。ANSI 间接负责，委托 SDO 制定，ANSI 负责审批。AF 两种方式相结合。标准及其出版物是主要的资金来源
认证工作	ANSI 负责 SDO 资格认证及行业、国家标准认证，其余三个标准机构均包含产品质量和体系认证。认证服务是资金重要组成部分
其他	均代表国家参与国际标准活动、培训教育等。另外，各机构为了生存发展，也扩展了其他相关业务。BSI 除标准制定拥有国家标准机构的职能外，标准技术信息提供、产品测试、体系认证和商检服务均同社会普通公司一样。DIN 拥有软件等多个下属企业。这些业务是资金重要组成部分

根据西方国家标准化机构业务范围可大致明确我国未来国家标准化机构业务。可结合实际情况，我国国家标准化机构可直接承担标准制定任务或间接委托其他机构负责，适当、灵活发展其他相关业务。

3、会员制度建立起了标准机构和企业、个人之间的桥梁。标准化机构提供平台，各会员通过平台制定包含自身利益的标准。广泛的会员与行业背景深厚的专家能够捕捉市场动向，保证标准质量。高质量标准提升机构权威性，高权威性又会吸引新会员加入。因此会员的重要性不言而喻，我国在构建非政府标准机构时，应将业内公司和资深专家利用会员制结合在一起，实现互利共荣的良性循环。

4、协商一致原则是西方标准制定过程中最重要的原则。我国在非政府机构建设中，标准制定不再由政府主导，而更多依赖于包含企业、制造商、供应商等各利益方在内的市场本身。企业作为标准使用者，要突出其主体作用，充分吸收其意见和建议，有助于提高标准适用性与合理性。

5、美国有 400 多家民间标准机构，为世界之最。虽有一定弊端，即政府过度放任，民间标准多而重复[7]。但利大于弊，其极大的丰富了标准内容，完善了标准体系。我国在构建民间标准成分时，一方面可将某些政府标准机构民间化，不受政府行政控制，先天优势是其职能划分已明确，只需在组织机构和运行机制加以改进。另一方面，鼓励民间成立标准机构，运行方式可借鉴ASTM等，政府可提供一定的财力、物力和技术支持，如美国政府层面通过美国国家标准与技术研究院借鉴（NIST）向民间标准机构提供技术支持。

参考文献

[1] 李凤云．美国标准化调研报告（中）[J]．美国标准化调研报告，2004

（04）：53-62.

[2] Product Certification. https://www.bsigroup.com.

[3] Financing Standards Work at DIN. http://www.din.de/cmd?level=tpl-home&contextid=din.

[4] 王金玉．发达国家标准化管理体系研究[J]．标准化动态，2002（04）：11-25.

[5] 上海市质量和标准化研究院．ASTM International 组织介绍．http://www.cnsis.info/index.jsp.

[6] Certification Process. http://www.api.org/.

[7] 李会光．欧美日中标准制定和管理机制的比较研究[D]．河北：河北工业大学，2007.

长输原油管道工程消防应急能力评估

周欢[1]，刘义[2]，赵东风[2]

(1. 中国石油大学（华东）机电工程学院，山东青岛 266580;

2. 中国石油大学（华东）化学工程学院，山东青岛 266580)

摘　要： 近年来，由于腐蚀、突发性自然灾害以及人为破坏等因素影响，油气管道突发事件频发。输油管道火灾爆炸事故是管道泄漏的次生事件，但其危害远远高于单纯的输油管道泄漏。建立长输原油管道工程消防应急能力评估体系，系统科学地评价管道消防应急管理水平，对提升长输原油管道消防应急能力具有十分重要的意义。本文对长输原油管道工程火灾爆炸危险性进行了分析，结合标准规范以及工程实践构建了一套长输原油管道工程消防应急能力评估体系。该体系选取的指标涵盖预防与应急准备、监控与报警以事故消防应急处置三个方面，利用层次分析法确定指标权重，采用主观打分的评价方法。最后，通过实例应用验证了评估体系的可行性，并对企业提出了改进建议。

关键词： 长输原油管道；消防系统；应急能力；指标；评估体系

Fire-fighting Emergency Capabilities Assessment of Long Distance Crude Oil Pipeline Engineering

Huan Zhou[1], Yi Liu[2], Dongfeng Zhao[2]

(1 Mechanical and electrical engineering college, China University of Petroleum Qingdao, Shandong, 266580, China;

2 Chemical engineering collage, China University of Petroleum, Qingdao, Shandong, 266580, China)

Abstract: Pipeline fire explosion accidents are secondary events of pipeline leakage,but its harm are much higher than that of pipe leaks. It is of great importance to set up an evaluation system for long distance crude oil pipeline engineering fire-fighting emergency capabilities assessment , which can help to evaluate pipeline engineering fire-fighting emergency capabilities thoroughly and scientifically. In this paper, fire and explosion hazard analyzes of pipeline engineering are presented, combined with standards and engineering practices to build a long-distance oil pipeline engineering fire emergency capability assessment system. The system indicators cover three aspects of emergency response. Than use AHP to determine the index weight and subjective evaluation method scoring. Finally, through an example application of the system to verify the feasibility of this assessment, and put some suggestion for improvement.

Key words: long distance crude oil pipeline; fire extinguishing system; indicator; emergency capability assessment

引言

目前，长输原油管道承担着我国约 70%的原油运输，是我国重要的能源基础设施。然而，由于受到腐蚀、突发性自然灾害，以及人为因素导致的意外事故和恶意的打孔盗油等因素影响，原油管道突发事件频发。长输原油管道输送的是易燃易爆介质，破裂泄漏时遇着火源极易导致火灾和爆炸事故，一旦发生，后果不堪设想。例如 11•22 青岛输油管道爆炸事件，由于输油管道原油泄漏进入市政排水暗渠，在形成的密闭空间暗渠内油气积聚遇火花发生爆炸。事故造成 62 人死亡、136 人受伤，直接经济损失 75172 万元。这次事故原因除了隐患排查不彻底外，还暴露出一个突出问题——应急处置不力。原油泄漏后，由于对事故风险评估出现严重错误，未及时启动应急预案，消防应急救援措施不到位，以至于造成严重后果[1]。因此，长输原油工程的消防安全极为重要，对长输原油管道工程消防应急能力进行评估，及时发现薄弱环节，对于提升长输管道消防应急能力水平具有长远意义。

本文进行了长输原油管道工程火灾爆炸危险性分析，依据标准规范和工程实际提出了一套长输原油管道工程消防应急能力评估体系，并通过实例验证了它的可行性。这套体系旨在找出管道消防系统薄弱环节，提出针对性的建议，提升管道消防应急水平。

1 管道工程火灾爆炸危险性分析

长输原油管道工程包括站场、阀室和管道 3 部分，输送介质原油属于甲B类物质，具有易燃、易爆、挥发性高、火灾爆炸危险性大等特点。站场和管道是最易引发火灾爆炸事故的区域，其中站场是主要生产场所和储罐装置集中区[2]。

1.1 站场火灾爆炸危险性分析

站场油罐区储存大量原油，并进行原油的收发作业，稍有不慎便会引发火灾爆炸事故。油罐区及工艺设备装置区发生火灾的原因主要有两方面：一方面是站场在日常运行维护过程中，原油的高挥发性以及油品的跑、冒、滴、漏蒸发形成易燃易爆蒸气，蒸气随风扩散、积聚到一定浓度遇明火可引起火灾或爆炸；另一方面是罐区及生产装置区含油污水排放口与排污口连接，外排口没有水封设施和无截断设置，易导致油气串通，遇明火引起火灾传递，严重时引发多米诺效应[3, 4]。

1.2 管道火灾爆炸危险性分析

长输原油管道输送的介质具有易燃易爆性，管道系统主要的火灾爆炸类型有：泄漏引起火灾爆炸、管道内爆炸性混合物爆炸、管道内超压爆炸、管道内堵塞爆炸、管道内发生自燃以及物料输送过程中遇多种引火源发生爆炸[3]。由于管道连接各种设备，一旦管道发生火灾爆炸事故就会发生联锁反应，使事故迅速扩大，对生产设备和建筑物造成严重破坏，还会危及人的生命财产安全。

2 构建长输原油管道工程消防应急能力评估体系

2.1 构建评估体系框架

在评价体系的建立过程中，本着科学、客观、系统的基本原则，参照管道工程设计相关规范以及管道工程消防系统应急管理现状进行指标的选取。评估体系建立的基本思路是：以管道工程消防应急系统的三个环节作为评价框架中针对总体评价目标的下一层级，三个环节中分别包含细化内容作为应急能力评价的第三层级和第四层级[5]。长输原油管道工程消防应急能力评估体系见图 1。

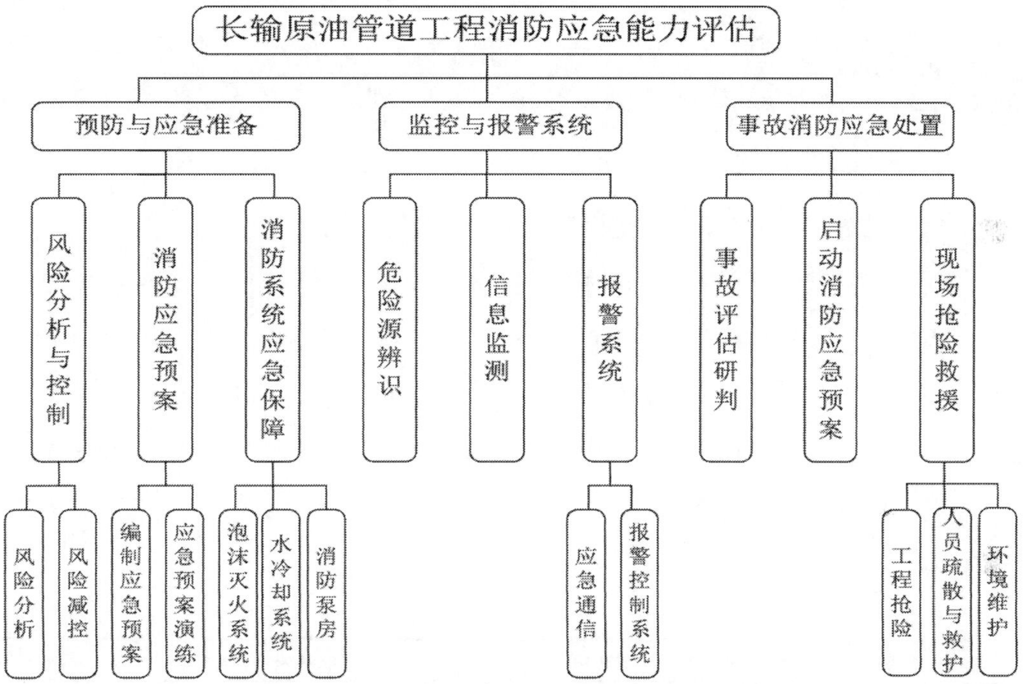

图 1 长输原油管道工程消防应急能力评估体系

Figure 1 Fire emergency capability assessment system of long distance crude oil pipeline engineering

预防与应急准备环节评价的主要对象包括：对管道火灾爆炸风险的全面有效分析、针对管道火灾爆炸危险性的减控措施；有效的消防应急预案以及定期演练记录；消防系统资源保障，满足规范要求的泡沫、冷却水储备量、泡沫泵房的设置以及其他辅助消防设施的配置。

监控与报警系统环节评价的主要对象包括：全面的火灾爆炸危险源辨识方法和标准；有效的火灾爆炸信息监测设备和人员，明确的监测内容；准确可靠的应急通信资源以及火灾报警控制系统。

事故消防应急处置环节评价的主要对象包括：火灾爆炸发生后对事故准确的评估把握；及时启动并响应消防应急预案；联合各方力量抢险救援：消防系统灭火，人员疏散，信息发布，各方支援，环境保护。

2.2 确定指标权重——层次分析法

层次分析法（简称 AHP）是由美国匹茨堡大学教授萨蒂在 20 世纪 70 年代提出的一种层次权重决策分析方法，是对非定量事件做定量分析的一种有效方法，符合确定应急能力评价体系指标权重的需求。层次分析法可有效地将定性问题进行定量计算，尽量避免人为误差造成的评价结果的偏离。使用层次分析法确定指标权重的步骤是：首先通过专家打分构建判断矩阵；其次是计算权重；最后一步是一致性检验。就本文建立的消防应急能力评估体系指标权重的确定具体操作如下。

第一步，构建判断矩阵。对于从属于上一层同一单元的同一层各个因素，相对于上一层某个单元的重要性进行两两比较，形成成对比较阵。1 表示两元素同样重要，3，5，7，9 分别表示一个元素相较于另一个元素稍微重要、明显重要、强烈重要和绝对重要；2，4，6，8 则表示相邻两个奇数刻度所表达意义的中值。第二步，计算权重。对每个比较阵计算其最大特征根和相对应的特征向量，经正规化作为元素a1，a2，…，an在准则G下排序权重。第三步，进行一致性检验。在此基础上进行一致性检验，如果一致性比例控制在 0.1 以内，则表示通过一致性检验。同理计算组合权向量并进行一致性检验。若检验通过，则可按结果进行决策[6]。基于层次分析法的以上三个步骤，制作AHP调查问卷请 8 位管道消防工程专家及现场人员进行评估，共收回 6 份有效问卷，问卷精简结果见表 1。

表 1 AHP 有效问卷权重统计表

指标	专家 1	专家 2	专家 3	专家 4	专家 5	专家 6
a1 预防与应急准备	0.422	0.395	0.478	0.431	0.416	0.372
b1 风险分析与控制	0.147	0.113	0.159	0.173	0.152	0.126
b2 消防应急预案	0.093	0.105	0.097	0.142	0.121	0.085
b3 消防系统应急保障	0.182	0.177	0.222	0.116	0.143	0.161
a2 监控与报警系统	0.217	0.237	0.194	0.275	0.208	0.264
b4 危险源辨识	0.065	0.093	0.067	0.082	0.054	0.072
b5 信息监测	0.071	0.048	0.047	0.095	0.079	0.087
b6 报警系统	0.081	0.096	0.080	0.098	0.075	0.105
a3 事故消防应急处置	0.361	0.368	0.328	0.294	0.376	0.364
b7 事故评估研判	0.104	0.095	0.082	0.116	0.099	0.113
b8 启动消防应急预案	0.128	0.136	0.097	0.073	0.098	0.085
b9 现场抢险救援	0.129	0.137	0.149	0.105	0.179	0.166

经过以上三个步骤确定了长输原油管道工程消防应急能力评估体系指标的权重如下表2所示。

表 2 长输原油管道工程消防应急能力评估体系指标权重表

指标	权重/本环节权重	指标	权重/本环节权重
a1 预防与应急准备	**41.9%**	b4 危险源辨识	7.22%/31.0%
b1 风险分析与控制	14.5%/34.6%	b5 信息监测	7.12%/30.1%
c1 风险分析	6.18%/14.7%	b6 报警系统	8.92%/38.3%
c2 风险减控	8.32%/19.9%	c8 应急通信	5.12%/22.0%
b2 消防应急预案	10.7%/25.5%	c9 报警控制系统	3.28%/14.1%
c3 编制应急预案	5.16%/12.3%	a3 事故消防应急处置	**34.9%**
c4 应急预案演练	5.54%/13.2%	b7 事故评估研判	10.2%/29.2%
b3 消防系统应急保障	16.7%/39.9%	b8 启动消防应急预案	10.3%/29.5%
c5 泡沫灭火系统	6.81%/14.7%	b9 现场抢险救援	14.4%/41.3%
c6 水冷却系统	5.93%/14.2%	c10 工程抢险	5.83%/16.7%
c7 消防泵房	3.96%/9.5%	c11 人员疏散与救护	4.98%/14.3%
a2 监控与报警系统	**23.3%**	c12 环境维护	3.59%/10.3%

从以上计算结果可以看出：预防与应急准备是管道工程消防应急管理三个环节中最重要的，它的权重占到41.9%，其次重要的是事故消防应急处置，占34.9%，监测与报警系统占23.3%。

2.3 确定评估方法——主观打分法

消防应急能力评估的目的是通过对长输原油管道工程消防能力进行评估，找出薄弱环节。本文对消防评估体系采用主观打分法，借鉴美国应急能力评价系统（简称CAR），根据被评价主体实际情况，对照评价条款进行打分评价。评价条款中评价标准分为1、2、3、NA四个等级："1"表示亟需加强改进；"2"表示基本符合要求；"3"表示完全符合要求；"NA"表示条款不适合实际情况。长输管道企业应急能力评价的总得分和每个环节能力的得分根据相应每项应急能力的得分与其所占权重乘积累加得出。针对最后的评价结果，设定评估结果标准如下：1~1.5（含）分，表示被评价对象亟需加强和改进；1.5~2.5（含）分，表示工作基本符合要求，但仍有较大改进空间；2.5~3分，表示工作水平很高，可以继续保持或稍加改进[7]。

3 实例应用

3.1 A长输原油管道工程消防应急能力评估

利用以上方法，将AHP问卷和打分表分发给相关专家和管道工程实践人员，汇总整理有效问卷后得出A长输原油管道工程消防应急能力评估结果，见表3。

表 3 A 管道工程消防应急能力评估结果

指标	分值	指标	分值
a1 预防与应急准备	**2.59**	b4 危险源辨识	2.85

b1 风险分析与控制	--	b5 信息监测	2.69
c1 风险分析	2.67	b6 报警系统	--
c2 风险减控	2.72	c8 应急通信	2.91
b2 消防应急预案	--	c9 报警控制系统	2.78
c3 编制应急预案	2.71	a3 事故消防应急处置	**2.75**
c4 应急预案演练	2.46	b7 事故评估研判	2.69
b3 消防系统应急保障	--	b8 启动消防应急预案	2.73
c5 泡沫灭火系统	2.50	b9 现场抢险救援	--
c6 水冷却系统	2.63	c10 工程抢险	2.84
c7 消防泵房	2.68	c11 人员疏散与救护	2.87
a2 监控与报警系统	**2.73**	c12 环境维护	2.63

3.2 评估结果分析及建议

从各项应急能力评价结果（表 3）来看，16 个应急指标中高于 2.50 分的有 14 项，说明 87.5%的应急能力很高。其余应急预案演练（2.46 分）和泡沫灭火系统（2.50 分）2 项具体应急能力基本符合要求，但有较大改进空间[8]。A长输原油管道工程消防应急能力最终得分是 2.68，表示目前的消防应急工作水平很高，可以继续保持或更加努力。针对薄弱环节，公司有关负责人员应提高警惕，重点改进。

根据评估结果，对于 A 管道工程消防应急工作提出以下建议：依据消防应急预案的要求定期进行应急演练，验收演练结果并及时改进，一方面提高员工的危机意识，另一方面提高消防实战应急水平；对于泡沫灭火系统，请有关专家和工程人员依据规范重新检查验收，对不符合要求的地方及时修正，定期进行泡沫系统检查，提高消防灭火系统的可靠性。

4 结论

根据相关规范和工程实践，构建了长输原油管道工程消防应急能力评估体系，利用层次分析法确定权重，采用主观打分法进行评估，建立起一套管道工程消防应急能力评估系统。利用这套系统对管道工程消防应急能力评估，能及时发现薄弱点，提出改进意见，对全面加强管道企业消防应急能力，切实提高突发事件应急水平具有深远的意义。

参考文献

[1] 何清本报记者. 青岛管道爆炸警示：中石化彻查辖下管道[N]. 21世纪经济报道, 2014-01-14.
[2] 查广平, 郭志强, 熊超. 长输原油管道工程消防设计[J]. 石油规划设计, 2013(05):35-37.
[3] 孙仪凯. 长输管道火灾爆炸危险因素分析与控制[J]. 当代化工, 2010(01):86-88.
[4] Tong S, Lo S, Zhang P, et al. Jet Fire Consequence Evaluation on the Natural Gas Transported by Pipelines[J]. Procedia Engineering, 2013,52(0):349-354.

[5] Liu X, Zhang Q, Xu X. Petrochemical Plant multi-Objective and multi-Stage fire Emergency Management Technology System Based on the fire risk Prediction[J]. Procedia Engineering, 2013, 62(0):1104-1111.
[6] 邓雪，李家铭，曾浩健，等. 层次分析法权重计算方法分析及其应用研究[J]. 数学的实践与认识，2012(07):93-100.
[7] Agency FEM, Association NEM. State Capability Assessment for Readiness（CAR）：2000.
[8] 郑明珠. 中石油油气长输管道企业应急能力评价[D]. 清华大学，2012.

基于数值模拟的天然气输送管道定量风险评估

袁长高[1]，赵东风[1]

（1. 美国化学工程师协会化工过程安全中心-中国分部，山东 青岛 266580）

摘 要： 由于天然气管道潜在危险性大且日常检测比较困难，管线一旦发生泄漏容易引发火灾、爆炸等重大事故，造成很大的人员伤亡和财产损失。因此，提高天然气管道的安全水平具有重大意义。采取定量风险评估的方法对天然气管道输送时的泄漏进行了研究。分析总结了天然气管道泄漏扩散的主要潜在危险因素；同时，采用了数值模拟的方法研究了天然气在不同风速、不同泄漏孔径下泄漏扩散的规律，并对其可能引发喷射火、闪火等造成的危害范围进行模拟分析，从而为天然气管道泄漏制定应急措施提供一定的参考依据。

关键词： 天然气管道；泄漏；风险分析；数值模拟

［中图分类号］ X937 ［文献标识码］ A **［文章编号］**

A Quantitative Risk Assessment Method of Natural Gas Pipelines Leakage Based on Numerical Simulation

Yuan Zhanggao[1], Zhao Dongfeng[1]

(1. Center for Chemical Process Safety-China Section, Qingdao, 266580, China)

Abstract: Natural gas pipelines are not only of large danger, but also difficult to be detected. Recently, leakage accidents of natural gas pipelines happened frequently, which resulted in enormous casualty and property loss. Therefore, it is significant to improve the safety of the natural gas pipelines. In this paper, quantitative risk assessment method is adopted to investigate the leakage of natural gas transmission pipelines. Firstly, it analyzes and summarizes the main potential factors of the leakage comprehensively. Secondly, under different wind speeds and apertures, a numerical simulation method is used to study the leakage and diffusion conditions of natural gas, as well as the possible consequences of jet fire, flash fire, etc. The results of this research can provide certain reference basis for gas pipeline leakage emergency measures.

Key words：natural gas pipelines; leakage; risk assessment; numerical simulation

引言 石油天然气管道是连接油气资源与市场的桥梁和纽带，中国目前有 70%的石油、99%的天然气通过管道运输，长输干线管道已达 8 万公里，海上油气管道近 5000 公里。到“十二五”末，中国长输油气管道总里程将超过 10 万公里，维护管道安全任重道远。然而，由于天然气管道潜在危险性大且日常检测比较困难，近年来管线泄漏引发的火灾、爆炸等重大事故频发，造成了重大人员伤亡和财产损失。根据统计，我国天然气管道的事故率为每千公里 4.3 次/a，是欧洲和美国的 6~8 倍[1,2]。因此，有必要对天然气管道泄漏进行定量的风险分析。采用数值模拟的方法研究了天然气在不同风速、不同泄漏孔径下泄漏扩散的规律，并对其可能引发喷射火、闪火等造成的危害范围进行模拟分析，以此防范于未然使得事故发生概率及事故损失得到有效降低。

1 天然气管道泄漏因素

天然气管道泄漏的原因很多，也很复杂，但归纳总结起来主要是 人为因素、管网设备设施因素及自然因素。天然气管道泄漏原因如表 1 所示。其中，腐蚀破坏是管道最常见的破坏因素。天燃气管网设备及设施的老化与腐蚀，会使管道还没达到服役年限就极易发生管壁穿孔而导致泄漏，酿成事故。管道腐蚀主要有一下三种腐蚀模式：一、外腐蚀。外腐蚀与防腐质量、土壤、保护方式、管道附近有无金属埋设物、管道附近磁场、电场有关。此外，外壁腐蚀不仅体现在埋地管道上，架空的钢管也会出现外壁腐蚀，所以架空的管道上面常常涂以保护类的东西进行防护。其中点蚀是外腐蚀的主要形式，占 90%左右。管外腐蚀主要表现为管—土壤体系的电化学腐蚀，这种腐蚀与防腐层质量、土壤性质、载荷等因素有关。二、内腐蚀。内壁腐蚀与电化学腐蚀的原理类似，因为天然气中自带的水分在管道上面形成的亲水层，为其形成电化学腐蚀提供了条件。此外，天然气中的一些化学物质与金属管道自身发生化学反应，形成腐蚀。所以内腐蚀与介质特性有关，其主要表现为均匀腐蚀，可以通过超声波测厚的方式确定其腐蚀速率；三、应力腐蚀。应力腐蚀开裂是埋地管道发生突发性破坏事故的主要危险之一。从腐蚀的类型看，点腐蚀和缝隙腐蚀造成的危害性相对较小，易于控制，而应力腐蚀开裂事故往往是在没有明显预兆的情况下，突然发生管道的灾难性破坏，其后果极其严重[3,4]。

表1　天然气管道泄漏主要原因分析

人为因素	设备设施因素	自然因素
1、施工质量：未按规范施工、焊接质量差等；	1、设备质量缺陷：管道设备未达到技术标准等；	1、地球运动；
2、误操作：操作不正确、施工操作不正确、使用管理操作不正确等；	2、腐蚀：外腐蚀、内腐蚀、应力腐蚀；	2、暴雨/洪水；
3、安全管理缺陷；	3、附件设施缺陷：垫片损坏、控制或泄压设备损坏、密封填料失效等。	3、气候变化；
4、第三方破坏。		4、其他自然力因素。

第三方破坏在整条管道的风险评估上占有重要位置，根据美国运输管道部(Department of Transportation 简称 DOT)统计，在 1971—1986 年的美国诸多管道事故中，第三方破坏则占 40%左右。我国情况也类似。尤其近年来在油、气管道上打孔、偷油、偷气事件每年都有若干起，有些造成了重大事故。在此处所指的第三方破坏，不仅限于管道破坏引起介质泄漏，一定程度上碰坏防腐层或给管道造成刮痕、压坑等亦属于破坏之例。管道防腐层的破坏会造成管道腐蚀，金属形成刮痕、压坑，造成应力集中形成疲劳裂纹扩展最终导致破坏，或造成应力腐蚀断裂[5-7]。

2 天然气管道泄漏数值模拟

2.1 SAFETI 软件简介

目前开发用于定量风险评价的软件很多，如挪威船级社 DNV Technical 公司的 SAFETI、美国环境署研发的 ALOHA 等等。SAFETI(Software for Assessment of Flammable ,Explosive and Toxic Impacts)安全评价软件在世界许多国家得到广泛运用，在国内被国家安全生产监督管理总局推荐使用，这个软件包包括 SAFETI、PHAST 和 LEAK 软件[8, 9]。SAFETI 软件是将 LEAK 的计算结果与 PHAST 的事故后果相结合，用于定量风险分析和危险性评价的软件，它对物料泄漏、扩散过程、事故后果及风险值进行模拟计算，给出可能的事故风险范围及指定点的风险值。它要求在熟悉可能发生的事故情形情况下，输入物质参数、工艺参数、气象参数及平面布置图等，对设置事故模型的影响范围及危险程度进行计算，并形成计算机模拟图表和报告。

2.2 管道泄漏后果数值模拟分析

天然气管道发生泄漏时，其过程具有高度的复杂性，主要表征为：天然气管道泄漏时在泄漏的发生、发展以及天然气扩散进入大气等一系列的过程之中，存在着很多的不可估性因素，泄漏燃气的温度、密度和大气的温度、密度的不同以及风速等等使天然气泄漏扩散变得十分的复杂；天然气为易燃易爆气体，泄漏后与空气混合，根据混合密度的不同，在遇到明火时易发生闪火、喷射火、爆炸等各种情况[10, 11]。

2.2.1 模拟参数

某天然气场站设计规模为输气自用燃料气量为 $100\times10^4m^3/d$，输气管道内部压力约为 4.0MPa，温度为 5～25℃。假设有一条管道发生泄漏，泄漏模型分为连续泄漏和瞬间泄漏，连续泄漏又分为 3 种，即泄漏孔径分别为 0.25in、1in、4in。具体模拟参数见表 2 所示。

表2　双天然气管道泄漏主要原因分析

泄漏模型	泄漏孔径	工作条件	潜在危险
连续泄漏	0.25in	压力：4.0Mpa 温度：25℃	喷射火，闪火
	1in		喷射火，闪火
	4in		喷射火，闪火
瞬间泄漏	灾难性破裂		喷射火，闪火

2.2.2 闪火危险性分析

闪火是泄漏后的可燃气体在扩散过程中，遇点火源在无约束空间产生的无爆炸性燃烧，其危害主要是热辐射。1in 泄漏造成闪火影响见图 1。

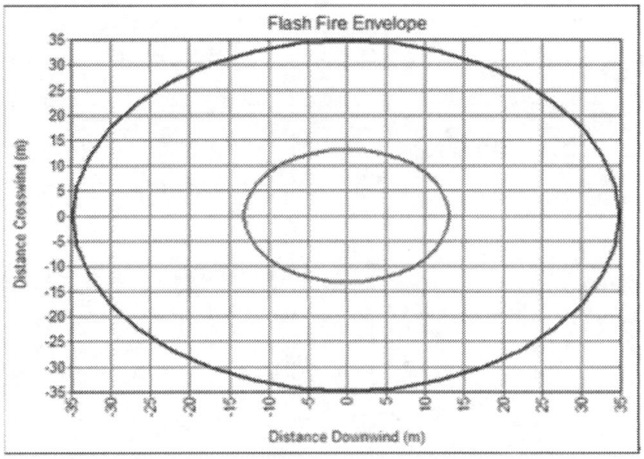

图 1　1in 泄漏（ C-5.5m/s）

Figure 1- The leakage aperture is 1in

由上图及模拟软件中得出的报告可得闪火造成影响见表 3 所示。

表3　闪火产生的热辐射影响

风速及大气稳定度组合		各泄漏孔径下泄漏点下风向影响距离/m							
		0.25in		1in		4in		瞬间破裂	
		44000 ppm	22000 ppm	44000 ppm	22000 ppm	44000 ppm	22000 ppm	44000 ppm	22000 ppm
1.5	A	3.68	6.54	15.6	37.7	91.9	161.8	200.6	255.4
(m/s)	F	3.81	7.07	16.3	42.2	96.9	181.0	218.8	275.2
5.5	C	3.52	5.39	13.2	34.8	93.4	183.1	237.9	303.5
(m/s)	D	3.37	5.68	13.5	36.2	94.5	187.7	238.8	324.6
10	C	3.01	4.78	10.8	29.9	94.1	194.2	262.3	314.4
(m/s)	D	3.09	4.83	11.3	32.0	95.2	201.8	263.9	312.2

分析上表数据：在泄漏孔径较小，泄漏速率较慢时，风速及大气稳定度的改变对结果无明显作用。当泄漏孔径较大或发生瞬间破裂时，风速及大气稳定度对结果有明显的影响。

2.2.3 喷射火危险性分析

带压的可燃气体从破裂口高速喷出，若被点燃就形成喷射火。其危害结果主要是火焰范围及热辐射引起的伤害半径，当喷射火焰小且表面热通量值小于一定程度时，则不会造成人员伤亡。热辐射影响造成的具体伤害见表 4 所示，1in 泄漏喷射火影响强度见图 2。

表4　辐射热通量造成伤害作用

入射通量 kW/m^2	对设备的损害	对人的损害
37.5	操作设备全部损坏	10 s，1%死亡；1 min，100%
25.0	在无火焰、长时间的辐射下，木材燃烧的最小能量	10 s，重大伤亡；1 min，100%
12.5	有火焰时，木材燃烧、塑料融化的最低能量	10 s，1 度烧伤；1 min，100%
4.0	—	20 s 以上感觉疼痛
1.6	—	长期辐射无不安全感

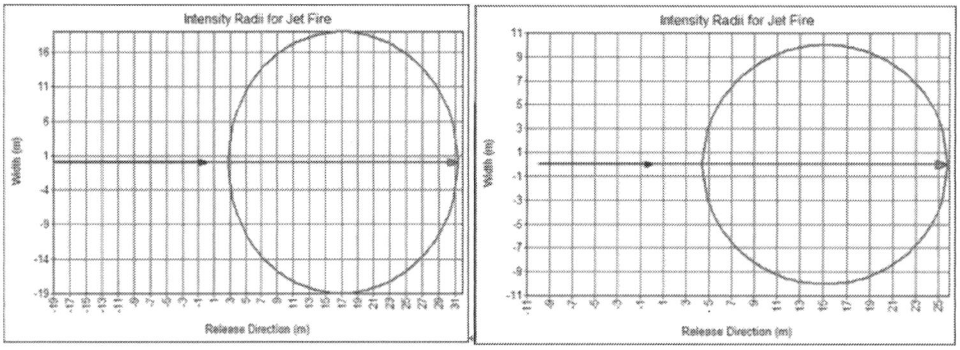

a- 1in 泄漏、辐射通量为 4KW/m2　　b- 1in 泄漏、辐射通量为 12.5KW/m2

图 2　1in 泄漏喷射火强度影响半径（ C-5.5m/s）

Figure 2- The intensity radii for Jet Fire

由上图及在软件报告中取得的详细数据可得下表 5 数据。

表 5　喷射火危害后果定量分析

泄漏孔径	**喷射火不同热辐射通量下的影响范围/m**					
	>4 KW/m^2			>12.5 KW/m^2		
	1.5(m/s)	5.5(m/s)	10(m/s)	1.5(m/s)	5.5(m/s)	10(m/s)
0.25 in	6.722	6.357	5.762			
1 in	31.43	31.42	31.26	24.84	25.96	26.82
4 in	123.75	122.61	119.36	92.77	94.72	97.85

分析上表数据，可得出以下几条结论：

（1）相同工况情况下，喷射火影响范围主要受泄漏孔径大小影响，而受风速影响较小，不受大气稳定度影响；

（2）不同孔径大小的泄漏源对周围影响严重程度不同，随着孔径的增大，泄漏速率增加，热辐射强度加强，影响的目标距离越大；

（3）在同一泄漏孔径下，风速在不同辐射通量下对其受热目标距离影响不一致，有促进有抑制。

3 结论

通过对天然气输送管道泄漏因素进行归纳总结，得出管道泄漏主要是 人为因素、管网设备设施因素及自然因素运用挪威船级社（DNV）的 SAFETI 软件对天然气输送管道泄漏情况进行风险后果模拟分析，从而得出管道一旦发生泄漏后天然气在不同风速、不同泄漏孔径下泄漏扩散的规律，以及可能引发的喷射火、闪火等对周边的影响范围，从而天然气管道输送制定安全措施及应急预案提供参考。

参考文献

[1] Morgan B. The importance of realistic representation of design features in the risk assessment of high-pressure gas pipeline. 5th International Conference and Exhibition Pipeline Reliability，Houston，Texas，1995 (9): 212—218.

[2] 王玉梅，郭书平. 国外天然气管道事故分析. 油气储运，2000，19(7)：5—10.

[3] 战丽华, 王为民, 戴旺旺, 等. 输气管道泄漏原因及检测方法述评[J]. 当代化工, 2013(09):1307-1309.

[4] 庞海艳. 天然气管道泄漏及其检测方法的几点分析[J]. 中国石油和化工标准与质量, 2013(17):110.

[5] 张增刚, 贾文磊, 田贯三, 等. 燃气管道泄漏原因及扩散影响因素分析[J]. 山东建筑大学学报, 2012(02):198-202.

[6] 李静, 汪彤, 吕良海. 城镇天然气管道泄漏原因分析[J]. 安全, 2005(04):33-34.

[7] 向启贵. 天然气管道泄漏扩散机理研究[D]. 西南交通大学, 2006.

[8] 银小兵, 周大为, 李明, 等. 运用 SAFETI6.3 软件模拟喷射燃烧[J]. 石油与天然气化工, 2004(02):142-144.

[9] 刘俊杰. SAFETI 软件在安全定量风险评价中的应用[J]. 油气田地面工程, 2008(12):46-47.

[10] 刘斐, 刘茂. 城市燃气管线的定量风险分析[J]. 南开大学学报：自然科学版, 2006,39(6):31-36.

[11] 李继兵. 有害气体运输过程中泄漏扩散的模拟与分析[D]. 西南交通大学, 2006.

Process Safety Management System Auditing

David A. Moore

President & CEO

AcuTech Consulting Group

1919 Gallows Rd. Suite 900

Vienna, Virginia, USA 22182

dmoore@acutech-consulting.com

Keywords: Process safety management, PSM, Auditing, Management System, Metrics

Abstract

Process safety management system auditing is becoming more of a business critical function, and as PSM evolves, it is receiving more attention. Auditing is a key opportunity to validate the process and improve performance to meet expectations for success. Auditing is more effective if aligned to organizational design and goals. Many auditing processes provide an incomplete approach, without consideration of the management system design or organizational goals.

Considerations such as management model, elements of the process, if the program is regulatory or performance driven, multiple performance objectives and key performance indicators should be key considerations when developing PSM management systems and the associated audit process.

Safety Assessment of L485 Gas Pipeline with Defects Based on Ansys Simulation

Li, G., Petro China Southwest Oil & Gas filed Company
Zheng, Z.,
LI, X.,

Safety Evaluation of L485 Gas Pipeline with Defects Based on ANSYS Simulation

LI Gen; ZHENG Zhi yu; LI Xiang rong;

(Petro China Gas Transmission Management Division of Southwest Oil & Gas filed Company,

Cheng du, 610213£©

Abstract: According to the gas pipelines¡¯ actual operating conditions of the Southwest Oil and Gas Field Company, and also based on the "fit for purpose" principle ,The finite element analysis and experiment were adopted to re-investigate the D813¡Á12 L485 gas pipeline under pressure with imperfection such as sunken, metal loss, unfitness of butt joint, undercut, etc. The comparison of the analysis and the actual measurement was made to verify the validity of simulation. Furthermore, based on the verified boundary condition, the utmost of the imperfections were carried out . In this paper ,a new method was adopted to safety evaluation of gas pipeline.

Study in ANN leak-detection method for natural gas pipeline based on acoustic signal

Qian haocheng
College of Pipeline and Civil Engineering in China University of Petroleum (East China)
No.66 Street Changjiangxi Qingdao Shandong Province 266555
631073144@qq.com

Li yuxing Liu cuiwei[1] Song zhiping[2]
College of Pipeline and Civil Engineering in China University of Petroleum (East China)[1]
Oil & Gas Storage & Transportation Company, PetroChina XinJiang Oilfield Company[2]
No.66 Street Changjiangxi Qingdao Shandong Province 266555[1]
Karamay , Xinjiang 834002[2]
Email address (es)
598269666@qq.com
chuyunlcw@163.com
242483547@qq.com

Prepared for Presentation at
American Institute of Chemical Engineers
2nd CCPS China Conference on Process Safety
Qingdao, China
August 28 – 29, 2014

UNPUBLISHED

Study in ANN leak-detection method for natural gas pipeline based on acoustic signal

Qian haocheng
College of Pipeline and Civil Engineering in China University of Petroleum (East China)
Qingdao Shandong Province 266555
pplailol@163.com

Li yuxing Liu cuiwei[1] Song zhiping[2]
College of Pipeline and Civil Engineering in China University of Petroleum (East China)[1]
Oil & Gas Storage & Transportation Company, PetroChina XinJiang Oilfield Company[2]

Keywords: leak detection signal processing artificial neural network program

Abstract

Introducing a leak–detection method of natural gas pipeline based on artificial neural network and acoustic signal. Taking the given acoustic characteristics of gas pipeline as different network's input , comparing the efficiencies and results (including network's complexity, training time and accuracy of their recognition system)of these networks, optimizing the parameters (training method ,epoch , performance , network's construct and gradient) and finding out the most appropriate network for leak-finder. Finally, we combine the selected network and characteristics to program the leak detection system of natural gas pipeline.

1. Introduction

Artificial neural network (ANN)[1] is an algorithm which can simulate human's neuron to process message. It has been widely used in field of forecast, classification, fitting, cluster and compress since 1980s. Error Back Propagation network, feedback neural network, self-organizing feature maps and radical basic network are most used in different area of message processing. Leak-detection of pipeline can be seen as a kind of modal identification, and the function of modal identification is an important feature of ANN. Therefore, Using ANN and acoustic signal to identify pipeline's leak-mode can be an efficient method to try.

2. Laboratory equipment

We experiment with pipeline's leak detection equipment of high-pressure gas in laboratory to acquire acoustic signal of each different working condition. Design of the laboratory equipment observes similarity laws, length of the circular pipeline is 251.5m, max operating pressure is 6.4MPa. Acoustic sensors settled in start and terminal of the circular pipeline, leak points are respectively 40.34m, 80.33m and 149.02m from start line. We take flange with external ball-valve and internal pore plate to simulate pipeline leak situation, use different aperture(0.45mm ,0.5mm ,0.6mm ,0.7mm and so on) to control leakage. At the same time, we also installed differential pressure transmitter, temperature transmitter, flow transmitter and pressure transmitter to monitor condition's parameter of fluid. Experimental fluid is compress air which simulate natural gas, the air flowed through compressor, refrigerated drier and filter to increase pressure and detach it's water and oil. Then it entered high pressure buffer tank and testing pipeline, in the end, it reached the terminal buffer tank and returned to the atmosphere. Flow chart and photo of the laboratory equipment is in the figures below.

Figure1: leak detection equipment of high-pressure gas

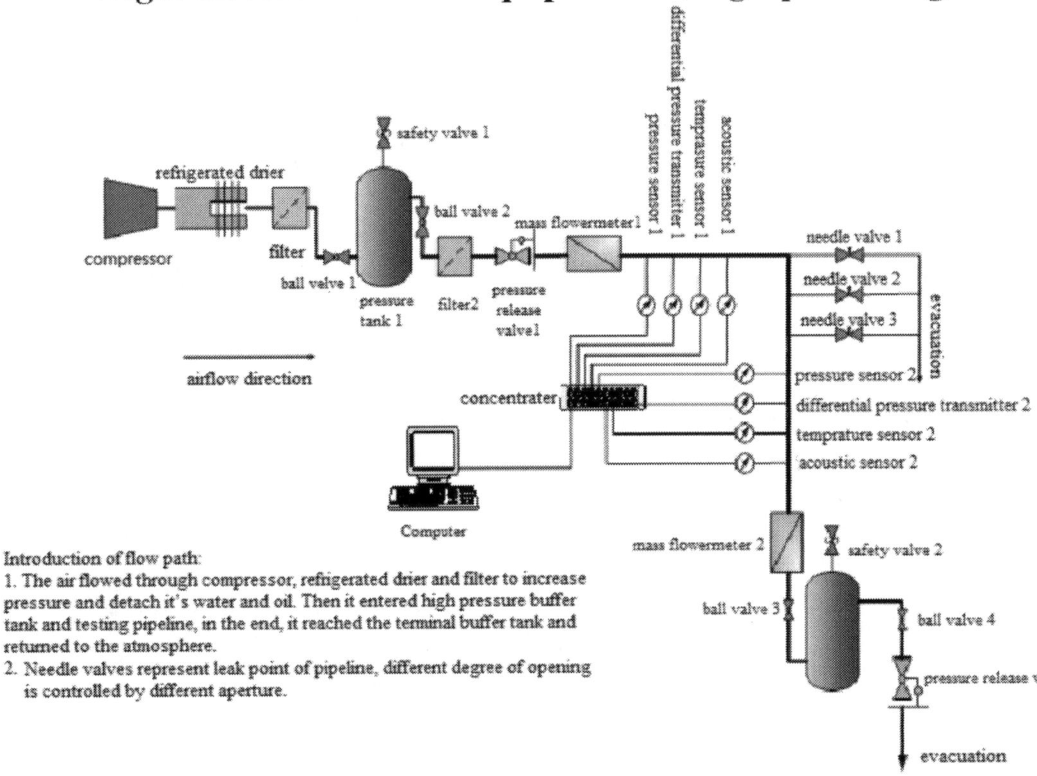

Figure2: Photo of laboratory equipment

3. Filtering of acoustic signal

We must take noise reduction measures to process acquired acoustic signal, because background noise always mix into the pipeline's acoustic signal in data acquisition. Expect that, preferable waveform of acquired signal is also essential, so , according to some documents and references , sampling frequency of hardware instrument must be 20~50 times more than detected signal[2].From analytical conclusion of leak acoustic signal's frequency spectrum and PSD(Power Spectral Density), we know that a majority of leak acoustic signal's energy gather in 50Hz or lower, so confirm sampling frequency of DAQ card (Data acquisition Card)is 1000Hz.

This article handled acoustic signal with wavelet transform to reduce its noise, we picked sym8 wavelet with 5 decomposition scales and rigrsure threshold selection rules to denoise the acquired acoustic signal.

4. Optimizing of acoustic signal's characteristics

To represent different operating mode's (compressor 's open and shut ,valve's open and shut and so on) character, we should describe it with appropriate characteristics.

4.1 Optimizing in time-domain

Invariable time-window[3] tends to reflect sudden change's grade, and be convenient for real-time detection. Taking a given leak acoustic signal with different time-windows(which has different width of time) to engage in a test of characteristics' calculating, we can conclude from the test that:

1. If time-window is too narrow (width of time is 10~100), sudden changes' grade of leak signal's root mean square, kurtosis and skewness are extremely low. We even can't distinguish it from background noise.
2. If time-window is too wide (width of time is 4000 or more), Leak signal's kurtosis and skewness become fluctuating, and their sudden changes are indistinct.

From the result of the test above, we take 1500 as the width of time-window.

As for finding out the most representable characteristics in time-domain, we took a series of leak detection experiment with 1mm leak aperture. After processing the acquired acoustic signal with wavelet de-noising program in Matlab, we compared every statistics of the acoustic signal, and the result shows the fallowing characteristics are more distinguished than others[4][5].

(1) Mean
Picking mathematical expectation or mean value to be the first statistical value of acoustic signal. As we discussed above, width of time-window is 1500. From their waveform, we can obviously see that sudden change of opening compressor's mean waveform is mild (-0.5~2.5kPa), and the sudden change through a longer time than others. Waveform of valve opening is more cliffy than the compressor's, and its sudden change's range is wilder (-60~20kPa). Compared with valve opening, Leak situation's waveform has narrower sudden change's range but sheerer peak.

(2) Kurtosis
Kurtosis is a statistic which is used to describe probability density curve's steepness of randomized data around its mean value. With kurtosis grows, the curve will be steeper. In their waveform, no matter valve opening or compressor opening's kurtosis curve has a dramatically decrease, their sudden change's range are between 1to 1.5kPa. However, leak condition's curve of sudden change is steeper than others, and its curve's peak is also 5~6 times higher than the other twos'. So, kurtosis can represent leak condition's sudden change efficiently.

(3) Root mean square
Root mean square can measure how much do the randomized data deviate from its mean value. Comparing leak condition's curve to other twos', we can conclude that leak's curve has the highest deviating grade; it is 10 times higher than valve's and 100 times higher than compressor's. At the same time, we noted compressor's waveform in an uptrend, when it turned to steady after a sudden change. That is a distinct difference between compressor opening and the other two. So, it is reasonable to say that root mean square can be used to distinguish three work conditions difference.

(4) Mean square value
In the processing of discrete random signal, mean square value represents signal's power and intensity. As we see in their waveform images, compressor's sudden change is about 10 times slower than leak's, and peak of valve's curve is hundreds times higher than others. So, mean value is also a representative characteristic for modal identification.

Table1 Discrimination of every characteristic value in different conditions

Type of condition / characteristic	Leak condition	Valve opening condition	Compressor opening condition
Mean	Low discrimination	High discrimination	High discrimination
Kurtosis	High discrimination	Low discrimination	Low discrimination
Root mean square	High discrimination	Low discrimination	High discrimination
Mean square value	Low discrimination	High discrimination	High discrimination

We can conclude that one characteristic in time-domain can distinguish 1 or 2 working condition from others from the table above. So, taking the four statistics as ANN's inputs, than training every working condition's corresponding weight, at last, we will get a specialized neural network to detect pipeline's leak condition in some complex situations.

4.2 Optimizing in frequency-domain

Processing leak and other two interfered conditions with fast fourier transform (FFT), we got each condition's frequency spectrum between 0 to 1500Hz, when its sampling frequency is 30kHz in 1.5MPa. At the same time, we picked hamming-window to get their power spectral density maps (PSD).

However, in the waveform images, frequency spectrum and PSD's trends of three operating situations are closed, and the difference between leak and interference conditions is very small in low-frequency stage, we can even hardly distinguish leak conditions from other two interference conditions and background noise only by sudden change's energy in low-frequency stage. So, frequency spectrum and PSD can just be the supplementary means in modal identification.

5. Optimizing of artificial neural networks

We picked three networks below for modal identification based on their advantages and characteristics.

5.1 Learning Vector Quantization

LVQ (Learning Vector Quantization)[6] inherits SOFM's (self organizing neural network) competitive mechanism, and can receive mentoring signals to go through a guiding study. The most difference between LVQ and SOFM is receiving mentoring signals. Learnlv1and learnlv2 which are functions in matlab can realize LVQ's training. Compared to learenlv1, learenlv2 can adjust two neurons simultaneously, therefore, its adjustment is more balanced and efficient. Using lvqnet (a function of matlab) to set up the network and taking learnlv2 as the training algorithm, we can get an impactful LVQ for modal identification.

5.2 Probabilistic Neural Networks

PNN (Probabilistic Neural Networks)[7] has simple structures and commendable practicability. We can directly utilize it without training, if the network has been set up. PNN is based on RBF (radial basis function network), it also adds bayesian decision theory and density function estimation to it, and that enhance PNN's ability of modal identification.

PNN consist of input layer x, hidden layer F, sum layer n and output layer y. The hidden layer F contains radial basis function, and calculates distance between input vector and modal center, then returns value to sum layer based on their modal. Finally, returned value get into competitive neuron from sum layer, competitive layer make a decision of output's modal based on the rule of "winner-take-all". The formula of radial basis function is:

$$F = \frac{1}{(2p)^{0.5} s^{d}} e^{\frac{(x- x_{ij})(x- x_{ij})^T}{s^2}}$$ [Eq.1]

x_{ij} in Eq.4 is No.j center of sample i, s is smoothing factor or diffusion velocity of network (It is 0.1 in common)

5.3 BP network

Although BP network [8]has a slow training, easy to fall into local minimizer, and its training result highly depend on initial samples, BP network (Error Back Propagation) is still a wildly used artificial neural network nowadays.

BP network is consist of input layer, hidden layer and output layer. In modal identification, number of input layer and output layers' neurons are certain, but the number of hidden layer's should be confirmed by Eq.5 and Eq.6.

$$M = \sqrt{n+ m} + a$$ [Eq.2]

$$M = \log_2 n$$ [Eq.3]

n is input layer's neurons, m is output layers' neurons, a is a constant which is between 1 and 10.

6. Result of ANN's test

To test how the selected ANN apply to identification of leak and interfered conditions, we took selected statistics of each working condition's acquired acoustic signal in 1MPa as samples for input, and picked another acoustic signal's characteristics as samples for test in the same pressure. We have to illustrate that each operating situation's sudden change will be intensified with pipeline's pressure increasing. When it comes to leak detection, we concern about leak missing alarm rate (leak can or not be detected) and leak false alarm rate (other interfered conditions are mistaken for pipeline's leak). The table below show leak missing alarm rate and false alarm rate in test condition.

Table2 Compare of different neural networks based on acoustic data in 1Mpa

Result / Type of network	Correct rate of leak detection
PNN（Probabilistic Neural Networks）	54.29%
BP（Error Back Propagation）	94.29%
PNN Dual-modal identification	87.14%
LVQ（Learning Vector Quantization）Dual-modal identification	85.56%

As we see in the table2, PNN's (Probabilistic Neural Networks) leak detection rate is very low (even less than 60%). However, when there are only leak condition and normal condition, its correct rate increased to 87.14%. We analyzed this situation form PNN's classified theory, and it should be caused by too many and overlap center of hidden layer in multi- modal identification. That could result in sum layer mistaking each modal's posterior probability, but leak condition has a minor sudden change and a relatively short time of change, so , its sample center of hidden layer is subject to absorbing of other modal conditions (just like valve opening, its range and time of sudden change are about 10 times more than leak conditions'). At the same time, we noted that the same situation happened in LVQ (Learning Vector Quantization), LVQ's correct rate of leak detection is very low in multi-modal, but when it comes to dual-modal identification(only leak and normal condition), LVQ is a definitely trustworthy algorithm for leak detection. That is a convincing evidence for hypothesis above, just like PNN, LVQ classify working conditions into certain types by calculating distance between input vector and modal center, the modal center should be similar in multi- modal identification, so, every modal field overlap and absorb each other, and that makes the condition of false alarm and missing alarm worse.

Except kind of working conditions, characteristics for input also affect the correct rate of leak detection. Picking an artificial neural network, we testing how is the network sensitive to one or more characteristics by changing the kinds of inputs. Table below shows the best result of every network with their specialized statistics.

Table3 Given characteristic value which made different networks to reach their max recognizable rate

Characteristic / Type of ANN	Mean	Root mean square	Mean value	Kurtosis	Correct rate of leak detection	False rate of leak detection
PNN（Multi-modal identification）	✓	✓			82.22%	2.18%

BP network（Multi-modal identification）	✓	✓	✓	✓	94.29%	1.36%
LVQ（Dual-modal identification）	✓			✓	96.67%	—
PNN（Dual-modal identification）	✓		✓		98.89%	—

Note: False rate of leak detection means other interfered conditions are mistaken for pipeline's leak

Comparing to table 2, PNN's correct rate of leak detection increased by 30% in multi-modal identification and increased to 98.89% in dual-modal identification, when we picked mean and root mean square for input. Just like PNN, the same situation came to LVQ, its correct rate of leak detection increased to 96.67% in dual-modal identification, when we picked mean and kurtosis as input. Therefore, we can conclude that every signal characteristic for input influences each other, we should optimize the type of input characteristics by the sensitivity of the statistics to the given network, not by the rule of the more comprehensive the better.

As we discussed before, in the problem of leak detection, we concerned about the leak missing alarm rate (leak can or not be detected) and leak false alarm rate (other interfered conditions are mistaken for pipeline's leak), so we should choose the network with lowest leak false alarm rate and highest leak detected rate. Observing that standard, we picked PNN and BP as the networks of leak detected system in multi-modal identification, because of their over 80% leak detected rate and lower false alarm rate.

7. Program of leak detection system based on acoustic signal

After optimizing the characteristics and artificial neural networks, we utilize Visual Basic and Matlab to mix-program the leak detection system based on acoustic signal according to the result of optimizing and the simulate instrument in laboratory. VB has a simple programming-language, and a programming-language with strong applicability. Matlab also has the advantages of numerous function interface and abundant toolkits. So, we programed interface of acquiring acoustic signal and system's UI (user interface) with VB language, and wrote signal processing & ANN's (which is used to identify each different mode) functions in Matlab. That realized a combining of two software's specialties. The figure below is the UI.

Figure3: User interface of leak detection program

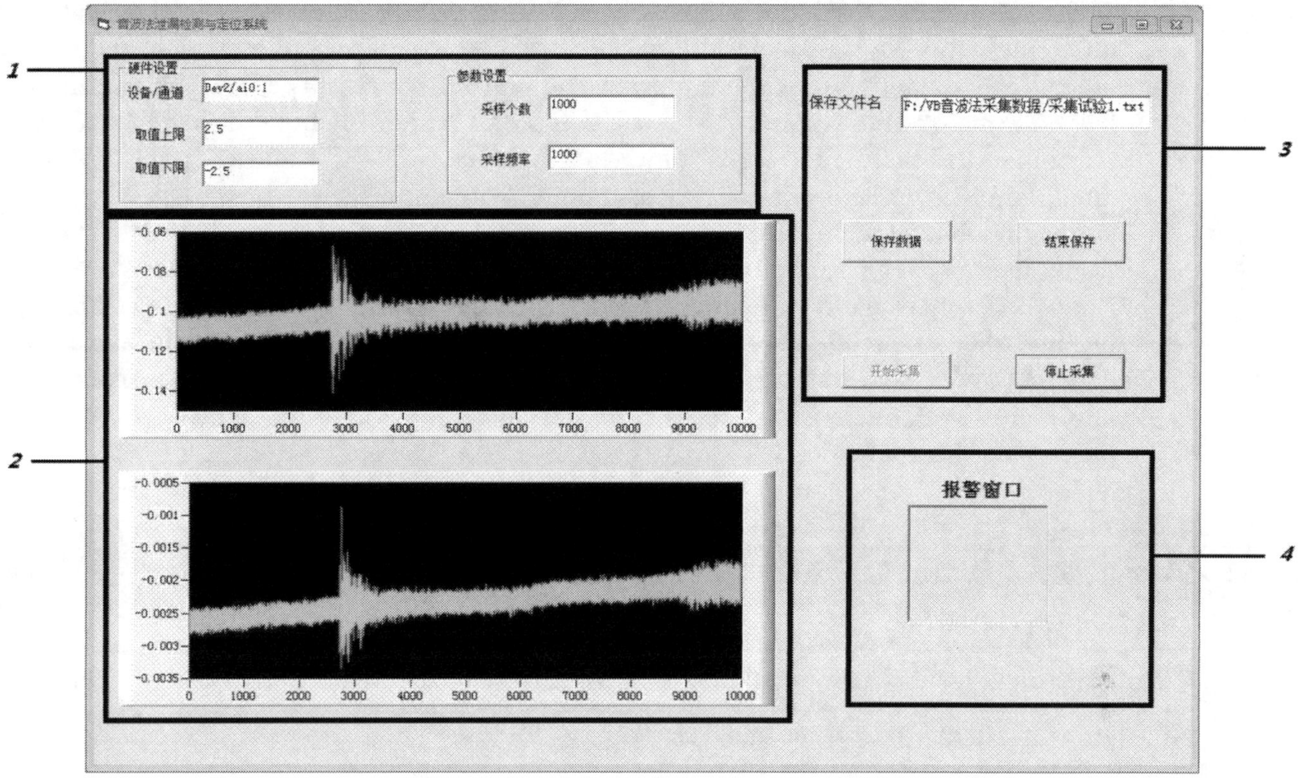

As we see in the picture above, the interface consist of four parts, they are respectively district 1, 2, 3 and 4:

- District 1- Parameter setting area, we can set the parameters of DAQ card
- District 2-Displaying of acoustic signal's waveform area, we can see the sudden change of waveform directly
- District 3-Controlling area, we can begin/end acoustic signal's acquiring and save the acquired data.
- District 4-Alarming area, we using the functions in VB to process acquired acoustic signal by the interface. In common situation, nothing will be happened in this area, but, when the leak condition is detected by the functions, It will return a signal to UI. When UI receives the signal, alarming window will turn to red and play a warning tone.

8. Conclusion and suggestion

As we tested above, picking the four characteristics as input of BP network can detect pipeline's leak on-line efficiently. Meanwhile, we noticed those below.

1. We took bayesian normalization training algorithm[9] in the test of BP network, and got a satisfactory application. Combining the character of the training algorithm to analyze, we can conclude that the algorithm solve the problem of optimal error adjustment rate effectively, help BP network avoid the trap of local minimum, and increase the correct rate of leak detection.

2. We always process the input samples with normalization function in Matlab, but this time, the procedure is unnecessary, because of the gap between each sample is dramatically huge(the maximum is even 100 or 1000 times more than the minimum). If we normalize these samples, the characteristics which fluctuate in lower level will be similar, and that will absorbs some working condition's specialities. So, we suggest taking samples without normalizing as input of network directly.
3. Due to the characteristics in frequency-domain can hardly distinguish leak condition from other interference, we only pick statistics in time-domain as the input of network. But as we actual use it in operating site, statistics in time-domain can't transmit too far, and they are easy to be absorbed by background noise. That will influence the correct rate of leak detection. So, propagation properties of characteristics are as important as characteristics' distinction for different condition, when we consider picking up characteristics.

9. References

[1]. Chen ming., "Theory of ANN and Application instance in Matlab," Beijing: Tsinghua University Press, 2013

[2]. Zhao guodong., "Filtering methods and Application in Signal processing," Hu bei: Huazhong University of Science and Technology, 2009

[3]. Liu chao., "Leak and Interference Signals' Characteristic Quantity Extraction for Gas Pipeline," Shan dong: China University of Petroleum(East China), 2011

[4]. Fu juntao, Li yuxing, Meng lingya, Liu chao., "Leak and Interference Acoustic Signals' Characteristic Quantity Extraction for Gas Pipeline," *Journal of Petrochemical University*, Volume 25, No.2, 2012

[5]. Fernando Velasco,Surendra P. Verma, "Importance of Skewness and Kurtosis Statistical Tests for Outlier Detection and Elimination in Evaluation of Geochemical Reference Materials,"*Mathematical Geology* . 1998 (1)

[6]. Cheng maoling, "Learning Vector Quantization Neural Network's Application in Discriminant analysis, " Mathematics in practice and theory, Volume 41, No.5, 2011

[7]. Du jixiang, Wang zengfu., "Automatic recognition of Plant leaves based on PNN," *Pattern recognition and artificial intelligence*, Volume 21, No.2, 2008,

[8]. Liu tianshu., "Improvement and Application of BP Network," Harbin, Northeast Agricultural University, 2011

[9]. Li ying, Wang aihua, Gu jianwei, Luo tian, "Application of BP network in Engineering project management and its Optimizing with Bias regularization ," *Value engineering*, Volume 8, No.31, 2009

The Soil Bioremediation for Oil Pipeline Leaked

Qingdong Zhang *
Xiaoshan Chen
Chaocheng Zhao
Department of Environmental and Safty Engineering, China University of Petroleum (East China)
zhangqd@upc.edu.cn
cxs910101@163.com
zhaochch@ upc.edu.cn

Keywords: oil; leak; soil; bioremediation

Abstract

Environmental restoration after the accidents is very important because pipeline leak causes a lot of carcinogens into the soil environment. There is a serious harm to the soil environment and human health when the oil spill happens. Bioremediation technology is better than traditional physical and chemical methods, because it is no secondary pollution, and has a good treatment effect. This study introduces the bioremediation and factors of affecting bioremediation, discuss the trends and prospects of bioremediation.

1. Introduction

Pipeline transportation has many advantages, such as low cost, fast speed and stable supply, it is suitable for oil and gas and other flammable substances. As the demand of energy advances, the strength of the land and subsea pipeline construction is also continuing to make progress. To the first half of 2012, total length of China's oil and gas pipeline has reached 93,000 km. Expect to 2015, the pipelines' total length will reach 150000 km. But with the increase of the pipelines' using time, pipeline leakage accidents happened frequently, the reasons include abrasion, aging, third-party damage and some other factors. 2013, a pipeline explosion happened in Qingdao Development Zone which caused by pipeline aging. Now, Our pipeline leak detection technology is still evolving, the emergency repair work cannot be carried out in time, there are 600000 tons of oil come into the environment because of escape or leakage, pollute the soil, groundwater and surface water[1]. Pipeline accidents always accompany environmental pollution, especially oil leak. Sludge contains many toxic compounds, teratogenic and carcinogenic. According to the United States environmental protection agency (epa) oil substances such as benzene, polycyclic aromatic hydrocarbons are the priority pollutants [2]. Oil pollution of the soil attracted the scholars' attention, and there are a lot of repair methods for contaminated soil. Traditional physical and chemical methods have some disadvantages, such as they cannot solve the fundamental problems, need high costs, and have a low effectiveness, but bioremediation technology has a good treatment effect and it is no secondary pollution. Bioremediation technology developed rapidly in recent years became the mainstream of the current technology.

2. The dangers of oil pipeline leak to the environment

2.1 The harm of soil

The main ingredients of oil are C_{15}~C_{36} alkanes, polycyclic aromatic hydrocarbons, olefins, benzene, phenol, etc, there are 30 kinds of priority pollutants in oil. In the soil, oil s difficult to degrade in a short time, it will change the ratio of carbon, phosphorus and nitrogen, affect the growth of plants and even killed them. Oil pollution also lead to the change of microbial communities, reduce the function and activity of indigenous microorganisms. At the same time, oil will enter the pores of soil, reduce the soil permeability, and destroy the soil structure [3].

2.2 The harm of sea

Oil brings a great harm to marine and the ecological environment. Firstly, a large number of marine animals will be trapped in the oil, result in their difficulty action or suffocate. Secondly, Oil can form a layer covering on the surface, block the exchange of O_2 and CO_2, destroy the cycle balance of dissolved gas in ocean, and oil also affect the photosynthesis [4]. Thirdly, there are some toxic substances in the oil, such as polycyclic aromatic hydrocarbon, it will destroy the healthy growth of marine organisms, and hydrocarbons through bioaccumulation and food chain will be further intensified. Finally, there is a more widespread range of influence when oil flow with water.

2.3 The harm of human

Oil contains polycyclic aromatic hydrocarbons (PAHS), they are carcinogenic, mutagenic, and teratogenic. Toxic pollutants will through the soil or ocean enters the food chain, and then enters the body, cause chronic poisoning and damage to the nerve center. Oil also has some volatile toxic substances, such as benzene and hydrogen sulfide, etc. Once inhaled concentration exceeds the critical value, the lesions will invade body's blood and bone.

3. Bioremediation technologies of contaminated soil

By microbial physiological processes of growth and development to absorption, degradation, or volatile the compositions in the oil, lower the harmful ingredients content in the oil [5]. Now bioremediation technology has become one of the most valuable and promising technology in eco-environmental protection field, which is recommended by United States Environmental Protection Agency (USEPA) as a preferred measure of the remediation of oil-contaminated soil[6]. Bioremediation can be divided into microbial remediation, phytoremediation and animal repair method. Phytoremediation and animal repair method are taking consumption of petroleum pollutants and environmental restoration as target, but phytoremediation is used of plant growth, animal repair method is the use of the cooking oil animals. This paper just introduces about the microbial remediation technology.

3.1 In-situ bioremediation technology

In-
situ bioremediation technology referres to a method that not move contaminated soil but directly repair it by microorganism, which needs to provide nutrients like nitrogen and phosphorus for in digenous microorganism or inoculate and domesticate strains so that improve the degradation an d achieve the purpose of consumption of petroleum hydrocarbons pollutants[7].

3.1.1 Culture adding method

This method is directly access exogenous degrading bacteria, and offer nitrogen, phosphorus, other nutrients and trace elements to microorganisms in the soil. Sanjeet Misshra.et al. [8] was successfully treated the contaminated land by microbial consortium on the carrier and nutrients. The proportion of carbon, nitrogen and phosphorus plays a very important role on the growth of microbial. An appropriate ratio is good for pollution control.

3.1.2 Biological incubation method

By optimizing the environmental conditions to improve indigenous microbial activity, so as to achieve the purpose that accelerates degradation of petroleum hydrocarbons pollutants. Regularly put nutrients and O_2 or H_2O_2 to the contaminated soil as an electron acceptor, and satisfy the needs of indigenous microbial growth and reproduction, improve the metabolic activity, finally, pollutants can be converted to CO_2 and H_2O.
Cultivate the indigenous microorganisms is better than introduce high efficient degrading bacterias of oil. Because indigenous microbes have adapted to the presence of contaminants, high efficient degrading bacterias' advantage can't keep longer than indigenous microorganisms [4]. Therefore, exogenous microorganisms can be added only if pollutants were not be effective degradation by indigenous microorganisms.

3.1.3 Biological aeration method

Biological aeration method is on the basis of soil vapor extraction technology, it's a successful example of the combination of physical method and biological method. By install the shafts in the contaminated soil, make air or O_2 into the contaminated area, and then extract the gas with the volatile organic compounds in the soil. At the same time, add nutrients to the contaminated areas, improve the microbial growth environment, promote its' metabolism. The advantage of this method is that less pollution, even if the intermediate product is a pollutant, we can install gas purification device at the outlet to avoid secondary pollution [9]. This method is through ventilation to provide a source of oxygen for microorganisms, so the soil permeability is one of the important factors affect the removal efficiency[10], and the soil moisture, pollution intensity, the ratio of the nutrients are all affect the removal.

3.1.4 Soil cultivation method

Soil cultivation method use indigenous microorganisms to degrade petroleum hydrocarbon. On the one hand, farming can make the pollutants and soil mixing, on the other hand, can make the soil aeration. The cultivation of regularly, can maintain good ventilation. If the depth of oil into the soil is small, can be processed directly, if the oil has penetrated above 60 meters in the soil, we will need to dig up the soil during treatment. In farming, we should also pay attention to adjust the ratio of nutrients, temperature, humidity and pH, because these factors have a great impact on microbial degradation [11]. Which the soil PH needs to be monitored on a regular basis, once the soil is acidic, can add limestone, make the increased to 7.8.

3.2 Ex-situ bioremediation technology

Ex-situ bioremediation technology is a method that concentrates the contaminated soil to a different place, and treats by biological method. It not only can prevent the proliferation and migration of pollutants, narrow the scope of pollution, also through various process controllers or bioreactor to generate conditions conducive to biodegradation.

3.2.1 Soil composting method

The addition of soil conditioners such as branches, straw, manure, peat and other composting materials, conduce to provide energy for microorganisms and improve soil quality. Adjust the pH value of the soil and fill with oxygen, and then degrade the pollutants by the microbial fermentation process [12]. This method is suitable for high hydrocarbon content in oil pollution, suitable for the area have a long winter, and the treated soil can be returned can be returned or used in agricultural production. Sarkar et al.[13] added fruit and vegetable and garden waste composting into diesel contaminated soil, the oil content of the soil decreased by 85% after 12 weeks. Gomez et al. [14] reported optimization of field scale biopiles for bioremediation of petroleum hydrocarbon contaminated soil at low temperature conditions. TPHs removal rates in the range of 74-82% was observed in the treatments setups where mature compost and microbial consortia were used simultaneously.

3.2.2 Biological mud method

Mix the contaminated soil and water then feed them to mud bioreactor. Add necessary nutrients and energy sources to the reactor for the growth and reproduction of microorganisms, through fully stir to achieve the purpose of removing pollutants. After reach the treatment goals, the soil can be shipped back to home position after dehydration, the treated effluent can be recycled and discharge directly or sent to the sewage treatment plant, which depends on the water quality. This method can accurately control the degradation condition, have a good processing effect and fast speed.

3.2.3 Prepared bed method

On platforms without leakage, cover with gravel and sand, then contaminated soil layer is on them, which keep the height between 15 to 30cm. Add nutrients, water, and some surfactants, periodically flip the soil to supplemental oxygen and ensure the microorganisms growth healthily. The leachate that flow during processing can recharge on the treated soil, in order to completely remove pollutants [15]. Zhang Jian et al. [16] established prepared bed treatment engineering covers an area of 2400 m^2 for shengli oilfield, treat oily sludge by bioremediation, 160 days later, the capacity up to 52.75%.

3.3 The influencing factors of bioremediation

3.3.1 Physical and chemical properties of oil

The nature of the petroleum hydrocarbons pollutants will influence repairing effect. The biodegradability of polymer compound is lower than low molecular weight compound [16]. Hydrocarbons biodegradation rate from high to low is: n-alkanes, branch alkanes, low molecular weight aromatic hydrocarbons, PAHs. And concentration of petroleum hydrocarbons also has a certain influence on the degradation rate, as the concentration increases, degradation rate of decline.

3.3.2 Microbe group

Microorganism is the most critical component in bioremediation, appropriate species and number are the important influencing factors of bioremediation. Different microorganisms for the different components of the oil have different degradation. A study shows that [17], mixed culture of bacteria for petroleum hydrocarbon degradation is superior to individual culture.

3.3.3 Environmental factors

Environmental factors affect the growth of microorganism, in order to improve the effect and the degradation rate of petroleum hydrocarbon, we need to adjust the environmental factors of contaminated soil to the optimal condition, as far as possible to create suitable environment for microbial growth. We also need to pay attention to PH, temperature, oxygen and nutrients of the soil and adjust them. Although there are so many kinds of Bioremediation technologies, but the root of them is the regulation of environmental factors.

4. Conclusions

China's economy has a rapid development, the demand for oil is growing, but environmental pollution is getting worse. Bioremediation is superior to the traditional physical rehabilitation and chemical remediation, but the degraded speed of petroleum hydrocarbon is slow, need a longer time. So strengthen the treatment technology is very important. The physical, chemical, biological methods can be combined with each other, play their respective advantages, make the repair system to optimization.

In nature, there are a large number of microorganism resources that have highly effective degrade ability, but the existing technology is not perfect to detect separation, therefore, to explore new strain screening methods and regulate the microbial growth environment are particularly important.

5. References

[1] Cheng Guoling, Li Peijun. "Phytoremediation and Microbial Remediation of Petroleum Contaminated Soil," *Journal of Environmental Engineering*, 2007, 06: 91-96.
[2] Mohammed N, Battikhi B. "Biodegradation of Total Organic Carbons (TOC) in Jordanian Petroleum Sludge," *Journal of Hazardous Materials*, 2005, 120(1/3): 127-134
[3] Zhang Zhenming, Lin Changhu, et al. "Analysis of Microbial Remediation of Petroleum Contaminated Soil," *Guizhou Science*, 2010, 03: 76-81.
[4] Fang Xi, Yang Wen. "The Current Status of Petroleum Pollution of the Ocean and the Prevention," *Environmental Science and Management*, 2007, 09: 78-80.
[5] Deng Shaoyun, Xu Xueyi, Qiu Qinghua. "Present Situation and Prospect of Research of Remediation in Oil-contaminated Soil," *Northern Horticulture*, 2012, 14: 184-190.
[6] Singh C, Lin J. "Isolation and Characterization of Diesel Oil Degrading Indigenous Microorganisms in Kwazulu Natal, South Africa," *African Journal of Biotechnology*, 2008, 7(12): 1927-1932.
[7] Shen Hongchen, Wang Jianxing, et al. "Risk and Emergency Treatment of Marine Oil Spill: An Overview," *Environment Engineering*, 2011, 06: 110-114+85.
[8] Sanjeet Misshra, Jeevan Jyot. "In situ Bioremediation Potential of an Oily Sludge-degrading Bacterial Consortium," *Current Microbiology*, 2001, 43(5): 328-335
[9] Liu Shasha, Dong Jiahua, et al. "Review on Bio-venting for VOCs-contaminated Soils," *Environmental Science and Management*, 2012, 07: 100-105.
[10] Yang Lewei, Li Xingang, et al. "Study on Soil Environment and Aboriginal Animalcule Condition of Bioventing System," *Environmental Chemistry*, 2007, 04: 479-482.
[11] Jiang Lanshu. "Oily Sludge Treatment with Soil Cultivation Method in the Oilfield," *Oil-gasfield Surface Engineering*, 2009, 01: 12-13.
[12] Zhao Yuxia, Yang Ke. "Review on Remediation Technology of Petroleum contaminated Soil," *Environmental Science and Technology*, 2009, S1: 60-63.
[13] Sarkar D, FergusonM, Datta R, et al. "Bioremediation of Petroleum Hydrocarbon in Contaminated Soils: Comparison of Biosolids Addition, Carbon Supplementation and Monitored Natural Attenuation," *Environm ental Pollution*, 2005, 136(1): 187-195.
[14] Francisco Gomez, Majid Sartaj. "Optimization of Field Scale Biopiles for Bioremediation of Petroleum Hydrocarbon Contaminated Soil at Low Temperature Conditions by Response Surface Methodology (RSM)," *International Biodeterioration & Biodegradation*, 2014, 89: 103-109.
[15] Bai Yun, Li Chuan. "Application and Development of Prepared Bed Restorating Oil-contaminated Soil," *Journal of Chongqing Technology and Business University (Natural Science Edition)*, 2009, 04: 373-377.
[16] Wilson S C, Jones K C. "Bioremediation of Soil Contaminated with Polynuclear Aromatic Hydrocarbons (PAHs): A review," *Environ Pollut*, 1993, 81: 229-249.
[17] Van Hamme J D, Ward O P. "Physical and Metabolic Interactions of Pseudomonas sp. Strain JA5-B45 and Rhodococcus sp. Strain F9-D79 during Growth on Crude Oil and Effect of a Chemical Surfactant on Them," *Appl Environ Microbiol*, 2001, 67(10): 4874-4879.

The Repair Technology of Oil and Gas Pipeline Accident

Xiaoshan Chen
Qingdong Zhang *
Chaocheng Zhao
Department of Environmental and Safty Engineering, China University of Petroleum (East China)
cxs910101@163.com

Keyword: oil and gas pipeline; leakage; tapping and plugging; intelligent repair

Abstract

Oil and gas pipeline accidents have uncontrollability. There are several serious damages after a leakage: fire and explosion, gas poisoning and environmental pollution. As much as possible to reduce the hazardous of leakage, this study describes technology of holing and plugging on non-stop pipeline with pressure, focuses on the welding tee, opening, plugging, under plugs, corrosion and other key aspects. It is found that opening rigs, plugging rig and under blocking device are developing toward large-caliber and automation. This paper also reviews the progress and benefits of inner tube intelligent repair technology. Discuss the development direction of repair technology and equipment in the future.

1. Introduction

Pipeline transport is the main means of energy transportation, it's an integral part of the petroleum and natural gas industries. To the first half of 2012, the total length of China's oil and gas pipeline has reached 93000 km. During the next National Twelfth Five-Year Plan period, China will vigorously develop the petroleum and natural gas industry. Expect to 2015, the pipelines' total length will reach 150000 km. But if oil and gas pipeline has a long time service, the third party damage, corrosion, pipe material, construction quality and some other reasons always lead to failure accidents[1], cause disasters such as fire, explosion, poisoning, great economic loss, casualty and environmental pollution. In recent years, pipeline accidents occurred frequently. 2006, Prudhoe Bay oil field in Alaska occurred a pipeline leakage accident, the leaky crude oil reached $76x10^4$L, polluted the permafrost and lakes [2]. Since 1970 to 2010, 1249 pipeline leakage accidents occurred in Europe, about 13500km a year on average which is 50% of European gas pipeline [3]. 2013, a pipeline leak explosion occurred in Qingdao Development Zone, the reason is that leaky oil flow into other channels, such as municipal sewage channel, power channel, communication channel and other network facilities, volatile oil mixed with air to form explosive gas, then encountered fire and exploded, causing many casualties[4]. Leakage of oil and gas pipeline is a probability event, cannot eliminate completely [5]. When the accident cannot be completely avoided, in order to resist the spread of hazards and reduce the accident harm to life and environmental consequences, repair technology plays a vital role.

2. The risk of oil and gas pipeline leakage

The characteristics of oil and gas are flammable, explosive and toxic. The leaky oil or gas will diffuse to soil, water and atmosphere, cause poisoning and environmental pollution. At present, the main method of pipeline leak detection is daily inspection, which unable to repair the leaking point accurately and timely, has a potential safety hazard.

2.1 Fire and explosion

(1) The leaky gas mixed with air, when the concentration in the combustion or explosion limit range, and meet fair, electric spark, or static electricity sparks, the pipeline will have danger. According to statistics of the European gas pipeline incidents, the rupture of pipeline may quickly release a large number of flammable hydrocarbon, it's easy to fire and explosion.
(2)The leaky liquid oil will flow around, and it's difficult to remove, when the oil meet fire will cause a large area of fire. The insoluble or slightly soluble light oil on the water can form a fire disaster on the surface and also not easy to put out.

2.2 Toxic substance

Oil and gas contains H_2S, toxic materials such as benzene, toluene, and the combustion will produce a large amount of SO_2, CO and CO_2 into the atmosphere. Once the toxic substances enter the food chain, either algae or mammals will suffer harm. Large areas of oil spill would lead to a large number of benzene volatiles be inhaled, cause acute poisoning.

2.3 Environmental pollution

After the leakage of oil and gas, toxic vapors or gases enter the atmosphere, will cause air pollution; Liquid oil flow into the water, will destroy the water quality, and lead to the ecological structural imbalance; liquid oil flow into the soil, especially arable farmland, it will cause land contamination. In addition, the smoke produced by a fire or explosion can also cause secondary pollution.

3. Technology of holing and plugging with pressure

Technology of holing and plugging with pressure is a safe, economical, environmentally friendly and efficient technique of in-service pipeline maintenance and repair, can be used for non-stop pipeline renovation and sudden accident emergency repair. The keys of this technology including construction preparation, welding tee, opening, plugging, corrosion and other aspects.

3.1 Welding tee

3.1.1 The choice of the tee

Tee can be divided into equal tee and reducing tee, tee is permanently retain on the pipeline, its design life should be more than the life of the pipeline. With the development of high pressure, large diameter pipelines, tee compressive ability also need to get further development. Jiang Nan [6-7] tested the large thick-walled welded tee, described the stress distribution of different parts which under the test pressure. Wang Ruili et al.[8] developed a new type of hot-pressing sealing tee for large-diameter pipeline, when the pressure among 2MPa to 10MPa, solved the shortcoming of low strength steel by the test of stress-strain.

3.1.2 Technology of welding tee

Technology of welding tee is the key to pipeline repair. Both at home and abroad have been accidents due to improper welding, burning through and hydrogen induced cracking are the main focus of welding. Wang Zhejiang[9] made a detailed introduction for its reasons. During the welding, the inappropriate temperature, flow rate of medium will cause burning through [10]. M.A. Wahab et al.[11] also analyzes the conditions of the temperature field and the cooling time which impact the welding result, through the simulation of welding process in service. The ambient humidity, welding materials, the types of welding wire are all have certain effect on hydrogen induced cracking, diffusible hydrogen which produced by flux cored wire is much higher than solid wire, it's more likely to produce hydrogen induced cracking[12]. And the quality if the pipe also affects the success rate [13]. Therefore, before us welding the tee, we need to pay attention to the environmental conditions and the basic parameters of pipeline, and then choose the appropriate welding technology.

3.2 Pipeline tapping

Due to the basic characteristics of oil and gas, tapping operation also has certain risks. In 2008, the defect of tapping machine and improper operation lead medium to spurt out, resulting in personnel casualties when a company was in the opening operation [14]. The safe and prefect opening require mutual cooperation in tapping equipment, tapping technology and operation status.

3.2.1 Research progress of tapping machine

In China, the study of land tapping machine is relatively mature, and the tapping machine for large diameter pipes, underwater pipeline is also gradually developed. With the increase of energy demand, the application of large diameter pipe is developing, and how to make an equal diameter hole on the large diameter pipe is a difficult problem. In order to satisfy the needs of repair operation, Li Guoxing[15] used hydraulic system as the power, to ensure the tapping machine has enough torque under the best speed. Opening tool is the key part of tapping machine, choose the appropriate opening tool conduce to safe and fast operation. Zhao Yangang[16-17] described the design and selection of the components on tapping machine, he also studied about the optimal material, angle and size of opening tools. Zhang Ruilin[18] solves the problem of low efficiency and multiple cutting by combined bimetal zonal saw blade with opening tool.

Underwater tapping machine is the focus of domestic research. Yu Zhongshan et al[19,20] researched the structure of tapping machine, and solved the problem of too big resistance on spindle by applied the spindle pressure balance loop to tapping machine, and cooperate with hydraulic control system. In order to solve the inconvenience brought by underwater environment for human action, Zhao Bingjie et al. [21] devised such as underwater automatic tapping machine, able to work according to the set itinerary automatic.

3.2.2 Operation points of pipeline tapping

(1) Before opening, to ensure that all equipment, pipeline are under normal pressure; When the media is flammable and explosive gas, to ensure that its concentration is out of explosive range; and test the sealing pressure of tapping machine, ensure its maximum working pressure, make sure the tapping machine is in good condition[14].
(2) A simulation test should be carried out before the first opening operation was programmed in different pipes, pipe diameter and working pressure on the gas pipeline.
(3) Before the tapping, the opening point should be clear coating, avoid the weld and the part corrode seriously. If unable to avoid them, appropriately polish the weld bead part, and the center drill of opening cutter cannot fall on the welding path [22].
(4) Under the working pressure, there will be a gaseous hydrocarbon or vapor of liquid hydrocarbon leak out, mixed with outside air and form the environment suitable for explosive [23].

3.3 Non-stop plugging

During the large-diameter pipe plugging, the weight of plug head is too big, resulting decline in seal ability; during the non-stop plugging, the pipe have high pressure and wide range of weld, lead to poor seal, and lower safety factor.
In order to improve the efficiency of large-diameter hole plugging, Liu Gang [24] optimized the configuration and structure of baffle-sac type pipeline closure, to improve the sealing pressure. Wang Xijuan et al. [25] developed a plugging agent, which can withstand the pressure less than 2.2MPa and the time it can assume more than 5h, avoid the seal failure due to exorbitant pressure. Li Bin et al. [26] assembled round wire to each plug of chain occlude, improved the sealing performance.

3.4 Corrosion protection

Weld of pipeline may have uneven tissue or micro-crack exists and so on, it is more likely to be corroded. Therefore after plugging, we should do corrosion protection for alterant tube (valve), tee and weld in accordance with requirements. The corrosion of T-Type ducts and short section can use stick projectile antiseptic cream, this kind of material is easy to use, have a strong cohesion, and a good capability of resistance to cathodic stripping [27].

4. Tube intelligent plugging technology

Since the 1990s, the tube intelligent plugging technology has a rapid development. 1999, PSI company successfully apply the tube intelligent plugging technology for the first time [28]. Our pipeline industry is progressing, especially the length of submarine pipelines increase quickly, and the demand of intelligent pipe plugging technology is more and more urgent.

4.1 Advantages of tube intelligent plugging technology

(1) Intelligent occlude enter into the tub from the serve end of go-devil, under the control of ultra-low frequency electromagnetic pulse signal (ELF), brake at the leaky pipe section and plug. After the plugging finish, ELF automatic discharge, plug flow with the media in tube and receive it out at the close ball end [29].
(2)Liu Jiehua et al. [30] made a detailed exposition for the working process of intelligent occlude, intelligent occlude can work several times in a pipe, greatly saves the time and tools.
(3) When the intelligent plug use for subsea pipeline maintenance, it can reduce the vent loss of the oil or gas; it can be applied to variety of situations, such as the seabed valve replacement and repair, riser, seabed pipe repair, build, pressure test and so on [31]; and tube intelligent plugging technology also can prevent seawater enter the pipe section, effectively improve the safety and operational efficiency of pipeline.

5. Conclusions

(1) China's oil and gas pipelines are constantly extending, pipeline construction and repair technologies are in constant progress, while the two technologies should be organic unify in together with accident emergency system, extend the service life of pipeline and reduce the risk of pipeline accidents.
(2) The features of pipeline in the future are long-distance, large-diameter, high pressure and high flow. Pipeline repair technologies and equipments should be toward intelligent, digital and easy. Furthermore, our deepwater pipeline construction and repair technologies are still not perfect enough. As for manual operation, deep water conditions are very harsh, so repair technologies should be more intelligent.

6. References

[1] Zheng Bin, Li Xin. "Risk Analysis and Respond Research of Oil Long-distance Pipeline Leak," *China Petroleum and Chemical Standard and Quality*, 2013, 01: 256.
[2] Stuart Kirsch. "Guest Editorial: Sustainability and the BP Oil Spill," *Dialectical Anthropology*, 2010, 34: 295–300.
[3] European Gas pipeline Incident date Group. "Gas Pipeline Incidents," *Nede-rlandse Gasunie: 8th Report of the European Gas Pipeline Incident Date Group*, 2011: 20-25.
[4] Zhang Chunsheng. "Reflection of "11 • 22" Accident," *Labour Protection*, 2014, 01: 60-61.
[5] Lai Jianbo. "Study on the Technique of Simulation and Its Leakage Risk of Gas Pipeline Network," *Tianjin University*, 2008.
[6] Jiang Nan, Zhen Liang, Wu Shengnian. "Test and Burst Test Study of Large Diameter Thick-walled Welded Tee Stress," *Pressure vessel Technology*, 2007, 06: 1-5.
[7] Cao Haibing, Jiang Nan. "Finite Element Analysis of the Stress of a Big-sized, Thick-wall, Equal Diameter, and Welded T-pipe," *Chemical Engineering & Machinery*, 2008, 04: 216-219.

[8] Wang Ruili, Gou Dongmei, et al. "Development of Hot-pressed Plugging T-joint for Oil and Gas Pipelines," *Oil & Gas Storage and Transportation,* 2013, 06: 620-622.
[9] Wang Zhiqiang. "Transport Appearance Bottom the Piping Seals to Block up to Weld the Craft Imitates to Experiment," *Petro-chemical Equipment*, 2009, 02: 29-30.
[10] P.N. Sabapathy, M.A. Wahab, M.J. Painter. "The Prediction of Burn-through during In-service Welding of Gas Pipelines," *International Journal of Pressure Vessels and Piping*, 2001, 7711.
[11] M.A. Wahab, P.N. Sabapathy, P.N. Sabapathy. "The Onset of Pipewall Failure during "In-service" Welding of Gas Pipelines," *Journal of Materials Processing Technology*, 2005, 168: 414-422.
[12] Wan Yating, Xu Yuelan, et al. "The Influencing Factors of Important Bearing Steel Hydrogen Induced Cracking," *Transactions of the China Welding Institution*, 2011, 08: 101-104+118.
[13] Cao Yongli, Huang Fuxiang, Wang Qin. "The Progress and the Development Direction Long-distance Pipeline Welding Technology in China," *Oil Forum*, 2012, 01: 7-11+68.
[14] Ou Yangxing. "Risks and Countermeasures of Petrochemical Enterprises Tapping Jobs," *Petrochemical Safety and Environmental Protection Technology*, 2011, 05: 31-33+68-69.
[15] Li Guoxing, Zhang Jiushan. "New Hydraulic Tapping Machine for Large-diameter Oil (Gas) Pipelines," *China Petroleum Machinery*, 2006, 06: 75-78.
[16] Zhao Yangang. "Research and Development on the Hot Tap Machine for Large-diameter Oil and Gas Pipelines," *Nanjing University of Aeronautics and Astronautics*, 2012.
[17] Zhao Yangang, Li Yanjun, Yang Jinpeng. "Design of Tapping Cutter for Large-Diameter Oil and Gas Pipelines," *Equipment Manufacturing Technology,* 2012, 05: 220-221+233.
[18] Zhang Ruilin, Zhao Laiguo, et al. "The Research and Development of the New Efficiency Electric Tapping Machine Used in Pipeline with Pressure," *Machinery Design & Manufacture*, 2010, 12: 128-129.
[19] Yu Zhongshan, Zhang Hong, et al. "The design of the Underwater Pipeline Tapping Machine," *China Mechanical Engineering*, 2010, 03: 268-270.
[20] Yu Zhongshan, Zhang Hong, et al. "The Design of Hydraulic Control System in Underwater Tapping Machine with Pressure Balance," *Machine Tool & Hydraulics*, 2007, 09: 127-128+132.
[21] Zhao Bingjie, Zhao Honglin, et al. "Design of Automatic Submarine Oil and Gas Tapping Machine," *Oil Field Equipment*, 2009, 07: 34-38.
[22] Su Huihui. "Research on Live Plugging Process Management Technology," *Shanghai Gas, 2*013, 04: 22-25.
[23] Ma Tianlin, Guo Wenkai. "Argue for the Security of Tapping and Plugging Technologies of Industrial Pipes with Pressure in Oil and Gas Chemical Industry," *China Association of Equipment Management, Brazilian Maintenance Association, European Union Maintenance Organizations*, 2008: 5.
[24] Liu Gang. "Research of In-service Baffle-capsule Pipeline Occludes," *Pipeline Technique and Equipment*, 2008, 05: 60-61.
[25] Wang Xijuan, Ma Chunwen, et cl. "Research of Large Diameter Pipelines Plugging Technology in Oil and Gas Fields," *Petrochemical Industry Application*, 2011, 05: 80-85+92.
[26] Li Bin, Liu Jialiang. "Development of a New Type of Chain Pipe Closure Device," *Machinery Design & Manufacture,* 2013, 12: 168-169+173.
[27] Zhang Qiang. "Construction Technological Points of Natural Gas Pipeline with Pressure Plugging," *Entrepreneur World*, 2013, 18:133.
[28] Aleksandersen J, Tveit E. "The Smart Plug: a Remotely Controlled Pipeline Isolation System," *Proceedings of the International Offshore and Polar Engineering Conference, Stavanger*, 2001: 192-199.
[29] Zhang Shimin, Li Yanmin, et al. "Intelligent Plugging Techniques in High-pressure Pipeline," *Oil & Gas Storage and Transportation*, 2009, 06: 59-61+79+87.
[30] Liu Huajie, Zhang Ce, Zhang Shimin. "The In-tube Pressurized Intelligent Plugging Robot," *China Petroleum Machinery*, 2013, 06: 115-118.
[31] Zhang Ce, Zhang Shimin, Geng Dai, et al. " FEM Analysis of Plug Packer Based on the Model of Mooney-rivlin," *Advanced Materials Research*, 2011, 201-203: 326-331.

基于工艺安全时间效应的火气系统减缓有效性研究

王海清
中国石油大学（华东）安全科学与工程系
山东青岛
wanghaiqing@upc.edu.cn
刘俊芳
中国石油大学（华东）安全科学与工程系
山东青岛
Liujunfang55555@126.com
黄景荣
中石油西南油气田 自贡西部燃气公司
四川自贡

Study on Effectiveness of F&G System based on Process Safety Time Effect Model

UNPUBLISHED

基于工艺安全时间效应的火气系统减缓有效性研究

王海清 *1 刘俊芳1 黄景荣2

（1.中国石油大学（华东）安全科学与工程系，山东青岛 266580）

（2.中石油西南油气田 自贡西部燃气公司, 四川自贡 643000）

摘要：火气系统 FGS（F&G System）的整体有效性取决于探测器覆盖率、火气联锁系统的可用性和减缓有效性等三个部分。目前的技术标准和研究成果主要涉及前两个部分，对于如何分析和评估 FGS 启动之后的减缓效果，尚鲜见有关报道。基于美国 ISA84.00.07 标准的 FGS 整体有效性框架，本文根据 FGS 作用情况下工艺安全时间的变化，实现了对 FGS 减缓贡献的定量刻画，进而建立有效性的计算方法。以液化天然气 LNG 接收终端为研究案例，通过仿真对比了 BOG 处理工艺区在有/无 FGS 作用情况下的工艺安全时间的变化，对所提出的减缓有效性的量化方法进行了验证。

关键词：火气系统；减缓有效性；工艺安全时间；安全联锁仪表

[中图分类号]：X937 [文献标识码]：A [文章编号]

Study on Effectiveness of F&G System based on Process Safety Time Effect Model

WANG Hai-qing[1], LIU Jun-fang[1], HUANG Jing-rong[2]

（1.*Dept. Safety Science & Engineering, China University of Petroleum, Qingdao* 266580, *China*）

（2.*CNPC Southwest Gas field Company, Zigong Branch, Zigong* 643000, *China*）

Abstract: The overall effectiveness of FGS (Fire &Gas System) depends on three parts: The coverage of detector, the reliability of SIS and the effectiveness of mitigation for the accidents. The current standard and research mainly involves the former two parts, it's hard to model the mitigation function after the FGS started. Based on the framework of the FGS overall effectiveness proposed in ISA84.00.07 standard, a new method is proposed to describe the mitigation function quantitatively based on the change of process safety time, as it varies with the mitigation effectiveness. At the end, fire in a LNG receiving terminal is simulated by FDS, by the comparison of process safety time when there is no FGS to that where there has FGS, the proposed method is verified.

Keywords: Fire and Gas system; Mitigation Effectiveness; Process Safety Time; Safety Instrumented System

*基金项目：山东省自然科学基金（ZR2013EEM030），中国石油大学（华东）拔尖人才工程（SD2013211153）

作者简介：王海清，教授，从事 SIS，HAZOP/LOPA，报警管理和可靠性 RAM 分析等研究。E-mail: wanghaiqing@upc.edu.cn

引言

火气系统FGS（F&G System）是石油天然气和化工等行业的重要安全保障设施之一，对降低由于可燃/易燃/有毒物质和气体的各类泄漏或暴露（即容纳失效/Loss of Containment）所造成的潜在火灾、爆炸灾害，具有重要意义。目前国内外火气系统的相关规范和技术标准，基本还是经验型的处方式（Prescription based）方法[1~3]。随着安全仪表系统SIS技术（联锁停车）的发展，火气系统也被归为SIS系统中的一类，并采用安全完整性等级SIL的概念进行设计，此类设计思想被称为基于风险（Risk based）的方法[4]。

然而英国健康安全委员会HSE的研究表明，在北海作业的油气设施发生的重大气体泄漏事故中，有高达 40%的事故没有被FGS系统探测到[5]。这是因为火气系统的整体有效性，不仅取决于系统本身的安全完整性（或可用性），还取决于探测系统设置的合理性（即覆盖率），以及减缓响应系统（通风、喷淋、消防灭火等）的有效性。近年来出现的以ISA 84.00.07 为代表的火气系统设计标准，属于强调以性能化（Performance based）为导向的设计方法[6]；然而该标准主要涉及火气系统的可用性和覆盖率，对减缓系统的有效性并未涉及，因此如何在性能化的框架下对火气系统进行描述与量化研究，对真正实现对整个火气系统的有效性研究具有重要理论和工程意义。笔者以工艺安全时间为衡量指标，量化了火气系统的减缓有效性，进一步得出了基于性能化框架下的火气系统整体有效性模型。

1. ISA84.00.07 框架的火气系统减缓有效性模型

2010 年美国ISA协会发布了ISA 84.00.07 标准（技术报告TR），给出了火气系统在设计阶段的生命周期，明确强调了影响火气系统有效性（Effectiveness）的 3 个要素[6]，如图 1 所示。该标准指出，火气系统主要是针对容纳失效（LOC）而导致的火灾或气体扩散灾害，而非某个具体工艺灾害，故火气系统具有整体和全局性考虑的特点。火气系统的目标是消除或减缓由于火灾或气体扩散所造成的灾害程度，避免其升级扩大；而非像预防型SIS那样是为了降低风险，因为此时容纳失效事故已经发生了（此为前提条件，即给定某个初始的LOC发生频率）。因此该标准通篇（包括标准名称）强调的都是有效性，而非风险（尽管其在案例分析中仍涉及到风险降低的概念）。

通过上述模型，即可将减缓系统的有效性和探测系统覆盖率、FGS可用性结合起来，实现对整体有效性E_{FGS}的计算。故本文延续ISA 84.00.07 的基本框架，采用事件树计算模型来研究减缓系统的有效性。根据火气系统的特点，探测器覆盖率D_C的取值范围为[0,1]，FGS可用性A_{FG}的取值范围为[0.9,1]，减缓有效性M_E的取值范围为[0,1]。其中前两个要素的计算方法在相关技术标准和文献中均有涉及[4,6]，而关于减缓有效性则鲜见报道，因此图 1 中取值ME=1 即简化认为减缓系统是 100%有效。

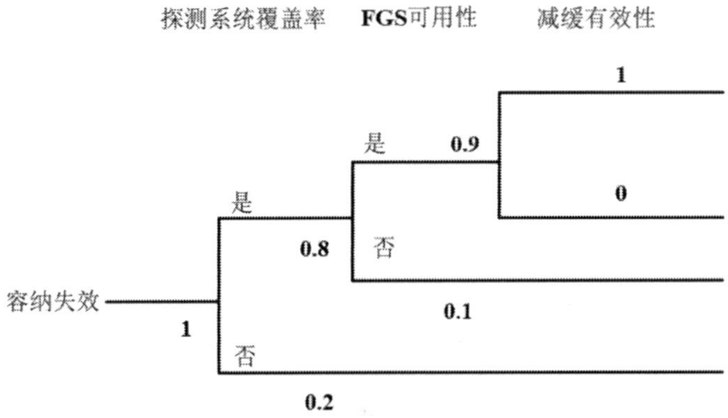

图 1 火气系统整体有效性的要素与模型

综上，提出如下的火气整体有效性E_{FGS}计算式如下：

$$E_{FGS}=D_C \times A_{FG} \times M_E \tag{1}$$

式中：D_C由火灾或气体探测器的覆盖率计算方法得到[6]；$A_{FG}=1-PFD_{avg}$，其中PFD_{avg}为火气系统联锁回路在检测周期内的平均不可用性，其计算方法见文献[6]。因此解决了减缓有效性参数M_E的计算问题，通过计算式(1)即可获得整个火气系统的有效性。进一步代入图 1 所示的事件树模型，还可计算各分支的风险降低程度[6]。

由于ISA 84.00.07标准发布的时间不长，一些研究仍将火气系统等减缓型的SIS系统（紧急泄放和火炬系统等），沿用基于风险降低概念的IEC 61508框架下进行量化计算。如日本JGC公司的Tanabe等提出了一种减缓型SIS系统的SIL定级方法[7]，但整个方法较为粗糙（见该文的式2），且无法考虑图1所示的其他影响要素；本文从操作人员可靠性和应急响应时间等两个方面，给出了新的量化计算减缓型SIS系统的SIL等级方法[8]。

2. 减缓有效性与工艺安全时间效应模型

火气系统的触发一般可分为2级：第1级为报警，由操作员确认后视情启动减缓系统，如1个光学火焰探测器超限（1ooN）或者气体探测器输出达到第1级设定值（对点型光学探测器通常为20-30%LFL）；第2级为自动触发减缓系统，如2个及以上的光学火焰探测器超限（2ooN）或者气体探测器输出达到第2级设定值（对点型光学探测器通常为40-60%LFL）。

当火气系统发出第1级报警后，至减缓系统启动的时间延迟记为T_d。影响T_d的因素包括探头的响应延迟时间、操作人员的确认与判断响应延迟时间、火气系统的响应延迟时间，以及减缓系统的响应延迟时间，这其中操作人员相关的延迟占据了主导地位。对于自动启动减缓系统的情况（第2级自动触发），则简化取$T_d = 0$。记火气探测系统触发报警为时间原点，火气系统之减缓响应过程的时序可以分为A-C共3个阶段，如图2所示。

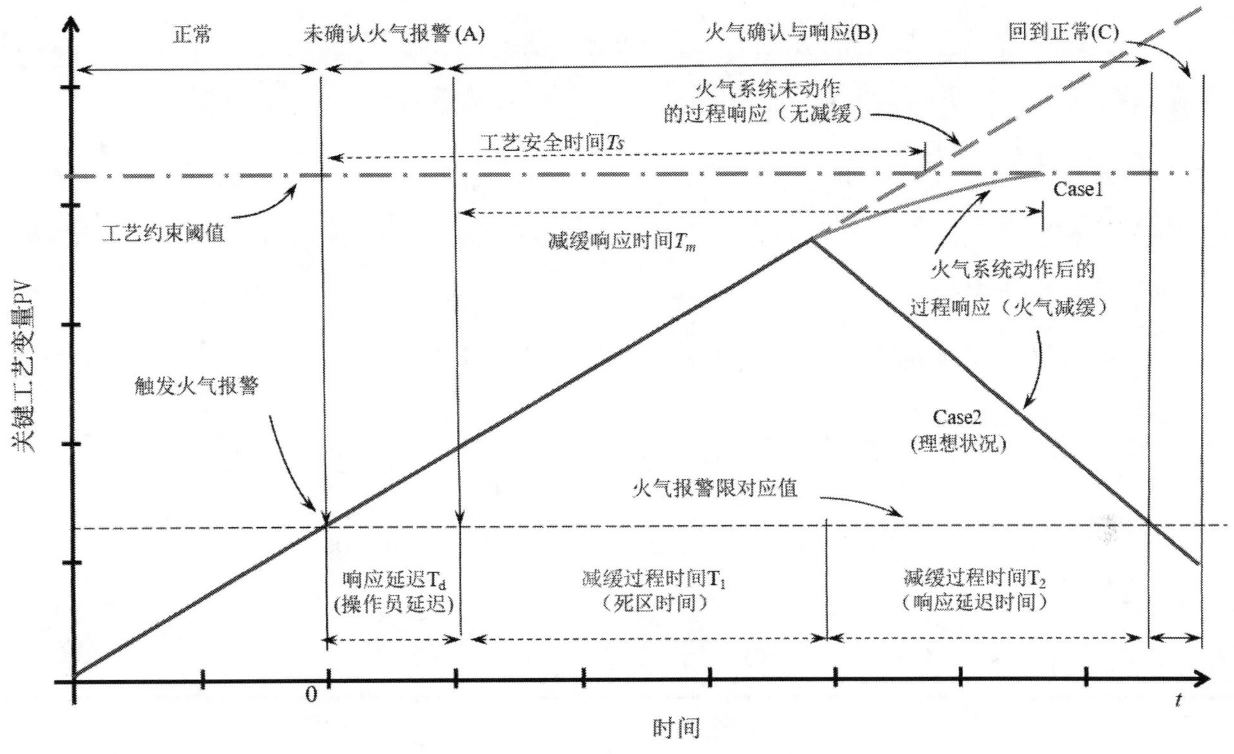

图 2 减缓有效性计算流程

记T_s为工艺安全时间，等于从探测到火气超限报警（火焰辐射或气体浓度）、但没有启动减缓系统的情况下，至PV变量超过工艺约束阈值的时间间隔；并记T_m为减缓响应时间，等于从减缓系统启动，至PV变量超过了工艺约束阈值的时间间隔。

下面按Case1和Case2两种情况分别考虑T_m的确定方法。其中Case1为减缓系统启动后，PV变量的变化趋势受到一定程度的减缓（这个假设的合理性是显然的），但经过一段时间之后仍然超过了工艺约束阈值，如图2所示，此时减缓系统的有效性M_E界于[0，1)之间，即存在一定的有效性（尽管未能避免LOC事故的最终扩大）；Case2为启动减缓系统后，至PV变量重新回到正常范围内，此时显然成立$T_m \to +\infty$，表明减缓系统达到理想状态，对应的减缓有效性为100%，即$M_E=1$。

当Case1的火气减缓效果越明显，则T_m将增大，从而使得有更充裕的时间和条件采取应急措响应施（即启动下一个保护层），防止LOC事故的扩大。当所采取的紧急关断、隔离等措施发生效果后，即从Case1情形转化为Case2情形。因此减缓系统的有效性，可以通过T_s（无减缓）和T_m（有减缓）之间的相互关系来刻画。因此本文提出建立如下的火气系统减缓有效性的计算式：

$$M_E = 1 - \frac{T_s}{T_m + T_d} \tag{2}$$

式中：$T_m \geq T_s$；且当$T_m \to +\infty$时，成立$M_E=1$。

由式3和图2可知，LOC事故的最终结果除了与火气系统有关外，与应急响应的有效性也有关。在工程应用中，通常有$T_m >> T_s - T_d$。考虑到火气系统的安全完整性（SIL1~4）实质上是数量级形式的[4]，故为了在精度上与图1所示的整体火气系统有效性匹配，同时简化工程判断，当满足下式时即可认为减缓系统是完全有效的，而无需满足Case2中的理论条件$T_m \to +\infty$。

$$M_E=1，当T_m \geq cT_s \tag{3}$$

式中：系数$c \in Z^+$，可根据具体工程需要来确定调整。考虑采取与IEC61508标准对应的SIL等级形式，本研究中取$c_{max}=4$。故根据M_E值的不同，本文提出将减缓有效性M_E按照SIL的形式在[1-4]内分为四个离散的等级，如表1所示：

表1 减缓有效性M_E等级划分表

M_E有效等级	1	2	3	4
M_E取值范围	0.1～0.3	0.3～0.6	0.6～0.9	≥0.90

综上式（2）和式（4），可建立火气减缓系统的整体有效性计算流程如图3所示。计算中涉及的各参数确定方法如下：

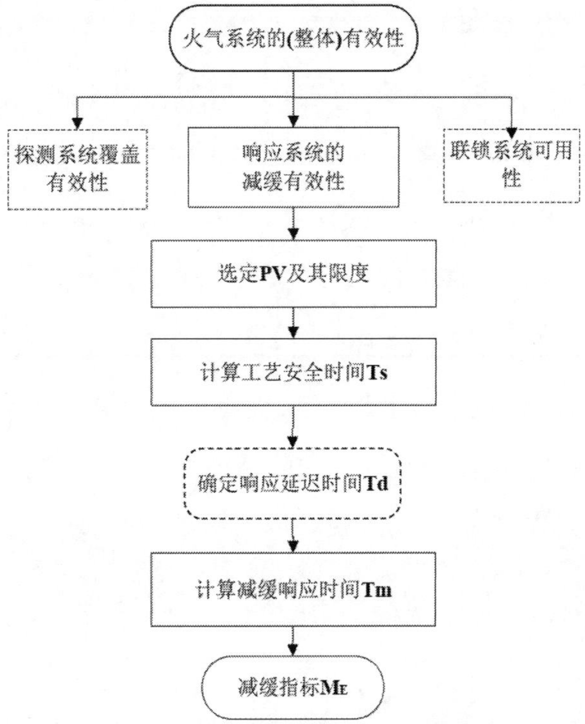

图 3 火气系统整体有效性计算流程

1. PV的选定：对于气体泄露探测，可燃气体可取爆炸极限LFL的百分比作为关键工艺变量参数；对于有毒气体可选择致死或严重伤害等后果的气体浓度作为关键工艺变量参数。对于火灾探测报警，选取热辐射量或关键工艺区域/设备的温度/压力等作为关键工艺变量参数。

2. T_s的确定：T_s为工艺安全时间，意为容纳失效触发一级报警后PV变量在安全范围之内的时间。对于可燃气体探测报警和有毒气体探测报警，T_s可根据漏孔尺寸、泄露源容器内压力、风向等参数，通过气体扩散模型计算得到；对于火灾探测报警，T_s可通过火灾仿真模拟求得。

3.响应延迟时间T_d的确定：如前所述T_d近似等于操作人员的响应延迟时间（对1级报警情况）。控制室人员的响应时间与操作人员探测、诊断和响应时间三部分有关。

4. T_m的确定方法：确定方法与T_s类似，区别在于模拟计算的过程中加入了火气系统动作之后的关断、隔离和喷淋通风等应急响应措施。

可见在其他条件不变的情况下，T_s和T_m是发生潜在LOC事故的工艺设施的平面位置参数的函数，因此与火气探测系统的覆盖率类似，通过本文所提出的计算模型和流程可以获得减缓系统的有效性映射图（Mitigation Mapping）。

3. LNG 接收终端案例研究

LNG 接收终端的工艺系统主要包括 LNG 卸船、LNG 储存、BOG 处理、LNG 气化/外输和火炬放空系统，按照工艺流程划分，可将其划分为四个区域：LNG 卸料区、LNG 储存区、闪蒸气（BOG）压缩与再冷凝区和 LNG 气化外输区，见图 4。卸料区通过卸料臂将 LNG 从 LNG 船舱输送至低温罐内；储存区的主要设备为 LNG 低温储罐与罐内泵；BOG 压缩与再冷凝区的主要功能是将储罐内的蒸发气经压缩机增压后，进入再冷凝器，与由 LNG 储罐泵出的 LNG 进行冷量交换，使蒸发气在再冷凝器中液化，再经高压泵增压后进入高压气化器气化外输，为 LNG 接收终端的核心单元；气化外输区则通过气化器将场内 LNG 气化，经加压、加臭、计量后输送至用户管道。

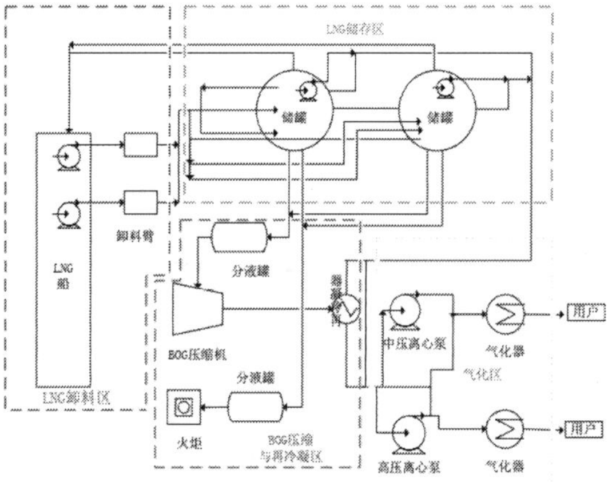

图 4 LNG 接收终端工艺流程及区域划分

LNG卸船、接收、储存及气化过程的火灾危险性均为甲类，需严格按照国标及行业标准规定的安全间距进行平面布置，还应配备足够数量的消防设备，除人员操作的固定式消防水炮及消火栓外，固定式水喷雾系统也应设置合理。假设所考虑的LNG接收终端现场的水喷雾系统配置[11]如表 2 所示：

表 2 LNG 接收终端的水喷雾系统

设备	喷雾强度/(L/min·m^{-2})	所在区域
高压输出泵	20.4	气化区
BOG 压缩机	20.4	BOG 压缩与再冷凝区
BOG 压缩机进口缓冲器	10.2	BOG 压缩与再冷凝区
BOG 冷凝器	10.2	BOG 压缩与再冷凝区
LNG 罐顶泵平台钢结构、阀门、管道等	20.4	LNG 罐区
码头火灾逃生通道	10.2	卸料区

针对 LNG 终端的不同工艺区域，可以采用本文提出的方法建立具体的减缓有效性映射图。在此选取 BOG 压缩与再冷凝区为火气减缓映射分析对象，其平面布局如图 5 示，并利用火灾动力仿真 FDS 软件对该区域仿真模拟，选取温度 T 为关键工艺参数变量 PV。为了得到不同位置的温度分布情况，将本区域根据关键设备所在位置分为以下几个区域：输出泵组区、缓冲罐区、压缩机房区以及再冷凝器区，仿真中分别在各个区域的近火源顶点与远火源顶点处设立 A1-E2 共 10 个温度模拟测点（黄色圈点）；火源中心位于靠近压缩机区域、距压缩机房中心 12.65 米处，此处火源仅作为热源考虑，见图 5。喷淋系统喷嘴均设置在各设备中心上方距地面高 5m 处，流量设置见表 2，无风环境。

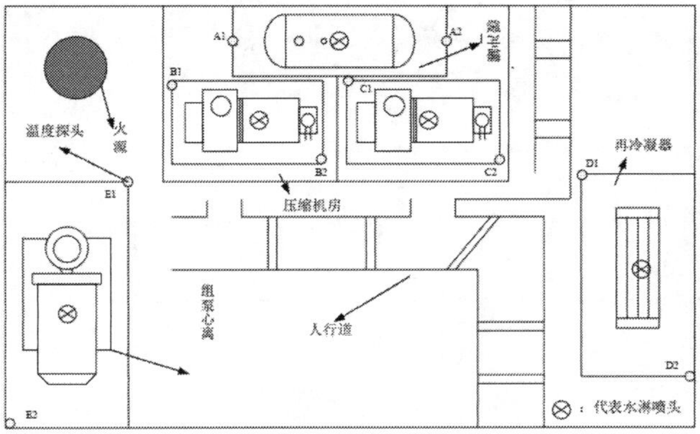

图 5 火源及 BOG 压缩与再冷凝区温度探测点布置

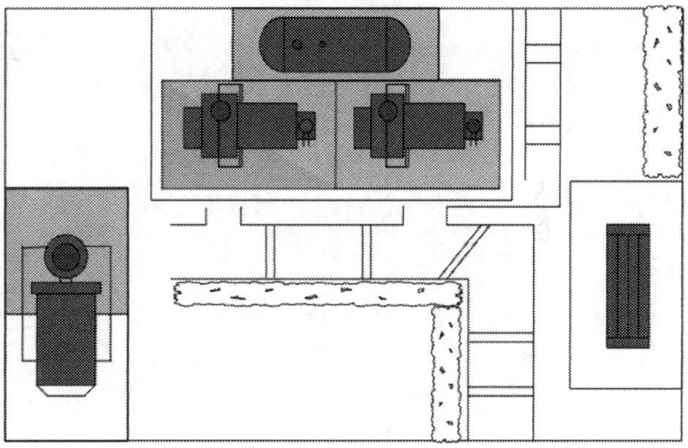
图 6 火气系统减缓有效性分布

欧美国家的LNG行业标准，如EN1473 和NFPA 59A等，规定了不同接收物体的临界热辐射强度要求[12]，在此基础上综合国内的有关研究，当环境温度高于 300℃时，会造成钢结构设备损坏，因此保守选定其临界值为 200℃。由FDS导出数据及图像结果，可获得有无水喷淋情况下A1-E2 各探测点温度随时间的变化曲线，从而得到各探测点工艺安全时间T_s与减缓响应时间T_m。假设本例中T_d为 32s（具体确定方法参考文[8]等），则通过进一步计算，结合本文给出的减缓有效性分级标准（表 1），可确定BOG压缩与再冷凝区域减缓系统的有效性。

仿真结果显示，在此火灾场景下，距离火源较近的缓冲器与压缩机区域所受影响较大。但由于处于无风环境，未造成火势的蔓延，且火灾造成的烟雾及热量能及时排放掉，位于压缩机泵房的设备温度未超过 100℃，故此火灾场景下不足以造成设备结构损害。但温度的升高必然会影响设备的工作性能，此时选取无喷淋情况下设备所能达到的最高温度作为 PV 极限值，仿真结果见表 4。

根据表 4 的结果，可得出 BOG 压缩与再冷凝区域的火气系统减缓有效性分布图，如图 6 所示（蓝色代表 3 级有效，绿色代表 4 级有效）。分析得出，在此火灾场景下，对于本区域火气系统的减缓有效性可达到 4 级。由于火源离工艺区域较远，且为无风环境，不会对设备造成结构性损坏，故水喷淋系统的喷淋强度可使各设备处于足够安全的环境状态中，故可达到较高水平的有效性。

表 4 仿真结果分析表

探测点	所在位置	温度限值（℃）	T_s(s)	T_m(s)	T_d(s)	M_E	M_E等级
A1	缓冲器	96.8	216	∞	32	1	4
A2	缓冲器	93.2	190	∞	32	1	4
B1	压缩机	50	5.04	5.20	32	0.86	3
B2	压缩机	98.2	217	∞	32	1	4
C1	压缩机	79.7	221	∞	32	1	4
C2	压缩机	112	248	∞	32	1	4
D1	冷凝器	200	--	--	32	--	--
D2	冷凝器	200	--	--	32	--	--
E1	泵组	74.9	216	∞	32	1	4
E2	泵组	200	--	--	32	--	--

注：“--”代表此火灾场景对该处无明显影响。

本案例表明在此火灾场景下，本区域火气系统的减缓响应系统设置合理，可满足该区域对减缓系统的减缓有效性要求。需要指出的是，以上 LNG 火灾仿真案例的目的在于验证所提出的火气系统减缓有效性的流程和计算方法，而不是为了研究案例本身（LNG 终端），因此仿真对涉及到的动力学、流体力学及其它的定量模拟条件进行了简化（因为这已属于另外的理论领域，与本文的主题无关）。仿真中对环境等有关参数进行了简化处理，在实际工程应用中需要根据具体情况而设置相应的仿真参数。

4. 结论

1. 探讨了基于 ISA84.00.07 框架的火气系统整体有效性模型与要素，首次提出了利用减缓时间与工艺安全时间的变化关系进行建模与定量分析。

2. 以火气系统减缓有效性为研究内容，确定了工艺安全时间是衡量火气系统减缓有效性的关键参数，并根据此指标将减缓有效性划分为 4 个离散的等级，实现与功能安全标准 IEC61508 对接。

3. LNG 接收终端案例采用火灾动力仿真 FDS 进行计算(也可采用其他更精确的模拟软件)，其目的在于说明如何应用所提出的减缓有效性模型与确定减缓有效性等级。FDS 本身的局限性和仿真参数的选取，对最终数值结果会有影响，但这并不影响本文模型与计算流程的有效性。此外对于不同的事故场景，应选取不同的工艺变量作为分析变量，以得到有效的工艺安全时间参数。

参考文献

[1]. API RP 14C: Recommended Practice for Analysis, Design, Installation and testing of basic Surface Safety Systems for Offshore Production Platforms [S], 2001.

[2]. GB 50493: 石油化工可燃气体和有毒气体检测报警设计规范[S], 2009.

[3]. GB 50116: 火灾自动报警系统设计规范[S]，2007.

[4]. ANSI/ISA-84.00.01 Part 3, (IEC 61511-3 Mod), Functional Safety: Safety Instrumented Systems for the Process Industry Sector—Part 3: Guidance for the Determination of the Required Safety Integrity Levels [S], 2004.

[5]. UK Health and Safety Executive: Offshore Gas Detection Siting Criterion Investigation of Detector Spacing [TR], OTO-93/002, 1993.

[6]. ISA-TR84.00.07：Guidance on the Evaluation of Fire, Combustible Gas and Toxic Gas System Effectiveness [S], 2010.

[7]. Masayuki Tanabe and, Atsumi Miyake: Risk reduction concept to provide design criteria for Emergency Systems for onshore LNG plants [J], *J. Loss Prev. Process Ind.*, 2011, Vol.24, 383-390.

[8]. Junfang Liu, HaiqingWang and FangLiu: A new SIL conception for emergency system and its application in LNG terminals[J],164,2013.

[9]. V. M. Fthenakis and D. N. Blewittt: Recent developments in modeling mitigation of accidental releases of hazardous gases [J], J. Loss Prev. Process Ind. Vol. 8(2): 71-77, 1995.

[10]. T.A. Roberts: Effectiveness of an enhanced deluge system to protect LPG tanks and sensitivity to blocked nozzles and delayed deluge initiation [J], J. Loss Prev. Process Ind. 2004, Vol.17, 151–15.

[11]. 郭揆常. 液化天然气 (LNG) 应用与安全[M]. 中国石化出版社, 2008.

[12]. 孙标, 郭开华. LNG 池火热辐射模型及安全距离影响因素研究 [J]. 中国安全科学学报, 2010, 20(9): 51-55.

Study on Filtering Methods of Acoustic Method of Leak Detection for Long Distance Natural Gas Pipeline

Cao, P.,
Li, Y., China University of Petroleum(Hua dong)
Liu, C.,

Acoustic method of leak detection has the advantage of accurate leakage location, high sensitivity, wide adaptability and short detecting time. Acoustic method has bright prospect in practical application of long distance natural gas pipelines. However, limited by the current stage of research, the acoustic leak detection method has not been practical used widely.

As the important link to the method, the filtering method for leaky acoustic wave needs a further study. This paper is mainly focused on the filtering method, summarizing filtering methods that have efficient effect on leaky acoustic wave. Meanwhile, in order to test the performance of all the methods, experiments are performed in the laboratory. Collecting acoustic data to compare the advantages and disadvantages of different filtering methods. Finally, give some useful advices in the filtering method of long distance natural gas pipeline.

The Study of Oil-Pipeline Vulnerability Assessment Index based on ISM

Abstract: In recent years the accidents of oil and gas pipeline have occurred frequently, by research we can get the feature of oil and gas pipeline on public safety and production safety. For getting a comprehensive risk assessment on pipeline, we refer to the international current comprehensive evaluation index system of the vulnerability and KENT evaluation method to determine the suitable evaluation index elements for oil and gas pipeline's vulnerability, then combing the ISM(Interpretative Structural Modeling) analysis method we transferred the assessment indicators into matrix multiplication to get the pipeline's vulnerability structure model. Using the vulnerability assessment system and the result of structure model, we set up a new applicable assessment system which is made up of the whole indicator tree.
Keywords: ISM analysis vulnerability capability index

1. 引言

"十二五"规划建设期间，我国油气管道网络已经基本上覆盖全国。2012 年 4 月份西三线实验段的提前开工建设，标志着我国油气管道建设在接下来几年内将迎来高峰期。随着油气管道网络的全面覆盖，其跨越的省份、区域越来越多，自然风貌、地理条件也愈加复杂，有些早年修建的管道所处地区随着经济发展已经成为人口相对密集经济相对发达的区域，公共基础设施与油气管道交错等问题都带来了新的安全挑战，油气管道安全问题已经不仅仅只是生产安全的问题，还具备了公共安全的特征。

在 2002 年，美国国家实验室联合提出了可识别和评价化工厂潜在威胁、风险和脆弱性的评价方法[1]，将脆弱性引入风险评价中，同时考虑了生产安全和公共安全 2 方面情况，改善了化工企业的安保状况。Calvin D Jaeger[2]和David A Moore[3-4]等都对针对化工行业非传统安全评估模型进行了研究。国内针对脆弱性指标体系的研究，灾害学领域中樊晓运等[5]结合以往研究成果，采用反推法、信息量法等多种方法对灾前承灾体脆弱性指标体系进行了研究，生态学领域中周嘉慧等[6]总结了已有的脆弱性评估方法，并对脆弱性指标体系进行了分类，主要有单一类型区域和综合型两类，其中综合型包括"成果及表现"、"压力-状态-影响"（PSR）和多系统评价指标体系，对于天然气管道，文献[7]从致灾脆弱性、后果脆弱性、应对能力脆弱性三个方面分析天然气管道脆弱性。刘松涛[8]用专家打分法对脆弱性和能力分别进行评估，然后运用层次分析法得到相应指标的权重，从而确定防范重点，在公共安全领域，朱正威等[9]提出了脆弱性-能力两个维度的评估框架，并且提到国际上现存单纯脆弱性评估、单纯能力评估和脆弱性-能力评估三类评估体系，其中泛美开发银行IADB建立的脆弱性和能力两方面的多灾害风险综合指标体系，可以更全面地分析各类灾害因子，达到展现事故连锁反应的评估目标。

通过分析比较上述研究，目前石油行业对于脆弱性评价是基于 API 发布的 SVA 评价方法，该方法在评价过程中主要是基于场景分析和表格记录，是一种跟 HAZOP 同样形式的方法，不同的在于其将脆弱作为风险发生可能性的一个修正因子，通过场景分析，赋予等级值后，根据其计算最终的风险值，该方法虽然流程简单，但是存在分析不完整，不容易组成专家组进行分析，影响因子不全面给并且评估过程缺乏量化，不利于进行不同区域的分析比较等缺点，而在生态学等其他领域，脆弱性分析主要采取的为指标分析方法，因此本文借鉴国际上现行脆弱性综合评价指标体系并结合 ISM 分析方法，确定适宜油气长输管道的脆弱性评价指标体系。

2. 解释结构模型

J.N.Warfield 在 1971-1973 年之间提出解释结构模型(Interpretative Structural Modeling，ISM)，用以分析复杂系统结构模型，专家根据经验和知识将某一复杂系统分解成若干要素，确定各要素之间的关系并转化为矩阵关系，通过矩阵的布尔运算确定系统内部多级结构模型，继而分析要素之间的相互关系。其原理如图 1 所示。

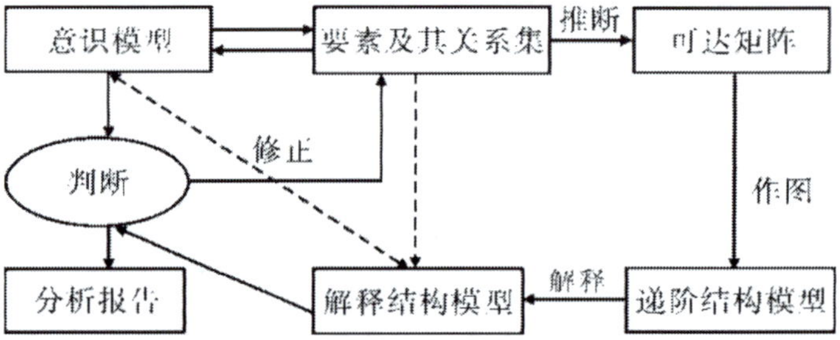

图 1 ISM构建指标体系原理[10]

2.1 ISM 模型构建流程

2.1.1 建立邻接矩阵A

邻接矩阵用以描述系统两两要素之间的直接关系，即作用关系、影响关系、因果关系等相互关系。在构建过程中，需要对可能存在的反馈关系。矩阵 A 构造如式（1）所示。

$$A = \begin{bmatrix} 1 & a_{12} & \cdots & a_{1n} \\ a_{21} & 1 & \cdots & a_{2n} \\ \vdots & \vdots & \ddots & \vdots \\ a_{n1} & a_{n2} & \cdots & 1 \end{bmatrix} \qquad (1)$$

其中：$a_{ij} = \begin{cases} 1, & \text{要素}s_i\text{对要素}s_j\text{有影响} \\ 0, & \text{要素}s_i\text{对要素}s_j\text{无影响} \end{cases}$，为了方便计算，要素对自身的影响视为 1。

2.1.2 确定可达矩阵M

邻接矩阵是专家根据个人经验和知识得出的，往往具有一定的主观和模糊性，为避免这种模糊性的影响，充分了解各个要素之间的关系，需要根据式（2）将邻接矩阵转化成可达矩阵 M。

$$M = (A + I)^k = (A + I)^{(k-1)} \neq (A + I)^{(k-2)} \neq \cdots \neq (A + I), (k \leq n - 1) \qquad (2)$$

可达矩阵表示系统各要素间直接或间接的影响关系，如在可达矩阵中，底层指标可通过中间层指标对上层指标间接产生影响，则底层指标对上层指标的影响为 1[11]。

2.1.3 可达矩阵的行级划分及结构模型建立

首先需要了解三个集合的定义，可达集R(Si)、前因集A(Si)和共同集C(Si)[12-13]：

（1）可达集 R(Si)：要素 Si 可以达到的要素集合定义为要素 Si 的可达集，由可达矩阵中第 Si 行中所有矩阵元素为 1 所对应列的要素组成；

（2）前因集 A(Si)：影响 Si 的因素集合，由可达矩阵中第 Si 列中所有矩阵元素为 1 所对应行的要素组成；

（3）共同集 C(Si)：为可达集 R(Si)与前因集的交集 A(Si)。

由 R(Si)=C(Si)进行行级间的划分，提取骨架矩阵，确定解释结构模型。

2.2 解释结构模型转化树形指标

ISM 得出的模型为一种网状结构模型，而许多综合评估方法需要使用树形结构指标体系，因此参考文献[11]和[14]方法，降低指标之间的相关性，将解释结构模型转化为一种树形指标体系结构。

3. 油气管道脆弱性评估指标体系构建

油气管道的脆弱性各个要素之间相互影响制约，运用解释结构模型（ISM）分析其相互关系，构建 ISM 模型，更好的了解各指标之间的关系。

3.1 油气管道脆弱性指标要素选取

本文认为脆弱性是指在威胁因素扰动下，因自身不利因素或缺乏应对能力而容易遭受损失、功能发生改变的内部固有属性。同时结合文献[15]中对国内外管道事故原因及类型的分析，并根据 SMART（S=Specific（明确性）、M=Measurable（可衡量性）、A=Attainable（可达成性）、R=Relevant（相关性）、T=Time-bound）（时限性））选取指标要素。

对于脆弱性主要从目标资产自身固有的脆弱性和受灾体脆弱性两个角度考虑，前者脆弱性包括管道设计、所运输的介质、介质理化特性和所蕴含的能量等，后者结合多米诺效应从人、机、环境三个角度考虑选取要素。在专家意见的基础上结合引言部分研究进展，本文选取指标要素见表 1。

表 1 油气管道脆弱性-能力指标要素

序号	指标要素	序号	指标要素
1	人员	9	管道运行条件
2	设施	10	管道主要功能
3	物料	11	介质的性质
4	环境	12	介质数量及能量
5	人员密度	13	土壤环境的影响
6	人员年龄结构	14	水环境的影响
7	人员安全意识	15	经济承载力的影响
8	人员安全培训程度	16	企业声誉的影响

3.2 脆弱性影响因素 ISM 模型构建

3.2.1 邻接矩阵的选取

分析表 1 中各个要素关系，经过调查和专家分析，确定邻接矩阵 A 如下：

$$
A=\begin{bmatrix}
0&0&0&0&1&1&1&1&0&0&0&0&0&0&0&0\\
1&0&0&0&0&0&0&0&1&1&0&0&0&0&0&0\\
0&0&0&0&0&0&0&0&0&0&1&1&0&0&0&0\\
0&0&1&0&0&0&0&0&0&0&0&0&1&1&1&1\\
0&0&0&0&0&0&0&0&0&0&0&0&0&0&0&0\\
0&0&0&0&0&0&0&0&0&0&0&0&0&0&0&0\\
0&0&0&0&0&0&0&0&0&0&0&0&0&0&0&0\\
0&0&0&0&0&0&0&0&0&0&0&0&0&0&0&0\\
0&0&0&0&0&0&0&0&0&0&0&0&0&0&0&0\\
0&0&0&0&0&0&0&0&0&0&0&0&0&0&0&0\\
0&0&0&0&0&0&0&0&0&0&0&0&0&0&0&0\\
0&0&0&0&0&0&0&0&0&0&0&0&0&0&0&0\\
0&0&0&0&0&0&0&0&1&0&0&0&0&0&0&0\\
0&0&0&0&0&0&0&0&0&0&0&0&0&0&0&0\\
0&0&0&0&0&0&0&0&0&0&0&0&0&0&0&0\\
0&0&0&0&0&0&0&0&0&0&0&0&0&0&0&0
\end{bmatrix}\qquad (3)
$$

3.2.2 可达矩阵M

根据式（2）求得可达矩阵 M：

$$
M=\begin{bmatrix}
1&1&0&0&0&0&0&0&0&0&0&0&0&0&0&0\\
0&1&0&0&0&0&0&0&1&1&0&0&0&0&0&0\\
0&0&1&0&0&0&0&0&0&0&1&1&0&0&0&0\\
0&0&1&1&0&0&0&0&0&0&0&0&1&1&1&1\\
1&0&0&0&1&0&0&0&1&0&0&0&0&0&0&0\\
1&0&0&0&0&1&1&0&0&0&0&0&0&0&0&0\\
1&0&0&0&0&0&1&1&0&0&0&0&0&0&0&0\\
1&0&0&0&0&0&0&1&0&0&0&0&0&0&0&0\\
0&1&0&0&0&0&0&0&1&1&0&0&0&0&0&0\\
0&1&0&0&0&0&0&0&0&1&0&0&0&0&0&0\\
0&0&1&0&0&0&0&0&0&0&1&0&1&1&0&0\\
0&0&1&0&0&0&0&0&0&0&0&1&1&1&0&0\\
0&0&0&1&0&0&0&0&1&0&0&0&1&0&0&0\\
0&0&0&1&0&0&0&0&0&0&0&0&0&1&0&0\\
0&0&0&1&0&0&0&0&0&0&0&0&0&0&1&0\\
0&0&0&1&0&0&0&0&0&0&0&0&0&0&0&1
\end{bmatrix}\qquad (3)
$$

3.2.3 可达矩阵的行级划分及结构模型建立

根据式（3）确定各要素的可达集 R(Si)、前因集 A(Si)和共同集 C(Si)见下表 2。

表 2 各要素的可达集 R(Si)、前因集 A(Si)和共同集 C(Si)

S_i	$R(S_i)$	$A(S_i)$	$C(S_i)$
1	1，2	1，5，6，7，8	1
2	2	1，2，9，10	2
3	2，3	3，11，12	3
4	3，4	4，13，14，15，16	4
5	5	5，9	5
6	6，7	6，7	6，7

7	7	7，8	7
8	7，8	8	8
9	9	9，10	9
10	10，11	10	10
11	11，13，14	11，13，14	11，12，13
12	12，13，14	12，13，14	12，13，14
13	13	13，14	13
14	14	14	14
15	15	15	15
16	16	16	16

根据表 2，提取骨骼矩阵，绘制 ISM 模型，如图 2 所示。

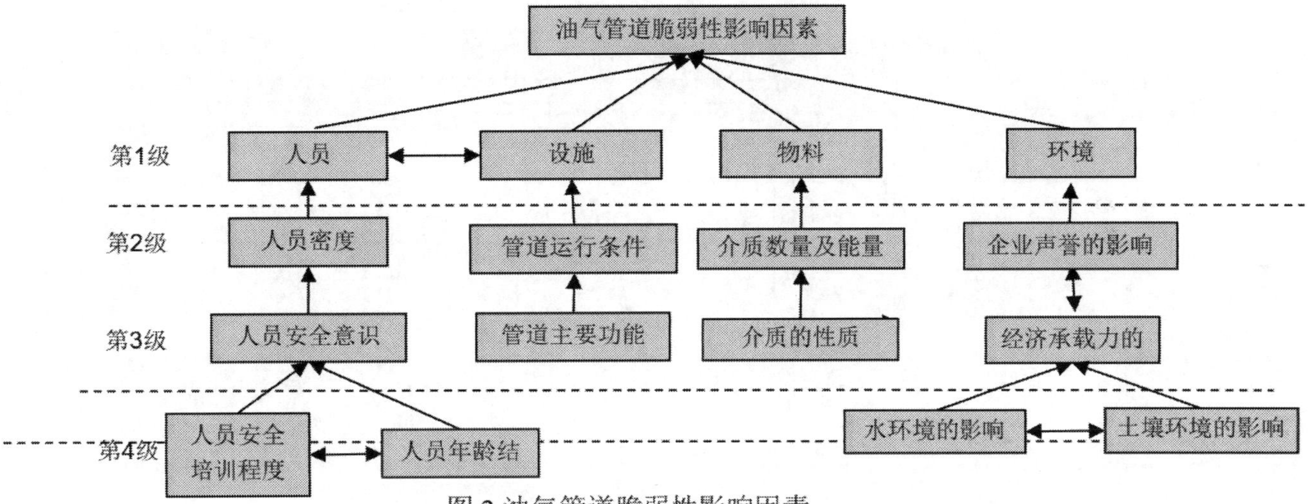

图 2 油气管道脆弱性影响因素

由图 2 可知，油气管道脆弱性影响因素体系是具有 4 级递阶结构的复杂系统，充分反映了影响因素之间的逻辑关系。笔者对模型的 3 个层次结构进行分析。

（1）表层直接影响因素包括人员、设施、物料和环境脆弱性。这 4 方面是影响油气管道脆弱性直接原因。需要对这 4 方面全面管理，才能有效降低油气管道的脆弱性。

（2）中层间接影响因素对油气管道脆弱性造成间接影响。人员密度和人员安全意识影响人员方面的脆弱性；根据管道主要功能确定管道运行条件继而影响设施方面的脆弱性；介质性质、介质数量和能量影响物料方面的脆弱性；企业的经济承载能力影响对事故处理能力，故对企业声誉也有一定的影响。中层因素通过表层影响因素对油气管道脆弱性产生影响，也反映了对表层影响因素的制约作用。

（3）深层根本影响因素包括人员培训程度、人员年龄结构，对土壤和水黄精的影响。人员安全教育培训主要影响相关的安全文化、安全沟通和员工安全意识，进而对油气管道脆弱性产生影响；而事故发生后，对污染的控制等对日后环境和企业声誉影响很大。

3.2.4 树形指标的转化

根据文献[14]，在对各个影响要素进行模糊性指标分解、相关性质指标合并、同类指标的聚类得到图 3 所示树形指标体系图。

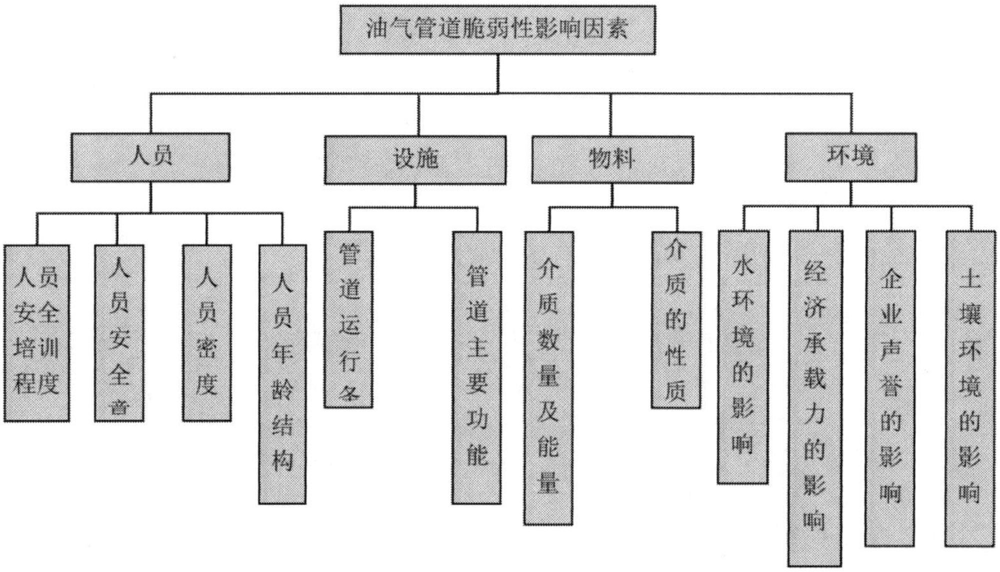

图 3 油气管道脆弱性估指标体系

4. 结论

（1）结合油气管道及常发事故的特点，经过专家分析，确定了 16 个对油管道脆弱性有影响的要素。

（2）利用 ISM 对各影响因素的关系进行分析计算，得出 4 级递阶有向图。员、设施、物料和环境脆弱性属于表层直接影响因素，人员密度和人员安全意识影、管道主要功和管道运行条件、介质性质、介质数量和能量中层影响因素、企业经济承载能力、企业声誉影响为中间层影响因素；深层根本影响因素主要是人员培训程度、人员年龄结构、水和土壤的承载能力。

（3）运用一种树形指标建立方法，结合 ISM 解释结构模型，从人员、设施、物料、环境 4 个方面，构建了油气管道脆弱性评估指标体系。

5. 参考文献

[1] American Petroleum Institute. Security vulnerability assessment methodology for the petroleum and petrochemical industries (second edition)[M]. Washington D C： API Publishing Services, 2004:1-142.

[2] C. D. Jaeger， Chemical facility vulnerability assessment project.[J]Journal of hazardous materials, 2003, 8(10)207-213.

[3] D.A.Moore,Security Risk Assessment Methodology for the petroleum and petrochemical

industries.[J].Journal of Loss Prevention in the Process Industries,2013,9(42):106-112.
[4] D.A.Moore, B. Fuller, M. Hazzan, J. W. Jones, Development of a security vulnerability assessment process for the RAMCAP chemical sector.[J]Journal of hazardous materials,2007,10(22):689-694.
[5] 樊运晓,罗云,陈庆寿.区域承灾体脆弱性评价指标体系研究.[J].现代地质, 2001,1(16)113-116.
[6] 周嘉慧,黄晓霞.生态脆弱性评价方法评述.[J].云南地理环境研究,2008,1(20)55-59.
[7] 赵秀雯，柴建设，城市埋地天然气管道系统脆弱性评估指标研究.[J].中国安全生产科学技术, 2011, 7(7),92-96.
[8] 刘松涛,刘文利,张向阳等.基于能力和脆弱性分析的火灾风险综合评价体系研究.[J].建筑科学, 2010,5 (26):44-46.
[9] 朱正威,蔡李,段栋栋.基于脆弱性-能力综合视角的公共安全评价框架:形成与范式.[J].中国行政管理, 2011,8(314):101-106.
[10] 胡俊锋,杨佩国,吕爱锋.基于ISM的区域综合减灾能力评价指标体系研究.[J].灾害学,2014,1(29):2014.
[11] 黄胜,郭海鹏,王超.ISM 方法在航空母舰评估指标体系构建中的应用.[J].中国舰船研究, 2014,1(29):.1-7
[12] 杨学营,李新春,杨学光.基于解释结构模型的风险预控管理体系建设分析.[J].煤矿安全,2014,5(45):234-237.
[13] 毛家鑫,高佳旭.基于ISM的全面质量管理影响因素分级研究.[J].中国管理信息化,2011,14(23):71-72.
[14] 张晓峰,裘杭萍,任正平.一种树形评估指标体系的建立方法.[J].火力与指挥控制,2007,6(32):77-79.

基于红外成像技术的高温管道泄漏检测研究

张丽珍 徐长航 陈国明 史焕地 艾素萍 韩国星 周乃望

(中国石油大学（华东），青岛，266520)

摘要：本文基于红外成像技术，开展了高温介质输送管道的泄漏检测实验，探索泄漏量、介质温度等因素对红外检测结果的影响。实验结果表明，通过管道表面红外热图，能够清晰识别泄漏源位置，并对不同工况条件下的红外热图进行特征提取，获得了不同泄漏量、介质温度条件下红外热图的差异性。在红外泄漏检测的基础上，制定了工程实用的管道严密性HSE管理体系，为现场人员开展高温介质输送管道安全保障工作提供理论依据。

关键词：管道泄漏；红外成像检测；图像处理；HSE 管理体系

The Detection of High-temperature Pipe Leakage by Infrared Thermography

Abstract: In this paper, based on the infrared thermography technology, carried out multiple sets of high -temperature pipeline leak detection experiments. It aims to search the fluent of different flow rate of leakage and temperature of medium. The results indicates that it can clearly identify the leak and locate the leak point by pipeline infrared thermal images. Then it can obtain infrared thermal image differences of different flow rate of leakage and temperature of medium through feature extraction. On the basis of test, we design the practical HSE management system which provides a theoretical basis for the personnel for pipeline security.

Keywords: pipeline leakage; infrared imaging detection; image processing; HSE management system

1. 引言

由于腐蚀、气候变化、外力破坏等因素的影响，管道很容易发生泄漏。管道泄漏不仅会造成财产损失和环境污染，甚至会进一步导致人员伤亡。所以及时识别泄漏并采取有效措施成为减少损害的当务之急。

红外成像技术区别于传统检测技术，以其设备便携、远距离检测、可靠性高在石油领域取得了越来越广泛的应用。Terumi Inagaki [1]提出管道垂直和水平置于水中两种实验方案，并通过改变温度和水深等变量来观测不同条件下的红外图像，研究管道泄漏的影响因素；Sadovnychiy S [2]首先分析管道泄漏温度异常变化，在此基础上设计出一套管道泄漏红外检测系统；Slobodan Dudic[3]以压缩气体管道泄漏为对象，开展了流量、压力、泄漏孔径变量条件下的管道表面温差变化规律研究。国内学者对红外成像技术也有一定的研究，李栋[4]等进行了红外成像埋地管道泄漏实验检测，并进一步研究了土壤热物性对红外检测结果的影响；范春利[5]等以船用保温层的内部和外部缺陷以及管道泄漏为检测对象，进行了红外成像技术的实验研究。

鉴于以上研究，本文以高温介质输送管道为对象，开展以泄漏量、介质温度为变量的红外成像检测实验研究，并在此基础上，制定适用于工程现场的管道严密性 HSE 管理体系。

2. 管道泄漏的红外检测系统设计

2.1 实验装置的建立

根据实验目的和实验要求，搭建了一套管道泄漏红外检测实验装置，该装置包括气体循环系统、管道泄漏检测系统。如图2-1所示，气体循环系统包括管道（直径25mm）、监控界面（可实现压力、温度、流量等工艺参数的监控）、压力调节阀（调节气体进口压力，保持气源压力稳定）、气体介质温度调控模块（可实现温度的自主调节）。如图2-2所示，管道泄漏系统外接于气体循环系统，该系统实验装置包括：红外热像仪-计算机系统、泄漏管道。本实验采用SAT HY6850 型红外热像仪作为检测仪器记录温度场，该仪器主要参数为：获取数据的最大帧速为 50 帧/秒，温度分辨率为 0.08℃，可以与计算机连接实现各种基本操作。通过调节实验系统，管道能模拟不同工艺参数下的管道泄漏。

图 2-1 气体循环系统

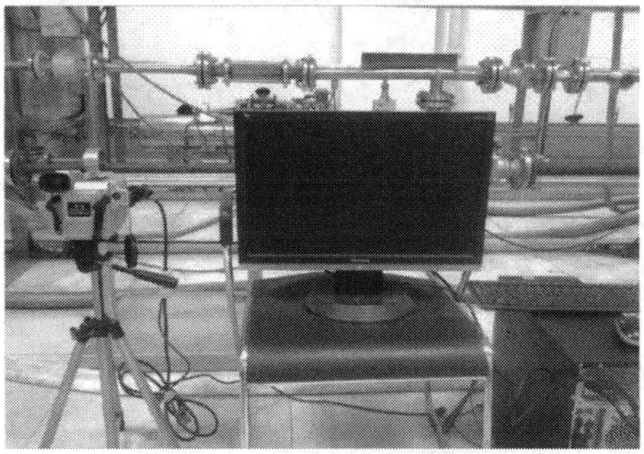

图2-2 管道泄漏检测系统

2.2 实验方案

实验样本和质量是进行分析和比较的前提，因此利用所搭建的气源稳定、工艺参数可调、气体介质温度自主调节、工况条件可监控的实验装置进行多组室内试验，将泄漏量以及介质温度作为变量，获得不同条件下的管道温度场。

2.2.1 泄漏量为变量的实验

打开气体循环系统，在气压稳定的条件下，调节气体介质温度调控模块，控制介质温度为60℃，然后通过控制阀改变管道泄漏量，红外热像仪-计算机系统实时观测管道泄漏前后的温度场，设置记录时间为3min。

2.2.2 介质温度为变量的实验

在气源稳定和泄漏量相同的条件下，调节气体介质温度调控模块，改变介质温度，红外热像仪-计算机系统设置如2.2.1，得到不同温度条件下的温度场。

3. 实验结果分析

3.1　管道泄漏源的识别和定位

图3-1、3-2、3-3分别为未泄漏、泄漏30s、泄漏90s的管道红外热像图，对比发现，从图中可以清晰地识别泄漏源位置。图3-1中低温区即为泄漏孔，当管道未泄漏时，泄漏孔温度明显低于管道温度。当泄漏30s后，由于热传导作用，泄漏孔温度升高，和未泄漏时对比明显。当泄漏90s后，泄漏孔温度接近管道的温度。

从图中可以看出，当管道发生泄漏时，红外热图可以清晰分辨出泄漏点的位置，并且泄漏时间越长，泄漏孔温度越高，稳定之后温度接近管道的温度。

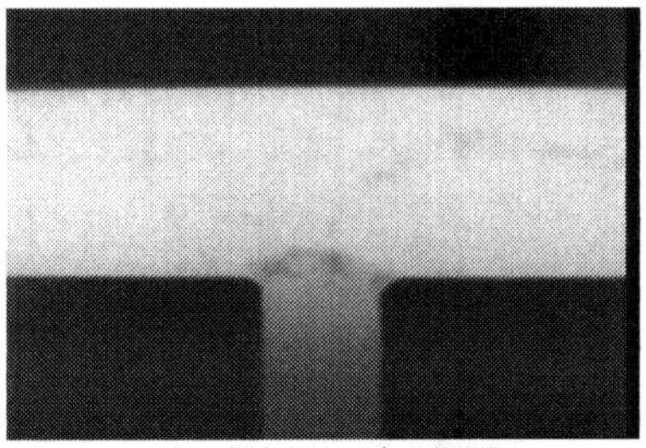

图3-1 未泄漏

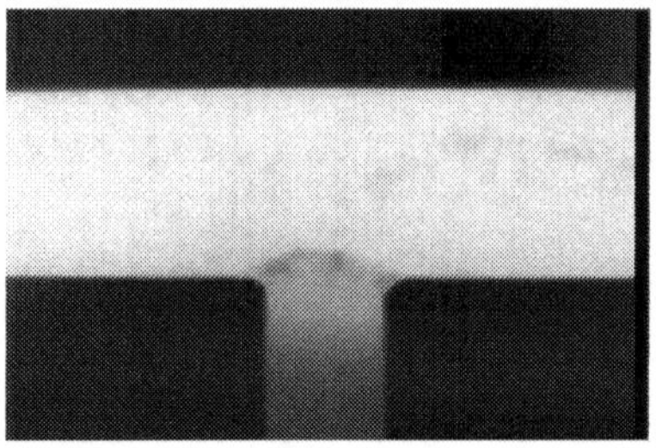

图3-2 泄漏30s

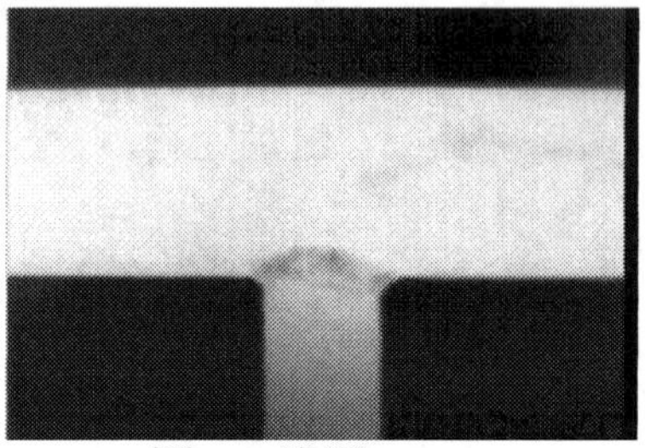

图3-3 泄漏90s

3.2　管道泄漏温度场定量分析

3.2.1　泄漏量为变量

分别对泄漏量为104L/min、115 L/min、123 L/min、143 L/min、154 L/min、173 L/min六组管道热图像进行温度特征提取，T1为泄漏点的温度，T2为泄漏孔前管道温度。

从图中可以观察到管道泄漏的温度场变化，并可总结出以下规律：

（1）管道发生泄漏后，泄漏孔和孔前管道温度场会发生明显的变化。前60s，温度上升速率较快，60s之后，温度变化比较平缓，最后趋于稳定。

（2）当泄漏一段时间之后，泄漏孔和孔前管道温差基本保持恒定。而且泄漏量越大，温差越小。

因此，当需要检测管道泄漏量的多少时，就可采用这种方法，一旦泄漏孔和孔前管道产生一定的温差变化，并呈现一定的规律时就可界定管道泄漏量的范围。

3.2.2　介质温度为变量

分别截取加热温度为60℃、45℃、30℃的管道泄漏30s的红外热图，研究介质不同温度条件下的管道泄漏温度场差异性。同样，对三组管道红外热图进行温度特征提取，T1为泄漏点的温度，T2为泄漏孔前管道温度。

从图3-10到3-15可以观察到不同介质温度下管道泄漏的温度场差异性，并可以得出：

（1）介质温度越高，泄漏孔与管道温度对比越明显。

（2）管道发生泄漏后，泄漏孔和孔前温度都会升高，在前60s温度增长比较快，60s之后温度增长相对缓慢，最后趋于稳定。并且，介质温度越高，泄漏孔和孔前管道温差越大。

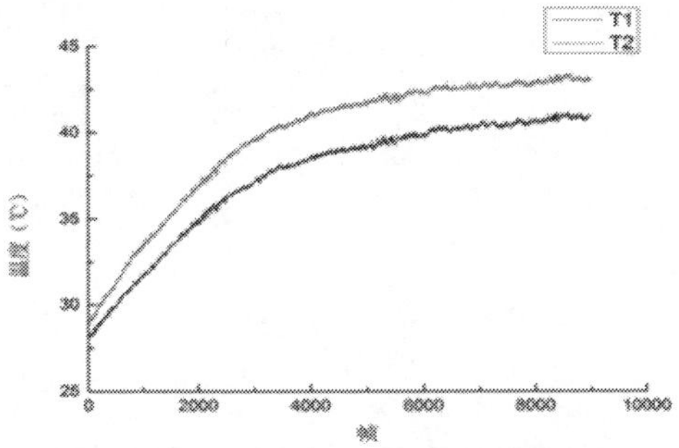

图3-4 泄漏量为104L/min

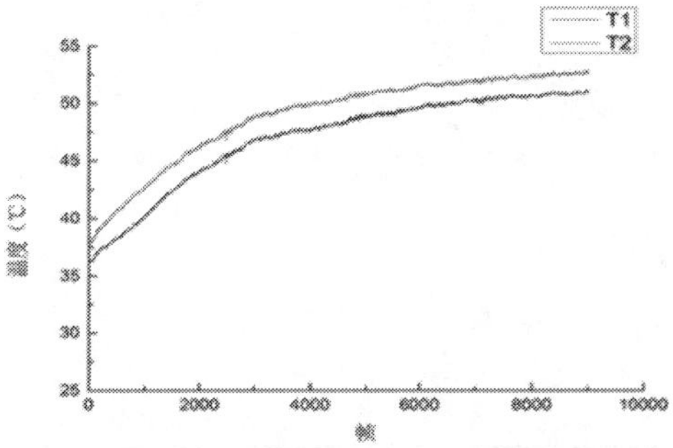

图3-5 泄漏量为115 L/min

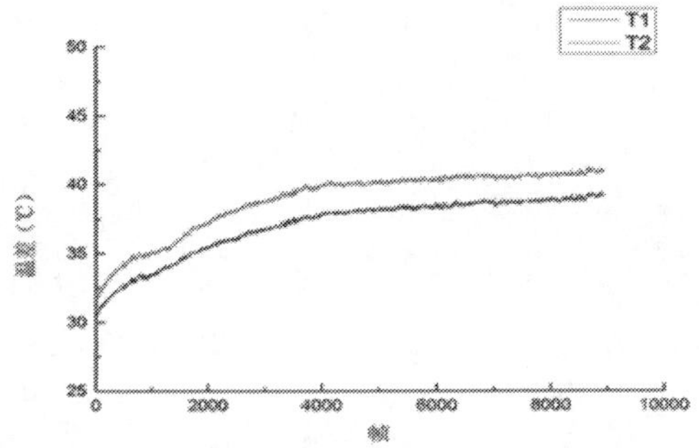

图3-6 泄漏量为123 L/min

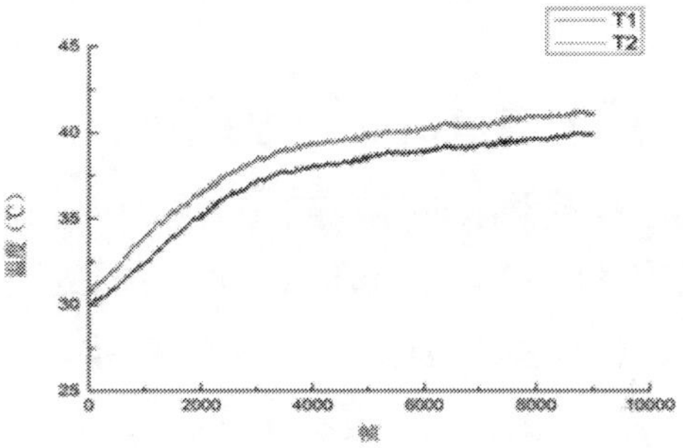

图3-7 泄漏量为143 L/min

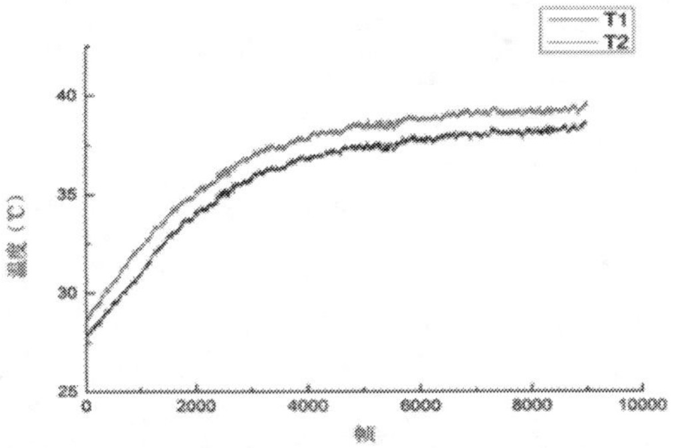

图3-8 泄漏量为154 L/min

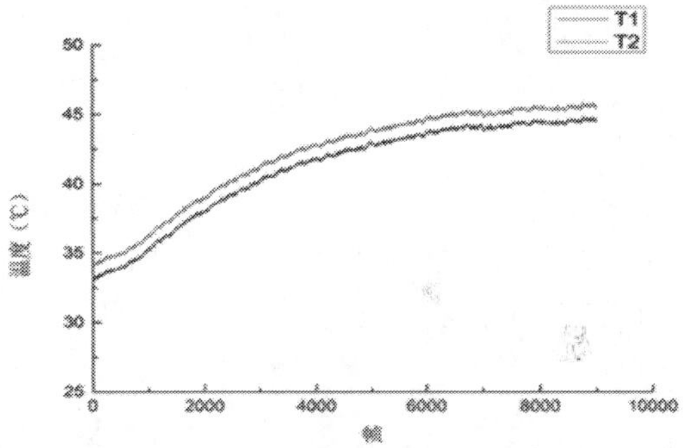

图3-9 泄漏量为173 L/min

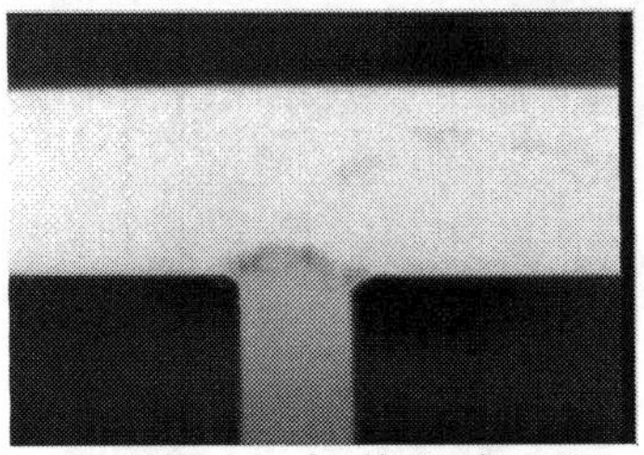

图3-10 加热温度60℃

图3-11 加热温度45℃

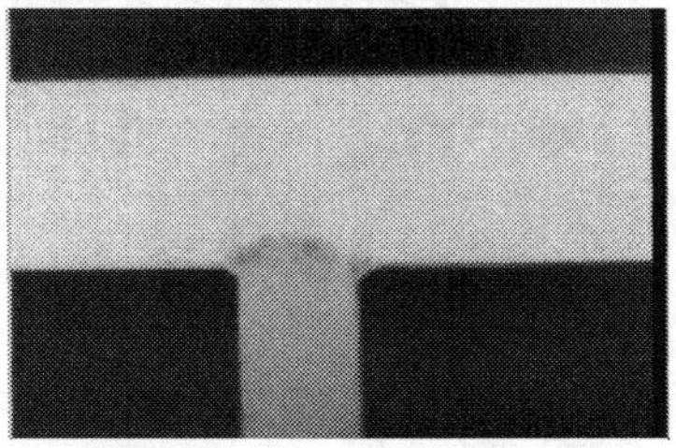

图3-12 加热温度30℃

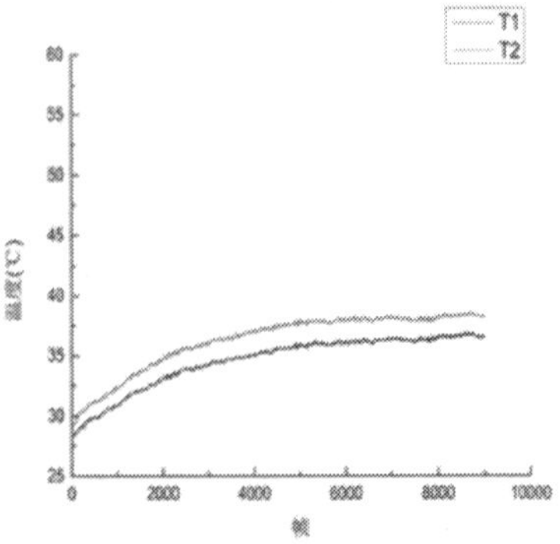

图3-13加热温度60℃

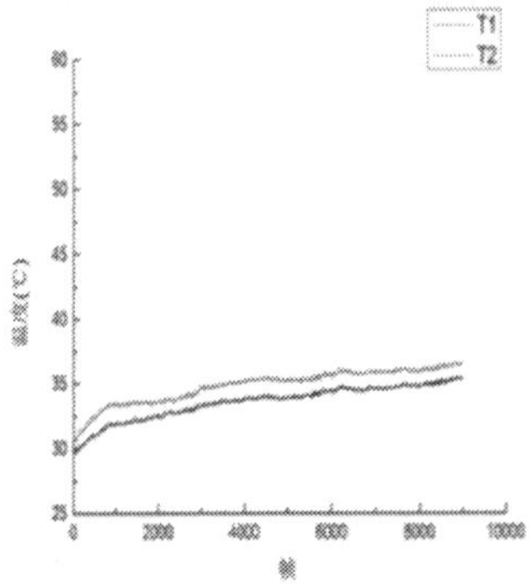

图3-14 加热温度45℃

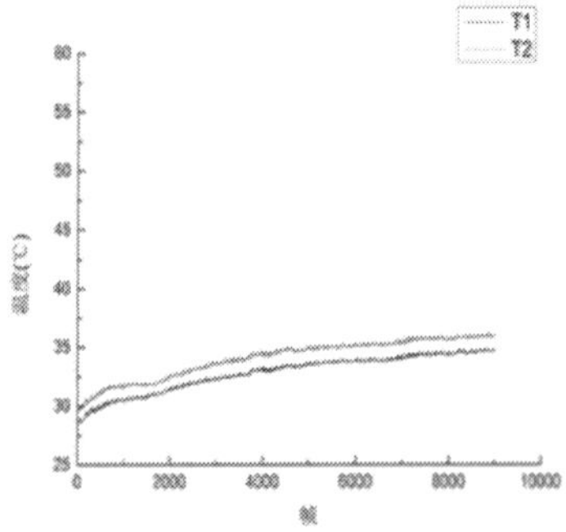

图3-15 加热温度30℃

4. 管道严密性 HSE 管理体系

4.1 HSE 管理体系的建立

从节约企业检测、维护成本、提升管道运行性能、控制事故损害后果出发，研究制定管道严密性HSE管理体系具有重要的工程价值。结合红外成像检测技术，HSE系统的制定应当以泄漏信息管理为基础，利用管道泄漏判定与泄漏定量识别的研究结果，综合企业人员应急培训进行制定。本文以HSE基本原则为指导，建立适用于工程应用的管道严密性安全管理措施。

图4-1是所建立的管道严密性HSE管理体系。领导承诺支持、参与是核心， 在对公司现状管理体系标准、计划实现管道安全保障目标总体认识的前提下，做出遵守法律法规、预防为主、持续改进的承诺，建立管道维护工作小组，明确责任，配备必要的资源（人、物、设备、工作制度），拟定管理实施计划表和进度安排表。在制度的支持下，工作小组合理开展管道严密性安全保障工作。

信息管理模块是以红外成像技术识别泄漏、定位泄漏孔位置、界定泄漏量和介质温度为检测基础，现场检测使用便携式红外检测设备，获得工作段管道的红外热图像，及时判断管道是否发生泄漏，并用相关软件对红外热图进行特征提取，得到管道温度场变化规律曲线，对泄漏情况进行界定，从而做出合理有效的控制措施。为加强工作效率和对今后工作的指导，对检测信息和分析数据进行记录，分别是分类编号、控制措施总结归纳。

结合信息管理模块，工作小组安全人员进行简单的风险评价，得出风险等级，进而安排下一步工作。应急管理是识别可能发生的突发事件和紧急事件，制定可靠的防范措施和应急预案，一般包括应急组织和职责、应急预案、应急预案演练、评估和修订、急救等。持续改进是强化HSE管理体系的过程，对体系文件进行不断修正完善，提高员工素质。

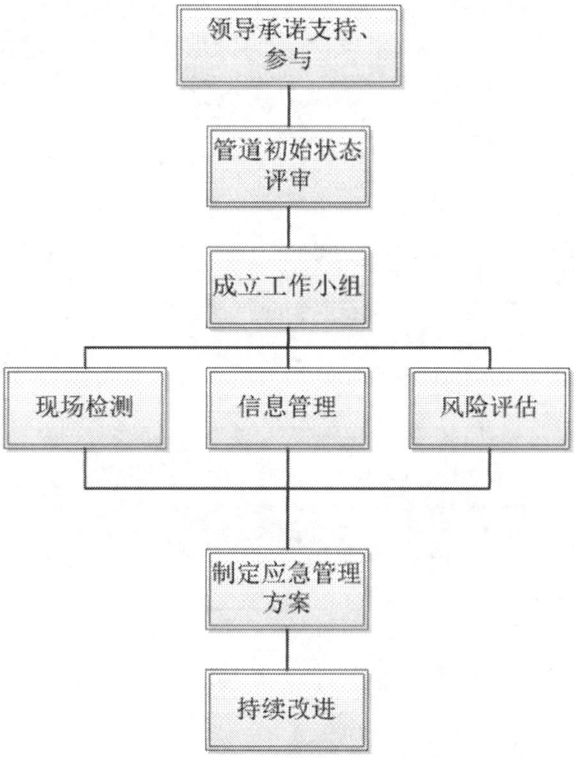

图 4-1 管道严密性 HSE 管理体系

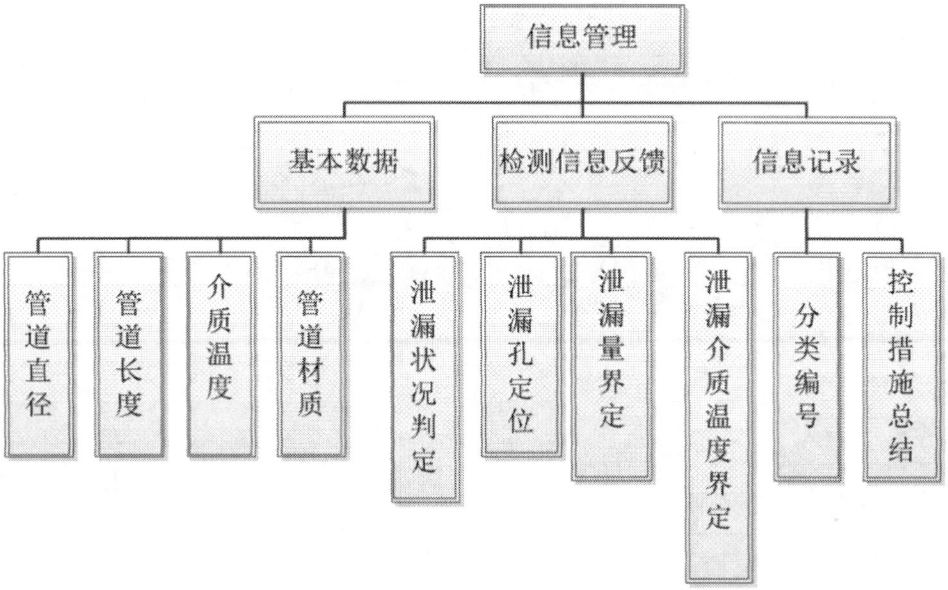

图4-2 信息管理模块

5. 总结

本文基于红外成像技术，开展了多组高温介质输送管道的泄漏检测实验，对比不同泄漏时间的红外热图像，可以清晰的识别泄漏并定位泄漏孔，进一步提取不同泄漏量、介质温度条件下的红外热图像特征，绘制温度曲线，观察曲线变化趋势发现，管道泄漏后泄漏孔温度和孔前温度会逐渐升高，并且泄漏量越小，介质温度越高，两点温差越大。实验结果表明，红外成像技术可以检测高温输送管道的泄漏并定量界定泄漏特征，对保障管道严密性具有重要意义。在检测的基础上，制定了工程实用的管道严密性 HSE 管理体系，为现场人员对管道进行安全保障提供了理论依据。

6. 参考文献

[1] Terumi Inagaki, Diagnosis of the leakage point on a structure surface using infrared thermography in near ambient conditions[J], NDT&E International, Vol. 30, No. 3, pp. 135-142, 1997.
[2] Sadovnychiy S, Ramirez A, Lopez M, et al. Pipeline leakage remote detection system[J]. WSEAS Transactions on Systems, 2004, 3(5): 1988-1992.
[3] Slobodan Dudic, Leakage quantification of compressed air using ultrasound and infrared thermography[J], Measurement ,45 (2012) :1689–1694.
[4] 李栋,吴国忠,李永柱,杨宏. 基于红外成像的埋地热油管道定位方法[J]. 管道技术与设备,2008,03:22-23.
[5] 范春利,沈立华,李孟良,杨立,孙丰瑞. 船用输热管道及保温层的红外热像无损检测研究[J]. 电子器件,2006,04:1367-1371.

初始温度对预混火焰传播特性影响仿真模拟研究

刘宇威[1]，刘义[2]

（1.中国石油大学（华东）机电工程学院，山东，青岛，266580；2. 中国石油大学（华东）化学工程学院，山东，青岛，266580）

摘　要：利用 Fluent 软件，探究管道内甲烷-空气预混气体的初始温度对其火焰传播的传播速度，火焰前阵面的变化，火焰传播过程中的最高温度，火焰传播过程中最大压力等特性的影响进行研究。研究不同初始温度下，管道内火焰的特性的变化情况。并以此为根据对不同温度下管道甲烷-空气预混气体传播的火灾危险性进行分析。

关键词：Fluent；管道；甲烷；火焰；传播。

[中图分类号] TE973　**[文献标识码]** A　**[文章编号]**

Initial Temperature of Premixed Flame Propagation Characteristics Affect Simulation Research

LIU Yu-wei[1], LIU Yi[2]

(1. College of Mechanical & Electrical Engineering, China University of Petroleum, Qingdao Shandong, 266580,P.R.China; 2.College of Chemistry, China University of Petroleum, Qingdao Shandong, 266580, P.R. China)

Abstract: : By using Fluent software, and to explore the initial temperature of the pipe inner - air premixed methane gas to the spread of flame propagation speed, the flame before the array surface, the maximum temperature in the process of flame propagation, the influence of the maximum pressure in the process of flame propagation characteristics were studied. Research under different initial temperature, the flame propagation characteristics of the change. Based on the pipeline - air premixed methane gas under different temperature the spread of fire risk analysis.

Key words: Fluent; The pipe; Methane; The flame; Transmission.

近年来输运天然气，油品等的管道频繁发生燃爆事故，且极易其造成非常巨大的经济损失。在实际的生产行为中，亦极其可能发生可燃气体与空气混合形成爆炸环境，若发生引燃事故，就会导致火灾、爆炸事故发生，会产生非常巨大的财产损失和严重的人员伤亡。2013年11月22日，位于山东省青岛经济技术开发区的中国石油化工股份有限公司管道储运分公司东黄输油管道泄漏原油进入市政排水暗渠，在形成密闭空间的暗渠内油气积聚遇火花发生爆炸，造成62人死亡、136人受伤，直接经济损失75172万元。2014年6月27日，印度南部安得拉邦一国有煤气公司的输气管道发生爆炸并引发大火，造成16人死亡，多人受伤。2014年8月1日台湾省高雄市发生天然气管道连环爆炸，已导致290余人死亡受伤。据统计，在煤炭、化工塑料、天然气等行业，火灾、爆炸事故在事故总数总所占比例分别高达46%，42%和 60%，而且每一次事故所造成直接经济损失也远远大于其它事故。本文使用CFD软件中的Fluent软件研究初始温度对于管道内甲烷空气预混火焰的传播特性的影响进，通过与各个文献中的数据资料对比和相关专业知识，证实了结果的有效性和合理性。

1 理论基础

1.1 数值模拟

1.2 CFD 简介

计算流体力学(Computational fluid dynamics，简称 CFD)是利用计算机数值计算并使用图像直观显示，对包含有流体流动、传热等相关物理现象的系统所做的数值计算分析。CFD的基本设计思想可以总结为：把实际问题中时间和空间上连续的物理量的场，如温度场和压力场，替代为一系列有限个离散点上的变量值的集合，运用一定的原则和方式建立起表征这些离散点上场变量关系的代数方程组，然后使用软件内计算模型，求解代数方程组，获得场变量的近似值。

1.3 Fluent 简介

Fluent 是目前应用较为广泛的商用 CFD 软件包，现已整合到 Ansys 软件中，它在美国的市场占有率将近 60%，凡是和流体和化学反应、热传递等有关的问题都可以使用 Fluent去解决。它具有丰富的物理计算模型、与时俱进的数值方法和相当灵活、适用的前后处理功能，在涡轮机设计、航空航天、汽车设计及相关实验，和石油天然气等方面都有着广泛的应用。Fluent 可以通过求解描述所有表征物质对流、扩散和反应源的质量，动力、能量守恒方程以及其它相关方程来模拟气体燃烧火爆炸，在算法上主要采用有限体积的数值方法。

1.4 本文中甲烷与空气相关化学常数

本文涉及的甲烷空气预混气体化学反映速率常数在表1中给出：

表 1 甲烷-空气初始混合物化学反应速率常数

CH_4+air	CH_4	O_2	CO_2	H_2O
化学反应当量比	1	2	1	2
速率指数	0.2	1.3	0	0
气体组份分子量 kg/mol	16.04	31.99	44.01	18.02
动力粘度 kg/mol·s	1.087×10^{-5}	1.919×10^{-5}	1.37×10^{-5}	1.34×10^{-5}
导热系数 W/m·k	0.0332	0.0246	0.0145	0.0261
标准状态焓 J/kg·mol	-7.489×10^{7}	0	-3.935×10^{8}	-2.418×10^{8}
标准状态熵 J/kg·mol·k	186040.1	205026.9	213715.9	188696.4

1.5 火焰传播机理

火焰前锋内剧烈的燃烧化学反应使其在边界上产生了很大的温度和浓度梯度，从而导致了强烈的热质交换。热质交换又引起了邻近的混合气的化学反应，由此形成了化学反应区在空间的移动，故火焰传播是一个复制的物理化学过程。按火焰传播过程中的气流的流动的状况，火焰传播形式可以分成层流火焰和湍流火焰。本文探究的是亚音速的燃烧过程中，初始温度对于火焰传播特性的影响。

2 模拟仿真

2.1 物理模型

模拟仿真的前提和关键之一是建立一个合理的物理模型，为了能够描述现象的主要特征并易于求解，采用了两端封闭的，长 100cm，直径 30cm 的碳素钢无缝管。圆柱容器封闭，而且内部没有各种障碍物。整个过程中的气体均为理想气体，燃烧反应为一步不可逆反应，整个容器与外界隔热处理。

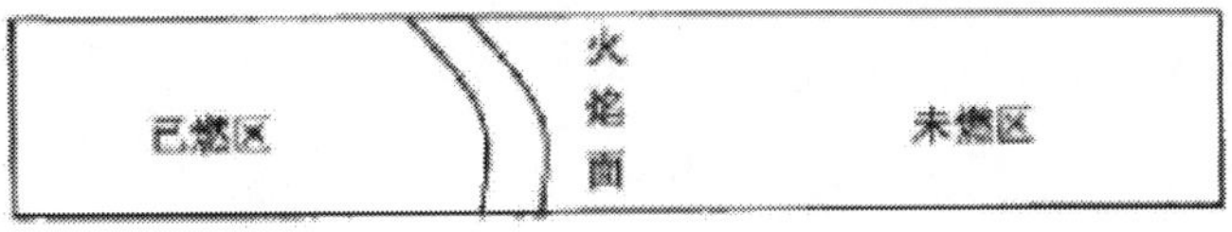

图 1 本文涉及的物理模型

2.2 本文涉及相关方程

模拟气体燃烧在管道内传播包括能量，动量，质量守恒方程，还需考虑组份方程，气体状态方程，湍流脉动方程以及燃烧过程中的化学反应方程。建立方程需做如下假设：不考虑热辐射换热，忽略体积力影响，粘性耗散作用为 0，也不甲酸 Soret 效应和 Dufour 效应，燃烧过程值考虑单步反应。在此鉴于受篇幅影响不对文中涉及方程做具体描述，具体方程内容参考相关文献所述。

2.3 模拟过程

首先使用 Gambit 软件进行建模和划分网格，然后在 Fluent 软件中对模拟问题进行初始化处理，确定边界条件、离散格式，并选用 SIMPLE 算法来处理压力/速度耦合问题，采用绝热燃烧模型，建立管道内预混甲烷/空气混合气体爆炸仿真处理。而后使用 CFD-Post 软件进行数据后处理。

3 模拟结果

3.1 不同时刻管道内火焰温度分布

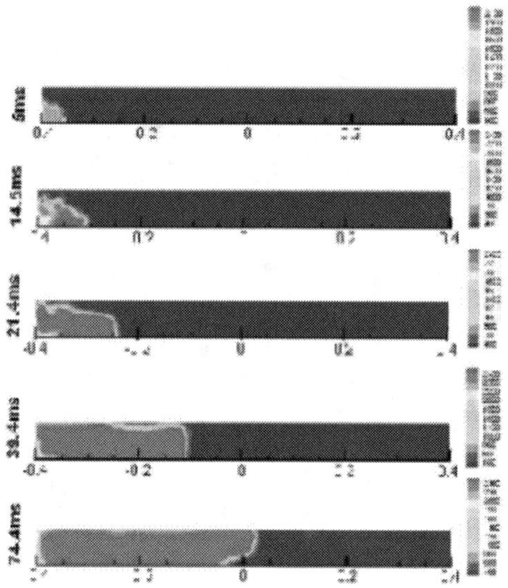

图 2 不同时刻管道内火焰温度分布

由图 2 可见，点火动作之后，点火点附近的温度迅速上升，同时火焰以相对缓慢的速度向四周扩散，火焰面呈表面比较光滑的半球形，说明 5ms 时的火焰是在以层流方式扩散传播。当火焰逐渐地靠近壁面之后，因为受到壁面的阻碍以及管端的约束，火焰改变方向，且与已燃气体相作用，在 14.5ms 时已燃区内发生湍流。壁面的约束加剧了层流向湍流的转变，在湍流漩涡的作用下，火焰开始发生变形、拉伸，火焰在 21.4ms~39.4ms 时进入加速传播阶段，而后形成传播较慢的火焰是中心轴线处火焰，速度较快的是沿壁面传播火焰，可以看出在 74.4ms 中心产生坍塌的火焰结构。这种坍陷的火焰结构是由于湍流火焰的加速传播机理所造成的，这是因为火焰坍陷实际上属于火焰面积增大的过程，而火焰面积增大就会使未燃区域的更多的可燃气体参与反应，从而使得燃烧反应速率加快，燃烧反应速率的增大，使反应释热速率增大，进而进一步导致了燃烧产物的膨胀加速及管道内压力的增大，管道内压力的增大从另一方面又推进了火焰前方可燃气体的流动，最终导致了火焰面附近的更高的湍流动能和燃烧速率，管道内甲烷-空气预混火焰正是在这种正反馈机制作用下使火焰加速传播的。

由图 3 可以看出管道内温度场分布，蓝色区域为未燃区，其温度较低，图中非蓝色区域是已燃区，其温度较高。火焰阵面在管道中传播时，已燃区和未燃区温度在不断的发生改变，具有温度梯度预热区域就是这样形成的，图中从接近燃烧区域的温度高于远离燃烧区的温度，这就说明燃烧火焰在传播的同时也在加热未燃气体，使邻近的未燃气体温度升高，图中由红色以右的部分就是燃烧火焰预热区。

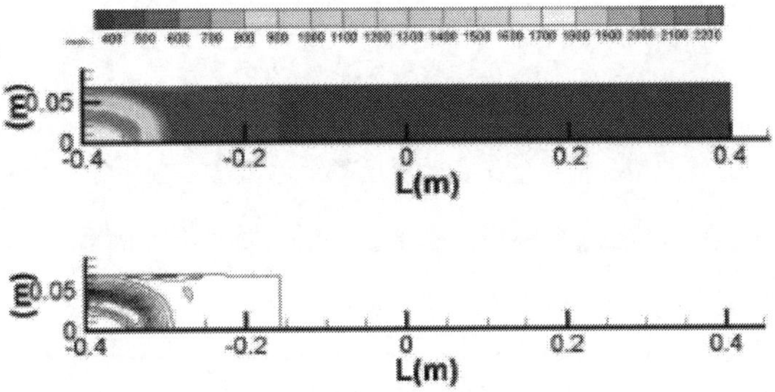

图 3 燃烧火焰预热区

从图 3 的下图预热区温度等值线图可以看出，管道内组份在预热区内进行大量的热量交换，靠近燃烧区附近的组份，其温度和能量变化更为剧烈，在图中就会呈现出密集分布的等值线。

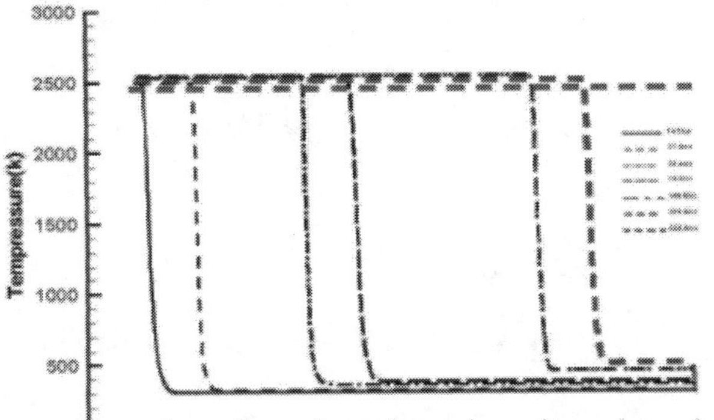

图 4 不同时刻温度-距离分布曲线

由图 4 不同时刻的温度-距离分布曲线可以看出，火焰阵面在燃烧区内形成了一个不断向外扩张的高温区；曲线 1~7 就表明了 7 个不同时间的管道内温度的分布情况，从其中可以观察整个燃烧过程中，管道内的出现温度峰值的位置向右移动的情况。当可燃气体完全燃烧以后时，气云区内的能量同时完全释放，此时整个管道内的温度达到一种平衡状态。

经模拟结果对比，发现初始温度对管道内火焰传播过程中最大压力的影响很明显。模拟中甲烷-空气预混的初始温度分别采用了 300K、600K、1000K，但是其它条件都保持不变。

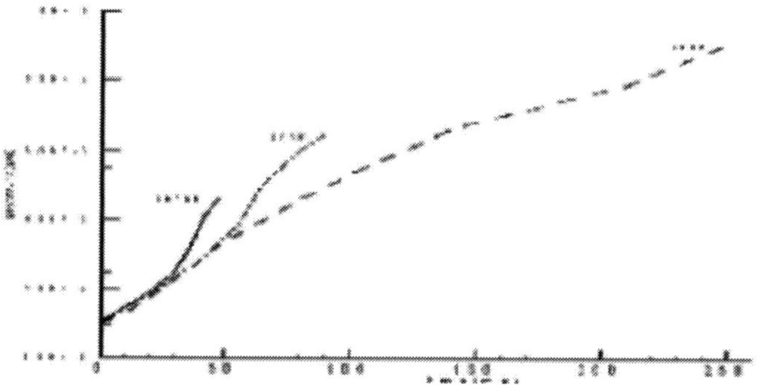

图 5 初始温度对气体爆炸压力的影响

从图 5 压力变化的情况中中可见初始温度越低，管道内的最高压力就越大。此外，也可以从图 5 中发现，初始温度对于火焰传播速度也有很大影响。初始温度越高，管道内组份彻底燃烧所需时间就越短。这是因为气体温度越高，气体活化分子越多，燃烧反应速度也就越快。

4 结论

由于燃烧，爆炸实验所需经费很高，且很危险，所以进行数值实验是很必要的，本文就是针对混合物初始温度对于管道内甲烷空气预混火焰特征的影响所做的一次数值实验。数值实验结果发现初始温度对于管道内火焰传播过程中的最大压力和火焰传播速度均有影响，其中最大压力与初始温度呈反比关系，火焰传播速度与初始温度呈正比关系。

参考文献：

[1] 谢烽，郭进，姚瑶，胡坤伦. 甲烷-空气预混气体燃烧特性研究[J]. 安全与环境学报,2014,02:5-9.

[2]徐艳冰. 燃烧火焰稳定性研究[D].浙江大学,2014.

[3]肖华华. 管道中氢—空气预混火焰传播动力学实验与数值模拟研究[D].中国科学技术大学,2013.

[4]张伟,龙江涛,刘洋. 甲烷-空气预混火焰稳定特性的实验研究[J]. 内燃机,2013,02:49-51.

[5] 王志荣,受限空间气体爆炸传播及其动力学过程研究[D]. 南京工业大学 ,2005

[6] 刘义,甲烷、煤尘火焰结构及传播特性的研究[D]. 中国科学技术大学 ,2006

[7] Shizhuo Zheng,Xin Zhang,Jian Xu,Biao Jin. Effects of initial pressure and hydrogen concentration on laminar combustion characteristics of diluted natural gas–hydrogen–air mixture[J]. International Journal of Hydrogen Energy . 2012 (17)

Numerical Simulation of in-Service Welding Residual Strength Size Effect Based on Sysweld

Guo, G.,

Abstract: The X70 pipeline steel high temperature tensile test and simulation that based on the welding software of SYSWELD were used to study the in service welding residual strength size effect. The results of tensile test shows that, when the temperature is 700°C, the strength of X70 pipeline steel is 250MPa, when the temperature rise from 800°C to 1300°C, the strength reduce nonlinearly from 102MPa to 19MPa The equivalent method of defect, the numerical simulation results and the high temperature mechanical properties research data were combined to establish residual strength burn through criterion of in service welding, and the influence of the pool size effect to the criterion was also be studied, and the results shows that, when the area of the equivalent defect increased to 1.1 times the bearing capacity decreased from 3.38MPa to 0.5MPa, which showed that the bearing capacity was decreased significantly with a slightly increase of molten pool area, the pool length had little effect on the residual strength, the pool depth and pool width have significant effect on the residual strength, and the change of penetration has a greater affected on the strength distribution of the high temperature area, the pool size effect is obvious.

Key words: in-service welding; gas pipeline; residual strength; size effect; numerical simulation

Introduction to API 570-Process Piping in-Service Inspection

Tong, D.,

Abstract: This presentation gives the general introduction of API 570 for in-service process piping inspection. It involves in the aspects on the codes history, scope, responsibility, design and construction codes, inspection classification and interval, damage mechanisms, RBI, FFS, etc. The main differences between Chinese process piping codes and API 570 is also discussed. The intended audiences will be inspectors, owner-users' engineers, QA QC, maintenance, operation personnel and project management.

Fire Risk Management and Emergency Disposal Technology of Pipe Racks in Chemical Industrial Park

Liu, X.,

According to the process characteristics of the pipe racks in chemical industry park, based on the statistical analysis of typical fire accident of chemical pipe racks, the leakage, fire and explosion accident risk and evolution rules of pipe racks in chemical industry park were studied in the paper. According to dynamic fire risk characteristics of pipe racks in chemical industry park, based on the on-line monitoring parameters of the pipe racks such as pipes pressure, flow rate, temperature, gas concentration and so on, combined with the utility pipes environment monitoring dates analysis and pipelines reliability analysis, the park pipe racks dynamic fire risk evaluation index system was established which can be used to realize the utility pipes accident risk early warning and real-time online monitoring. And based on the intelligence and Internet information technology, the accident risk control strategy of chemical material conveying pipe galleries and accident emergency disposition key technology were put forward, to realize early warning, intellisense and rapid emergency disposal function of the pipeline leakage and fire accident, which to avoid the material leakage and fire & explosion disaster accidents, effectively to improve the capacities of petrochemical enterprise accident emergency disposal technology.

Numerical Research of Pipeline Leakage Sound Field Based on Acoustic Emission Technique

Shi, H., China University of Petroleum(East China)

Pipeline leak is an important factor for affecting the safe operation of the oil and gas gathering and transportation system. Acoustic Emission (AE) technique is an effective means of pipeline leak detection. Analysis of the characteristics of the sound source and the transmission mechanism of AE signals caused by leakage can be one of the important conditions that ensure the efficiency and accuracy of detection. The paper was based on the Lighthill aeroacoustic theory, combined the numerical simulation with experimental study, used computational fluid dynamics software Fluent and acoustic software LMS Virtual Lab to simulate the flow field and the sound field of pipeline leakage, and analyzed them in time domain and frequency domain to provide guidance for the acoustic emission pipeline leak detection. First of all, the dipole and quadrupole source at the pipe leak hole were respectively selected. The pipeline leak fluid velocity and noise directivity pattern were analyzed to determine the sound source type. And then the comparative analysis of pipeline leak sound field distribution with pipe wall vibration and no pipe wall vibration was conducted. The AE experiment data verified the necessity of acoustic vibration coupling on the calculation model. Finally, taking the pipe wall as a boundary, the inside and outside pipe acoustic model was established to study the pipeline leakage sound field distribution when the medium outside the pipe was respectively air, water and soil. The AE technique was used to detect pipeline leakage to verify simulation results.

The Analysis of the Difference Between China and American Oil & Gas Pipeline Safety Inspection Standards

Qingdao, China
August 28 - 29, 2014

Min Qi, Shuai Lu, Dongfeng Zhao

China University of Petroleum(East China)

CATALOGUE

- **Introduction**
- **Overview of pipeline safety inspection standards between China and US**
- **Comparison and analysis on typical standards**
  - Inspection method
  - Testing location
  - Inspection cycle
  - Inspection operator requirement
- **Conclusion and suggestion**

Introduction

- Until now, the whole length of oil and gas pipeline is 93000km.
- The inspection is the effective measure to guarantee pipeline operation, Pipeline maintenance decision and ensure safety and integrity of pipeline.

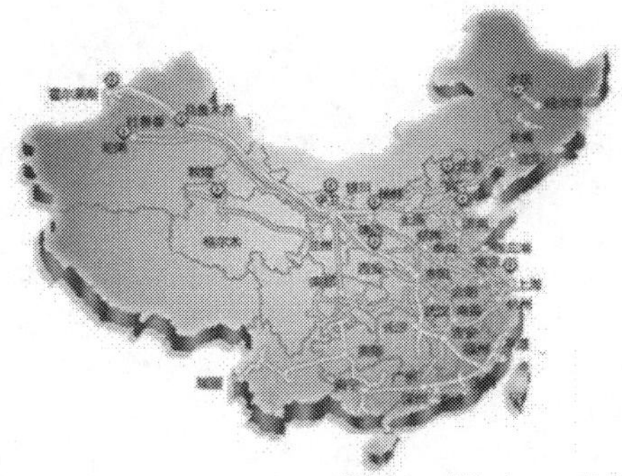

China University of Petroleum(East China)

Overview of pipeline safety detection standard between China and US

China(21)	U.S(19)
SY 6186-1996石油天然气管道安全规程	ASTM G 62Standard Test Methods for Holiday Detection in
SY 5737-1995 原油管道输送安全规定	Pipeline Coatings
SY 6457-2000含硫天然气管道安全规程	ANSI/API RP5A 5-2009 Field Inspection of New Casing,
CJJ33-2005城镇输气输配工程施工及验收规	Tubing, and Plain-End Drill Pipe, 7th Edition
GB50251－2003输气管道工程设计规范	ASTM A1047/A1047M-05(2009) Standard Test Method for
GB/T20801.5-2006压力管道规范 工业管道 第5部分：	Pneumatic Leak Testing of Tubing
检验与试验	API RP 574—2009 Inspection Practices for Piping System
SY/T 6553-2003管道检验规范在用管道系统检验、修	Components
理、改造和再定级》	API 577—2013Welding Inspection and Metallurgy
SY/T 5918-2004埋地钢质管道外防腐层修复技术规范	API 578—2010Material Verification Program for New and
Q/SY 93-2004天然气管道检验规程	Existing Alloy Piping Systems
SY/T 0087.1埋地钢质管道外腐蚀直接评价	NACE RP 0102—2010 In-Line Inspection of Pipelines
SY/T 0087.2-2012钢质管道及储罐腐蚀评价标准 埋地	API 1163—2013 In-line Inspection Systems Qualification
钢质管道内腐蚀直接评价	Standard
SY/T 0087.3-2010 钢质管道及储罐腐蚀评价标准钢质	NACE pub 35100—2000 In-Line Nondestructive Inspection of
储罐腐蚀直接评价	Pipelines
GB/T 15830-2008 [illegible]	NACE SP 0502—2010 Pipeline External Corr[illegible]

Overview of pipeline safety Inspection standard

U.S

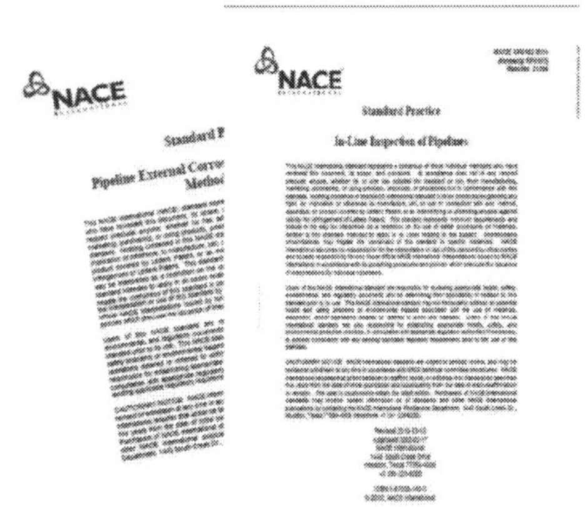

China

Comparison and analysis on typical standards

Here we selected four typical aspects to compare and analyze the difference between two countries. We will discuss and focus on inspection methods, testing location, inspection cycle, inspection operator requirement specifically.

(1) Inspection method

Category		Ch	
Online testing	Type of failure	corrosion of in... pi...	...sses of
	Main method	magnetic flux leakage(MFL), ultrasonic, eddy current testing and etc.	flux leakage(MFL)、 ultrasonic、 Magnetic tomography、 ultrasonic guided wave testing and etc.
	indicator	Thickness Variation, pitting	Thickness Variation
pressure testing	Type of failure	Failure depended on time	Failure depended on time
	Main method	strength test or leak test	strength test or leak test
	indicator	Thickness and crack of the pipeline	Thickness and crack of the pipeline
direct assess-ment	Type of failure	corrosion of inside pip... corrosion out of pip...	...n of inside pipe, and ...osion out of pipe
	Main method	ECDA、ICDA	ECDA、ICDA
	indicator	The maximal defect size	The maximal defect size

American new technology and method are not introduced into China

Basically the same

- The major difference: the new technology and method were not introduced into China's standards.
- The main reason: the standards are not updated on time.

China University of Petroleum(East China)

(2) Testing location

comparison	China（SY/T 0087）	US（NACE SP 0502）
pre-assessment	The partition of pipeline is according to the material of pipeline, construction factors	The partition of pipeline is according to the material selected.
Indirect detection	Choosing measured spacing small enough to satisfy assessment requirement, and the spacing should ensure the detection result available.	The same as China
Direct detection	Excavation points: according to 3 categories of detected points to decide excavation points; no requirement on non-detected points.	Excavation points: according to 3 categories of detected points to decide excavation points; if non-detected points are ensured, pre-assessment makes sure corrosion is most likely to be in ECDA area, at least one point needs to be tested. Direct testing location should be in ECDA area during pre-assessment.
later stage evaluation	Detection point choosing is not involved	Detection point choosing is not involved

difference

- The main difference: NACE SP 0502 stipulates the whole process of detection location in details with more consideration.
- The main reason: NACE SP 0502 updated regularly, but SY/T 0087 did not, which cause differences in detail.

China University of Petroleum(East China)

(3) Inspection cycle

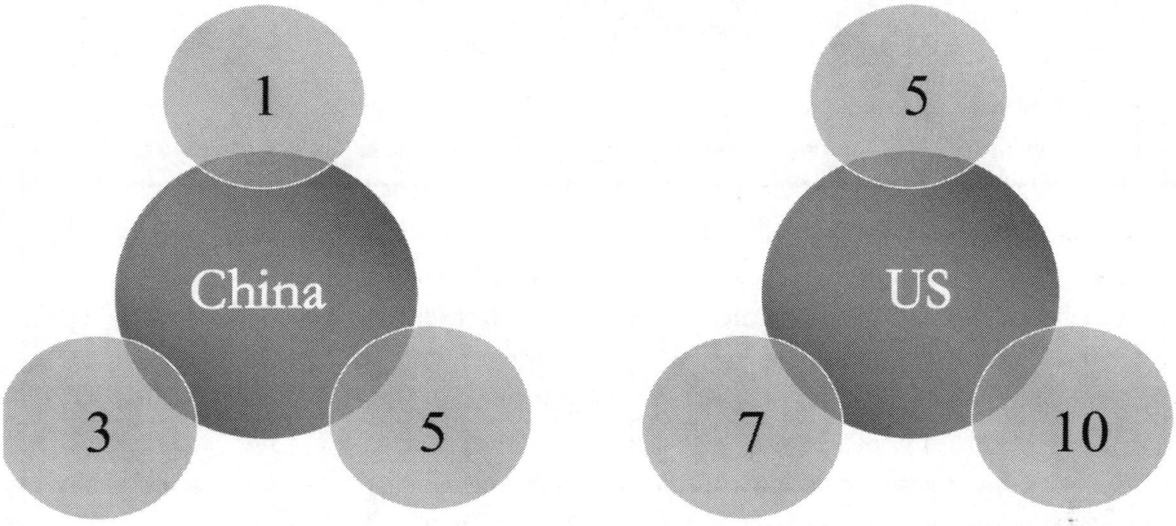

China University of Petroleum(East China)

- The main difference: China's inspection cycle is shorter than US's, the requirement is stricter.
- The main reason: this difference results from the gap of pipeline management level and pipeline production level between China and US and the actual state of pipeline operation.

China University of Petroleum(East China)

(4) Inspection operator requirement

Higher requirement

ECDA	GB/T19285	SY/T0087
The provisions of this standard should be applied under the direction of competent persons who, by reason of knowledge of the physical sciences and the principles of engineering and mathematics. Such persons may be registered professiona l engineers or persons recognized as corrosion specialists or cathodic protecti on (CP) specialists by organizations such as NACE or engineers or technicians with suit able levels of experience.	1)full-time staff appointed by owner of buried steel pipeline are responsible for the corrosion record of the corrosion and protection system 2)Overall inspect of the corrosion and protection system are conducted by qualified unit and professional staff .	Corrosion Engineers should be Involved in the testing, and the primary detection should be finished by corrosion and protection professional team. .

- The main difference: US standard has higher requirement on professional knowledge and ability .
- The main reason: the different background of experts who made standards.

Conclusion and suggestion

China has develop oil & gas pipeline inspection standards for several years, but compared with the United States this gap still exists.

CCPS CENTER FOR CHEMICAL PROCESS SAFETY

Conclusion and suggestion

China should update the content of safety detection standard timely, and adopt new technology and method so that the engineering application can offer powerful guarantee.

CCPS CENTER FOR CHEMICAL PROCESS SAFETY

Thank you for your attention!

境外石油管道项目公共安全预评估方法探讨

安聪[1]，赵东风[2]，陈爽[1]，胡苏[1]

（1. 中国石油大学(华东)机电工程学院，山东 青岛，266580；2. 中国石油大学(华东) 化学工程学院，山东 青岛，266580）

摘 要： 本文在定义海外管道项目脆弱性概念的基础上，将公共安全风险程度延伸为管道项目脆弱性等级程度。以脆弱性概念的两个维度结合项目本身计划投入、人员配备等情况以及项目所在地的环境脆弱性例如政治、经济、自然因素和项目因素等方面，建立脆弱性鱼骨图，全面辨识境外项目相关风险因素。最后，以苏丹境内中资企业实际管道案例分析脆弱点，辨识关键脆弱性因素，为境外管道项目安全决策提供依据。

关键词： 公共安全；脆弱性；石油管道；安全预评价

［中图分类号］ X937 **[文献标识码]** A

Security Risk Pre-assessment Method for Overseas Oil Pipeline Projects

Cong An [1], Dongfeng Zhao [2],Shuang Chen[1],Su Hu[1]

(1 China University of Petroleum (Huadong), Department of Safety Technology and Engineering ,Qingdao,Shandong,266580,China;

2 China University of Petroleum (Huadong), Department of Environmental and Safety Engineering ,Qingdao,Shandong,266580,China)

Abstract：In this paper, the security risk level is connected with the condition of the pipeline project vulnerability. Based on the identified vulnerability concept, a fishbone diagrams is proposed in two dimensions, which are sensitivity and capability, meanwhile each dimension is divided into social, economic and the project several aspects to identify the risk-associate factors completely. In the end of the paper, a virtual overseas oil pipeline project in Sudanese is chosen to demonstrate the model application to find out the vulnerability items, identify the key factor influenced the project security and mitigate the security risk.

Key words：security; vulnerability; oil pipeline; pre-assessment

随着国民经济快速发展，中国原油对外依存度逐年攀升。近年来中国积极开展国际化项目，上游勘探业务逐年攀升，原油输送业务需求不断加大。境外原油运输主要依赖海运和管道运输两种形式，然而海上运输容易受天气、政治、人为等未知因素影响，一旦发生事故对周边海域生态造成毁灭性打击，风险巨大。管道运输具有高效、低耗、连续输送和自动化程度高等优势，风险相对较小。因此中国正积极建设境外输油管道，为国内现代化建设运转保驾护航[1]。

中国原油进口来源区域主要分布在沙特、伊朗、伊拉克、苏丹等社会动荡的中东和西非等政治敏感地区，境外管道敷设不论在项目建设、管道运输以及后期维护方面仍存在风险。在项目实施前中国企业首先进行风险评估，但对于境外公共安全风险评估仍停留在对当地政府、政治局势等宏观把握上，并未深入、细致分析境外管道公共安全风险因素，准备不足，不能很好的支持项目进展[2-4]。因此，本文提出基于目前公共安全领域公认的脆弱性概念以及评估方法，在项目实施前详利用鱼骨图分析方法尽评估管道项目脆弱性影响因素，计算各环节脆弱度，评估境外输油管道风险。

1. 中国境外管道建设情况分析

1.1 中国近年来原油贸易

中国《能源发展"十二五"规划》以及中国统计年鉴（图 1）[5, 6]，指出由于粗放式发展导致我国能源需求过快增长，石油对外依存度从本世纪初的 26%上升至 2012 年的 58%。在面对日益增长的原油消费量，日趋稳定的原油生产量已不能满足需要，原油缺口逐年增大，海外原油进口势在必行，海洋运输影响因素繁多，风险较大，相对稳定的路上管道运输成为了更好的选择。因此境外原油管道建设对国内经济发展、社会稳定意义深远。

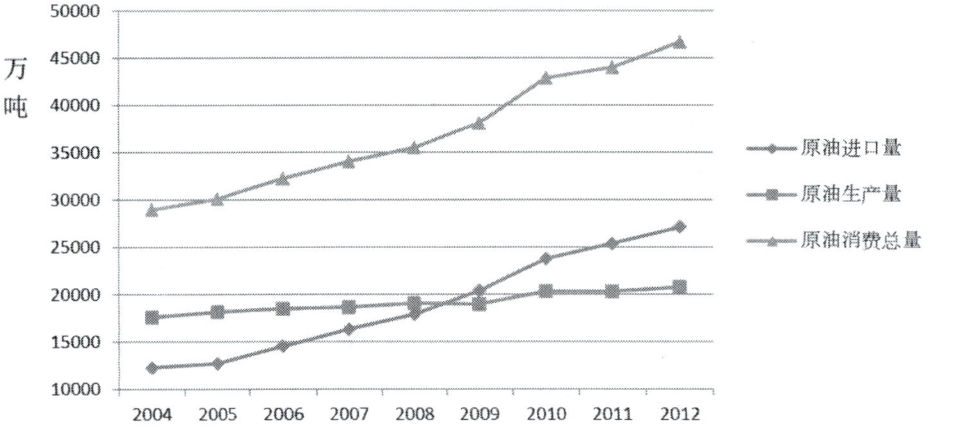

图 1：中国原油能源统计图

2.2 中国石油进口渠道

中国近年来实施油气进口多元化战略，建设了三大管运通道和海上航运通道形成中国油气进口的四大战略通道，如图 2 所示，即西北、中哈原油管道和中亚天然气管道；东北，中俄原油管道以及西南中缅原油和中缅天然气管道，降低马六甲海峡运油的风险。

同时中国境外勘探项目中，同样需要在本地建设配套输油管运至港口，输送往中国境内。因此在未来几年内，中国不论是境外跨国原油输送管道建设，还是境外项目区域管道建设都将进一步扩大，满足国内石油能源巨大需求。

3. 境外石油管道项目公共安全预评估

3.1 管道脆弱性评估模型

脆弱性最初源于自然灾害的研究，美国学者G. F. White1994 年在《Natural Hazards》一文中首次提出脆弱性概念。多个文献将脆弱性评估方法应用于公共安全领域，并取得了很好的效果[7-11]。本文将管道系统脆弱性定义为由于外界扰动使得管道项目引发事故的易损程度和控制事故进一步发展的能力匮乏，由此脆弱性评估蕴含着两个方面，一是在事故发生前薄弱环节的存在，事故发生的可能性；二是在事故发生后，能力不足导致的事故后果的严重性。公共安全事故中可能性与后果严重程度相互交错不能独立考虑，脆弱性能更好的展开评估。脆弱性评估也是风险评估的一种，将事故的可能性与后果的严重性的综合考量，以脆弱性程度表现出来。

管道系统中薄弱环节指管道项目周边环境以及本身特性使项目进展受阻的程度；能力匮乏指项目各方投入无法保证发生扰动后管道项目仍继续开展的能力。本文中新上管道项目脆弱性包括项目实施各个环节所处的环境脆弱性即各个区域社会、经济脆弱性以及项目本身脆弱性[12, 13]，如图 3 所示。脆弱性评估将以易损性和能力两个维度结合环境脆弱性和项目脆弱性两方面展开评估。

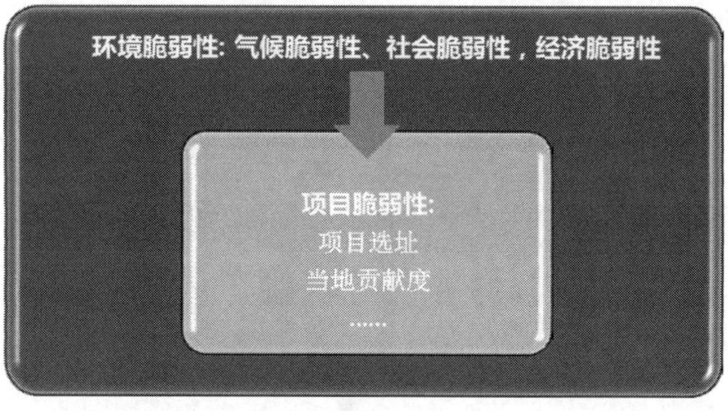

图 3 管道项目因素脆弱性分类

3.2 管道公共安全影响因素

中国石油石化行业海外项目所在的国家中有 29 个被外交部、国际专业安保机构评定为高风险国家，这些国家不同程度地存在着政局动荡、宗教冲突、恐怖袭击、绑架劫持、部落纷争、治安犯罪等公共安全风险。管道项目脆弱性预评估包括两部分评估：首先是依靠此管道进口原油供应链脆弱性分析其次是管道项目在建设、运行阶段脆弱性分析。本文中脆弱性分析采用管理方面广泛应用的鱼骨图分析方法，以脆弱性的两个维度即易损性和能力，在外界环境脆弱性和项目本身脆弱性等多个方面展开评估，综合考虑项目实际情况，找出管道系统的脆弱性影响因素。

3.2.1 供应链脆弱性分析

境外原油通过管道直接输送至中国或运输至所属地码头继而海运至中国过程中，发生意外导致管道供应链永久或者暂时中断[14, 15]。

管道项目供应链影响因素错综复杂，包括项目内外信息、物质以及资金等交错关系，必然会受到一系列不确定因素的扰动，有可能导致供应链系统的直接崩溃，因此在管道项目开展前，评估原油管道供应链的脆弱性程度是管道输油得以运行的基础，供应链鱼骨图脆弱性分析如图 4 所示。

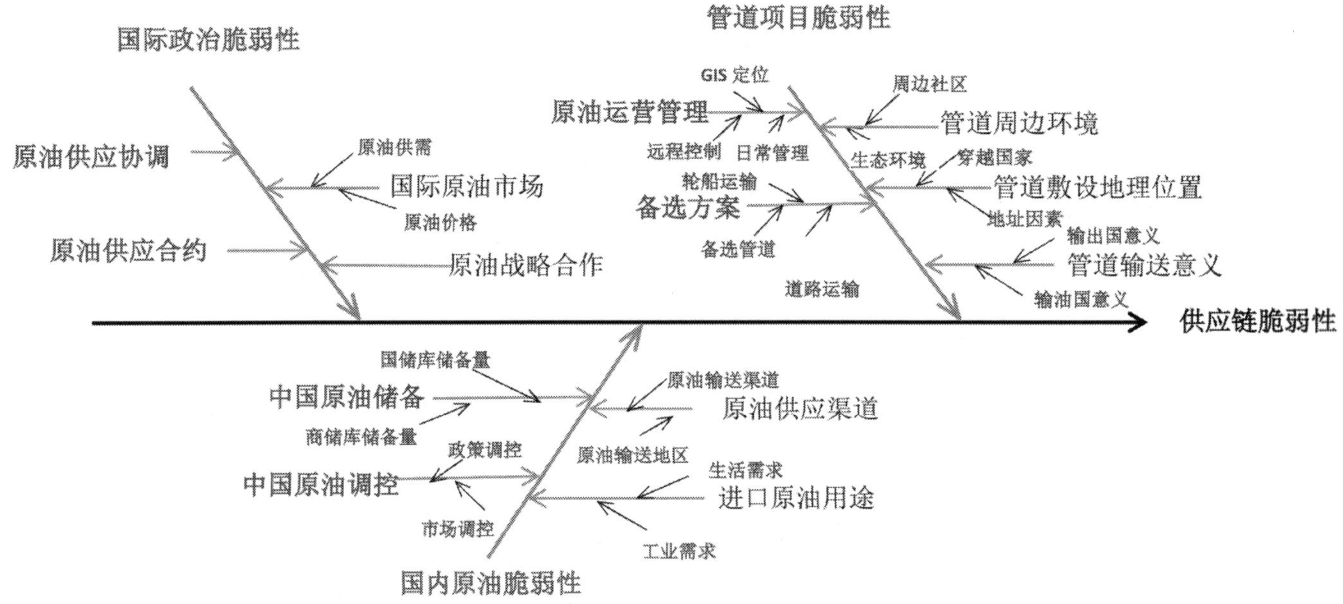

图 4：供应链鱼骨图

3.2.1 境外管道项目建设、运行脆弱性

境外管道项目建设往往需要投入大量的人力、物力、财力，项目施工区域大多在远离城市的荒漠、丛林中，道路交通差、自然条件恶劣、医疗卫生保障困难，同时这些地区政治局势、宗教势力错综复杂。项目建设过程中面临自然灾害、传染病甚至是极端恐怖袭击、绑架以及部落冲突等突发事件风险。同时输油管道项目在运行后面临着偷油、蓄意毁坏输油管道、袭击管道维护人员等事件[16, 17]。预先评估管道项目脆弱性程度可指导合理配备后续安保力量以及应急能力投入，保障原油顺利输送至中国，同时对保障境外人员、设备以及财产安全有重大的作

用。其中管道项目公共安全预评估中本文只重点考虑影响管道项目可顺利开展的脆弱性因素，尚未建设的管道项目应对能力方面脆弱性影响因素仅考虑目前可控因素。管道项目鱼骨图脆弱性分析如图 5 所示。

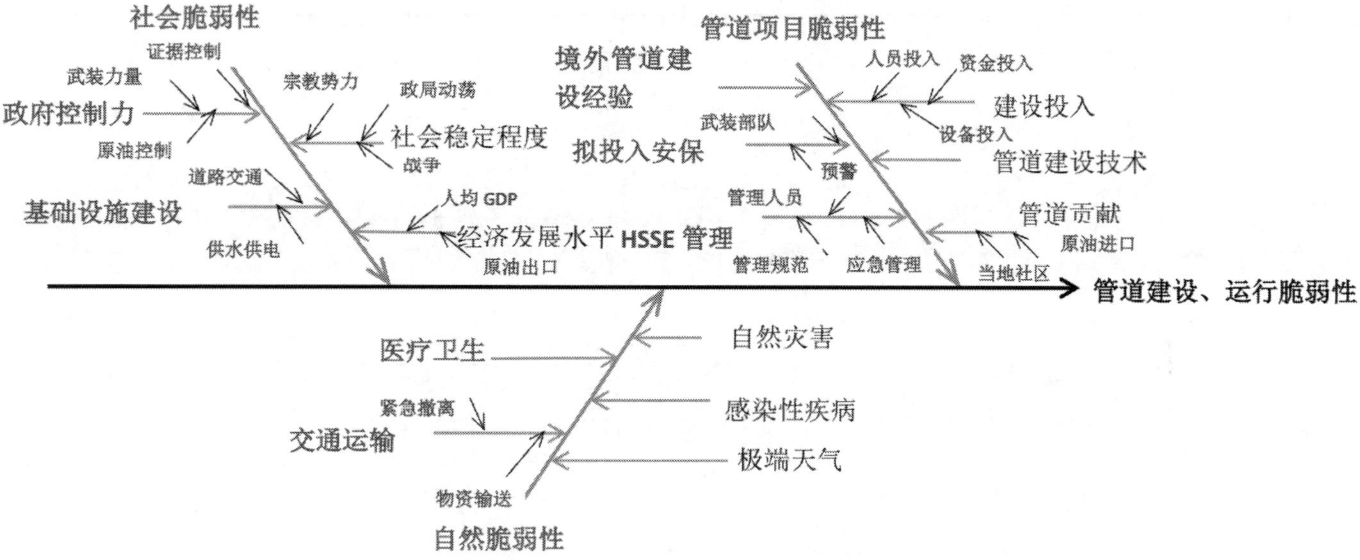

图 5：管道建设、运行脆弱性

3.3 境外石油管道定量评估

项目预评估阶段，资料较少，难以直接量化评估易损性以及能力，本文采用模糊层次分析法量化公共安全脆弱度，计算公式如下。脆弱性打分等级分为五级，打分标准如表 1 所示。在脆弱性鱼骨图辨识的脆弱性相关因素基础上，采用层次分析法分别计算脆弱性中各个因素的权重并对其脆弱性程度打分，计算各因素脆弱度，最终综合得出整个项目脆弱度，判断是否在可接受范围内，项目是否能够开展。其次对项目中各个影响因素脆弱度进行排序，辨识在项目运行过程中脆弱点，为管道项目顺利进行提供风险决策。

表1. 脆弱性分级标准

等级	描述	分值
I	脆弱性程度极低	0-20
II	脆弱性程度较低	20-40
III	脆弱性程度一般	40—60
IV	脆弱性程度较高	60—80
V	脆弱性程度极高	80—100

管道公共安全评价模型计算表达式为：

$$V_i = C_i \times W_i \qquad [公式. 1]$$

$$V = \sum_{i=1}^{n}(C_i \times W_i) \quad [公式. 2]$$

其中：

V_i 表示为各管道项目脆弱性影响因素脆弱度

C_i 表示为影响因素脆弱性等级打分

W_i 表示为影响因素权重指标

4. 苏丹管道案例分析

20 世纪初，中国石油业务在非洲蓬勃发展，中石油与苏丹签订合同（产品分成协议），原苏丹生产的原油主要出口到中国，目前苏丹为中国第七大原油进口国，管道输送业务量巨大。然而自 2011 年原苏丹分裂成南北两国，中国原油输送不定性增大，同时苏丹自然气候恶劣、传染性疾病感染率高，因此苏丹管道项目存在巨大公共安全风险。本文采用前文分析方法，结合苏丹本国政治、经济状况，自然气候以及公共卫生现状，预先评估苏丹管道项目公共安全脆弱性。结合管道项目建设方原有的项目经验采用上文鱼骨图综合评估辨识公共安全风险类型，类型汇总见表 2，对各类型影响因素易损性程度以及能力程度采用模糊评估方法计算脆弱度。

表2. 苏丹管道项目风险

风险种类	描述
重大事故	管道项目有导致当地生态环境破坏、原油泄露、爆炸等重大
社会安全	各类社会治安、恐怖袭击、战争和武装冲突、政局动荡等事
公共卫生	疟疾、霍乱、黄热病等流行性疾病风险
自然灾害	雷电、沙尘暴等自然灾害风险以及高温、暴雨、洪水等极端

根据海关总署统计数据、中国统计年鉴中国原油进口量以及苏丹原油进口量，苏丹政治局势相关资料，中国石油、中国石化公开发表管道建设以及和苏丹合作资料[18, 19]，结合鱼骨图相关因素辨识情况，确定脆弱性分值以及计算各因素权重，主要影响因素及说明列于表 3、4 中。

表3. 供应链脆弱度

影响因素	脆弱度	描述
中国原油需求	50*0.24	南北苏丹内战影响，中国苏丹原油进口量大幅下降，需求降低
管道运营管理	20*0.18	管道建设、维护技术成熟，偷油、漏油概率较小
管道敷设地理位置	60*0.15	南北苏丹政治局势不稳定，管道炸毁、破坏风险巨大
中国原油调控	70*0.11	原油弹性系数持续下降，调控能力较低

表4. 管道建设、运行脆弱度

影响因素	脆弱度	描述
管道投入	70*0.27	管道项目投入大量人员、资金以及技术，极易成为袭击、绑架、偷到事件目标，一旦发生事故后果较为严重
社会稳定程度	90*0.20	南北苏丹局势紧张，政局及其不稳定
医疗卫生	65*0.15	公共医疗卫生建设较差，管道项目
政府控制力	70*0.1	管道辖区内政府对局势控制力有限

通过计算得出，苏丹管道项目供应链脆弱性为III级，中苏两国原油供应不发

生突发违约停供事件，供应链脆弱性仍处在可控范围内，重点关注苏丹局势，选取其他相对稳定国家供油，满足国内原油需要。苏丹管道建设、运营项目由于目前局势紧张，战争频发，脆弱性为Ⅳ级，项目展开后加强安保投入以及日常管理，保证境外作业人员安全。

5. 结论

错综复杂的石油格局下，国内原油进口急剧不平衡，境外原油管道项目预评估显得尤为重要和迫切。本文在脆弱性评估基础上采用鱼骨图分析方法以及模糊层次量化计算，针对我国境外管道项目进行脆弱性预评估，将公共安全风险程度延伸为脆弱性程度。重点对境外管道项目脆弱性含义进行定义，以易损性和能力两个维度，针对外部环境以及项目本身两个方面辨识境外管道项目中脆弱性影响因素，提出采用模糊层次分析法评估各因素以及管道项目脆弱度，并选取苏丹管道项目预评估脆弱性程度，本评估方法为管道项目公共安全辨识以及安全管理决策提供了依据。然而本文在脆弱性量化分析上完全依靠事故经验和专家评估，评判上仍有不足，笔者将在后续预评估方法研究中重点改进。

参考文献

[1] 王玉生, 中国石油海外管道发展历程及启示[J]. 石油规划设计, 2010. 21(4): 第15-19页.

[2] 潘红磊, 境外企业社会安全事件调查分析方法研究[J]. 中国安全生产科学技术, 2012(05): 第121-124页.

[3] 辛辉, 境外项目公共安全管理[J]. 石油化工建设, 2013(03): 第52-54页.

[4] 王锐, 中国石油企业境外公共安全管理探讨[J]. 经营管理者, 2013(17): 第50页.

[5] 能源生产、消费总量和构成,中华人民共和国国家统计局[EB/OL].http://data.stats.gov.cn.

[6] 进出口主要国别总值,中华人民共和国海关总署[EB/OL]. http://www.customs.gov.cn.

[7] 吕稚知, 公共安全脆弱性评价现状探析[J]. 现代商贸工业, 2012(01): 第269-270页.

[8] 朱正威, 赵欣欣与蔡李, "脆弱性-能力"综合视角下的区域公共安全机理研究——以广东省为例[J]. 中国行政管理, 2013(04): 第113-118页.

[9] David A Moore. Security Risk Assessment Methodology for the petroleum and petrochemical industries[J]. Journal of Loss Prevention in the Process Industries, 2013, 26:1685-1689.

[10] Ebert. A. N. Kerle and A. Stein, Urban social vulnerability assessment with physical proxies and spatial metrics derived from air- and spaceborne imagery and GIS data[J]. Natural Hazards, 2009. 48(2): p. 275-294.

[11] Christian Delvosalle, Pipart C F A. ARAMIS Project A Comprehensive Methodology for the Identification of Reference Accident Scenarios in Process Industries[J]. Journal of Hazardous Materials, 2006,130:200-219.

[12] 王岩, 方创琳与张蔷, 城市脆弱性研究评述与展望[J]. 地理科学进展, 2013(05): 第755-768页.

[13] Smit B, Wandel J. Adaptation, Adaptive Capacity and Vulnerability[J]. Global Environmental Change-Human and Policy Dimensions, 2006,16(3):282-292.

[14]　宁钟，供应链脆弱性的影响因素及其管理原则[J]. 中国流通经济，2004(04). 第15-18页.
[15]　刘中伟，国际局势对中国石油安全的影响及相关战略分析[J]. 国际关系学院学报, 2011(05): 第79-88页.
[16]　李阳，我国国际工程EPC总承包项目风险管理研究[D], 2009, 长沙理工大学: 长沙.
[17]　戴联双，油气管道恐怖袭击风险评估方法与应用[J]. 油气储运，2010. 30(5): 第337-339页.
[18]　岳松伟，浅析中国石油在苏丹地区的市场环境[J]. 现代经济信息，2009(19): 第3页.
[19]　吴广义，苏丹油田安全管理与应急预案[D], 2010, 中国石油大学.

炼化企业机械完整性管理研究

李宇静[1]，刘义[2]

（1. 中国石油大学（华东），山东省，青岛市，266580；2. 中国石油大学（华东），山东省，青岛市，266580）

摘要：机械完整性管理已被世界各大石油、石化企业和危险化学品生产企业应用。炼化企业属于典型的连续型大规模生产类企业，生产过程复杂，资产和设备的投入高。本文介绍了设备完整性管理技术在炼化企业的应用现状，即采用技术改进措施和规范设备管理相结合的方式，以保证整个装置中关键设备运行状态的完好性。简要介绍了炼化企业现阶段的现代化管理技术，分析其优缺点，并提出了将机械完整性管理与现代化设备管理技术相结合的理念。

关键词：炼化企业　机械完整性管理　现代化管理

Refinery Mechanical Integrity Management Research

Li Yujing[1]，Liu Yi[2]

（1. China University of Petroleum，Shandong Province，Qingdao，266580，China；2. China University of Petroleum，Shandong Province，Qingdao，266580，China）

Abstract： Mechanical integrity management has been applied by The world's major oil and petrochemical enterprises and dangerous chemicals production enterprise. Refinery business is a typical continuous large-scale production enterprises, whose production process is complex and investment of assets and equipment is high.This paper describes the application status of equipment integrity management technologies in refining enterprises, which combination of technical improvement measures and standardize equipment management to ensure the integrity of the entire device operating status of key equipment. It introduces the refinery's current modern management technology briefly, analyzing its advantages and disadvantages. And it proposes the concept that combines mechanical integrity management with modern equipment management technology.

Key words： petroleum refinery　Mechanical integrity management　modernization management

目前，全国各地炼油化工基地正在不断的建立，要建立新的生产装置，其很重要的一部分是选用符合生产运行条件的机器设备，且这些设备在企业各方面的发展中起到关键性作用。设备是人们在生产和生活上所需的机械、装置和设施等，可以支持长期使用，并在使用中基本保持原有的功能和实物形态的物质资料和劳动资料[1]。我国的炼化企业均具备良好的设备管理传统，在生产运营过程中积累了丰富的管理经验，拥有一支技术过硬的管理和维修队伍。随着企业现代化设备管理技术的全面发展，信息化管理、状态监测应用等的普及，企业在设备管理方面又有了新的发展与进步。

1 机械完整性的基本描述

机械完整性也称设备完好性，是指从设备首次安装起一直到其使用寿命终止的全生命周期内，采取管理措施保持工艺设备处于可满足其特定服务功能的状态[2]。它是确保设备正常运行的重要手段。机械完整性属于预防性维护，它可以确保可靠的、无故障的运作。对机械完整性中所涉及的设备是一个广义的概念，其内容涵盖了设备设施、管道线路、电器仪表、安全预防监控系统；如高能动设备、泄放和通风系统、紧急停车系统、气体检测报警系统、消防设施、防雷防静电系统、自动启动、关闭、停止装置等，一旦该设备失效或故障，会引起过程安全事故[3]。

机械完整性是风险管理中的一个重要环节，它贯穿于设备从设计、制造、安装、使用、维护保养、报废整个生命周期[3]。科学的管理设备的生命周期，即便设备达到生命周期依然能够正常运转也应报废，因为在该状态下的设备发生故障的概率很大且很难预测和控制，从设备出发保障全过程的安全。机械完整性管理的意义在于防止因设备故障致使工艺过程由于超压、超限、泄露等引发事故。机械完整性的增强是保证设备利用率的根本所在。机械完整性的提高是防止和减少工艺过程安全事故的基础。

2 机械完整性的应用现状

2.1 机械完整性应用的国外现状

Reynolds J在1996年提出在石油工业中基于风险检验方法的机械完整性合理改进。19个石化公司赞助的项目，开发基于风险检验（RBI）的机械完整性方法应用于炼油和石化行业。发展行业标准描述并给出了RBI方法的更新进度，RBI方法允许所有公司通过技术来获取利益。第一个试点项目回顾运用RBI验证方法的几个现有炼油装置。压力设备的故障的处理单元中，可以有多个不良影响。API RBI评估如何失效的压力边界的潜在后果，并评估故障的可能性[4]。

F. Walter Pinto在2008年提出了描述对于石化设备中静设备机械完整性要素。文中涉及了地址管理系统、工程实践、预防/预测性维护/检验系统、性能指标和资源。共享了利安德化学公司的设备机械完整性，且大量的最佳实践开发出一部分可行性计划。同时共享一些成功故事和经验教训[5]。

2.2 机械完整性应用的国内现状

陈炜，庄力健，朱建新等在2012年提出构成石化系统的三个重要组成部分是静设备、动设备及仪表联锁。静设备、动设备与仪表联锁三个体系中某一环节的非正常动作都可导致系统的失效，最终致使了压力类设备的失效，因此，简单地对设备进行一个个的评估，不一定能检测到系统中全部的失效可能。提出了一种基于石化装置系统RBI－SIL分析的压力设备完整性评估技术的构思，可以更有效、更准确地评估系统存在的风险，查找系统出现风险的起源，制定更有效、优化的降险措施，最大可能地预防这种风险传递而发生压力设备失效事故[6]。

苏建峰，张建军，李智勇等在 2012 年针对电气设备管理和预防性工作的现状，提出了建立基于状态评估的电气设备完整性管理体系。通过对预防性维护数据和设备运行、监测数据进行分析，构建设备的状态评估系统，依据设备状态评估、风险评价、检维修策略建议和状态检维修决策结果，尝试把设备周期性检维修转变为状态检维修，实现对电气设备的完整性管理[7]。

刘雪峰在 2014 年提出了实施管道完整性管理减小线路风险的举措，探讨了运用管道完整性管理理念和手段控制管道线路风险的具体做法：实施管体缺陷修复，加强管道巡护，保证管道运行压力平稳调整，做好应急准备，预防打孔盗油案件发生，加强对第三方施工管控[8]。

这些现状表明，机械完整性管理技术在国外企业运用的已经相当成熟了，而在我国，机械完整性管理技术的推进才起步不久，虽然已经应用到一些设备或流程的管理当中，但还需要进一步广泛推广，可以将其余企业现有的设备管理技术相结合，制定出适合各企业具体情况的机械完整性管理技术。

3 炼化企业现代化设备管理技术

炼化企业的设备数量和复杂程度都不是其它行业可以比较的，特别是伴随设备的自动化、智能化的发展趋势，设备管理日益细致，传统的管理模式已经满足不了现代设备管理的需求，设备的现代化管理技术营运而生，并在我国得到了广泛的应用。

在信息化管理建设方面起步较早的以中国石油、中国石化为代表的炼化企业，计算机网络及硬件配置已基本满足实施信息化管理的要求。近年来，石油、石化两大公司所属各炼化企业依据自身实际需要应用或开发了许多设备管理方面的软件系统，如ERP、EAM、PMIS、特种设备安全管理、RBI评估软件、机组状态监测系统等等，这些管理系统在企业全方面的设备管理中起到了重要作用[9]。

3.1 现代化设备管理技术内容

通常的现代化设备管理系统一般涵盖以下几个方面。

（1）企业设备固定资产和相关技术：设备管理的基础是建立企业设备信息库，实现设备选型，设备购买和安装调试，备件管理和定期检维修以及设备报废流程的自动化。同时需要记录各设备全生命周期的运行状态，包括技术参数状态、检维修和保养记录等[10]。

（2）设备故障和设备事故处理：记录企业生产设备发生过的故障，同时必须跟踪设备的维修和故障处理的流程[10]。

（3）设备档案管理：完整的记录和整理设备在生产过程中全生命周期活动相关的档案信息[10]。

（4）检维修管理：制定设备定期的检维修计划，并跟踪记录计划流程的执行情况，建立报告和任务提醒[10]。

（5）编制设备维修计划：可根据设备运行的记录和监测当前状态，自动制定设备维修计划，实现以设备的关键度条件为依据来确定设备的维修计划[10]。

（6）备件管理：建立设备备件的详细信息库，并能够制定备件计划和备件安排流程，根据备件数量自动生成采购安排，并设置设备与备件的关联关系[10]。

（7）设备维修及成本核算：根据使用设备的人员、工时、消耗的原材料等计算设备的产出与成本的数据关系[10]。

（8）设备缺陷分析管理：建立完整的设备故障分析体系，记录每次发生故障的信息，并分析原因，生成方便以后使用的故障分析数据库和专家信息库[10]。

（9）报表管理：实现自动生成满足各管理人员需求的设备生产过程的各种报表。统计报表可完成查管询功能，能够生成设备的检维修统计和故障分析报表等[10]。

3.2 现代化设备管理的实际意义

（1）建立整个中国炼化企业范围内的设备管理体系。实现在企业范围内设备的数据共享，以便提高企业设备的运行维护能力，实现企业各部门及时进行数据上报，为企业管理决策提供有效的技术支持[11]。

（2）控制和跟踪设备检维修流程。在企业范围内实施检修维护体系，实现对检维修项目的费用、备品配件、进度的控制和跟踪，通过网络技术手段，实现各检维修信息的无保留和即时传递[11]。

（3）支持设备管理的持续改进与持续发展，有效增强设备管理标准化、规范化的运行，为累积设备管理经验、使用更先进的管理技术、优化管理流程和持续改进管理提供技术支持[11]。

（4）提高企业的综合分析和决策支持能力，对数据进行记录、整理和加工，形成设备完整的信息库，再对所得的信息进行分析、判断和整合，得出有价值的基础数据，进而得到直观的分析图表、曲线，对正确决策提供有效的支持与帮助，并确保各项指标的及时性、准确性和有效性[11]。

（5）实现企业业务流程的自动化管理，用储存的电子信息替换储存的纸质信息，以此实现设备数据的网上提交、审批、反馈、分析等功能，有效提升相关部门的工作效率，改善相关部门之间的协调合作[11]。

3.3 炼化企业现代化管理存在缺陷

（1）组织机构设置不合理，不利于调动员工的积极性[12]。炼化企业的设备管理组织未有效地将员工职责方面的能力运用到实际的工作中去，仅是一个执行的组织。

（2）对设备管理重要性认识不足，员工的设备管理意识薄弱[12]。关于工艺、操作人员，对于设备相关的技术知识培训工作进行的不彻底，当设备出现异常和发生故障时，无法及时发现、分析判断和积极的应对处理；缺乏系统的、完整的改革设备管理流程，与其相关的考评体系也很不完善，无法提高设备管理在企业管理中重要作用。

（3）目前，我国炼化企业现代化设备管理不规范、标准不统一，各系统间数据不能及时共享，限制了管理系统功能的发挥；大多数管理系统没有先进的设备管理模式和管理方法的指导，系统开发商缺少了解和研究炼化企业设备管理的特点与要求，仅将设备管理流程计算机化，很难从本质上提高企业设备综合管理水平[11]。

（4）检维修计划编制与事实不符[12]。在炼化企业设备检维修管理中，对设备的检维修计划不完善，造成一些设备的综合效率很低，配件供应和检维修保养非常不易，故也使维修人员的工作量、维修费用及设备的故障率加大了，最后减弱了设备保障生产的能力。

4 机械完整性管理与现代化设备管理相结合

由于炼化企业现代化设备管理体制还不够完善，本文提出将先进的设备管理方法—机械完整性管理与现代化设备管理方式相结合，使得炼化企业设备管理更加全面，更加可靠。

4.1 完整性管理体系建设

实现炼化企业设备全生命周期的管理，提供全面的设备信息支持，构建规范化的设备管理系统，实现设备从审查设计、定货、采购，到使用、维护、检修、报废等全过程管理。机械完整性的增强是确保设备利用率的基础，设备可靠性的提升是防止和减少安全事故的基础。

（1）设备的选择[3]

设备的选择是机械完整性管理体系的首要方面，所选的设备需要符合国家相关的法律法规，经济合理，仪器仪表、安全监控系统、有毒有害物、泄露探测装置、火情探测和灭火系统、照明及通风系统等齐全。

（2）人员培训及资格考核

企业各级设备管理部门有计划地开展人员培训工作，以确保培训工作有效实施，保证设备操作、维修人员的岗位技能，提高他们的业务水平和综合能力。所有设备操作、维修人员在培训后，均需通过各种考核，以确定培训的有效性或资格的认定。

（3）检查检测和预防性维护

为了使设备符合应有的要求而进行检查检测和预防性维护，提出作业流程和规划。作业流程分为执行和监控两部分的内容[3]。对所有设备制定预防性维护计划，可以根据设备关键性或设备类型将预防性维护计划进行分类管理。确定作业频率和时间安排及执行的顺序。

（4）机械完整性作业规范

机械完整性管理体系包括规范的编制、指导意见及修改说明等且应与企业自身的管理规范保持一致。形成文档的作业程序明确机械完整性管理计划；相关部门人员的职能和职责；质量保证规程；维护维修规程；检查测试规程；预防性维护规程等内容[9]。

（5）缺陷管理[3]

通过检查检测一定会找出有缺陷的设备，依据设备各参数规定正常设备的可行性标准；定期对各设备运行情况进行评估；分析缺陷设备的缺陷状态；制定和实施解决缺陷的方案对策；将设备缺陷和相应的措施通知可能受到影响的人员；及时排除设备缺陷并对其效果进行跟踪。

（6）故障原因分析

分析设备故障原因，找出问题所在，采取预防措施，避免再发生。原因分析也是设备完整性管理体系PDCA循环的一个重要组成部分[13]。

4.2 现代化技术结合机械完整性管理体系

炼化企业现代化设备管理技术结合机械完整性管理体系，运用机械完整性管理技术统一制定企业的标准化规范，再用信息化的手段对设备管理进行分级管理，实现对设备的生命周期各阶段的数据和管理功能的需求分析，设备的数据录入、编辑、浏览、查询、数据的导入和导出、历史查询、设计图纸与技术资料文档存储和管理等功能。

管理体系是以面向全生命周期管理为原则的综合管理体系，即通过一定的技术、经济、组织措施，对设备的规划、设计、制造、选型、购置、安装、使用、维护、修理、改造、更新直至报废的整个生命周期进行综合的科学管理。在此基础上，再融合先进的风险分析技术软件，RBI分析软件、RCM分析软件和SIL分析软件等，分别对静设备、动设备和电气仪表控制系统进行风险分析，继而实现设施设备的检验检测和维护的计划制定、审评、检验检测和维护的实施、验收和后期跟踪等流程管理及相关数据和文档资料的存储和管理。通过现代化设备管理信息系统化技术，再运用机械完整性管理的先进技术，提高设备使用效率，减少设备的维修的次数、投入人员和维修资金，也使设备风险降到可接受的范围，更全面实现设备的机械完整性现代化管理。

5 结论

炼化企业进行机械完整性现代化管理之后，实现强化计划管理、定额管理，统一调配，合理储备。机械完整性现代化管理可以系统科学的选择将要购买的设备，并合理地安装设备，按照规程来运行设备，确保能够定期对设备进行检查维修，实施备品配件管理，能全面完整地收集和分析数据，及时共享多项信息，增加管理透明度，降低设备的故障概率，良好地生产运行。能够减少库存费用，节省采购费用，大大降低了生产成本。增强了数据的权威性、合理性、准确性、及时性，很大程度上地提高了工作效率和质量。通过机械完整性现代化管理，炼化企业有可能实现设备管理自动化，还可以大大减少人工操作的时间，提高操作的准确性，是炼化企业设备管理的主要发展趋势。

参考文献

[1]沈晓飞，夏翔鸣，朱晓东. 炼化企业设备管理模式研究[J]. 石油化工管理工部学院学报，2004（3）：13-18

[2]刘义. 机械完整性管理——过程安全管理的重要因素[J]. 劳动保护，2013（3）：102-103

[3]袁树平. 浅析机械设备完整性管理是企业安全的重要保障[J]. 科技资讯，2013（23）

[4]Reynolds. Mechanical integrity compliance improvement with risk-based inspection methodology in the petroleum industry[A]. API Safety and Fire Protection Conference，1996

[5]F Walter Pinto. Mechanical integrity—stationary equipment reliability[J]. Process Safety Progress，2008，27(2)：105-111

[6]陈炜，庄力健，朱建新. 基于（石化装置）系统RBI—SIL分析的承压设备完整性评估技术[J]. 压力容器，2012，29（9）：43-49

[7]苏建峰，张建军，李智勇. 基于状态评价的电气设备完整性管理[J]. 油气储运，2012，31（5）：397-399

[8]刘雪峰. 实施管道完整性管理消减线路风险的措施[J]. 石油工程建设，2014，40（1）：88-90

[9]吴刚. 炼化企业设备管理信息化建设的现状及建议[J]. 当代化工，2007，36（1）：101-103

[10]蔡畅. 企业设备信息管理系统[D]. 成都：电子科技大学，2013

[11]殷启超，张永海. 炼化企业设备信息化管理综述[J]. 科技信息，2012，（15）：472

[12]马亚丽. 浅谈当前石油企业设备管理中存在的问题及对策[J]. 经营管理者，2012，17：133

[13]范旭潮. 浅析设备完整性管理体系的建立与实施[J]. 机电信息，2013（6）：167-168

2nd CCPS
China Conference
On Process Safety

Effective Off-site incident emergency response plan development

Qingdao, China
August 28 - 29, 2014

Allan Su

AcuTech Consulting Group

CCPS CENTER FOR CHEMICAL PROCESS SAFETY

Index

Typical process to develop the Emergency Response Plan(ERP)

Difficulties to develop off-site ERP

Each step samples

Key learning

CCPS CENTER FOR CHEMICAL PROCESS SAFETY

Typical ERP procedure

- Collect related information for facility, process, material, resources, impact area
- Develop the most severity scenarios could happened
- Base on different scenarios and available resources develop the various emergency response plan for different cases
- Consolidate the plans for different group as response actions.
- Drill the plan to ensure the effectiveness and efficiency
- Regular audit to ensure ERP is workable and effective

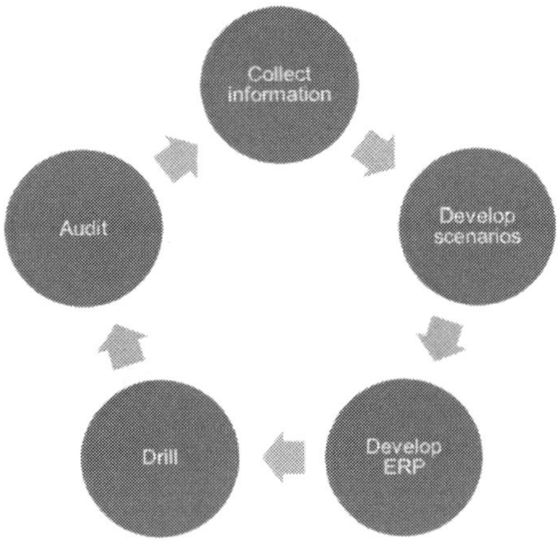

CCPS CENTER FOR CHEMICAL PROCESS SAFETY

Why off-site ERP is difficult

- Involve multi-parties
- Public impact
- Information not sufficient or clear
- Resources (hardware and knowledge) split in multi-parties

CCPS CENTER FOR CHEMICAL PROCESS SAFETY

ERP develop consideration

- Base on the correct information
- Base on professional knowledge
- Rapid response from multi-resources
- Effective Drilling
- Auditable

CCPS CENTER FOR CHEMICAL PROCESS SAFETY

Information collection

- Authority control the information registration for all off-site facility and piping.
- Company could apply for updated information related with off-site facility/piping
- Regular update the piping network information to related company

- Authority (government/ management department) take more responsibility on collecting off-site facility information due to off-site is the “sharing area” for multi-parties.

CCPS CENTER FOR CHEMICAL PROCESS SAFETY

Worst scenario development

- Scenario should developed base on both Safety and Security Risk Assessment
- Mainly initiated as leakage of chemical
- Severity should be developed base on Quantitative Risk Assessment (QRA) and consequence modeling.

Risk Assessment should be responsible by company due to they have most knowledge of hazard resources/process/control method.

Authority should guide on methodology used for risk assessment.

Stricter risk tolerable level should be used for public related event

CCPS CENTER FOR CHEMICAL PROCESS SAFETY

Develop the plans

- Plan should be stick with scenarios and consequence modeling results.
- The Plan will be adjusted base on real incident situation (weather, wind direction, etc.), the plan should give the guidance upon these adjustment? The help-decision tools should be developed if it is necessary.
- Un-available resources should be provided by company ， authority will coordinate different sites to share these resources.
- The Scenarios and Plan registered in authority and related parties (fire-fighting team, hospital, etc…)

Effective communication

- Code the company, scenario and response plans for clearly communication.
- Setup the standard data collection sheet for information collection and communication
- Set the clear threshold for further actions (Evacuation, Area Isolation)
- Plans need to consolidated into immediate actions for each departments after incident.
- Capable Emergency Management Center (communication, field command)， provide the decision-making tools and knowledge.

ERP Drilling

- Table drill for communication and different level threshold
- Actual drill for special skills (could be in other area)

- Authority will take the main responsibility for arrange the drill.
- Table drill will be the major method for verify the plan's applicability by multi-parties

Audit

- Audit will be conducted for company who will keep all related information/documents
- Audit to make sure the ERP is developed and drilled following the guidelines
- Audit to make sure the ERP is up to date.

Key Learning

- The ERP plan should be developed by the party who have most knowledge for hazard resources.
- The communication difficulty could be solved by special plan made and pre-defined communication method. (scenario code, information collection sheet, clear response actions, threshold)
- Table drill could be used to test the plan by multi-parties.

Thanks

Lessons Learn from Ptw (Permit To Work) Failure Incidents and Best Practices Sharing on Improvements

Su, A., AcuTech Consulting Group

Allan Su
Senior Engineer AcuTech Consulting Group
Asu@acutech-consulting.com

The permit to work program is the key defense for chemical site abnormal operational incident. However it involves many related parties, abnormal operation, needs to conduct the complex isolation process, etc, the incident caused by permit to work procedure failure still happened frequently in chemical site.

The article shares 13th incidents which related with Permit to Work procedure failure and discuss the most easy failure parts of PTW procedure and share the best practices of preventing these failures.

The Studies of Flame Propagation Simulation in the Pipeline Base on Fluent Software

HAN Feng-lei[1], XIONG Hao[1]

(1 China University of Petroleum, Qingdao, Shandong, 266580, China)

Abstract: The flame is dangerous, and flame propagation can create more damage. However, the generation and propagation of the flame is conditional and it's easier to realize in a closed or half-closed space such as pipeline. That '11.22' Huangdao pipeline explosion accident showed damage of the flame in a pipeline. By studying the flame of methane-air premixed in underground drains using Fluent software for numerical simulation and analyze that results. This article draw the flame propagation state of methane-air premixed in underground drains in different conditions and the impact on external space.

Designing a Pipeline Asset Management System Using a Continual Improvement Life Cycle Approach ñ Applying Standards, Practices & Guidelines to Prevent Process Safety Incidents

Sun, J., ATKINS
Jones, D. A., AYKINS
Sapkota, B.,

Assurance of safe and reliable operation of pipelines for hazardous materials can be achieved by implementing an asset management system based upon industry standards, practices and guidelines. A well-designed asset management system includes provisions for continual improvement of the life cycles of the management system itself (design/development, implementation, auditing), the pipeline assets (design, construction, commissioning/start-up, operation, maintenance), and the control of risks to people, the environment, assets and delivery of products (before, during, and after incidents).

To minimize the risks of pipeline incidents and to maximize the performance of the pipeline assets, continual improvement must be sustained by managing the quality of resources and activities throughout each life cycle. Specifically, management processes are necessary to assure the competence of people, the availability of reliable information, and the integrity of pipeline assets, as well as to assure the control of work activities.

Based on proven quality management principles, this paper will present a plan-do-check-act life cycle approach to designing a risk-based asset management system aimed at preventing and mitigating process safety pipeline incidents. This paper will show how principles from available international standards, industry recommended practices, and technical society guidelines can be incorporated into the design of the asset management system.

This paper will present core processes for: (1) managing process safety to prevent the occurrence of incidents, (2) managing emergencies to prevent the escalation of incidents, and (3) managing investigations to prevent the recurrence of incidents. These processes will address risk assessment and control, pipeline integrity management, human factors, root cause analysis, and management of change. This paper will also show how supporting management processes are integrated into the asset management system to sustain its effectiveness: leadership and management commitment, resource planning and organization, communications, document and records control, performance monitoring and evaluation, and corrective and preventive actions.

Corrosion Monitoring and Controls Technology

Lisa C. Hutto
Chilworth Technology
113 Campus Drive, Princeton, NJ 08540, USA
Lisa.Hutto@Dekra.com

Ethan Zhang
Dekra China
800 Dongfang Road, Pudong, Shanghai 200122, China
Ethan.Zhang@Dekra.com.cn

UNPUBLISHED

Controls Monitoring and Controls Technology

Lisa C. Hutto
Chilworth Technology
113 Campus Drive, Princeton, NJ 08540, USA
Lisa.Hutto@Dekra.com

Ethan Zhang
Dekra China

Keywords: Corrosion Control, Corrosion, Catastrophic Failures, Monitoring Technology

Abstract

Equipment and pipeline corrosion monitoring and controls technology is of paramount importance for all chemical and refining operations. Maintaining equipment integrity is essential for not only safety but also for operating a business to the best potential possible. There are many options available for corrosion monitoring and controls in each industry. While each process may be different, it is up to the facility to determine their needs and which type of technique works best for their process. There are many options available for corrosion measurement to include, non-intrusive, intrusive, in line inspection, chemical testing, and electrochemistry. The proper procedures and monitoring can be the determining factor on how the facility can minimize downtime as well as preventing catastrophic equipment failures. This paper illustrates the various corrosion monitoring techniques available and the need for control monitoring techniques. The options for the facility are clearly defined as to advantages and disadvantages along with a clear description of what each method can offer. This paper, therefore, seeks to emphasize that there are many options for a facility to choose from in corrosion control and how to utilize available monitoring techniques.

1. Introduction

Failure to manage corrosion of plant equipment is often the main cause of catastrophic failure of this equipment. There are several reasons for these failures often ranging from lack of understanding, costs, risk management, and use of equipment that is not compatible with the process. Corrosion measurements are necessary to determine corrosion rates in a facility. Once this data is generated the facility can use it to estimate how long the process may operate safely. Remedies can include temporary repairs or a complete shutdown if the failure is on the edge of catastrophic. Measuring and monitoring the corrosion rate allow the facility to determine the most cost-effective way to control costs associated with plant operations. The results from corrosion inspection and measurement can be used to prevent catastrophic failure and maintain safe, reliable operation of their industrial equipment.

A facility must manage corrosion monitoring and control, take the effort to determine timelines for testing, methods for inspection and determine how to control the mechanical integrity of their process with the wide availability of options that are available.

2. What is corrosion monitoring and why should a facility implement it?

Corrosion measurements include a wide variety of techniques to determine how corrosive an environment is, and the rate at which metal loss is occurring. Corrosion measurement also evaluates the effectiveness of corrosion control and prevention techniques and provides the data regarding the effectiveness of a particular corrosion control and prevention method.

3. Why perform corrosion control?

On August 6, 2012, a large fire and explosion erupted at about 6:15 PM due to the release of flammable vapor at the Chevron Refinery in Richmond, California. **(Figure 1)** After eight months of interviewing, reviewing records and investigation, the US Chemical Safety Board (CSB) released their preliminary report. The fire was sparked by a pipe that had been thinned by corrosion from the heavy crude the refinery was utilizing. The piping was recommended to be replaced in 2002 when it was discovered, but Chevron continued to operate with the degraded piping after discussing the risk hazards. The extent of the 2012 incident was potentially made worse because the refinery continued to operate for another two hours after the leak was discovered and the operators and supervision debated what to do about it.

The Chevron Richmond incident encouraged the California division of Occupational Safety and Health to give Chevron the maximum allowable fine, also the largest in state history, for 11 willful violations. OSHA found that Chevron disregarded obvious hazards and continued to operate. The CSB cited the Chevron Richmond Refinery for chronic failures to replace aging equipment. This finding also initiated a request to overhaul regulatory control of the industry to prevent such accidents from happening again.

Figure 1 - Commuters step out of their cars to take pictures of the fire raging within the Chevron oil refinery on Aug. 6, 2012

Every company knows that inspections can be expensive. Inspections take time and usually require production shutdown that results in lost profits. Then, there's the mindset that, "we can make it until the next turnaround", or "it's really not that bad of a leak" or "why do we need to inspect, it looks fine on the outside." This type of thinking is why we have seen industry incidents such as Chevron Richmond – August, 2012, NDK America Inc. – December, 2009, Carbide Industries – March, 2011 and Jaipur Oil Depot Fire, August, 2009.

Companies should perform corrosion control to prevent failures that can lead to leaks, spills, fire, explosion, injuries and fatalities. Based on past incidents, internal corrosion has proven to be the leading cause of pipeline failures. It is a known fact that pipelines are difficult to inspect which often results in a failure to conduct an inspection. Problems have to be addressed in a proactive manner, rather than waiting until the equipment or piping completely fails.

4. What are some ways corrosion monitoring techniques can help a facility?

If a company does take the time to implement a good mechanical integrity program that includes performance of corrosion control and monitoring several benefits can be expected to include:

- Providing early warnings that failing process conditions exist due to corrosion related failure.
- Allowing an analysis of process parameter changes and their effect on the corrosivity.
- Diagnosing and identifying the cause and rate of the parameter changes.
- Evaluating the corrosion control effectiveness and providing information relating to the ongoing condition of the facility.

5. Monitoring Technology

There are many different options for a facility when it comes to corrosion monitoring techniques. Standard methods used in industries today:

- Corrosion Coupons
- Electrical Resistance (ER)
- Linear Polarization Resistance (LPR)
- Galvanic (ZRA)

5.1 Corrosion Coupons

Corrosion Coupons are one of the better known and simplest of methods for monitoring. The technique exposes a sample of material (the coupon) to the process environment for a predetermined time, then the sample is removed an analyzed for weight loss. The total weight loss from the sample during the specified time determines the corrosion rate. Coupon monitoring is most effective in areas where corrosion rates do not significantly change over a long period.

Advantages:

- Visual inspections are complete.

- Easy to observe and analyze any deposits.
- Corrosion rate easily calculated.
- Very simple test, sophisticated instruments not required

Disadvantages:

- If a corrosion upset occurs during the testing period, the coupon alone cannot identify the time of occurrence of an upset.
- Does not always give corrosion rate values with the exact pipe wall loss. Coupons only offer an estimate and not real metal lost.
- As pipe surface deposits increase, the actual corrosion rate versus what the accuracy of what the coupon measures significantly decreases.
- Coupons need flow to work. Measurement is not possible on no flow piping such as bypass lines, out of service equipment, or lines redundant system piping.
- Unable to detect pitting due to the coupon being a new and smooth sample and the test subject have some wear or pitting.

5.2 Electrical Resistance Monitoring (ER)

Electrical Resistance Monitoring (ER) probes are considered to be an "electronic coupon." ER probes provide the same basic measurement of metal loss just as Corrosion Coupons, however, the difference is that ER probes can measure metal loss at any time during the testing. It is the only on-line, instrumented technique applicable to almost every type of corrosive environments. ER probes are available in variety of types and sizes and allow different options for mounting so that the operator does not have to completely reconfigure the probes for different test objects. The probes also have a sensitivity range that allows the operator to select the best response for the process. The different element types available are shown below in **Figure 2**.

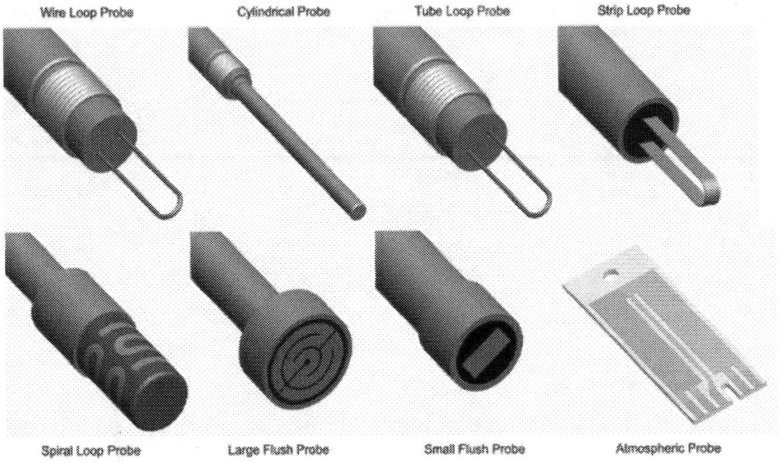

Figure 2 - ER Sensing Elements

The ER technique measures the change in electrical (Ohmic) resistance of a corroding metal element exposed to the process stream. Corrosion on the surface of the element produces a decrease in its cross-sectional area with a corresponding increase in its electrical resistance. This increase in resistance can be related directly to metal loss. The metal loss, as a function of time, is by definition, the corrosion rate.

Advantages (All the advantages of coupons, plus):

- Response time is much shorter than for "weight loss" coupons.
- Direct corrosion rates are quickly obtained.
- Probe remains installed until operational life has been exhausted.
- Response to corrosion upsets is quick and can be used to trigger an alarm.

Disadvantages:

- Unable to detect pitting (Corrosion must be uniform in order for the ER to be able to measure)
- If there is solid corrosion, precise results can only be obtained on a new element once the probe has been conditioned.

5.3 Linear Polarization Resistance Monitoring (LPR)

The electrical conductivity of a fluid is associated with its ability to cause corrosion. Linear Polarization Resistance (LPR) Monitoring applies a small amount of voltage to an electrode in a liquid solution. This technique has been used successfully for many years, in almost every type of water-based, corrosive environment.

LPR monitors the relationship between the current from charged electrons and the electrochemical potential, which results in the information needed for the calculation of the corrosion rates. To conduct LPR, two to three electrode probes are put into a liquid solution where the electrodes are isolated electronically. The current needed to maintain a particular voltage shift (typically 10 - 20 mV). Linear polarization is the applied potential ratio with the current level resulting from the process that is directly related to the corrosion on the surface of the electrode in the solution. The measurement of this current allows the corrosion rate to be calculated.

Advantages:

- The corrosion rate measurement is instantaneous.
- A more accurate tool compared to coupons or ER.

Disadvantages:

- LPR requires testing in relatively clean, aqueous electrolytic environments.
- LPR does not work in gas or water/oil emulsions because it can cause fouling of the electrodes.

5.4 Galvanic/Potential Monitoring

Galvanic monitoring, (also known as 'dissimilar metal corrosion' or wrongly 'electrolysis') is another electrochemical measuring technique. In this method, two electrodes of dissimilar metals are coupled in a corrosive electrolyte. Once immersed in an electrolyte solution, a natural voltage (potential) difference exists between the electrodes. The current generated relates to the rate of corrosion. A galvanic couple forms as a result of one of the metals in the couple becoming an anode and corroding faster than it normally would alone, while the other becomes the cathode and corrodes slower than it normally would alone. However, if there is contact with a dissimilar metal, the self-corrosion rates will change. The dissimilar metals will cause corrosion of the anode to accelerate, and corrosion of the cathode will decelerate or possibly stop. Galvanic coupling is known to be the foundation to many corrosion monitoring methods. Galvanic monitoring is applicable to the following major modes of corrosion:

- Bimetallic corrosion
- Crevice and pitting attack
- Corrosion assisted cracking
- Corrosion by highly oxidizing species
- Weld decay

Galvanic monitoring is widely used in water injection systems where dissolved oxygen concentrations are a primary concern. Oxygen leaking into such systems greatly increases galvanic currents that result in increased corrosion rates of steel process components. Galvanic monitoring serves as an indicator that oxygen might be in injection waters due to leaking gaskets or deaeration systems.

Advantage:

- Provides rapid, continuous method for assessing corrosive conditions in aqueous flows.

Disadvantage:

- Foreign metals in the system can result in chemical reactions taking place and interfering with the measurement.

6. Corrosion Control and Protection

For corrosion management, good mechanical integrity programs utilize measurement elements offered by four combinations. The four basic methods used for corrosion control and protection are:

- Using materials resistant to corrosion
- Applying protective coatings
- Utilizing cathodic protection
- Using corrosion inhibitors - (i.e., modify the operating environment)

Effective corrosion control can be obtained by combining two or more of the above methods. Well managed mechanical integrity programs and methods can be used to analyze the data and develop corrective action, plans and conclusions on how to operate the facility and manage the corrosion rates. The purpose of developing a good mechanical integrity program and utilizing the basic corrosion protections are to control catastrophic failures, downtime and costs while still operating in a safe manner. It is beneficial for companies to look at the way the process is designed prior to implementation and the best way to monitor and protect the pipelines and process equipment.

7. Conclusion

Of the four techniques listed above, corrosion coupons, ER, and LPR form the core of industrial corrosion monitoring systems. Specialized applications may use other methods.

These corrosion monitoring techniques are proven to be successful and accurate due to:

- Easy to understand and implement.
- Equipment reliability validation had completed through operational success.
- Results are apparent.
- Measuring equipment can be made intrinsically safe for hazardous area operation.

Understanding the options available and usefulness of corrosion monitoring and control should be a priority of all industries. Monitoring can serve to reduce the potential of catastrophic failures and unexpected releases in a facility. A facility must be proactive with their mechanical integrity programs and be vigilant about their inspection programs for the protection of life and facility. Corrosion monitoring and control offer a facility the best chance of protecting the facility and reducing damaged equipment if the program is utilized as an essential tool to manage the facilities risk. There is a financial benefit to utilizing corrosion monitoring and control methods due to reduced plant downtime and plant life extension, as well as a safety benefit for the company and the neighboring communities.

8. References

[1] "A year after a refinery explosion, Richmond, Calif., is ..." (n.d.). Retrieved from http://grist.org/climate-energy/a-year-after-a-refinery-explosion-richmond-cali-is-fighting-back/_br

[2] Advanced Metallurgical Services Ltd in OLDHAM, Lancashire. (n.d.). Retrieved from https://1stdirectory.co.uk/companies/16315507_advanced-metallurgical-services-ltd.htm (accessed Jul. 22, 2014)

[3] "Corrosion Monitoring, Corrosion Coupons, and Corrosion Coupon Racks." http://www.gwt-inc.com/resources/corrosion-monitoring/ (accessed Jul. 22, 2014)

[4] Corrosion Monitoring and Corrosion Coupons - Global Water. (n.d.). Retrieved from http://www.gwt-inc.com/resources/corrosion-monitoring/_br (accessed Jul. 18, 2014)

[5] "ER Probes and Instruments." (n.d.). Retrieved from http://www.anodeengineering.com/er-probes-instruments.html_br (accessed Jul. 18, 2014)

[6] Industrial Equipment News. Keeping Pipelines Safe from Internal Corrosion: http://www.ien.com/article/keeping-pipelines-safe/4252 (accessed Jul. 22, 2014)

[7] "Introduction to Corrosion Monitoring" - alspi.com. (n.d.). Retrieved from http://www.alspi.com/introduction.htm (accessed Jul. 18, 2014)

[8] Metal Samples Company. Introduction to Corrosion Monitoring: http://www.alspi.com/corrosion-intro.htm (accessed Jul. 18, 2014)

[9] Synergy Services. Corrosion monitoring in oilfield, pipeline, and industrial processes: Retrieved from http://www.greenesenergy.com/Images/Interior/download%20center/synergy/corrosionmonitoring.pdf (accessed Jul. 18, 2014)

[10] TECHNICAL BULLETIN - Greene's Energy Group. (n.d.). Retrieved from http://www.greenesenergy.com/Images/Interior/download%20center/synergy/corrosionmonitoring.pdf_br

S8000 检测系统在聚乙烯关键机组的应用

郭磊 杨卫 张毅

（中国石油独山子石化公司乙烯厂聚乙烯车间，新疆独山子 833699）

摘要：介绍了 S8000 旋转机械在线状态检测与分析系统的原理与图谱，针对独山子石化公司聚乙烯装置流化气压缩机 21C400 振动升高的状况，借助故障诊断软件，利用频谱分析法，对引起压缩机振动的可能原因进行了详细分析，排除了转子不对中、轴弯曲、摩擦，最后找出振动升高原因是转子不平衡，并造成了轴瓦轻微磨损，最后进行了验证。

关键词：S8000 系统；流化气压缩机；振动；频谱图；转子不平衡；轴瓦；磨损

Application of S8000 System in the Polyethylene Key Unit

Abstract: This paper introduces principles and patterns of S8000 rotating machinery online Condition monitoring and analysis system, conducts a detailed analysis on the possible causes for the libration on fluidization gas compressor 21C400 in polyethylene unit in Dushanzi Petrochemical Corporation in view of the libration increase in the unit by using malfunction diagnosis software and spectrum analysis. Through analysis, the paper excludes the causes of rotor misalignment, shaft bending, friction for the increases of libration and finds out the reason is the imbalance of rotor, which causes the abrasion of bearing. And the result is tested and verified.

Keywords: S8000 system; fluidization gas compressor; Libration; Spectrogram; Imbalance of rotor; Bearing; Abrasion

独山子石化公司乙烯厂聚乙烯车间流化气压缩机 21/22C400 是公司关键机组，该机组由日本 EBARA 公司制造，型号 180TC，工艺介质为流化气，入口压力 1.91MPa，出口压力 2.14 MPa，入口温度 60℃，出口温度 68℃，转速 2985r/min，功率分别为 2100Kw 和 2600Kw。流化气体的流量由压缩机入口的导向叶片控制。在压缩机的密封系统中，叶轮入口采用阶梯式梳齿密封，出口采用直通型梳齿密封，轴封采用直通型梳齿密封和单端面机械密封相结合的密封方式。其工艺作用如图 1 所示，聚合反应在 D400 中连续进行，S400 将循环气中分离出来的细小颗粒通过喷射器 J400 返回反应器，流化气经过流化气冷却器 E400 初级冷却后，由流化气压缩机 C400 增压打循环，经过最终冷却器 E401 冷却后进入反应器底部。C400 为气相法聚乙烯流化床反应器提供动力，补偿循环气流通过流化床及冷却器所产生的压降，并保证一定的气速使反应器床层处于流化状态。

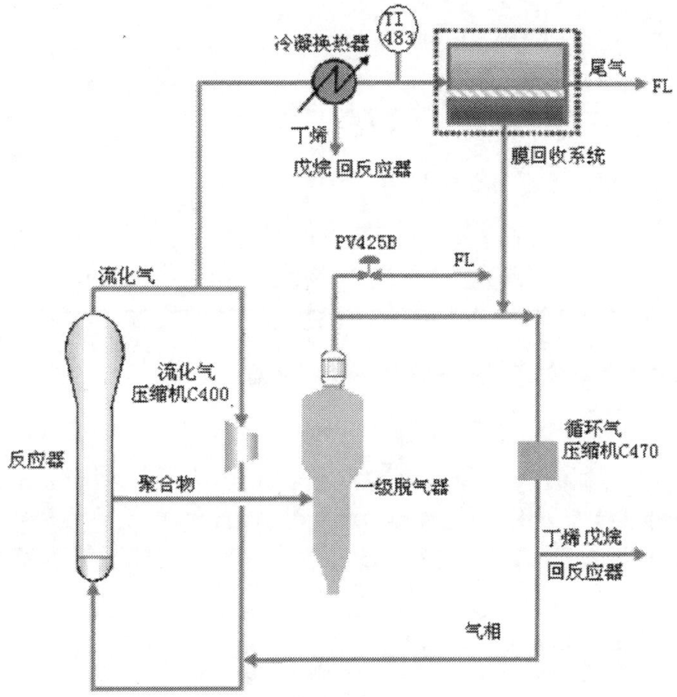

图 1 流化气压缩机工艺流程简图

1. S8000 旋转机械在线状态检测与分析系统

S8000 旋转机械在线状态检测与分析系统为状态检测和设备管理、运行提供丰富、专业的机组状态分析图谱，通过分析图谱可以方便掌握机组运行的状态。S8000 旋转机械在线状态检测与分析系统能够实现对不同采集设备采集的数据的浏览、分析与处理，使得在一个单独平台能同时浏览分析不同采集设备上传的数据。

1.1 FFT 频谱分析法

目前，频谱分析方法是在计算机上用快速傅立叶变换来实现的，因此又称为FFT分析法。频谱图是频谱分析方法提取诊断信息的一种表达方式。频谱图有：幅值谱、相位谱、功率谱等，以下主要介绍幅值谱分析方法。即[1]

$$X(f) = \int_{-\infty}^{+\infty} x(t) e^{-12pft} dt$$

式中：$x(t)$——振动信号；

$X(f)$——振动信号 $x(t)$ 的傅立叶变换谱。

一般情况下 $X(f)$ 为一复变函数，令

$$X(f) = R(f) = iI(f) = |X(f)| e^{ip(f)}$$

式中：$R(f)$ ——实谱

$I(f)$ ——虚谱

即：

$$|X(f)| = \sqrt{R(f)^2 + I(f)^2}$$

$$\varphi(f) = arctg\left[I(f) / R(f) \right]$$

式中：$|X(f)|$——幅值谱或 FFT 谱，它表示信号中各频率成分的幅值大小沿频率轴的分布情况；

$\varphi(f)$——相位谱，它表示信号中各频率成分的相位沿频率轴的变化状况。

目前幅值谱在机械故障诊断领域中的应用相当普遍。幅值谱可以提供以下诊断信息：

1、振动信号中主要由哪些频率成分及谐波分量所组成；

2、组成的谐波分量中哪些成分的幅值最为突出，这提示着和故障的某种联系。

1.2 S8000 旋转机械在线状态检测与分析系统图谱

S8000 旋转机械在线状态检测与分析系统分为常规图谱和启停机图谱，常规图谱有总貌图、综合分析图、单值傍图、多值棒图、波形频谱图、频谱图、多频谱图、振动趋势图、过程振动趋势图、轴心轨迹图、多轨迹图、全息谱图、三维全息谱图、全频谱图、轴心位置图、极坐标图及工艺量频谱瀑布图共计 17 种，启停机图谱有转速时间图、Nyquist 图、Bode 图、频谱瀑布图和级联图共计 5 种。

2. S8000 检测系统在聚乙烯关键机组的应用案例

2.1 关键机组 21C400 出现的问题

如图 2，流化气压缩机 21C400 在 S8000 检测系统的总貌图，总貌图展示了转子测点分布位置。取该轴左、右端互相垂直的两个方向和轴向共六个点作为检测点，其中 21CVE400A、21CVE400B、21CVE401A、21CVE1401B 为径向振动测点，21CXE400A、21CXE400B 为轴向测点。

2012 年 10 月 11 日，聚乙烯 21 线按计划转产，停 21C400 压缩机，转产前 21C400 运行正常，轴振动均在 40μm 以下。10 月 14 日，启动 21C400 压缩机，在转产过程中，对 21C400 仪表探头进行预防性检查，校验探头，紧固连接处。钳工对压缩机转子进行对中检查，地脚螺栓进行预防紧固处理，未见异常。启动压缩机后，轴振动 21CVE400B、21CVE401A 振动值经过短暂波动后，降低至 40μm 以下，未超过停机前振动值，21CVE1401B 启动后振动值较大，超过 60μm 后，缓慢降低至 40μm 以下，也未超过停机前振动值，21CVE400A 启动后振动值较大，波动较大，振动值居高不下，最大超过 75μm ，逼近到振动报警值，如图 3，严重威胁着设备的安全运行和装置的稳定生产。21C400 轴振动增加后，使用 S8000 旋转机械在线状态检测与分析系统对振动数据采集，进行频谱分析，通过对采集数据的分析，可以加快对振动原因的判断。

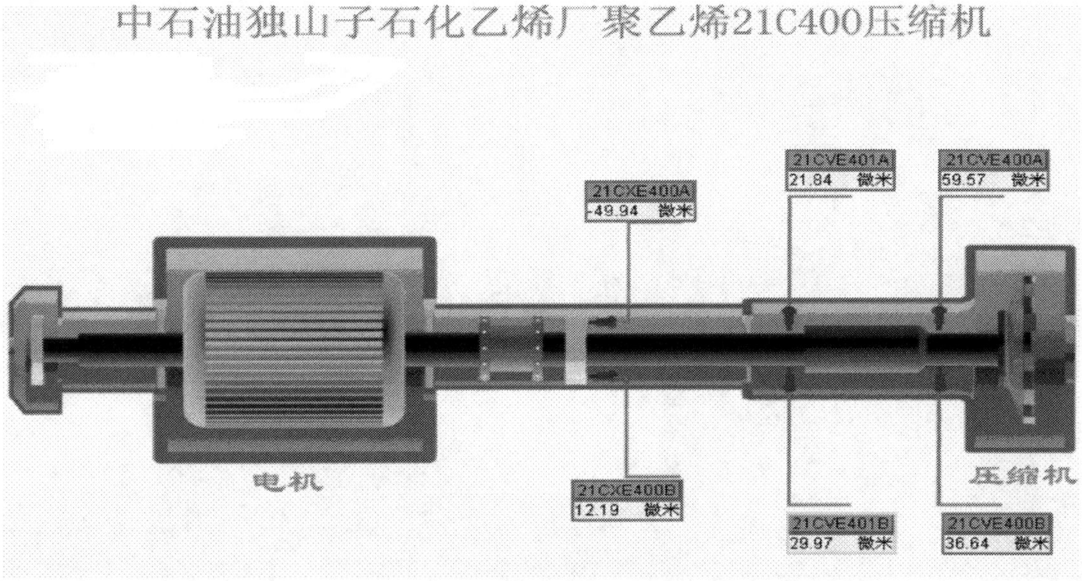

图 2 流化气压缩机 21C400 总貌图

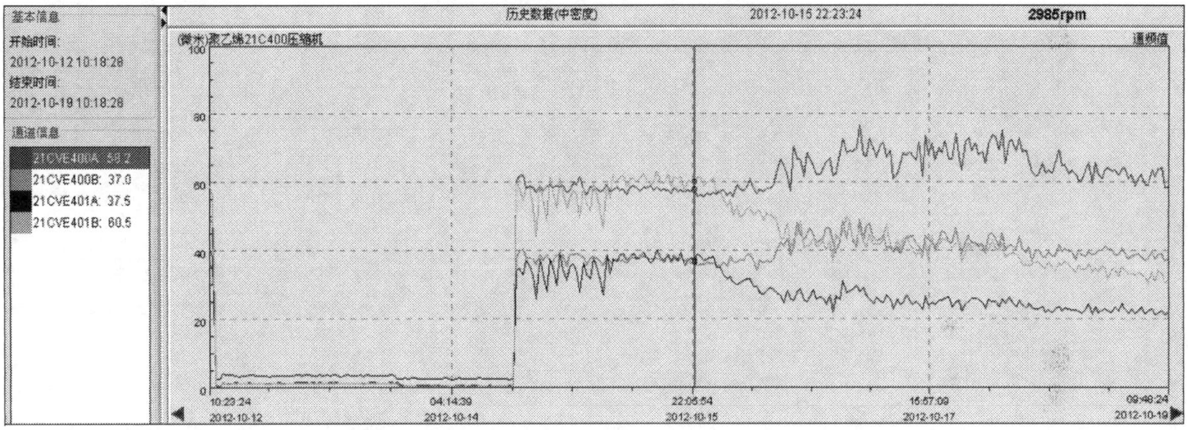

图 3 流化气压缩机 21C400 启动振动趋势图

2.2 关键机组波形频谱图

波形频谱图是反应机器振动幅值和转动频率的关系曲线，横坐标是转动频率，纵坐标是振幅的峰-峰值。2012 年 10 月 15 日利用 S8000 旋转机械在线状态检测与分析系统测得 21C400 轴振动波形频谱图，如图 4，分别是压缩机 21C400 四个测振点 21CVE400A、21CVE400B、21CVE401A、21CVE1401B 的波形频谱图。

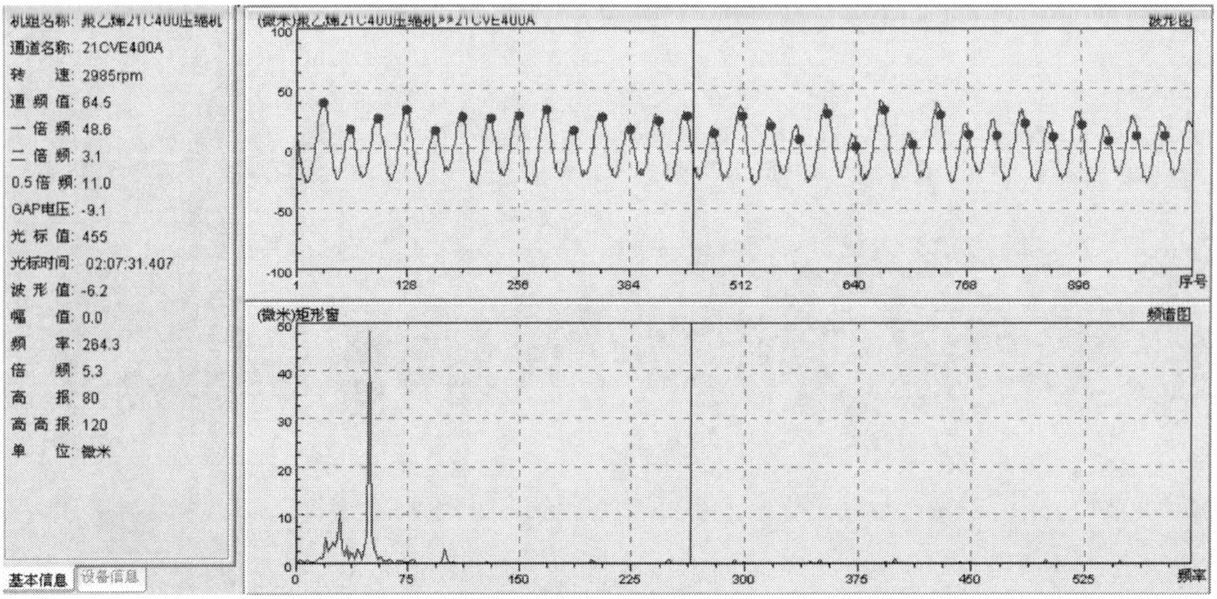

(a) 21CVE400A 波形频谱图

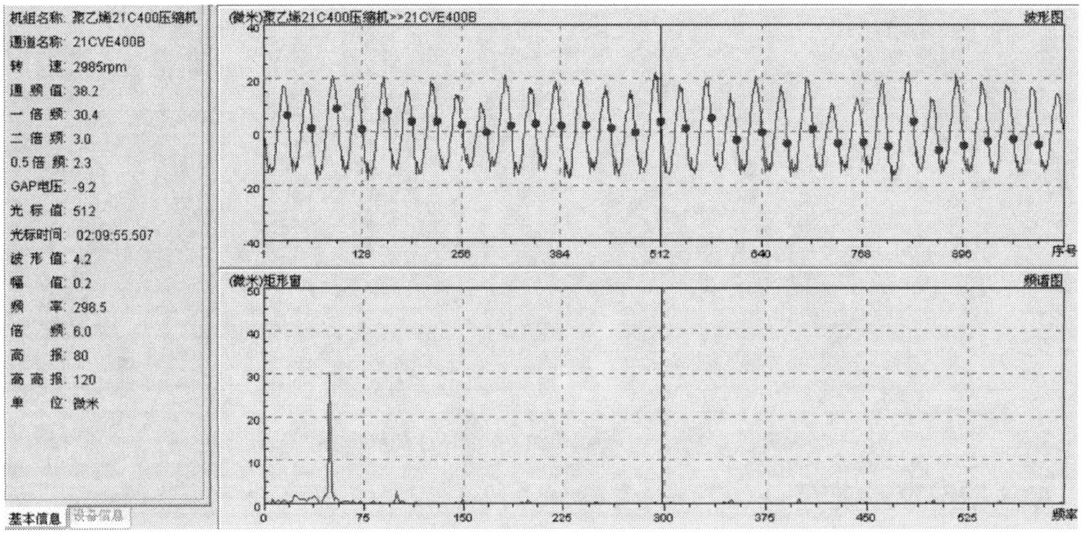

(b) 21CVE400B 波形频谱图

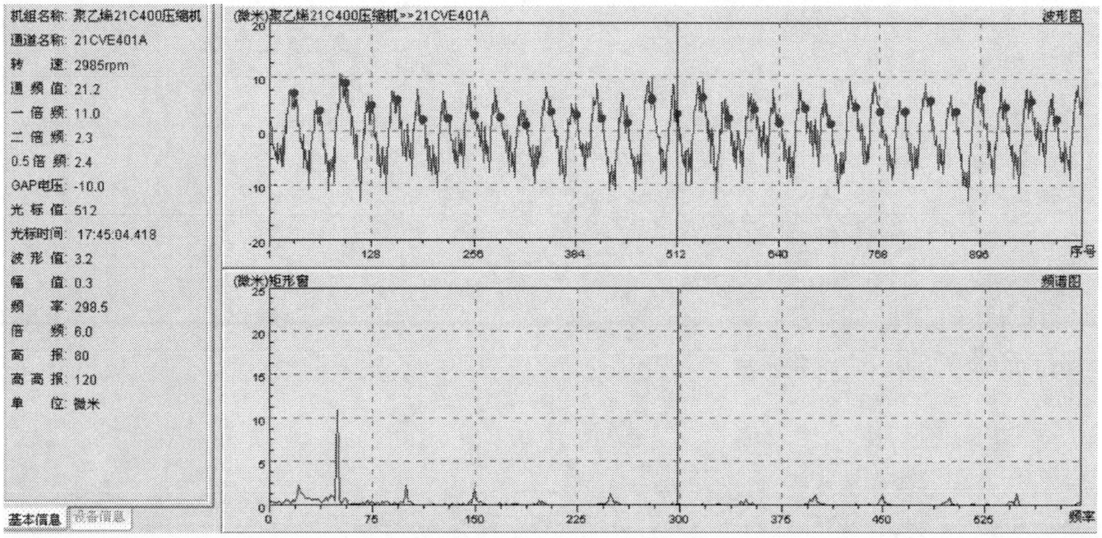

(c) 21CVE401A 波形频谱图

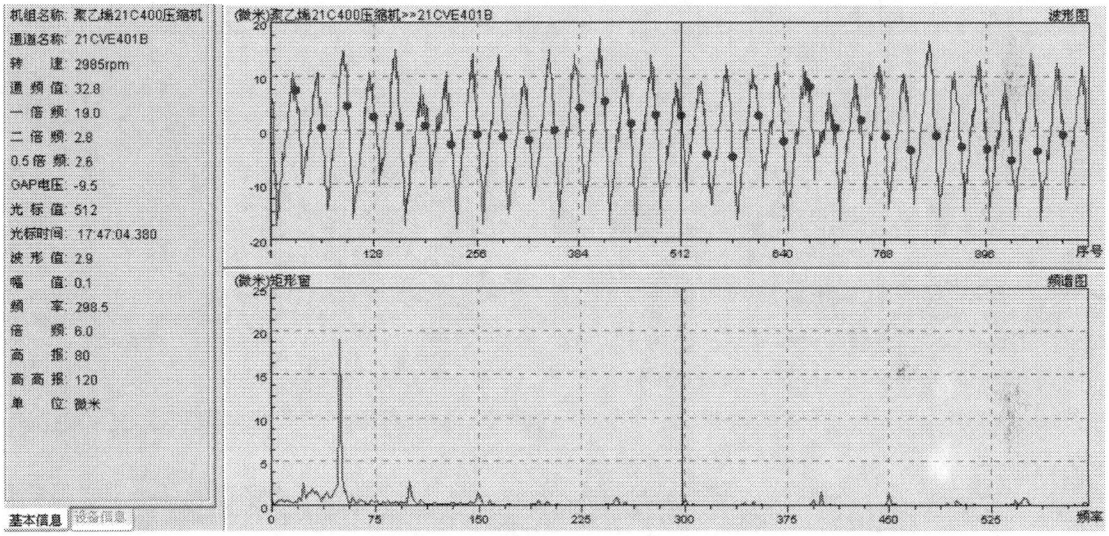

(d) 21CVE401B 波形频谱图

图 4 21C400 波形频谱图

流化气压缩机 21C400 工作转速为 2985r/min，转动频率为约 50Hz。由 21C400 波形频谱图（如图 4）可以看出，21C400 四个测振点在 1 倍频处的振动幅值分别为 48μm、30μm、12μm、18μm，轴振动幅值在 1 倍频（工作频率）50 Hz 处最大，即轴振动以 1 倍频为主，21CVE400A 出现明显 0.5 倍频，2 倍极小。21CVE400B、21CVE401A、21CVE1401B 轴振动以 1 倍频为主，0.5 倍频、2 倍极小。由此判断，振动峰-峰值增长的原因是由基频振动分量增长引起的。

2.3 流化气压缩机 21C400 振动原因分析

2.3.1 振动原因分析

根据过去设备运行经验，流化气压缩机 21C400 振动原因有：

1、探头失效，使测试数据不准；
2、转子不对中；
3、轴弯曲；
4、机械松动；
5、摩擦；
6、压缩机内部气流不稳；
7、旋转失效与喘振；
8、轴瓦磨损或失效；
9、油膜涡动或油摸振荡；
10、转子不平衡等共约 10 种原因。

经检测和分析可以直接排除 5 种因素：
1、探头失效，使测试数据不准；
2、机械松动；
3、旋转失效与喘振；
4、压缩机内部气流不稳；
5、油膜涡动或油摸振荡；

下面对其余 5 种振动原因逐一进行分析。

2.3.2 转子不对中

流化气压缩机转轴与电机主轴之间，由联轴器联接构成轴系，由于机器安装误差、承载后的变形及机器基础的松动等，会造成轴系平行位移、轴线角度位移或综合位移等轴系对中变化误差，统称为转子不对中。当出现不对中故障时，会产生一系列有害于设备的动态效应，如机器联轴器偏转、轴瓦早期损坏、油膜失稳和轴挠曲变形等，导致机器发生异常振动和噪声，危害极大。其振动特征为：

（1）频率特征。转子不对中的型式不同，频率表现也有些差别。平行不对中主要激起 2 倍频，即为$2fr$，角度不对中则表现为同频振动突出，它们的共同点都会产生多倍转频振动，如$1fr$、$3fr$、$4fr$、$5fr$等高次谐波；

（2）振动方向特征。与不对中的型式有关，当存在平行不对中时，径向振动较大。

由上述波形频谱图可知，流化气压缩机 21C400 轴振动以 1 倍频为主，而转子不对中的主要特征共同点都会产生多倍频振动，所以排除转子不对中的可能性。

2.3.3 转子弯曲

转子弯曲分为永久性弯曲和临时性弯曲两种类型。转子永久性弯曲是指转子的轴呈永久性的弓形，它是由于转子结构不合理、制造误差大、材质不均匀、转子长期存放不当而发生永久性的弯曲变形，或是由于热态停车时未及时盘车、转子的热稳定性差、长期运行后主轴的自然弯曲加大等原因造成的。转子临时性弯曲是指转子上有较大预负荷、开机运行时的暖机操作不当、升速过快、转子热变形不均匀等原因等造成。

转子弯曲在频谱图和时域图反映的特征是：

（1）产生较大的二倍频分量；

（2）在低转速下转子的工频振动幅值，比在开车时相同转速下的振动值要大很多。

由上述波形频谱图可知，流化气压缩机 21C400 振动以 1 倍频为主，而转子弯曲的主要特征应是产生较大的 2 倍频分量，同时在低转速下转子的工频振动幅比在开车时相同转速下的振动值要大很多，所以排除轴弯曲的可能性。

2.3.4 摩擦

摩擦产生分为两种情况：第一种是部分摩擦，此时转子仅偶然接触静止部分，同时维持接触仅在转子进动整周期的一个分数部分，这通常对于机器整体来说，它的破坏性和危险性相对较小；第二种是整周的环状摩擦，也称为“全摩擦”或“干摩擦”，它们大都在密封中产生。摩擦一般会产生更多的分数谐波振动分量，即$\frac{1}{2}$倍频、$\frac{1}{3}$倍频、$\frac{1}{4}$倍频等，这与流化气压缩机 21C400 的频谱图不一致，故排除摩擦现象。

2.3.5 轴瓦磨损或失效

轴瓦磨损或失效在频谱上的反应，应该是以$\frac{1}{2}$倍频、1 倍频为主，2 倍频以上也有，这与流化气压缩机 21C400 的波形频谱图反应较一致，（如图 5），并且联轴器侧径向轴瓦温度 21TI410 和止推轴瓦温度 21TI407、21TI411 在压缩机停机前与启动后无变化，而叶轮侧径向轴瓦温度 21TI409 在压缩机启动后比停机前上升 2℃，因此存在很大轴瓦磨损或失效的可能性[1]。

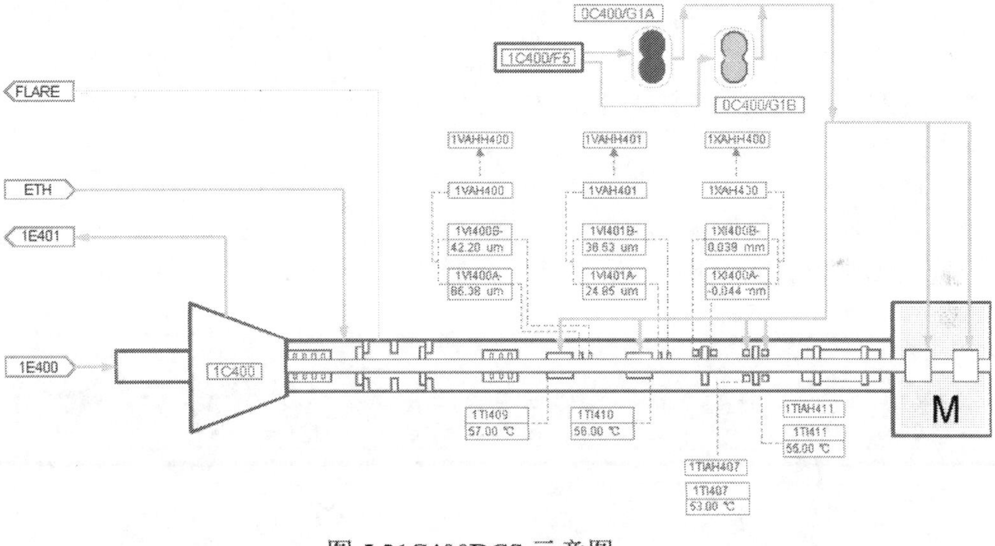

图 5 21C400DCS 示意图

2.3.6 转子不平衡

转子不平衡包括转子系统的质量偏心和转子部件出现缺损。转子质量偏心是由转子的制造误差、装配误差、材质的非均匀性等原因造成的，称为初始不平衡；转子部位缺损是指转子在运行中由于腐蚀、磨损、介质结垢及转子受疲劳力的作用，使转子的零部件局部损坏、脱落，碎块飞出等，造成新的转子不平衡。转子质量偏心及转子部件缺损是 2 种不同的故障，但其不平衡振动机理却是相同的。当出现转子不平衡时，其振动特征为[2]：

（1）振动频率 不平衡振动的频率成分单一且明朗，主要表现为转子的基频等于转子的旋转频率，即工作频率 $fr = \frac{n}{60}$Hz，n 为轴转速，除此之外，不平衡振动还会激起其它一些弱小的频率成分，如 $\frac{1}{2}fr$、$2fr$ 等谐波；

（2）振动方向特征 主要是径向振动较大。

通过对上述可能引起流化气压缩机 21C400 振动原因的分析以及排除，初步判断压缩机振动超标的原因可能是转子不平衡引起的，根据 21C400 波形频谱图，如图 4，可以得出流化气压缩机 21C400 振动特征频率为 1 倍频（$1fr$），振动稳定性是渐变的，即振动值逐渐增大，轴向振动较小，振动主要集中在径向垂直和水平位置，是转子不平衡产生振动的一个特征。这是因为转子由于腐蚀、磨损、结垢，零部件脱落等原因具有偏心质量，在高速旋转时会产生偏心惯量，引发转子失衡，振动主要表现在径向位置。所以最终可以确认转子不平衡是导致压缩机振动超标的真正原因。

2.4 验证流化气压缩机 21C400 振动原因

经过车间频谱分析，认为 21C400 可能存在压缩机轴瓦磨损、转子不平衡的隐患，2012 年 10 月 20 日停机进行解体检查。压缩机解体后，发现叶轮侧五油楔可倾瓦磨损，如图 6。

如图 7，压缩机解体后，发现叶轮表面聚合物结垢较严重，同时发现表面结垢脱落一大块，这是因为 21C400 在转产停机前生产 Cr 系聚乙烯 HD5420，这种聚乙烯熔融指数非常低，因此很脆。在压缩机开、停机过程，叶轮振动突然增大，造成一大块附着在叶轮表面的聚合物剥落，引起转子不平衡。当转子旋转时，质量不平衡将激起转子的振动，这就是造成转子不平衡的主要原因。转子不平衡，轴振动较大，造成轴瓦轻微磨损。车间更换一套新轴瓦，使用高压水彻底清理压缩机叶轮及叶轮流道聚合物结垢后做动平衡重新开车，开机后轴振动、轴瓦温度正常。

图 6 解体后叶轮侧五油楔可倾瓦

图 7 解体后 21C400 叶轮

3. 结语

本文利用 S8000 旋转机械在线状态检测与分析系统对流化气压缩机的振动故障进行分析。通过对 S8000 旋转机械在线状态检测与分析系统的振动趋势图、振动频谱图进行分析，列举了可能导致压缩机振动超标的 10 种原因，经检测和分析直接排除了 5 种原因，然后结合波形频谱图，利用特征信号分析技术，对其余 5 种原因进行分析。通过对比分析，排除了转子不对中、轴弯曲、旋转失速与喘振，最终得出引起压缩机振动超标的真正原因是转子不平衡，转子不平衡又造成了轴瓦轻微磨损。最后停车对压缩机进行解体检修，根据检修情况及检修后压缩机的振动情况验证了我们判断，引起压缩机振动超标的真正原因是转子不平衡。

4. 参考文献

[1] 沈庆根 设备故障诊断［M］第1版 北京 化学工业出版社，2006：135-148.
[2] 李录平 汽轮机组故障诊断技术［M］第1版 北京 中国电力出版社，2002：49-56.

2nd CCPS CHINA CONFERENCE ON PROCESS SAFETY™

Pipeline Safety

Lisa C. Hutto
Chilworth Technology
113 Campus Drive, Princeton, NJ 08540, USA
Lisa.Hutto@Dekra.com

Ethan Zhang
Dekra China
800 Dongfang Road, Pudong, Shanghai 200122, China
Ethan.Zhang@Dekra.com.cn

UNPUBLISHED

Pipeline Safety

Lisa C. Hutto
Chilworth Technology
113 Campus Drive, Princeton, NJ 08540, USA
Lisa.Hutto@Dekra.com

Ethan Zhang
Dekra China

Keywords: Pipeline Safety, Internal Corrosion, Catastrophic Failures

Abstract

The safe transportation of hazardous materials to facilities and consumers is a valid concern today as we see more pipelines being built and more hazardous material transportation in these pipelines. To ensure pipelines are a minimum threat to people and the environment, regulations have to be followed and continually developed to ensure the safe classification, handling and packing of these hazardous materials. In the United States, the Pipeline and Hazardous Materials Safety Administration (PHMSA) oversees the design, construction, operation, maintenance and spill response planning for some 2.6 million miles of natural gas and hazardous liquid pipelines. This paper discusses some of the pipeline regulations in the US, significant safety incidents that have occurred from pipelines, internal corrosion, and new pipeline safety and routing concerns. This paper also discusses PHMSA applying the integrity management program (IMP) requirements, or some elements of IMP, to areas beyond current high consequence areas (HCAs).

1. Introduction

Pipeline regulations and enforcement are essential in the prevention of incidents related to pipeline safety. Not only are there an abundance of older pipelines, but there are new pipelines being built which is creating a greater concern for pipeline safety due to their routing locations. As pipelines multiply, there is a greater need to improve pipeline inspections, maintenance and security. The number of pipeline safety incidents over the years have increased and brought to light more deficiencies and the need to enhance the current programs. Numerous pipeline incidents are due to poor mechanical integrity inspection plans. Many of the pipeline failures which were considered significant were due to a lack of internal pipeline inspections.

Another issue that has led to numerous pipeline safety incidences is the lack of security around pipelines. The inability to patrol entire pipelines and the cost to provide security for the entire pipeline causes companies difficulty in complying with a high level of security that pipelines require.

2. Significant Pipeline Safety Incidents

In the United States, PHMSA defines Significant Incidents as those incidents reported by pipeline companies when any of the following specifically defined consequences occur:

- fatality or injury requiring in-patient hospitalization
- $50,000 or more in total costs, measured in 1984 dollars
- highly volatile liquid releases of 5 barrels or more or other liquid releases of 50 barrels or more
- liquid releases resulting in an unintentional fire or explosion

Since 2001, natural gas pipeline explosions and other accidents have resulted in the loss of at least 45 lives and many more serious injuries, usually from burns. In just the year 2012, natural gas high-pressure transmission lines accounted for more than 80 explosions and fires. Of these 80 incidents, there were seven injuries, no fatalities, and more than $44 million of damage. The smaller, natural gas distribution lines that operate under low pressure and bring gas directly to residential buildings and commercial facilities added another 71 incidents with nine fatalities and 21 injuries. Some Pipeline Safety Incidents that are considered significant are the following:

- On August 19, 2000 in the early morning hours, near the Pecos River in New Mexico, a 30-inch El Paso Energy natural gas pipeline exploded. Twelve campers who were nearby were killed by the explosion. The fireball was seen as far as 20 miles from the incident area. After the investigation was complete, it was determined that over 70% of the internal pipe wall had been eaten away due to severe internal corrosion near the rupture point. The explosion occurred underground on the east side of the river 200 to 300 yards from the campers around 5:30 a.m. The explosion left a crater 86 feet long, 46 feet wide and 20 feet deep. The pipeline was installed in 1950.

- September 7, 2000 - A bulldozer ruptured a 12 inch diameter NGL pipeline on Route 36, south of Abilene, Texas. A police was killed as a result of the explosion was killed.

- July 30, 2004 - A high-pressure natural gas pipeline buried below the ground surface exploded in Ghislenghien, Belgium resulting in the death of 24 people and leaving 122 wounded. Most of the dead were police and fire-fighters responding to reports of a gas leak. Early indications suggest that construction work may have damaged the pipeline.

- August 21, 2004 - A natural gas explosion destroyed a residence located in DuBois, Pennsylvania. Two residents were killed in this accident. The NTSB determined that the probable cause of the leak, explosion, and fire was the fracture of a defective butt-fusion joint. **See Figure 1**

Figure 1 - DuBois, Pennsylvania – Natural gas explosion

- September 19, 2005 - A pipeline pumping station employee was killed in Monroe, Ohio, when leaking propane was ignited and exploded by an arcing pump. Flames reached 300 feet high in the following fire.
- December 26, 2006 - Abule Egba, Nigeria pipeline explosion - An elevated pipeline carrying petroleum products was punctured by armed vandals at about 12 midnight. An armed gang punctured the underground pipeline in order to siphon fuel into road tankers. The vandalized oil pipeline exploded and up to 500 people were killed.
- On July 26, 2010, Enbridge Energy Partners LLP (Enbridge), reported that one of their 30-Inch pipelines burst in Marshall, Michigan. Enbridge estimated over 800,000 US gallons of crude oil leaked into Talmadge Creek. However, the EPA reported that over 1,139,569 gallons of oil had been recovered as of November 2011. In 2012, the NTSB later cited known but unrepaired cracks and external corrosion as the cause.
- September 9, 2010 – The San Bruno pipeline explosion occurred at 6:11 pm PDT in San Bruno, California. A 30-inch diameter, steel, natural gas pipeline burst into flames in the Crestmoor residential neighborhood 2 miles west of San Francisco International. Pacific Gas & Electric owned the pipeline. Eyewitnesses reported the initial blast "had a wall of fire more than 1,000 feet high." Residents reported a loud "boom" and shaking which led some residents, first responders, and news media to initially believe that it was an earthquake or that a large jetliner had crashed. It took crews nearly an hour to determine it was a gas pipeline explosion. The death toll was eight people.
- December 19, 2010 - an oil pipeline containing high-pressure crude at a pumping station in San Martín Texmelucan de Labastida, Mexico exploded, killing at least 27 people and injuring more than 50. The explosion is believed to have been caused by attempts to puncture the pipe to steal oil.
- February 9, 2011 - Five people were killed, and eight homes were destroyed, in a gas explosion and fire in Allentown, Pennsylvania. The NTSB had warned UGI about cast iron gas mains needing replacement after the 1990 gas explosion in that city. UGI was cited in 2012 for several safety violations, including lack of valves on their gas system.

- December 11, 2012, a 20" buried natural gas transmission line installed in 1967 ruptured near Sissonville, WV damaging nine homes and leaving a badly burned section of highway 77.

Figure 2 – Sissonville, WV – Natural gas transmission line explosion

The investigation proved that severe corrosion, which reduced the pipe wall to around 30% of its original thickness, was the cause and this could have been prevented with proper inspections, which had not been done since 1988.

- November 22, 2013 a Sinopec Corp oil pipeline exploded in Huangdao, Qingdao, Shandong Province, killing 62 people. The investigation determined that an oil pipeline leaked, caught on fire and exploded.

3. Pipeline Regulations

3.1 Pipeline Safety Improvement Act of 2002

In 2002, the United States government enacted the Pipeline Safety Improvement Act (PSIA) due in part to the El Paso Energy - New Mexico natural gas pipeline explosion. It mandated significant changes and requirements that the `natural gas industry had to ensure the safety and integrity of pipelines. The PSIA applies to natural gas transmission pipeline companies.

The PSIA requires pipeline companies to develop and implement an "integrity management program." The purpose of the program is to require pipeline operators to identify any "high consequence areas" (HCA) on their pipelines, perform risk assessments, and baseline mechanical integrity inspections on the entire pipeline utilizing approved inspection methods.

All pipeline companies were required to have all HCA's on their pipelines identified and submitted by December 17, 2004. The other portion of the PSIA required companies to complete the inspections on the identified HCA's and remediation plans developed if required. Going forward, all pipeline segments must be re-inspected every seven years with only minor exceptions.

3.2 *Pipeline Safety, Regulatory Certainty, and Job Creation Act of 2011*

President Obama signed the Pipeline Safety, Regulatory Certainty and Job Creation Act of 2011 ("Pipeline Safety Act" or the "Act") into law on January 3, 2012. The purpose of this Act is to examine and improve the regulation of pipeline safety. The Act also gives enhanced safety review authority to the Department of Transportation ("DOT"). The new rules stemmed from pipeline safety incidents in San Bruno, California, Allentown, Pennsylvania and Marshall, Michigan.

The Pipeline Safety act was designed to examine and improve the state of pipeline safety regulation. The purpose of the Act is to ensure the following:

1. Reapprove and enforce all of PHMSA's federal pipeline safety programs through fiscal year 2015
2. Provide regulatory leadership to pipeline owners and operators for developing infrastructure investments and creating new job positions
3. Ensure pipeline transportation improvements are made by strengthening enforcement of current laws and making improvements to existing laws as necessary
4. Implement a balanced regulatory approach to improving safety while also applying cost benefits
5. Ensures key rulemaking are not finalized until Congress has an opportunity to act

4. New Pipeline Safety & Routing Concerns

For the vast majority of new pipelines, the pipeline company decides on a general route they prefer for their proposed pipeline as well as some alternative routes. Once a company has completed studies on the pipeline location and are satisfied with the results, they begin a more formal process with various government agencies. If the proposed pipeline is going to cross international or state boundaries (interstate lines), it requires a much different approval process than the one that is going to stay within just one state (intrastate lines).

4.1 *Pipeline Siting*

Once the pipeline company has a pipeline proposal, and route determined, they must apply to the Federal Energy Regulatory Commission (FERC) for approval. If approved, the pipeline company receives a "Certificate of Public Convenience and Necessity" from the FERC. The FERC performs an environmental impact review that is quite extensive and includes many opportunities for landowners to become involved. A valid argument that routinely occurs during the process is when FERC is asked about their mission to provide energy to consumers across the nation sometimes trumps individual property owner's concerns and protection of the environment. To date, there are no known denials of the Certificate of Public Convenience and Necessity for an interstate gas transmission line.

4.2 Permits

There is no comprehensive federal permitting process for the siting of new interstate hazardous liquid pipelines. If a pipeline crosses an international border (Canada or Mexico), then a permit is required from the State Department. The foreign country then takes the lead on the proposal in a process similar to the one described for FERC. If the pipeline does not cross an international border then, the responsibility for approval of the pipeline route falls on the individual states. If the state has no agency in charge of pipeline siting, the responsibility falls to the regular land use authority of local governments along the proposed route.

Depending on where the new pipeline is proposed, there are a variety of other permitting processes that may apply. Some examples are Section 401 Clean Water Act water quality certificates; Section 404 dredge and fill permits from the Army Corps of Engineers, NPDES discharge permit (if applicable), Coastal Zone Management Act determination of consistency, Endangered Species Act consultation, Historic Preservation Act consultation and impact determination, and a variety of state and local permits. Each of these permits has its process for involvement. The federal permits require compliance with the National Environmental Policy Act, which ultimately means some environmental review.

5. Pipeline Internal Corrosion

Not only is internal corrosion the leading cause of pipeline failure, it is also the most difficult to detect. These types of failures have led to catastrophic injuries and destruction. Due to the failures, integrity management requirements were placed on pipeline companies/operators by the U.S. Department of Transportation as an aid to compliance. It can be very difficult and expensive for a company to try and maintain their pipelines and keep them safe from internal corrosion. To date, some industries spend several billion dollars to maintain the integrity of their pipelines.

5.1 Causes of Internal Corrosion

Internal corrosion can be caused by impurities in the natural gas, crude, and refined products being transported. One of the biggest and most costly impurities is moisture. Moisture occurs because it easily combines with salts, like chlorine and sulphur compounds. Water can chemically react to form very corrosive liquids that collect in low-lying spots.

It is not unheard of for corrosion to destroy a pipeline wall thickness at a rate of 2-3 mils/year, but it can happen much more quickly in the upstream area of operations.

It is up to the company to determine which inspection method works best for the pipeline. There are many options available including intrusive, non-intrusive, inline and permanently mounted dataloggers.

6. Integrity Management Program (IMP) Requirements beyond High Consequence Areas (HCA)

While maintaining good pipeline conditions, inspection plans and correcting pipeline abnormalities are essential, there are still other elements that should be taken into consideration for pipeline safety. Companies must take a more hands on approach to improving the management and analyses technology available to successfully determine pipeline risks with pipeline segments in HCAs. To avoid more catastrophic failures companies need to evaluate and implement more advanced damage prevention programs, and additional risk control measures than what is already required by Part 192. Some options to meeting these requirements are:

- installing computerized monitoring and leak detection systems
- replacing pipe segments with pipe of heavier wall thickness
- providing additional training to personnel on response procedures
- conducting drills with local emergency responders
- implementing additional inspection and maintenance programs

The Pipeline Safety, Regulatory Certainty, and Job Creation Act, Section 5 requires that the Secretary of Transportation complete an evaluation and issue a report on whether integrity management requirements should expand beyond high consequence areas (HCAs) and whether such expansion would mitigate the need for class location requirements. Based on the long list of pipeline disasters over the years, many people think more mitigation requirements should be forthcoming in the future.

7. Conclusion

In conclusion, there are options available to companies in improving pipeline safety. The public has valid concerns about the existing and new pipelines being built around the country. Updated regulations need to be implemented to protect the community and companies must budget for increased pipeline inspections. There are options for inspections and monitoring equipment that could greatly reduce catastrophic pipeline failures. In the future, if PHMSA begins to apply more elements of IMP to areas beyond current high consequence areas (HCAs) significant advances in safety should be made.

8. References

[1] Baldonado V. El Paso Natural Gas Company - FindLaw. (n.d.). Retrieved from http://caselaw.findlaw.com/nm-supreme-court/1044166.html_br (accessed Jul. 15, 2014)

[2] Briefing Paper Feb 05, 2014 - US Department of Transportation. (n.d.). Retrieved from http://www.phmsa.dot.gov/staticfiles/PHMSA/DownloadableFiles/Pipeline/Technical%20Advisory%20Committees/4_Brief_for_TPSSC_Class_Location_02_05_2014_MI.pdf_br (accessed Jul. 15, 2014)

[3] "Keeping Pipelines Safe from Internal Corrosion" http://www.ien.com/article/keeping-pipelines-safe/4252 (accessed Jul. 15, 2014)

[4] "Key Pipeline Accidents" - Philly.com. (n.d.). Retrieved from http://www.philly.com/philly/news/special_packages/inquirer/marcellus-shale/Key_Pipeline_Accidents.html_br (accessed Jul. 15, 2014)

[5] List of pipeline accidents - Wikipedia, the free encyclopedia. (n.d.). Retrieved from http://en.wikipedia.org/wiki/List_of_pipeline_accidents_br (accessed Jul. 15, 2014)

[6] MDCH - Kalamazoo River Oil Spill - Michigan. (n.d.). Retrieved from http://www.michigan.gov/mdch/0,4612,7-132-54783_54784_54984_59961---,00.html_br (accessed Jul. 15, 2014)

[7] National Transportation Safety Board – Pipeline Accident Brief – DCA-04-MB-006. PAB-06/01

[8] Natural Gas Fire Leaves Homes, Highway Destroyed. CDL Life. http://cdllife.com/2012/top-trucking-news/natural-gas-fire/leaves-homes-highway-destroyed

[9] Natural Gas Pipeline Accidents in the United States Since 2001. (n.d.). Retrieved from http://fracdallas.org/docs/pipelines.html_br (accessed Jul. 15, 2014)

[10] Part 192 Transp of Natural Gas by Pipeline. (n.d.). Retrieved from http://www.viadata.com/manualsdemo/desktop/mergedProjects/192cb/html/_192_935_what_additional_preventive_and_mitigative_measures_must_an_operator_take_.htm_br

[11] PHMSA - Home - The Pipeline Safety Act - Track PHMSA's Progress, http://www.phmsa.dot.gov/ (accessed Jul. 21, 2014)

[12] PHMSA - NPRM and ANPRM - 78 FR 46560. (n.d.). Retrieved from http://www.phmsa.dot.gov/portal/site/PHMSA/menuitem.ebdc7a8a7e39f2e55cf2031050248a0c/?vgnextoid=65dcc4a1c0cf0410VgnVCM100000d2c97898RCRD&vgnextchannel=2dd0d95c4d037110VgnVCM1000009ed07898RCRD_br (accessed Jul. 21, 2014)

[13] PHMSA: Stakeholder Communications - Pipeline Risk Management. (n.d.). Retrieved from http://primis.phmsa.dot.gov/comm/reports/safety/sigpsi.html_br (accessed Jul. 21, 2014)

[14] Pipeline Explosions Since 2001 http://fracdallas.org/docs/pipelines.html (accessed Jul. 15, 2014)

[15] Pipeline Safety: Class Location Requirements https://www.federalregister.gov/articles/2013/08/01/2013-18286/pipeline-safety-class-location-requirements (accessed Jul. 15, 2014)

[16] Pipeline Safety Regulations http://www.mrsc.org/subjects/pubsafe/pipesafetyregs.aspx (accessed Jul. 15, 2014)

[17] Pipeline Safety Trust. (n.d.). Retrieved from http://www.pstrust.org/docs/PST_Briefing_Paper_09_1.pdf_br (accessed Jul. 21, 2014)

[18] SOS Catastrophes - European Network Of Disaster Victims. (n.d.). Retrieved from http://www.soscatastrophes.org/Explosion-de-gaz-de-Ghislenghien_br (accessed Jul. 21, 2014)

[19] West Virginia Pipeline Blast in 2012 Blamed on Corrosion. Insurance Journal. March 12, 2014. http://www.insurancejournal.com/news/southeast/2014/03/12/323070.htm (accessed Aug 1, 2014)

新疆乌鲁木齐市天然气输配站场 HAZOP 分析

白元，陈勇，赖小林

（新疆维吾尔自治区安全科学技术研究院，新疆，乌鲁木齐 830002）

摘 要： 城市燃气门站是城市燃气输配系统的起点，对在役天然气站场进行 HAZOP 分析可以辨识出装置可能存在的设计、操作及管理缺陷，分析设备故障、作业过程中的人员失误等可能带来的各种后果及验证保护措施。实例应用表明通过 HAZOP 分析对提高设施安全、可靠性，为天然气输配站场系统风险评价工作有着积极的作用。

关键词： 输配站场；HAZOP 分析；风险评估

［中图分类号］TE832　［文献标识码］A

HAZOP Study for Natural Gas Station in Urumqi, Xinjiang, China

Bai Yuan, Chen Yong，Lai Xiaolin

(Xinjiang Uygur Autonomous Region Academy of Safety Science and Technology, Urumqi, 830002, China)

Abstract: City gas gate station is the starting point of urban gas pipeline system, HAZOP could identify and rank the hidden pipeline station hazards, analysis the equipment failure, human error in the process of the possible consequences and bring forward the suggested measure which helps the enterprises to regulate the hidden hazards. The application results showed that HAZOP could improve the security of station equipments and operation and take good effect on risk assessment.

Key words: Natural Gas Station; HAZOP analysis; Risk assessment

随着城市的发展和人口的增加，天然气管网不断扩大，随之而来的是天然气站场不断增多与越来越靠近人口稠密区域。天然气站场的泄露和爆炸事故会给人们带来严重的生命和财产损失，同时还会对周边环境造成巨大的污染。研究城市燃气门站的风险，实时对城市燃气门站进行风险评价，对促进天然气工业的健康发展，具有非常重要的作用。

国家安监总局高度重视、积极倡导推广应用危险与可操作性分析工作，先后发布了一系列文件，对在危险化学品领域开展危险与可操作性分析工作提出了明确要求。开展《危险与可操作性分析（HAZOP分析）应用导则》(AQ/T3049-2013)和《化工企业工艺安全管理实施导则》等标准规范的学习宣贯和培训工作，《中华人民共和国安全生产法》和《危险化学品安全管理条例》的颁布，进一步推动了我国风险评价技术向更广、更深的方向发展[1]。

HAZOP (Hazard and Operability Study)中文的意思是“危险性和可操作性分析”，是一种被工业界广泛采用的工艺危险方法，也是有效排查事故隐患，预防重大事故和实现安全生产的重要手段之一[2]。目的是识别危险和潜在的危险事件序列，通过分析生产运行过程中工艺状态参数的变动，操作控制中可能出现的偏差，以及这些变动与偏差对系统的影响及可能导致的后果，找出出现变动或偏差的原因，以及针对这些偏差，系统现有的安全措施，并提出相应的附加安全设施或操作程序，从而减少事故发生的频率及可能的后果，保障系统安全、高效运行[3]。

1 HAZOP 分析技术

1.1 HAZOP 原理

HAZOP分析是对危险与可操作性问题进行详细识别的过程，由各专业技术人员组成分析小组，以“分析会议”的形式进行。HAZOP分析包括辨识潜在的偏离设计目的的偏差、分析其可能的原因并评估相应的后果。会议期间，在分析小组组长的引导下，使用一套核心引导词，对系统的设计进行全面、系统地检查，识别对系统设计意图的偏差。该技术旨在利用系统的方法激发参与者的想象力，识别系统中潜在的危险与可操作性问题。HAZOP应视为一种基于经验的方法，用于完善设计，而不是要取代其他的经验方法[4]。本文HAZOP分析采用南京安元科技有限公司的“安元企业工艺安全管理 HAZOP系统软件”完成，软件HAZOP分析流程如图 1 所示。

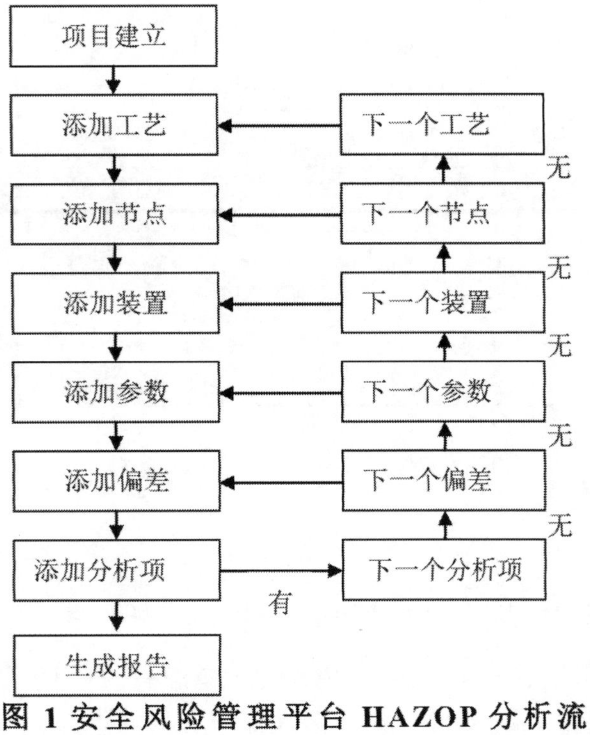

图 1 安全风险管理平台 HAZOP 分析流程图

Figure1- Hazop analysis flow chart of the system

1.2 HAZOP 分析步骤

首先，界定，在HAZOP分析组长的协助下，明确系统边界，以及系统与其他系统和周围环境之间的界面，不关注与分析范围和目标无关的区域。将HAZOP分析的焦点严格地集中于辨识危险，能够节省精力，并在较短的时间内完成。HAZOP分析组长应明确规定HAZOP小组的分工和职责，对HAZOP小组的人员应进行HAZOP培训，使HAZOP小组所有成员具备开展HAZOP分析的基本知识，以便高效地参与HAZOP分析。其次，准备，获得系统信息，包括管道仪表流程图（P&ID）、工艺流程图（PFD）、操作规程、安全连锁系统因果逻辑图、控制程序逻辑、平面布置图、设备数据表、操作规程、维护手册和事故记录等。最后，分析，将系统分解为若干部分，选择某一部分并明确设计目的，对每个要素使用引导词确定偏差，识别原因和后果，确定是否存在重大问题，识别保护、检测和显示装置，确定可能的补救/减缓措施，对建议措施达成一致意见，依次对每个要素重复以上步骤，然后对系统每个部分重复以上步骤[5]。

1.3 HAZOP 风险评价

风险是事故发生可能性和后果严重度的函数，是对事故的量化描述。根据后果所发生的概率和严重性的评估，计算出风险分值，按照风险矩阵(表 1) 确定风险等级。

表1 风险评分矩阵表

危害程度					可能性(增加)				
					A	B	C	D	E
级别	人员 P	财产 A	环境 E	声誉 R	在行业内未听说过	在行业内发生过	在公司内发生过	在公司内每年多次发生	在基层经常发生
1	无、轻微伤害	无、轻微损失	无、轻微影响	无、轻微影响	一般风险	一般风险	中度风险	高度风险	高度风险
2	小伤害	小损失	小影响	有限影响	一般风险	中度风险	高度风险	高度风险	高度风险
3	重大伤害	局部损伤	局部影响	很大影响	中度风险	高度风险	高度风险	严重风险	严重风险
4	一人死亡	重大损失	重大影响	全国影响	高度风险	高度风险	严重风险	严重风险	严重风险
5	多人死亡	特大损伤	巨大影响	国际影响	高度风险	严重风险	严重风险	严重风险	严重风险

2 燃气输配系统 HAZOP 分析

2.1 工艺系统简介

该站场为城市燃气管道首站，天然气来源于油田天然气长输管线，门站高压进口压力：P＝8.0Mpa，门站高压出口压力：P＝4.0MPa，设计流量为 50000Nm^3/h。天然气由高压管道输送到门站总进口阀门，经调压室调压系统将压力由 8.0MPa调到 4.0MPa的高压后、过滤；计量后经过加臭装置进入地下输气管道，分别向调压计量间和储罐供气；调压计量间天然气进入汇管 1，再经四路（两开一备一旁通）过滤、调压计量进入汇管 2；最后进入城市中压管道。其工艺流程图见图 2。

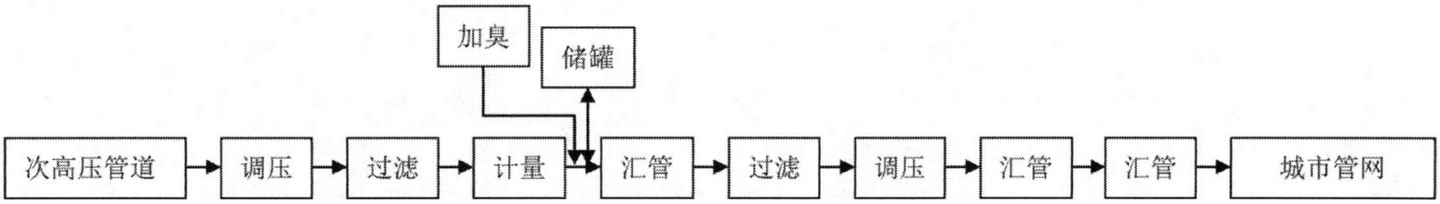

图 2　站场工艺流程示意图

Figure2- Flow chart of the natural gas station

2.2 分析对象及步骤

在完成前期准备工作的基础上，组织召开了 HAZOP 分析会议，站场工程师对分析对象的工艺流程和原理进行详细的讲解，分析评价组进行了该站现场数据分析验证，运用 HAZOP 分析对天然气输配站场进行了风险分析。根据现场工艺概况和工艺流程图划分了分析节点(表 2)。

表2　站场节点划分表

节点	节点类型	设计意图
1	进站系统	控制整个站场安全，对燃气进出站场进行总控制。
2	过滤系统	对油田分输站接来的天然气中的固态、液态杂质进行去除，提高进入管网中天然气的清洁程度，降低对管网设备的损坏。
3	计量系统	燃气计量
4	调压系统	降压、稳压
5	加臭系统	给燃气人为添加气味，提高燃气泄漏时的警觉性。
6	水浴系统	用于站场内各管线保温伴热。
7	放散系统	将站场内各段因安全、调压需要，多余天然气引至安全空地实行集中排放。
8	排污系统	将站场内液、固态废料收集、处理后进行集中排放。
9	全站控制系统	对全站具有自控设备的装置进行自动控制。

2.3 分析结果与建议

HAZOP 分析根据系统的工艺流程、工艺设备或操作程序的特殊点，站场使用的主要设备有电动球阀、汇管、气体成分含量分析仪、绝缘法兰、过滤器、超声流量计、调压器、储罐、流量计、伴热带、放散塔、放散管线、排污池、排污管线、控制系统等。表 3 列出了天然气站场系统进行 HAZOP 分析时，所有考虑的设备、引导词及参数，总计有 62 个偏差。

表3　站场主要设备及HAZOP引导词

设备	引导词	参数	设备	引导词	参数	设备	引导词	参数
电动球阀	过量	压力	调压器	过量	压力	伴热带	减量	温度
电动球阀	减量	压力	调压器	减量	压力	放散塔	过量	流量
汇管 1	过量	压力	调压器	减量	温度	放散塔	减量	流量
汇管 1	过量	液位	储罐	过量	流量	放散塔	相反	流量
汇管 1	减量	温度	储罐	减量	流量	放散塔	过量	压力
气体成分含量分析仪	异常	仪表	储罐	相反	流量	放散塔	减量	压力
绝缘法兰	过量	泄漏	储罐	过量	压力	放散塔	减量	温度
过滤器	过量	流量	储罐	减量	压力	放散管线	过量	流量
过滤器	减量	流量	储罐	过量	温度	放散管线	减量	流量
过滤器	过量	压力	储罐	减量	温度	放散管线	减量	温度
过滤器	减量	温度	储罐	过量	液位	放散管线	过量	泄漏
过滤器	过量	液位	储罐	减量	液位	排污池	过量	液位
超声流量计	过量	流量	储罐	过量	温度	排污池	减量	液位
超声流量计	减量	流量	储罐	减量	温度	排污池	减量	温度
超声流量计	过量	压力	流量计	过量	流量	排污池	过量	温度
超声流量计	减量	压力	流量计	减量	流量	排污管线	减量	温度
超声流量计	过量	温度	储药罐	过量	泄漏	排污管线	过量	压力
超声流量计	减量	温度	伴热带	过量	流量	排污管线	过量	流量
调压器	过量	流量	伴热带	减量	流量	排污管线	过量	泄漏
调压器	减量	流量	伴热带	伴随	流量	控制系统	过量	流量
调压器	相反	流量	伴热带	过量	温度			

表 4 所示为天然气输配站场的危险与可操作性（HAZOP）分析部分记录过程。对站场 HAZOP 分析工作中，针对现有的工艺系统提出近以下主要建议，并给出相应的风险等级评价。(1) 电动球阀定期检验安全附件；定期检查应急电源完好情况；建议按照有关标准重新核查装置区内可燃气体报警装置探头是否能够覆盖所有区域；每年定期对管线进行吹扫，去除施工等遗留的杂质。(2) 绝缘法兰采用具有弹性密封环结构的螺纹连接。(3) 过滤器定期检验安全附件，建议将电伴热温度信息传送至空中系统。(4) 超声流量计选择合适量程范围的温度表和温度变送器。(5) 调压器在调节阀前后增加压力变送器。(6) 储罐建议进行场站作业时，对自动放空阀门的阀位状态进行重点监控；建议出站阀门设置联锁，防止反串，引起下游管道超压；场站作业时，对自动放空阀门的阀位状态进行重点监控。(7) 伴热带建议将水浴温度信息传送至控制台。(8) 排污池建议用专用排污罐替代排污池进行回收处理。

表4 天然气输配站场HAZOP 分析部分记录表

偏差	原因	后果	风险矩阵			安全措施	建议措施
			后果	可能性	风险		
流量	下游管道或设	环境污染	1	2	I	压力检测	

过大	备破裂/泄漏						
	下游管道或设备破裂/泄漏	天然气损失/设施损坏	2	2	II	进、出站 ESD 阀	
	下游管道或设备破裂/泄漏	火灾或爆炸/员工伤害	4	1	III	工业电视	
	下游管道或设备破裂/泄漏	管道停输/降量	3	1	II	定期巡检	
	进气流量偏高	设施损坏	2	2	II	流量检测	
	上游管道或设备破裂/泄漏	管道停输/降量	3	1	II	定期巡检	
	出站 ESD 阀门误关断	管道停输/降量	3	1	II	压力检测、阀位检测、定期巡检	
	进气流量偏低	降量	1	2	I	流量检测	
	分站阀门/出站阀门关断	管道停输/降量	3	1	II	压力检测、阀位检测、定期巡检	
流量过小	放空阀门误开	天然气损失	1	3	II	定期巡检、阀位检测、阀门动作报警	建议进行场站作业时，对自动放空阀门的阀位状态进行重点监控。
	上游管道或设备破裂/泄漏	天然气损失/设施损坏	2	2	II	进、出站 ESD 阀	
	上游管道或设备破裂/泄漏	火灾或爆炸/员工伤害	4	1	III	工业电视	
	上游管道或设备破裂/泄漏	环境污染	1	2	I	压力检测	
逆流量	与善乌线相连的站场排气管线阀门误打开	下游直排管道超压	4	2	III	设置进出站	
	与善乌线相连的站场排气管线阀门误打开	站场无法进行正常检定工作	2	2	II	设置进出站 ESD 阀门	建议出站阀门设置联锁，防止反串，引起下游管道超压。
压力过低	放空阀门误打开	天然气损失，环境污染	1	3	II	定期巡检、阀位检测、阀门动作报警	建议进行场站作业时，对自动放空阀门的阀位状态进行重点监控

	下游管道或设备破裂/泄漏	管道停输/降量	3	1	II	定期巡检
	下游管道或设备破裂/泄漏	天然气损失/设施损坏	2	2	II	进、出站 ESD 阀
	下游管道或设备破裂/泄漏	环境污染	1	2	I	压力检测
	下游管道或设备破裂/泄漏	火灾或爆炸/员工伤害	4	1	III	工业电视
温度过高	进气温度偏高	设备损坏	2	2	II	选择合适量程范围的温度表和温度变送器
温度过低	来气温度低	下游调压装置冰堵	4	2	III	温度检测
液位高	加臭剂添加过量，药罐出口堵塞，控制阀失效	溢流、影响正生产	1	2	I	加臭药罐远传液位计，定期巡检
液位低	加臭剂添加不足，药罐泄漏，现场液位计不准	设备损坏	1	1	I	远传液位计，定期巡检
温度过高	夏季温度高	罐内压力升高、达到闪点温度，有爆炸危险	1	1	I	远传温度，定期巡检，避阴保存。
温度过低	冬季温度过低	易混性降低，雾化效果不佳。	1	1	I	远传温度，定期巡检。
流量过大	控制器失效，加药量过大	管线内腐蚀	1	1	I	流量计流量远传自动控制
流量过小	排液泵未开、堵塞，流量计失效	设备损坏，加臭失效	1	2	I	流量计流量远传自动控制
过量	储罐密封失效，人为破坏	环境污染，影响正常生产。	2	2	II	储罐定期检定，定期巡检。

3 结论

本次 HAZOP 分析针对天然气站场现有的工艺系统划分为 9 个节点，提出 62 个偏差，描述了现有的安全保护措施，一般风险 51 个，中度风险 58 个，高度风险 14 个，提出了 8 条主要建议措施，并给出相应的风险等级评价。应用实践证明，HAZOP 分析技术适用于天然气输配站场的工艺安全分析，HAZOP 分析可以辨识出装置可能存在的设计、操作及管理缺陷，分析设备故障、作业过程中的人员失误等可能带来的各种后果及提高设施安全可靠性，为天然气输配站场系统风险评价工作有着积极的作用。

参考文献

［1］ 危险与可操作性分析（HAZOP 分析）应用导则[S].中华人民共和国安全生产行业标准 AQ/T:3049—2013.

［2］ 吴重光.危险与可操作性分析（HAZOP）应用指南[M].中国石化出版社，2012：1-10.

［3］ 冯文兴,贾光明,谷雨雷,等.HAZOP 分析技术在输油管道站场的应用[J].油气储运，2012，31(12)：903-905.

［4］ 赵宏振.长输管道天然气站场 HAZOP 分析实践[J].天然气与石油,2010,28(4):1-4.

［5］ 付建民,王新生,徐长航,等.操作规程 HAZOP 分析技术原理与应用分析[J].中国安全生产科学技术,2013,9(5):111-117.

作者简介：白 元：(1986-)，男，新疆乌鲁木齐，主要从事城镇燃气、安全科学方面研究。
E-mail: baiyuan1215@sina.com

2nd CCPS CHINA CONFERENCE ON PROCESS SAFETY™

The Applied Research on Quantitative Risk Assessment Models and Software for High Pressure Gas Pipeline Network in Urumqi, Xinjiang,China

Chen Yong
Xinjiang Uygur Autonomous Region Academy of Safety Science and Technology
111 Jianguo Road, Urumqi, Xinjiang
316910929@qq.com

Bai Yuan
Xinjiang Uygur Autonomous Region Academy of Safety Science and Technology
111 Jianguo Road, Urumqi, Xinjiang
472362673@qq.com

Lai Xiaolin
Xinjiang Uygur Autonomous Region Academy of Safety Science and Technology
111 Jianguo Road, Urumqi, Xinjiang
25180988@qq.com

UNPUBLISHED

The Applied Research on Quantitative Risk Assessment Models and Software for High Pressure Gas Pipeline Network in Urumqi, Xinjiang,China

Chen Yong
Xinjiang Uygur Autonomous Region Academy of Safety Science and Technology
111 Jianguo Road, Urumqi, Xinjiang
316910929@qq.com

Bai Yuan
Xinjiang Uygur Autonomous Region Academy of Safety Science and Technology
111 Jianguo Road, Urumqi, Xinjiang

Lai Xiaolin
Xinjiang Uygur Autonomous Region Academy of Safety Science and Technology
111 Jianguo Road, Urumqi, Xinjiang

Keywords: Gas pipeline; Quantitative risk assessment; Software, Application

Abstract

In recent years, as the project of changing fuel from coal to natural gas progress in Urumqi of China, a large number of high-pressure gas pipeline enter into the center city, bring huge risk to city public safety. In order to quantitatively assess the risk of gas pipelines, the quantitative evaluation index system, which include inherent risk index, causation index, consequence index and the weight of each index was established. Based on the quantitative evaluation index system, the gas pipeline network qualitative risk analysis Software was developed. The Software can calculated each pipeline quantitative risk value based on the input data automatically, and the weight of each factor can also be modified according to the actual needs. More than 50 high pressure, (>1.6MPa) gas pipelines were conducted quantitative risk assessment in Urumqi using the qualitative risk analysis Software. The assessment results can provide data support for the management of city gas pipeline network, and also can provide basis for establishing forecasting systems and mechanism of disaster prevention and mitigation.

1. Introduction

Since 2010, Urumqi, taken adjustment of heating energy structure in order to improve air quality as a permanent cure, began implement the "coal to gas" project. At the end of 2012, the total Urumqi mileage of natural gas transmission and distribution pipelines is about 3908.46 km, which the high-pressure pipelines of is 67.94 km, secondary high-pressure pipeline is 191.53 km the low pressure pipeline is 3589 km, the total gas pipeline length of the network increased 1042.47 kilometers compared with the end of 2009.With the increasing of Urumqi urban size and the gas pipeline network mileage, the gas fire, explosion, poisoning and other accidents have occurred, bring great negative impact to public security and social harmony and stability. Gas pipe network accident has become the major risk factor after traffic accidents and industrial accident[1-3]. Because of the high pressure gas pipeline network with a high security risk, a large degree of harm, it is necessary to assessment the security risk level of Urumqi high-pressure gas pipeline network.

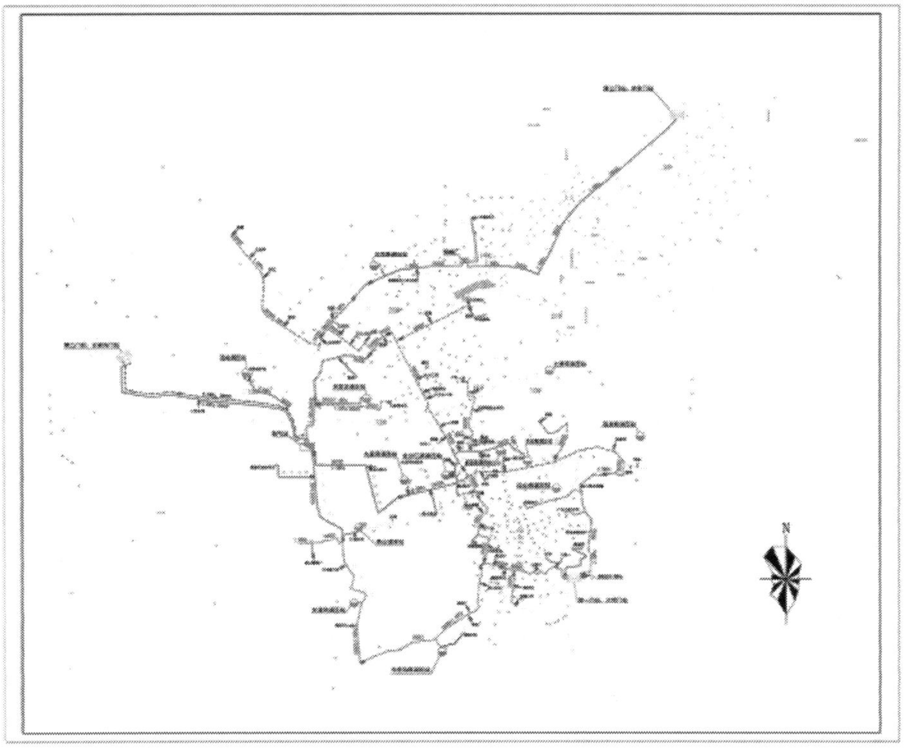

Figure 1. The Schematic of Urumqi Downtown Natural Gas Transmission and Distribution Mainline

2. Method

2.1 Quantitative risk analysis model building

In this paper, an integrated quantitative risk analysis method for natural gas pipeline network is proposed. The quantitative evaluation index system is composed of network vulnerability index, accident incentive index, accident consequences index and the weights of each index[4-6]. Network vulnerability index can be divided into network operation and line pipe laying construction, including pipeline gas flow, pipeline gas pressure, the minimum depth, pipe diameter, wall thickness, pipeline life and so on. Accident incentive index is divided into external interference index corrosion index, design flaw index, misuse index, geological activity index, including the ground activities, electrochemical corrosion etc. Accident consequences index is composed of leakage hazard index and accident impact index, including population density, large-scale distribution of public facilities and buildings, medium risk, property distribution, urban lifeline distribution .The quantitative risk analysis index can describe both the likelihood and severity degree of the accident, taken into account the effects of hazards factors mutual coupling to the integrated risk.

2.2 Qualitative risk analysis software development

Based on the quantitative evaluation index system, the gas pipeline network qualitative risk analysis Software was developed. Measures for risk management of gas pipeline network can be given automatically from the software based on the calculate risk level. The software can calculated each pipeline quantitative risk value based on the input data automatically, and the weight of each factor can also be modified according to the actual needs.

3. Results and analysis

More than 50 high pressure, (>1.6MPa) gas pipelines were conducted quantitative risk assessment using the qualitative risk analysis software.

Table 1. Urumqi Downtown Natural Gas Transmission and Distribution Mainline Security Risk Value

Pipeline No.	Vulnerability Value(A1)	Incentive Value(A2)	Consequences Value(A3)	Security Risk Value
1	4.84	5.89	4.37	124.40
2	3.13	5.39	6.55	110.48
3	3.81	6.21	7.49	177.17
4	3.13	6.09	7.49	142.76
5	2.70	5.55	5.93	88.88
6	2.82	4.71	10.86	144.50
7	2.97	5.90	7.49	131.18
8	2.83	5.85	7.49	124.15
9	2.45	4.58	1.87	21.00
10	2.26	4.82	5.93	64.49
11	2.90	4.85	3.74	52.70
12	2.26	4.92	3.43	38.08
13	3.15	4.36	4.68	64.10

14	2.93	5.49	11.85	190.22
15	2.90	4.30	5.62	70.17
16	2.90	5.18	8.42	126.84
17	2.50	4.37	11.85	129.39
18	2.75	4.94	10.37	140.63
19	2.23	5.11	11.85	135.05
20	2.48	5.61	1.87	26.00
21	2.48	5.96	1.87	27.65
22	2.82	4.75	10.37	138.93
23	2.61	5.93	2.81	43.44
24	2.82	4.68	7.49	98.96
25	3.51	5.41	7.17	136.14
26	3.51	5.19	7.49	136.13
27	3.51	5.23	7.17	131.59
28	3.91	4.53	3.43	60.72
29	3.13	5.65	11.85	209.89
30	2.51	4.20	11.85	125.11
31	2.89	4.16	13.33	160.38
32	3.30	4.34	13.33	191.23
33	2.89	5.30	14.32	219.28
34	2.89	5.03	5.62	81.67
35	2.89	4.75	4.68	64.21
36	3.03	5.25	7.49	118.87
37	2.89	6.18	4.99	89.11
38	2.89	5.06	13.33	195.06
39	2.89	4.69	6.55	88.85
40	3.70	5.90	1.87	40.83
41	2.89	5.54	0.94	14.98
42	2.89	5.08	7.17	105.40
43	3.00	6.22	5.62	104.94
44	4.08	6.21	5.62	142.27
45	3.14	2.63	1.87	15.44
46	2.89	4.63	10.37	138.68
47	2.89	4.71	11.85	161.16
48	2.89	5.54	6.86	109.92
49	2.89	4.65	1.87	25.17
50	2.93	4.68	6.55	89.68
51	4.70	6.22	10.37	303.35
52	2.58	5.13	10.86	143.62
53	2.58	5.20	11.85	158.60
54	2.73	4.64	4.06	51.29
55	4.03	4.65	5.62	105.27
56	4.06	6.21	2.81	70.75

57	3.63	6.32	7.49	171.93

The above risk assessment results shows: the average risk value of the 57 gas pipeline is 112, One pipeline risk assessment value is greater than 300, having a high degree of risk; Two gas pipeline risk assessment values are between 200-300, having a certain risk, Fifty four gas pipeline risk value is less than 200, they are relatively safe. Urumqi's high-pressure gas pipeline is overall safe based on the following main reasons: First, Urumqi's vast majority of high-pressure gas pipeline were builded after 2010, which having higher construction standards and material specifications, the gas pipeline having high intrinsic safety. Secondly, there is no flooding, landslides and mudslides, river alluvium and other geological activity in Urumqi, which may affect pipeline external security. Finally, Due to the relatively low population density of Urumqi, property distribution is not concentrated, there are not sensitive targets of the vast majority of pipeline crossing area, the severity of the consequence of accidents is relatively low.

4. Conclusion

The results show that the quantitative risk assessment models and software can be used in practical application. Results of the assessment values are consistent with Urumqi's high-pressure gas pipeline true risk level. The assessment results can provide data support for the management of city gas pipeline network, and also can provide basis for establishing forecasting systems and mechanism of disaster prevention and mitigation.

5. References

[1] Montiel H, Vilchez JA, Amaldos J., " Historical Analysis of Accidents in the Transportation of Natural Gas," *J. Hazardous Materials*, Volume 51, NO. 1-3,1996

[2] European Gas Pipeline Incident Data Group. 7th Report of the European Gas Pipeline Incident Data Group. 2008

[3] China Fire Statistical Yearbook[M]. Ministry of Public Security Fire Department of China.BeiJing: Chinese People's Public Security University Press, 1996-2005

[4] Han Z Y, Weng W G., "An Integrated Quantitative Risk Analysis Method for Natural Gas Pipeline Network," *Journal of Loss Prevention in the Process Industries*, Volume 23, NO. 3, 2010

[5] Han Z Y, Weng W G., "An Overview of Gas Pipeline Network Quantitative Risk Analysis," *China Safety Science Journal*, Volume 19, NO. 17, 2010

[6] Jonkman S N, van Gelder P H A J M, Vrijling J K., "An Overview of Quantitative Risk Measures for Loss of Life and Economic Damage," *J. Hazardous Materials*, Volume 99, NO. 1,2003

2nd CCPS China Conference On Process Safety

Study of Special Hazop Technology for SIL Determination

Qingdao, China
August 28 – 29,2014

Tang Bin
CNOOC Safety Technology Service Co., Ltd

Content

This PPT will introduce the following contents:

1. Principle of process safety accident occured
2. Prevention and mitigation of process safety accident
3. SIS introduction
4. SIF and SIL introduction
5. SIL determination
6. LOPA introduction
7. Normal Hazop
8. How to convert information from Hazop to LOPA
9. Special Hazop for LOPA to determine SIL
10. Conclusion

Route of process Safety Accident Occured

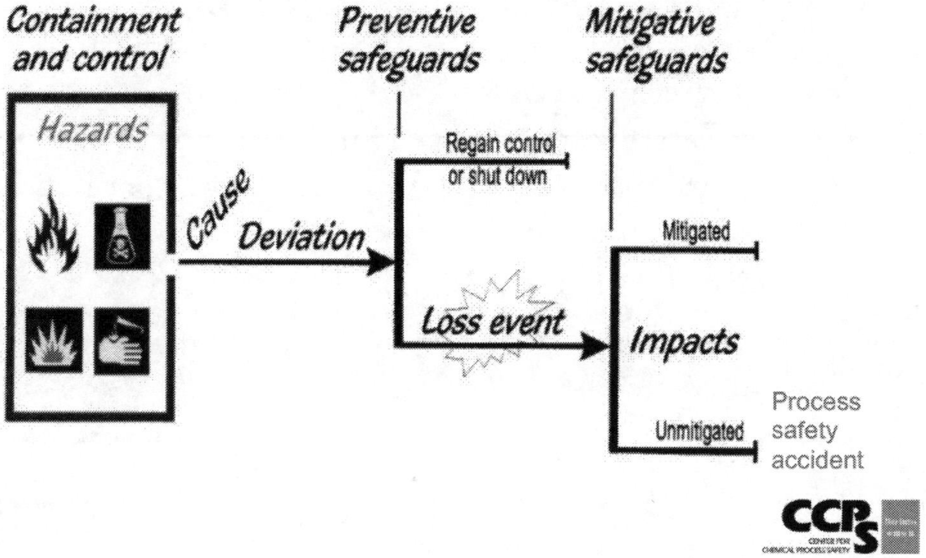

Control Model of Material and Energy

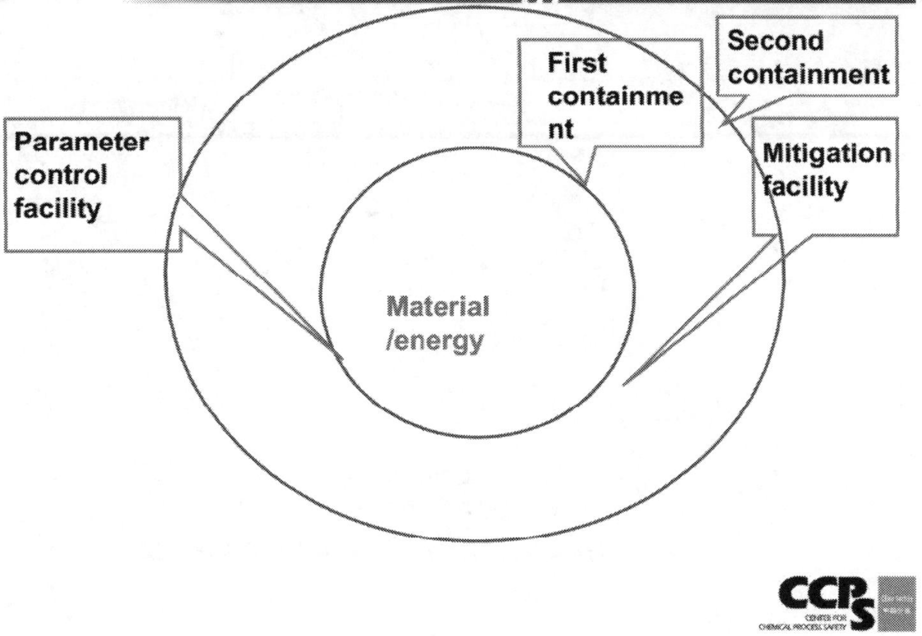

Control Model of Material and Energy(2)

During normal production, two following conditions must be met to prevent material leakage from first containment:

- The parameter below design limit
- First containment integrity , i.e. first containment must can contain material well when no parameter deviation

CCPS

Onion model

Shown in Order of Activation

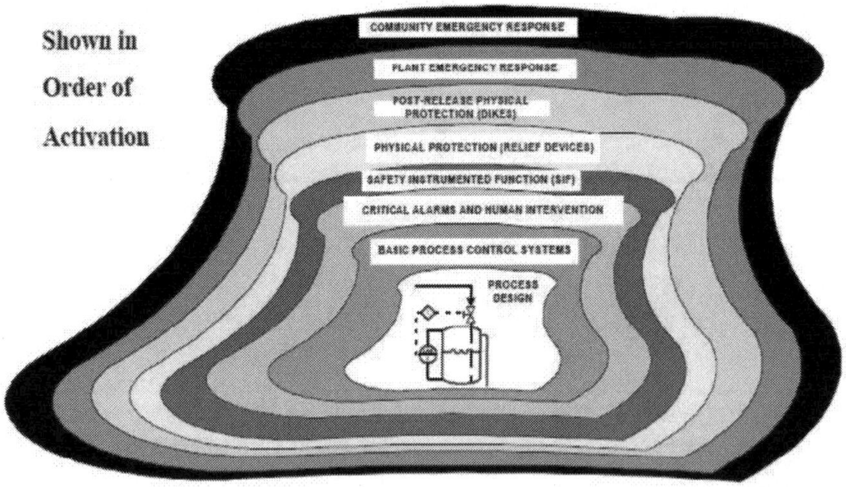

CCPS

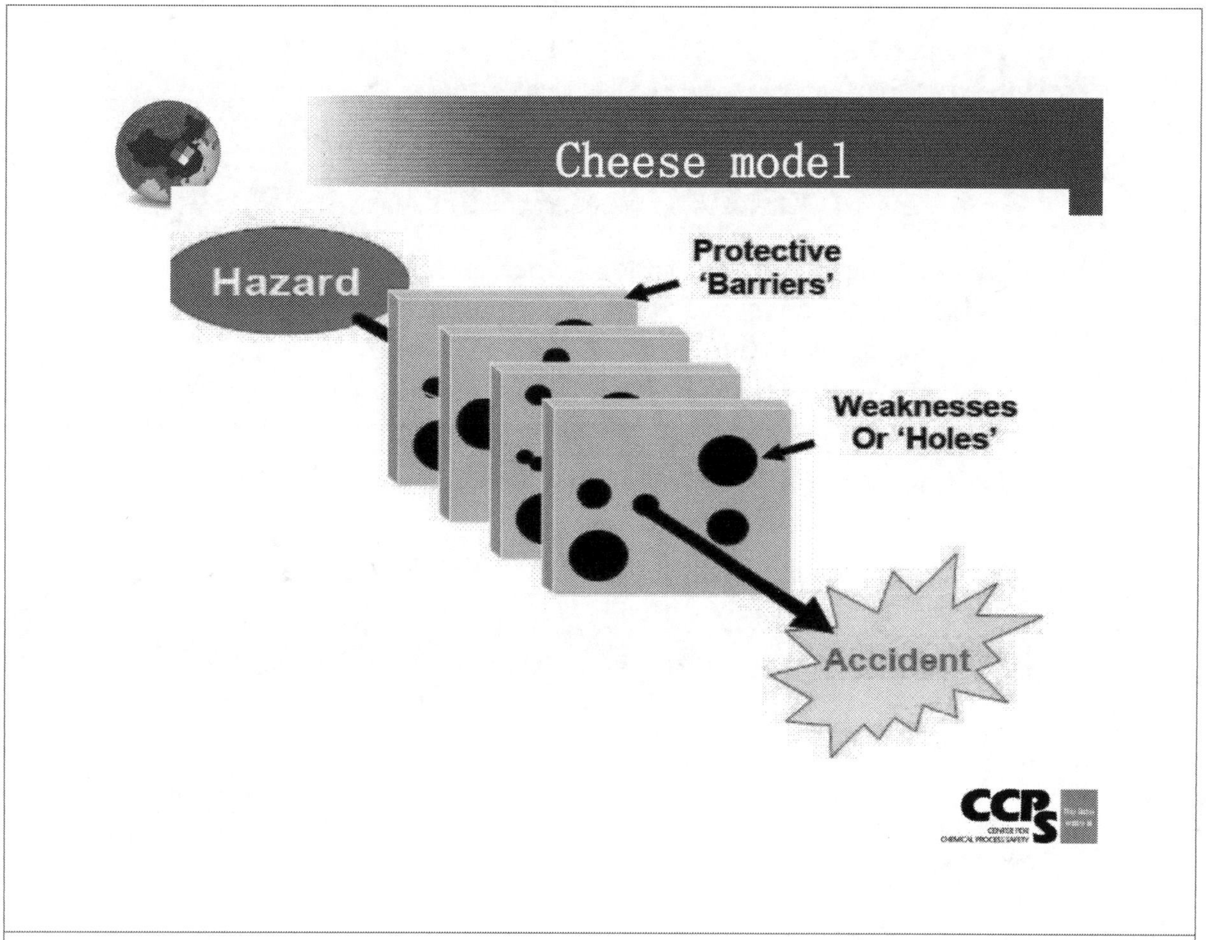
Cheese model
Hazard
Protective
'Barriers'
Weaknesses
Or 'Holes'
Accident
CCPS
CENTER FOR
CHEMICAL PROCESS SAFETY

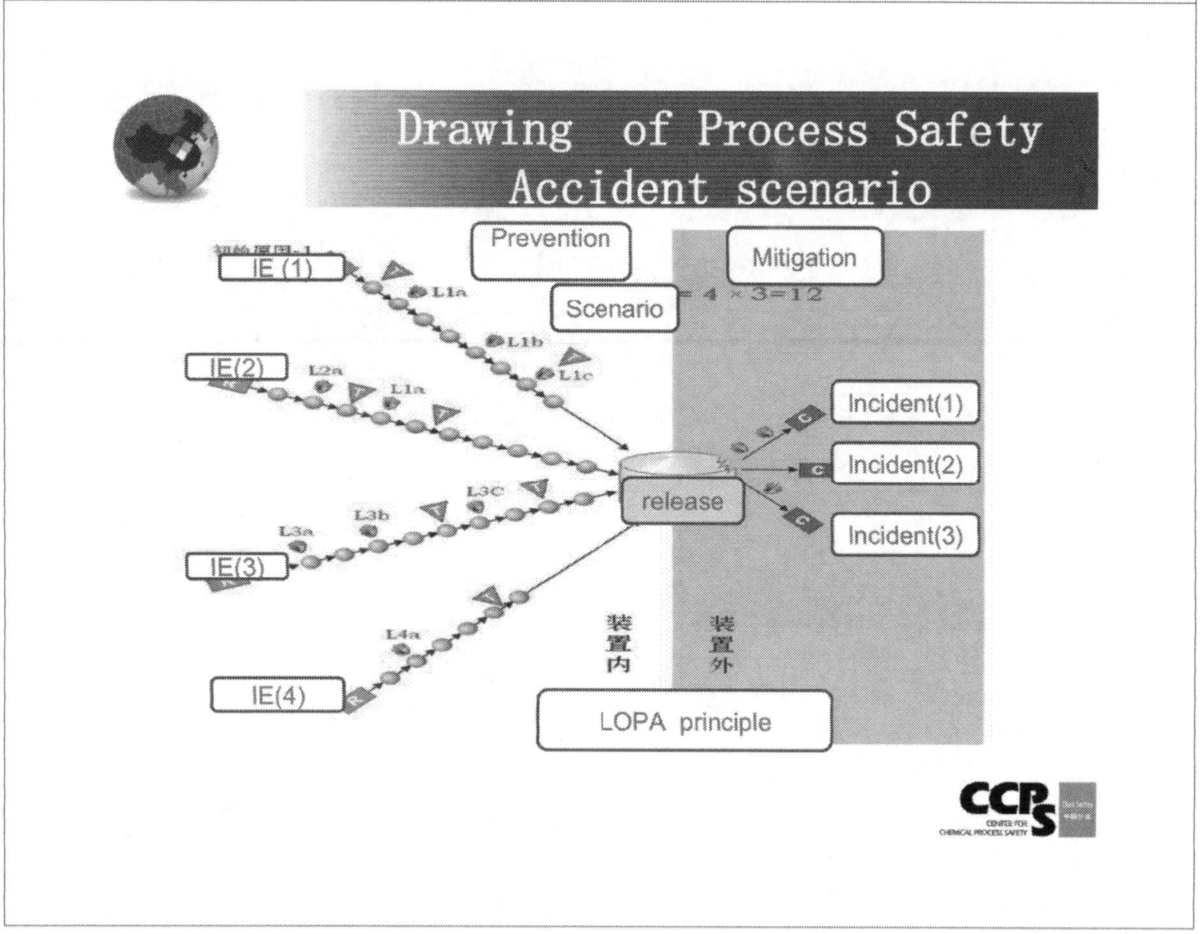
Drawing of Process Safety
Accident scenario
Prevention
Mitigation
Scenario
IE (1)
L1a
L1b
L1c
IE(2)
L2a
L1a
Incident(1)
Incident(2)
release
L3C
L3b
L3a
Incident(3)
IE(3)
L4a
IE(4)
装置内
装置外
LOPA principle
CCPS
CENTER FOR
CHEMICAL PROCESS SAFETY

How to Reduce Process Safety Accident

According to onion model, cheese model and drawing of process safety scenario, we can get the following conclusion:

1. Enough protection layers needed to reduce frequency of process safety accident
2. Average Probability of failure on demand (PFD) is as lower as possible
3. No common cause failure between different layer-independent
4. Protection layer includes: process design, BPCS, key alarm plus human intervention, SIF, physical protection(relief devices)

CCPS CENTER FOR CHEMICAL PROCESS SAFETY

When need SIF(1)

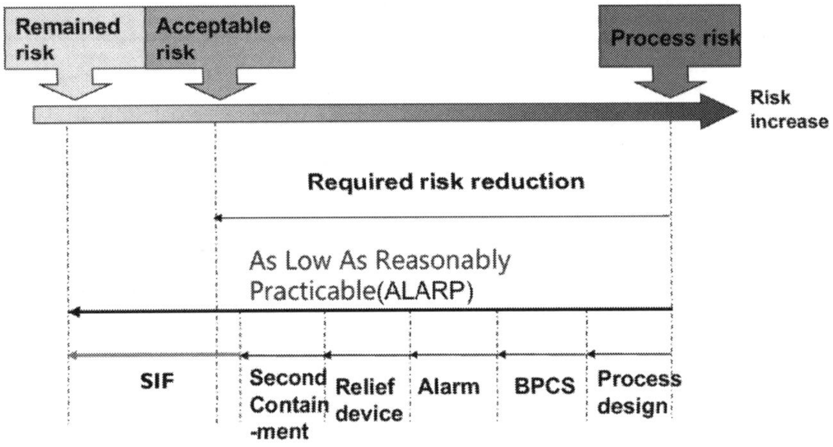

CCPS CENTER FOR CHEMICAL PROCESS SAFETY

Introduction of SIS ,SIF, SIL (1)

•

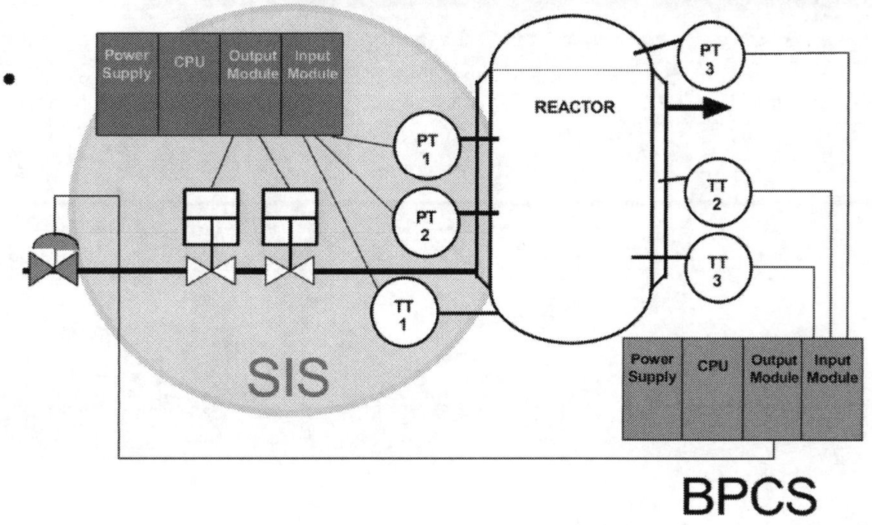

Introduction of SIS ,SIF, SIL(2)

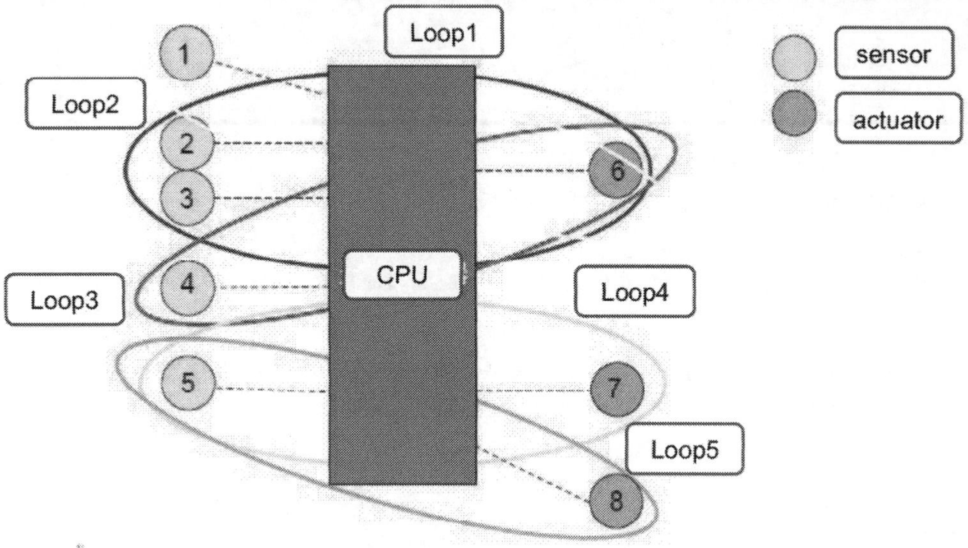

Introduction of SIS ,SIF, SIL(3)

SIL (Safety integrity level)	PFD (Probability of failure on demand)	RRF (Risk reduction factor)
4	$\geq 10^{-5}$ to $< 10^{-4}$	> 10,000 to ≤ 100,000
3	$\geq 10^{-4}$ to $< 10^{-3}$	> 1000 to ≤ 10,000
2	$\geq 10^{-3}$ to $< 10^{-2}$	> 100 to ≤ 1000
1	$\geq 10^{-2}$ to $< 10^{-1}$	>10 to ≤ 100

CCPS CENTER FOR CHEMICAL PROCESS SAFETY

Introduction of SIS ,SIF, SIL(4)

SIF is one of protection layer to reduce process risk , i.e. one piece of onion ,but this piece of onion has below characters:

- Compared with other protections, it is more expansive , so only consider to use it when other protections cant reduce the risk below acceptable level;
- We can select a PFD(changeable) of SIF, but PFD of other protection can be selected.

CCPS CENTER FOR CHEMICAL PROCESS SAFETY

When need SIF(1)

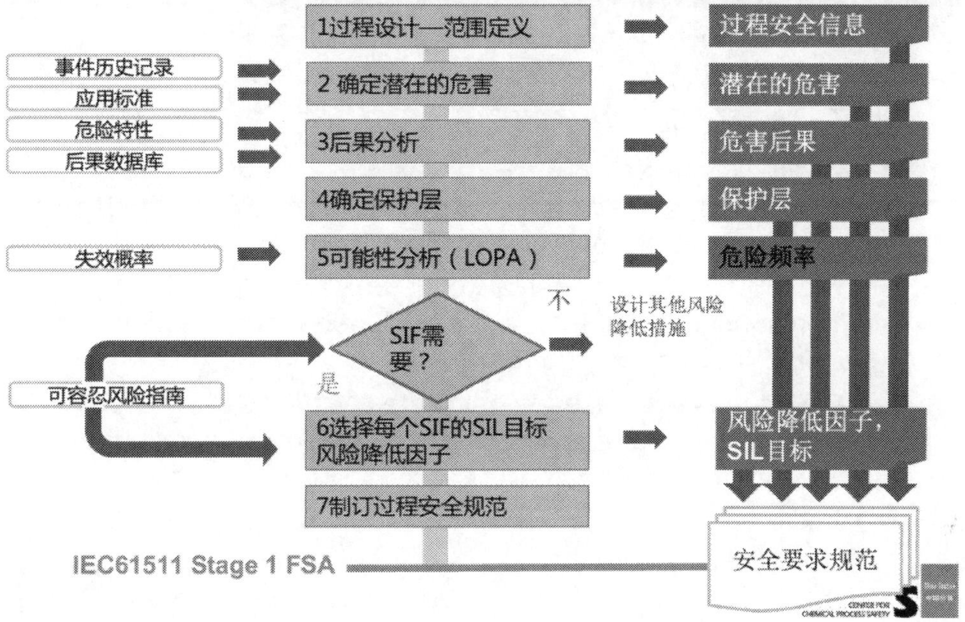

SIF is Set Really

SIF set really: At primary design stage (basic design stage) , no PHA and LOPA calculation, SIF and other protection layers have been designed, but mostly, SIL of SIF is not determined, which will lead to :

1. Lower SIL results in less protection, which can not reduce risk below acceptable level;

2. More protection layers results in remained risk much less than acceptable level, which leads to more investment and higher maintenance cost later

In order to overcome above shortcoming, LOPA method should be used to determine SIL

Basic Principle of LOPA

- LOPA can be used to calculate PFD of SIF

$$PFD_{SIF} = \frac{F_{\text{acceptable}}}{F_{\text{initial event}} \times PFD_1 \cdots PFD_N}$$

- If PFD>1, no SIL required, but SIF function is normally suggested to be maintained. If PFD< 1, SIL can be determined according to form of 13th page;
- Initial event(IE) for a special scenario should be determined when conducting LOPA , and all independent protection layers(IPL) following the IE as well.
- LOPA can only be used to calculate frequency of process safety accident, and can't be used to determine accident scenario (IE and IPL)
- In order to overcome the shortcoming of LOPA, HAZOP is needed as supplement
-

Principle of Hazop

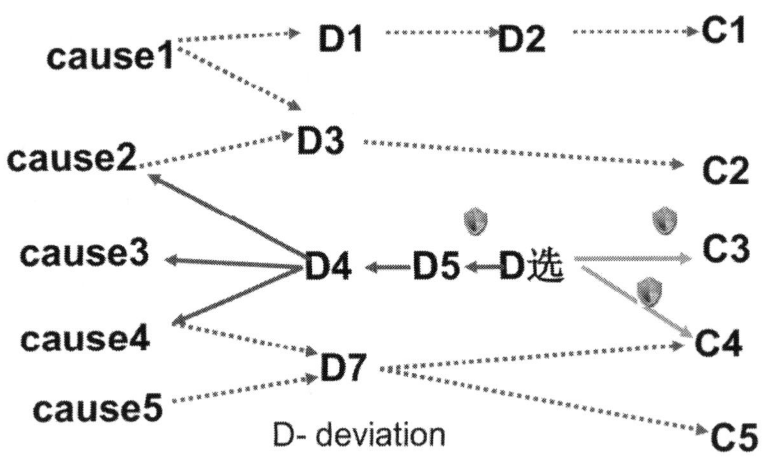

D- deviation

C= consequence

=Safeguard

18

Information of HAZOP and LOPA

LOPA need information	Information from Hazop
Scenario including SIF	Deviation
SIF description	N
Consequence	Consequence
Tolerable risk(frequency)	N
Initial event(IE)	cause
Frequency of IE	N
Enabling event	N
Frequency of enabling event	N
Independent protection layer(IPL)	safeguard
Failure probability of IPL	N
Description of outcome modification （ignition, probability of personnel in affected area, personal injury）	N
Probability of outcome modification	N

Cant Get information from HAZOP for LOPA (1)

Deviation : May lack of parameter deviations protected by SIF in HAZOP form, which result scenarios cant be got, which is protected by SIF from HAZOP

Cause: LOPA need use IE, but cause of Hazop study may not be IE

Enabling event/condition: LOPA may need this information , but no enabling event/condition in Hazop form, resulting in no enough information for LOPA

Consequence: LOPA need the final outcome which has a acceptable risk , but consequence of Hazop may not have a acceptable risk

Independent protection layer: LOPA need IPL to calculate frequency of a scenario, but safe guard of Hazop may not be IPL

常规HAZOP无法和LOPA对接（2）

- **Outcome modification:** When conducting LOPA, sometimes, outcome after material leakage may need to be modified, but no modification information in Hazop
- 5 differences mentioned above, may result we can't get enough information from Hazop for LOPA

Requirement for Special Hazop

- **Deviation :** Deviation of HAZOP must include parameter deviations which are protected by SIF
- **Cause:** Cause of Hazop should find direct cause, i.e. initial event, which has a frequency
- **Consequence :** Consequence of Hazop should find final HSE or property lost.
- **Safe guard :** Current safe guard should divided into non-IPL and IPL
- **Scenario:** Each scenario should be descripted in Hazop form

Example of Special HAZOP

NO	Deviation	Cause	Consequence	Non-IPL	IPL(1)	IPL(2)	IPL(3)	COMMENTS
1.1	Test seperatorW HPB-V-1301 high high pressure	Anyone of downstream valves closed	Rupture of test separator, and material leakage result in potential fire	1higher pressure alarm 2.safety valve PSV1301	Safety valve V1302		N	
		Malfunction of upstream well choke	Rupture of test separator, and material leakage result in potential fire	higher pressure alarm	Safety valve V1302	N	N	
1.2	Test seperatorW HPB-V-1301 low low pressure	Rupture of test separator and associated pipe due to external affection	material leakage result in potential fire	Lower pressure alarm	N	N	N	

CCPS CENTER FOR CHEMICAL PROCESS SAFETY

LOPA Calculation with Information from Special Hazop

- **Example of SIL determination**

Microsoft Word 文档

CCPS CENTER FOR CHEMICAL PROCESS SAFETY

Conclusion

- SIF is a special protection layer, whose SIL can be selected according to different risk
- LOPA is a good method ,which can be used to determine SIL
- HAZOP is a good method, which can be used to determine accident scenario
- Normal HAZOP need to be adjusted for special HAZOP, the information can be input to the LOPA for SIL determination

Thanks!

加强消防安全管理，构筑聚乙烯装置生产防火墙

郭运涛[1]马媛[2]贺新佳[3]司广顺[4]

（1.2.3.4 中国石油独山子石化分公司 新疆维吾尔自治区克拉玛依市独山子区，邮编：833699）

摘　要：聚乙烯装置生产流程比较长，动静设备较多，好多设备都是在高温高压下运行，又是可燃性气体，容易发生火灾的系数较高。所以聚乙烯装置很注重这方面的管理与经验总结。结合聚乙烯装置近年来在消防安全管理的一些做法，本文着重介绍聚乙烯装置如何在车间管理中实施这些做法。通过这些做法的实施，来构筑聚乙烯装置安全生产的防火墙。

关键词： 消防 安全 聚乙烯

Strengthen Fire Safety Management, to Build a Production Firewall Polyethylene Device

Yuntao Guo[1] Yuan Ma[2] Guangshun Si[3] Xinjia He[4] ,

（1.2.3.4 PetroChina Dushanzi Petrochemical Company Dushanzi Karamay City the Xinjiang Uygur Autonomous Region, postcode：833699）

Abstract: Polyethylene deice has a long and complex production process, It contains so many rolling equipments, most of them run in a condition of high pressure and high temperature, more crucial is that process fluids are inflammable, so it is easy to catch fire. Most Polyethylene deices take those management and experience seriously. Concluding those fire protection measures that have been adopt in Polyethylene deice , this paper focus on some novel methods, putting them into practice will construct a reliable firewall to keep the safety production of Polyethylene deice.

Key words: fire protection safety Polyethylene deice

引言：

独山子石化公司乙烯厂聚乙烯装置是以乙烯为主要原料，丁烯-1 或己烯为共聚单

体，H_2为分子量调节剂，经气相流化床反应器反应，以生产聚乙烯树脂为最终产品的化工生产装置。设备数量已经超过 700 台，有 500 多个控制阀，200 多个仪表控制回路，包括溶剂回收、化学品储存、催化剂制备、活化单元、预聚合和聚合等 12 个生产工序，采用的是DCS计算机控制系统。聚乙烯装置和其它生产装置相比，具有工艺流程长、工艺复杂、影响因素多和操作难度大的特点；设备上具有种类繁多、单台无备用设备多、专用设备多和设备管理难度大的特点；安全上经常使用的有毒、有害及危险化学品就达 20 种之多，安全责任重大[1]；发生火灾的系数也就比较高了，所以车间特别注重防火防爆的方面的管理，每年都进行相关方面的业务培训，提高车间防火防爆的能力。主要通过以下几个方面来进行实施。

1 加强对聚乙烯操作人员的业务培训增强主人翁意识提高他们应对突发问题的能力

火灾事故的发生一般都具有突发性的共性，所以在操作的过程中，严禁设备超温超压超负荷运行。外操在巡检的过程中，要对其巡检的区域进行仔细的检查，防患于未然。一旦发生泄漏，要迅速的赶到现场，查看泄漏的情况并迅速的切断物料的来源，同时启动应急处理预案，对可燃性气体进行稀释。在处理时要保持冷静的心态，切不可慌乱。内心的恐慌往往会造成事故的进一步的扩大。

增强员工的主人翁意识有助于车间的消防安全管理，如果一个泄漏点能及时的被发现，就有可能避免一场火灾的事故的发生。车间在这方面下了很大功夫，先后发现一些大的泄漏点，如抽出线管线震裂泄漏，乙烯进料线自动控制阀门密封失效的泄漏，都是由于员工的及时发现而避免了灾难的发生。

这方面的培训不能松懈，要经常抓，使他深入到每位员工的心中。这样才能确保生产的安全进行。

2 加强对聚乙烯装置检修工作尤其是动火作业管理

在检修的过程中由于某些措施未到位，往往成为火灾发生的隐患。例如在作业中未使用防爆工具，未对检修周围环境进行烃类气体含量检测，就盲目的施工。所以检修作业要进行作业前风险评估，开具作业票，现场要有监护人。施工人员要有工作资质，车间监护人要对其进行现场作业指导，确保整个作业过程处于受控状态。如果是动火作业，不但要对作业点进行测爆，更要对周围的环境进行测爆，在作业现场要放置灭火器，应对发生的初期火灾。在动火的过程要严格的遵守“四不动火”的原则即“没有批准的作业风险评估许可作业票不用火；没有防火措施或用火措施不落实不用火；没有用火监护人不用火；动火部位、动火时间与作业风险评估许可作业票不符不动火。”加强对检修工作尤其是动火作业的管理，降低发生火灾的几率，提高车间对动火作业管理的能力。

3 适时的进行聚乙烯装置的防火防爆的应急演练

聚乙烯装置每个月都要举行防火防爆的应急演练，以此来检验员工的处理问题的能力。在演练的过程中要学会如何安排人员到现场去处理、如何向上级汇报、如何采取有效措施来避免事故扩大化，通过一次次的演练使我们应急预案牢记在员工的心中，万一发生这种情况使员工有章法可依，不至于慌乱。演练前有讲解，演练结束后有讲评，让大家认识到自己的不足之处，不断对我们的应急预案进行改进，使它更加贴合我们实际生产。这样才能达到我们演练的目的。

4 对照近年来其他单位发生的火灾事故来查找自己的不足

通过不断学习他们的事故案例，来查找我们存在的不足之处，对我们的不足之处做一个修改，把隐患消灭在萌芽状态。例如茂名石化公司因雷击发生火灾的事故，通过对照发现我们装置上的避雷设施，年久失修，好多都已经失效，如果被雷击中岂不是一样会发生事故，亡羊补牢犹未晚也。还有发生在我厂某车间由于玻璃管液位计冻裂而发生的大面积泄漏事故，而我车间也存在类似的玻璃管液位计，通过积极排查消除安全隐患，保证了装置的平稳运行。所以事故案例一定要认真的去学习，分析其中的来龙去脉，得出自己的东西，更要避免犯类似的错误。

5 加强对消防器具使用的培训

冬季和夏季是化工企业发生火灾事故的高发期，所以车间每年都在入冬前和入夏前，组织员工对装置内的消防灭火器具进行实际演练，通过实际的灭火演练让员工掌握各种消防灭火器具使用方法，并达到熟练地应用。在灭火的过程中消除员工对于火灾的恐惧心理，提高了员工应对火灾事故的能力。也让员工知道了如何在火灾中保护自己不受到伤害。使车间整体处理应急事故的能力也得到了提高。

6 及时的检查消防设备是否处于良好的备用状态

每个月都要对装置上的消防设备进行检查，包括干粉式灭火器、消防泡沫车、消防栓炮以及消防地井。确保他们时时处于备用状态。干粉式灭火器的检查主要查看它的压力是否还在工作状态，如果压力偏低就要及时的充压；消防泡沫车的检查主要查看它的发泡剂是否还有效，如果失效了就要及时的更换；消防栓炮要及时的对其加脂，确保它转动自如。无跑冒滴漏，冬季要把里面的存水放净，并用工厂风进行吹扫，防止冻裂；消防地井冬季要加盖保温井盖防止冻裂管线。冬季下雪后要及时的打扫积雪。

7 装置现场各种探测器的正确运用

我装置共有各种探测器近 200 个，其中包括可燃气体探测器、热感探测器和烟雾探测器，还有分布在装置的摄像头，他们就相当于我们员工的耳朵和眼睛，时时刻刻监控装置的生产，他们的作用不可忽视，裂解车间的高速离心泵轴密封失效导致丙烯大量泄漏，最先发现泄漏的就是现场的可燃气体探测器。所以我们认真对待他所发出的报警，一定要安排人员到现场进行查看确认。这样就避免了事故的发生。

总结：

通过几条措施的实施，进一步的加强了聚乙烯装置消防安全管理，保证装置的长周期的平稳运行，消防安全的任务任重而道远，需要我们长期不懈的努力，才能构筑我们车间安全生产的防火墙。

参考文献

[1] 梁永刚 王继文. 聚乙烯装置操作手册

循环水场氯气的安全管理与使用

于海波 陈静

（中国石油新疆独山子石化公司乙烯厂安质环处，克拉玛依市 833600）

摘要：运用工艺工程危害分析，通过安全评价方法对目前循环水场加氯系统安全半部进行分析，找出风险，并提出风险控制方法措施。

关键词：加氯；工艺危害分析；风险；控制措施

Safety Management and the Use of Circulating Water Yards Chlorine Gas

Abstract: The process hazard analysis and safety evaluation method are applied to circulating water yards chlorination system security part analyzing to find out the risk, and puts forward measures of risk control methods.

Key Word: chlorination; process hazard analysis; risk; control measures

氯气作为一廉价高效的杀菌剂，多年来以其可靠的杀菌性能一直被工业循环水场广泛使用。氯气是我循环水场主要的水处理剂之一，每年用量较大，同时氯气又是危险化学品，根据国家《危险化学品名录》氯气被列为剧毒气体，采用严格的强制控制措施，购买、使用、贮存、运输都必须严格执行规定。因此，我们在搞好循环水场正常供水生产的同时，必须加强对氯气的管理和安全使用。

本文从氯气工艺安全管理过程入手，结合我厂的工作实际，依据风险分析与控制原理，识别氯气使用过程中的风险，判定其风险度，探讨合理可行的风险控制措施，提高的风险管理和事故防范能力。

1. 加强氯气风险分析与控制的必要性

（一）循环水水处理系统概况

工业循环水在运行过程中会发生因腐蚀、结垢、及微生物滋生等导致水质恶化的现象，其直接危害就是对系统设备产生腐蚀、结垢，影响换热器的使用寿命和换热效率，从而影响工艺生产运行，为此必须进行连续的水质稳定处理。乙烯厂公用工程联合车间利用氧化杀生剂和非氧化杀生剂交替杀菌灭藻，以控制设备的腐蚀结垢和微生物滋生。

乙烯厂公用工程联合车间目前共有4套循环水系统，其日常使用（投加）氯气情况见图1。

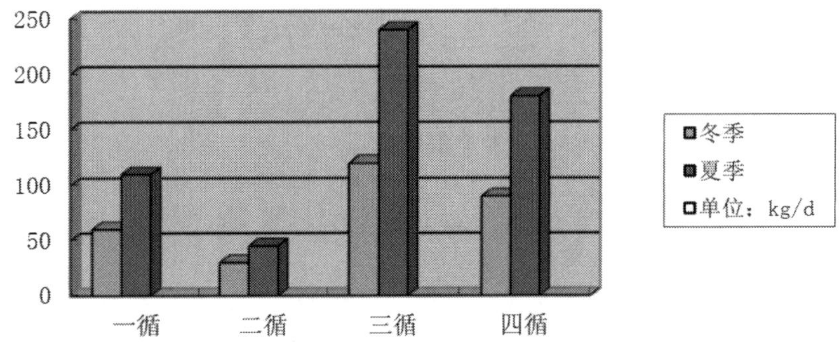

图 1 氯气日使用（投加）量

运去和现在的氯气投加方式及人员接触频次见表1。

表 1 投加方式及操作频次对照表

项目	时态	投加方式	存储方式	人工操作频次
液氯（氯气）	过去（2009 年 3 月前）	人工装卸氯瓶，加氯机投加氯气，人工操作	净重 500 公斤的单液氯钢瓶	1 次/周，并随时人工调节
	现在（2009 年 3 月）	人工装卸氯瓶，自动加氯	净重 500－1000 公斤的液氯钢瓶并联	1 次/周，人员减少氯气接触时间

从表1可以看出，经过不断改进，现在员工接触和操作氯气的频次已大大减少，但不等于危害已经消除，下面对危害做进一步分析。

（二）氯气的物理、化学性质

氯气在常温下为黄绿色并带有刺激性臭味的剧毒气体，常温下7.09×105Pa以上压力时为液体。液态氯为金黄色。氯气在空气中不燃烧，它属于助燃性气体。与氢、甲烷、等可燃性气体混合形成具有爆炸性的混合气，在日光、灯光、热源或火星作用下会发生爆炸。遇水能生成次氯酸和盐酸，而次氯酸又可分解成盐酸和初生态氧。也能与许多有机化合物、某些金属粉末发生剧烈反应，造成爆炸或生成爆炸性产物。

盛装于气瓶中的氯气是气态氯与液态氯共存，且处于饱和状态，瓶内压力就是氯的饱和蒸气压，它随温度升高而增大，0℃为0.367MPa，60℃为1.799MPa。1L液氯可气化成484L气态氯，由于其比重比空气重，泄漏后，便沿地面扩散，对人、畜、庄稼、树木等都会造成毒害。对人的眼睛、鼻腔、咽喉粘膜有极强的刺激作用，主要通过皮肤和呼吸道黏膜进入体内，引起流泪、咽干、咳嗽、胸闷等症状。进入呼吸道会引起支气管炎和肺水肿等症，严重者可引起肺肿，高浓度吸入者可致人死亡。

氯气一旦泄漏到空气中，会对大气造成污染，严重破坏生态环境。氯气属于危险化学品，必须严格遵守国家有关条例、规程和规定以及根据这些法令和法规规订的安全操作规程，对操作人员必须进行专业培训和安全技术考核持证上岗操作。

2. 工艺安全分析及风险评价

（一）泄漏故障（事故）统计

据统计，从2001年8月至2014年6月共13年间，发生一般氯瓶、加氯机泄漏有33起（指影响半径在50米内的泄漏），轻微毒气伤害（指不须到医院处理，自行休息就能康复的）3起，较严重毒气伤害（指须到医院处理的轻伤）2起。

1、泄漏原因：

氯瓶泄漏。绝大多数为氯瓶的角阀泄漏（因腐蚀或角阀本身质量问题，还有一种是在开阀过程中的瞬间泄漏）。

加氯机、加氯管线泄漏。多为腐蚀穿孔，也可能是因机械碰撞，安全设备管线不熟练引发的泄漏。泄漏点多为接头等静密封点，铜管本身也可直接被腐蚀穿孔。

2、中毒原因：

操作者或监护者未正确佩戴空气呼吸器等防毒面具。

（二）分析评价

用HAZOP分析法对加氯危害进行风险评价，见表2。

表2 加氯操作危害识别与风险评价表

设备	参数	偏差	原因	后果	S	L	R	安全措施
加氯机、氯瓶	流量	过高流量过大，或过低，流量过小	角阀操作不当，开太大，或角阀开太小，或加氯管线有堵塞，或加氯系统有泄漏	流量过大，水质指标超标，加氯管线易泄漏；流量过小，杀菌不受控，管线易堵塞和腐蚀穿孔	2	2	II	工业视频监控，加氯间单独调节流量，转子流量计检定，加氯管线使用柔性紫铜管，人员做好个人安全防护进行操作
	压力	过高液位高，或过小	P2 线堵，水流不畅	压力过大，水质指标超标，加氯管线易泄漏；压力过小，管线易堵塞和腐蚀穿孔	2	4	II	工业视频监控，加氯间单独调节流量，使用耐腐蚀膜盒压力表，加氯管线使用柔性紫铜管，人员做好个人安全防护进行操作

3. 氯气使用中的风险控制措施

通过上述分析可见，氯气在循环水场使用各个环节中存在的风险数量集中，等级普遍较高。因此，必须把风险控制工作做足、做细，做扎实。根据上述对氯气使用过程中的风险分析与判定，我们应该采取如下控制措施：

（一）氯气运输采购中存在的风险控制：

（1）对氯气生产厂家进行资质审查，对运输车辆加强管理，必须车况良好，有危化品运输资质并签订运输合同。对运输人员加强管理。所有运输、押运人员必须掌握危化品性质、事故应急处理等知识，经考核合格后持证上岗，并定期对其进行安全识教育。

（2）氯气钢瓶的附件完好，安全技术说明书MSDS资料齐全。

（二）氯气钢瓶装卸时的风险控制

（1）严格执行吊葫芦、叉车操作规程。对起重设备按要求定期检测。

（2）对起重和叉车操作人员加强管理，严禁无关人员开动吊车、叉车。

（3）设置钢瓶意外坠地缓冲装置。

（三）钢瓶存放时存在的风险控制

（1）氯瓶库区明确划分，严格区分氯气空瓶、满瓶存放区域。满瓶、空瓶分开存放。不能靠近火源或高温处，不得用蒸汽直接对钢瓶升温。液氯瓶在使用时不要把氯气全部用完，要留5～10kg余量。

（2）加强出入库管理，并按危险化学品管理规定执行库房大门双锁管理。

（3）存放氯瓶有加氯区必须严格执行门窗紧闭。

（4）设置警示标志和隔离措施，限制无关人员进入毒区。

（四）氯气使用时存在的风险控制

（1）编制氯气使用的专项操作卡，操作时按操作步骤执行，并步步确认，严格执行安全规程。

（2）在加氯操作中，应用10％氨水检查氯瓶是否漏气，如发现有白色烟雾时表示有漏氯现象，应处理好再使用。

（3）安装工业视频监控，全天候监控氯瓶库区及加氯区。

（4）在氯气使用区安装气体（氯气）漏报系统。目前，公用工程联合车间共在四个氯气使用区安装固定式氯气报警仪14台。每半年对报警器探头进行一次计量检定。

（5）对操作人员巡检和使用操作做规定，进行氯瓶使用区域必须强制通风，双人操作，携带便携式氯气报警器，并佩戴空气呼吸器方可进行作业区进行操作。

（6）建立应急救援专业队伍，按规定配置应急救援器材、氯气防护器材和人员中毒现场救治药品，配备应急救援物质和防护器材，如重型防化服、空气呼吸器、碱液回收系统等。公用工程联合车间每半年进行一次氯气泄漏的专项在急演练。

（7）高处设置风向袋或风向标，在厂区常年主导风向两侧设立安全区域用于人员疏散或集结，应急疏散路线和安全集结区域应有明显的标志

（8）操作人员和检修人员必须经过培训，并经常性、制度化加强危化品安全生产知识培训。目前公用工程联合车间每年不少于2次气防知识培训，2次危险化学品及职业卫生防护培训。

目前已有安全防护措施：

表 3 安全防护措施一览表

项目	机械通风	报警设备	泄漏处置设施	警示标识	个人防护	管理措施
分布	每个操作间 1—2 个通风扇	每个操作间不少于 2 个固定式报警器	在氯瓶间均设有碱液回收系统	注意通风、防止中毒、必须戴防毒面具，并进行警示色和硬隔离栏杆	空气呼吸器、氯气报警仪	操作规程、应急预案，应急操作卡齐全

4. 改进措施与方向

（一）根据杜邦理论，更强调加氯设备的安全完善性和仪表自动化

（1）建立碱液回收系统。一旦氯气发生泄漏，碱液回收系统自动启动，切断对外通风，吸收塔工作，将泄漏的氯气吸收到碱液罐内进行中和处理。

（2）确保氯气检测仪正常使用。对操作人员配备量达50%以上。

（二）追求原料的本质安全

（1）氯气始终是一大安全隐患。应寻求杀菌替代物，以低毒或无毒药剂代替剧毒物液氯。

（2）为消减风险，保障现场操作人员的安全和健康，目前乙烯厂公用工程联合车间第二循环水场氧化杀生剂由氯气改为次氯酸钠溶液，并进行配套设备设施的改造。通过对氯气改为次氯酸钠杀菌的改造，可以避免氯气泄漏对现场操作人员及周边人员造成伤害，降低作业安全风险。并通过对次氯酸钠杀菌效果观察后，将其它循环水场氯气杀菌进行工艺革新。

5. 结束语

虽然循环水处理系统经过多年不断的改进，安全性已得到较大的提高，员工的操作环境得到了较好的改善，同时在长期生产实践活动中，积累了很多利用氯气、控制氯气的宝贵经验，但离本质安全还有一定的距离，这是今后的努力方向，特别是对于高危害氯气的使用，还要进一步改进，直至实现水处理系统的本质安全。

6. 参考文献

[1] GB 5138 工业用液氯[S].
[2]GB1984－89氯气安全规范[S].
[3]GB18218－2000 重大危险源辨识[S].
[4]彭力，李发新，危险识别与风险评价技术[M]北京石油工业出版社.

氢气长输安全及风险管理浅析

黄玖来 林任远 薛东

（厦门熙宝源化工技术有限公司，福建厦门 361000）

摘要：介绍了某氢气长输管道及分输站的安全管理及运行监控技术。应用 HAZOP 等方法对该系统进行了深入研究，对发现的风险提出了建议措施。应用研究表明，该分析技术很适合长输可燃气体管道的危险和可操作性分析，为该项目风险识别、消除安全隐患提供了强有力的手段，为完善可燃气体长输管道设计、安全运行及风险管理提供了借鉴，值得推广应用。

关键词：氢气；长输管道；分输站；HAZOP；过程安全；风险管理；建议措施

The Process Safety & Risk Management Analysis in the Hydrogen Long-distance-transportation

Abstract: This paper introduces the process safety & risk management analysis in the hydrogen long-distance-transportation pipeline and sub transmission station. Recommendation measures of risks are analyzed and provided. The application research shows that, the HAZOP analysis technique is very suitable for long distance combustible gas pipeline hazard and operability analysis, provides a powerful means for the project risk identification, to eliminate potential safety problems, in order to improve the combustible gas to provide a reliable reference for perfecting long-distant pipeline design, safe operation and risk management. It is worthy of promotion application.

Keywords: hydrogen; long-distance-transportation pipeline; sub transmission station; process safety management（process safety management; risk management; recommendation measures

氢气是主要的工业原料之一，也是最重要的工业气体和特种气体，在石油化工、电子工业、冶金工业、食品加工、浮法玻璃、精细有机合成、航空航天等方面有着广泛的应用[1]。

作为烃类化合物中两大组成元素之一的氢元素，在炼油化工过程中，可以起到改善油品分子结构、替换杂原子的重要作用。进入21世纪以来，随着人们对自身生存环境的越来越重视，对燃料清洁性的要求越来越苛刻，汽柴油质量升级步伐不断加快。目前，中国全国范围内国IV汽柴油标准正在抓紧推行，部分地区如北京、上海、广州等已先行跨越推行国V标准汽柴油。为降低硫含量、烯烃含量及芳烃含量，因此炼化企业加工过流程从部分产品加氢逐步过渡到全加氢，加氢工艺得到越来越广泛的应用，相应地氢气需求也迅速增加。优质、稳定的高品质氢或纯氢已成为新世纪炼化企业提高轻油收率、改善产品品质不可缺少的基本原料，很多炼化企业已将氢气系统纳入如氮气、仪表风等一样的公用工程的模式来管理。同时，在化工领域，高纯氢气作为苯加氢、己内酰胺精制等必须的原料得到了大量应用。

氢气作为最轻的气体，无色无味，易燃易爆，在空气中爆炸极限为4.1~75.0%（体积）。在加热或有催化剂存在的条件下，氢气非常活泼，很容易与多种物质发生化学反应。因此，如何做到安全稳定产氢、供氢、用氢已经成为与氢气密切相关人们时刻关心的话题。一般而言，为安全起见，国内制氢、供氢及用氢装置距离设计优先考虑相对较近，长距离设计时主要考虑优化区域资源利用。本文以某氢气长输管道及分输站为例，应用HAZOP、现场风险查勘等过程安全的科学方法对该系统进行了深入研究，为该系统的安全管理及运行监控提供了建议。

1. 氢气长输系统概况

1.1 工程概况

A公司与B公司同属某石化集团公司的下属企业。A公司以化工生产为主，B公司以炼油生产为主，两家企业具有原料互供、互补的需要，且两家单位单元装置可以形成经济规模，具有良好的经济效益。

因干气或天然气存在缺口，B公司干气或天然气制氢装置难以满足新增氢气需求。故统筹兼顾两公司，考虑新建一条A公司-B公司的氢气输送管线和一套8.4×104Nm³/h（6万吨/年）的变压吸附（PSA）装置，变压吸附装置设置在A公司气化事业部，将A公司氢气供应给B公司和A公司本身需氢的其他各事业部。

项目主要分为站场部分和管线部分。站场主要功能为：过滤、计量、清管、分输，本项目站场为A公司气化事业部首站、A公司分输站、B公司末站。管线主要分为三大段：A公司气化事业部首站至A公司分输站段，A公司分输站至B公司末站段及A公司本身需氢的其他各事业部，前一段以埋地为主，后两大段全部架空走管廊。

1.2 输送管线简介

根据输送压力（起点压力2.65MPa（表压）、输量及规范要求安全流速，A公司气化事业部首站至A公司分输站段管径为DN450，主要以埋地为主，管道长度约23km，采用符合国标要求的20#无缝钢管。A公司分输站至B公司末站段管径为DN400，利用原有管架架空敷设，管道长度约20km。A公司分输站至A公司本身需氢的其他各事业部的管道主管管径为DN400，分支管管径根据不同事业部实际需要为DN350、DN200、DN100不等，均利用原有管架架空敷设，管道长度约7km。输送管道设计压力4.0MPa（表压），管道总长约50km。

输送管道设有A公司气化事业部首站、截断阀室、A公司分输站、B公司末站段。输送管线含有站、阀室及线路。线路的日常工作是巡线，操作主要在分输站及阀室，其中以A公司分输站工作最多。气化首站建在气化事业部，由气化PSA岗位操作人员兼顾，截断阀室由A公司经管部巡线人员监管，A公司分输站由其经管部管道车间氢气班管理，B公司末站由B公司管理。

1.3 系统公用工程

氢气输送管网及分输站所需要的公用工程有蒸汽、生产水、生活水、仪表空气、氮气、燃料气和电等公用工程依托现有设施改造。

1.4 自动化控制

A公司分输站新建控制室1个，控制室内设有PLC，用于对分输站工艺参数的监控。同时该控制室也完成对整个氢气输送管线运行的监控。由A公司经管部调度室负责氢气输送管线运行的日常管理调度。

输送管线A公司气化事业部首站和A公司分输站的清管收、发球筒的球通过清管指示器由设备商成套提供。

2. 氢气长输系统安全措施及风险管理分析

运用HAZOP[2]工具对该系统进行全面的危险和可操作性分析研究。整个分析过程中结合系统现场风险查勘，主要参考了本项目相关的以下资料：管道和仪表控制流程图（P&ID）、工艺流程图（PFD）、工艺技术规程或操作法（含管道数据、安全阀、设备数据表及报警联锁）等等。

在常规石油、油品、化工原料、天然气及油气的储运过程中，经常存在很多风险，可能会发生很多问题。例如：静电和地震破坏、人为破坏、管道腐蚀和老化造成储运管道内介质泄漏等问题，严重时就会着火爆炸，如2013年“11.22”青岛东黄输油管道泄漏爆炸特别重大事故和2014年“8.1”台湾高雄气爆事故。

氢气长输过程中一样存在类似风险，因此分析、知晓氢气长输系统存在的可能风险，落实风险管理控制和事故预防措施是否充分和足够，对确保该系统安稳长周期运行十分重要。

2.1 工艺设备管道

根据GB 50177-2005标准[3]要求，氢气管道应采用无缝金属管道，禁止采用铸铁管道，管道的连接应采用焊接或其他有效防止氢气泄漏的连接方式。本项目采用符合国标要求的20#无缝钢管，管道采用焊接及法兰连接方式。管道上选择的阀门都采用球阀、截止阀，如管道球阀XV3301/02/03/04/05/06型号为Q347F-300 LBF，埋地管线球阀001型号为全焊接 300LB RF R-PTFE（蜗轮），阀门材料均为A105，符合标准表12.0.3的有关规定，法兰及密封垫垫片的选择符合GB 50177-2005 中表12.0.4的有关规定。首站至分输站埋地输送管道安装时已按设计要求做好电化学保护措施，吹扫前开展了通球处理。建议电化学保护每年检测一次并存档备案。

本项目工艺设备管道施工质量满足规范要求，为日常安稳长周期运行奠定了硬件基础。

2.2 电气专业设施

氢气传输过程中，容易产生静电并积累。当氢气在空气中的体积分数为4%-75%时，遇到火源或静电火花，可引起爆炸。因此氢气输送全过程中相关的工艺设备及管道等要重点做好防静电积累，在项目设计及现场要严格落实有关设施，做好危害告知。图1为本系统部分输送系统管道阀门流量计等防静电措施。图2为氢气危害告知牌。

图 1 氢气输送系统管道阀门流量计等防静电措施

图 2 氢气危害告知牌

本工程项目采用TN-S 接地系统。系统内接地干线采用镀锌扁钢-40×4。接地分支线采用镀锌扁钢-25×4。所有电气设备正常不带电金属外壳、工艺设备管道、电缆桥架等均可靠接地。电气设备的保护接地、工艺设备的防静电接地、人体静电消除装置的接地、建筑物的防雷接地及仪表接地共用一套接地装置，该接地装置的接地电阻不大于10Ω。经现场风险查勘验证，均符合GB 50177-2005标准第9章防雷及接地的有关要求。

为消除人体静电对现场设施及操作可能带来的危害，需要在现场适当增设静电消除桩。图3为氢气分输站内的日常操作巡检使用的静电消除桩。

图 3 静电消除桩

在日常巡检、操作及维护过程中，要认真检查有关静电接地的牢固及可靠性，需要定期做好系统所有静电接地点电阻检测，每年至少检测一次并做好台账记录。在拆卸阀门、流量计调校、检修之后恢复运行时一定要恢复防静电设施如法兰铜片或连接导线等。

2.3 仪表专业设施

本系统涉及的物料氢气和燃料气为易燃易爆介质，仪表均选用本安防爆或隔爆仪表，本安防爆全达到不低于ExiaⅡCT4 防爆要求，隔爆全达到不低于ExdⅡCT4 防爆要求。

为了防止信号传输中的电磁干扰，仪表信号电缆采用屏蔽电缆，电缆敷设沿架空电缆安全技术措施。

本系统合理设计安装在线式可燃气体及氢气检测报警器等设施。气体检测报警控制器放置于监控室内，对各监测点进行监测控制，氢气检测器安装于气体最易泄漏的地点，其核心部件为气体传感器。氢气检测器将传感器检测到的氢气浓度转换成电信号，通过线缆传输到气体检测报警控制器，气体浓度越高，电信号越强，当气体浓度达到或超过报警控制器设置的报警点时，气体检测报警控制器发出报警信号，并可启动电磁阀、排气扇等外联设备，自动排除隐患。在线式氢气检测报警器可有效防止爆炸事故的发生。

在项目建设及工程竣工验收过程中要全面落实本系统设计仪表安全等级，严格按设计要求进行采购、安装及验收。从有关项目中交、工程竣工验收资料及现场DCS运行实况可以看出，在线式氢气及可燃气体等检测报警器能正常工作，本系统仪表专业设施及控制室内PLC系统是安全、可靠、有效的，仪表点校、及联校台账等符合设计及有关法律法规的要求。

2.4 氢气长输系统日常风险管理

2.4.1 氢气长输系统日常风险查勘

因本系统管线长度约50km，氢气长输系统及管线日常巡线风险查勘非常重要。每一位操作及技术管理人员都要经过上岗考试，合格后方可上岗。日常巡线风险查勘时须穿戴劳保用品带三件宝，同时要携带好有效可靠的便携式氢气检测仪（携带式可燃气体检测报警仪应定期校验），按时对系统所有设备及管线巡检重点要做好以下检查或风险查勘工作。

（1）管线及阀门静密封点、法兰连结处有无泄漏。

（2）管线上有压力表的，要检查压力是否正常。

（3）管线阴极保护测试桩是否完好。

（4）管线的防腐、保温是否完好，伴热管线是否泄漏，伴热阀门是否按要求开关。

（5）管托、管架状况是否完好，周围的杂草树木及杂物是否影响管线安全。

（6）管线标识是否一直保持清晰，埋地管线的标识墩是否完好。

（7）管线附近或埋地管线上是否有新增的可能影响管线安全运行的在建构、建筑物，巡线时要随时注意管线经过的居民区、道路、沟渠等周边环境等有无开挖或破坏埋地管线现象。

（8）其它影响管线安全运行的隐患。

2.4.2 日常风险及问题处理

原则上，日常巡线或生产运行监控维护过程中对发现的风险或问题应该按以下方式进行处理。

(1)各单位巡线过程中一旦发现泄漏和正在危及运行管线的安全隐患，首先要确定管线名称、泄漏地点、部位或造成隐患的因素后，宜立即报生产调度室，由调度安排应急处理，管线巡线管辖单位、有关生产车间、相关职能科室参与应急处置。

(2)各单位巡线发现的其它问题，按照管线巡线职责划分，报告相关单位落实整改。

(3)巡线人员应根据外管线巡线及整改情况，建立相应外管线巡线、整改记录，记录要求至少保存3年以上。

(4)每天早调会上，各相关单位要汇报管道巡线及发现问题的整改情况。

2.5 突发性事故与应急处理

为应对突发事故和氢气排放安全燃烧，本项目设计了火炬系统。其最主要的功能是保证输送管线系统在正常、事故、特别是紧急等工况时，能够将排放的可燃性气体安全、可靠地进入此火炬装置中烧掉，并满足环保排放的要求。火炬系统现场图见图4。火炬系统的开工投用、停工及事故应急等需要严格按火炬操作法及技术规程进行操作及维护。

图 4 火炬系统

日常出现突发事件或事故时，根据具体情况要采取以下主要应急处理措施：

（1）管道巡视或检修人员在发现氢气泄漏时，若泄漏点为法兰面且漏点不大，则采取喷水等防火措施后，立即通知检维修公司用防爆工具进行紧固，同时及时通知有关部门阻断漏点附近交通，在漏点周围设置警戒线与警示标识。根据需要是否按规程截断有关阀门，并采取必要的措施，防止发生爆炸，使事故扩大。

（2）在氢气发生泄漏可能出现不能控制的趋势时应立即通知氢气首站关闭供气阀门，通知下游用氢部门及时关闭相关阀门。通过火警电话报警，专职消防队到场后，应采取隔离措施，设置警戒区，利用水枪驱散氢气，使氢气在空气中含量降至1%以下。尽可能切断泄漏源，根据需要通氮气置换。将管道内剩余的氢气向火炬系统排放并保证安全燃烧。

（3）如燃烧爆炸已经发生，专职消防队应首先采取抢救人员措施，同时采取利用水枪驱散氢气等措施，防止二次爆炸。

（4）防止火势蔓延，主要利用消防水冷却设备。

（5）迅速将人员撤离至泄漏污染区上风处，并进行隔离，严格控制人员及车辆的出入。切断火源。建议应急处理人员戴自给正压式呼吸器，穿消防防护服。合理通风，加速扩散。如有可能，将漏出气用排风机送至空旷地方或装设适当喷头烧掉。漏气管道要妥善处理，修复、检验后再用。

2.6 其他安全风险或隐患

当用“压力”或“流量”作参数来研究分析该氢气长输系统时，因A公司气化事业部制氢车间仅一条生产线，一旦制氢车间出现事故从源头减少或停止供氢，将导致氢气压力低或流量低甚至氢气输送管道无压力或无流量，进而导致下游炼油加氢或化工加氢装置停车。而下游需氢部门特别是大炼油或大化工需氢装置必须要求氢气作为公用工程一样需要持续供应，否则整个炼油或大化工装置生产平衡将被打破，无法完成生产计划，经济损失巨大。

为此，提出建议措施，为降低该类风险或隐患，尽快新增一条备用氢气生产线，确保源头氢气首站做到持续供氢。同时建议下游炼油装置的原有小制氢装置等能在长输管道较长时间不能供氢时（如主管、氢气分输站等发生爆管等异常事故）随时启用以弥补最低或基本需氢装置的供氢需要，防止大炼油因氢气长输管线供氢停止面临全面停工，以减少大炼油装置的经济损失。

3. 结论

通过应用HAZOP分析及现场风险查勘等方法，对某氢气长输管道及分输站的过程安全管理及运行监控技术进行了深入研究，对发现的风险或隐患提出了建议措施。研究表明，该氢气长输管道及分输站安全设计总体符合有关规范要求，运行安全监控及风险管控持续有效，应急措施得力，值得同行借鉴和推广应用。

4. 参考文献

[1] 李大东. 加氢处理工艺与工程. 北京：中国石化出版社，2011年6月.

[2] 吴重光，张广文，黄玖来等. 危险与可操作性分析（HAZOP）基础及应用. 北京：中国石化出版社，2012.

[3] 陈霖新，章光护等. 氢气站设计规范（GB 50177-2005）.中华人民共和国建设部第330号公告，2005年4月15日.

化工厂厂外事故应急处理程序的制定，审核和演练

Allan Su
（Senior Engineer, AcuTech, Consulting Group）

摘要：厂外事故应急响应制度由于涉及非私有区域，涉及多方利益，涉及公众，资料不完备等因素往往不能如同厂内应急响应制度一样建立包含预设事件，应急设备，应急人员，应急演习等元素组成的完备计划。本文通过讨论在厂外事故应急响应制度的各个阶段中如何更好的制定计划，并通过多种不同的演习方式和审核来保证制度的有效性。
关键词：外部事件；应急预案；有效钻孔；PSM 审计；PSM

Establishing, auditing and drillings effective off-site incident Emergency Response Plan for chemical site

Abstract: The emergency response procedure is the basic arrangement in chemical site to reduce the damages of the incident. But due to multi-parties involvement, public related, and not having the sufficient information in public area, unclear responsibilities, the off-site incident emergency response procedure are not as efficiently set up as on-site procedure. The article discuss the difficulties in building up, drilling and auditing the off-site emergency response procedure and share the best practices from many sites on how to use appropriate risk assessment tools and effective audits and multi-style of drills on building up and maintain an effective off-site emergency response procedure.
Keywords: off-site incident; emergency response plan; effective drill; PSM audit; PSM

1. 概述

现代工业更加趋向于规模化，集中化和自动化的生产方式，为了提供更加稳定的上下游产品传输，现代工业也采用了大量的管道连接，为了实现更加稳定的上下游连接，有些管道也采用架空或埋地的方式将连接管建造在厂区以外。这些厂外管道在为生产提供了便利以外，同时也将危险通过管道引出了厂区，对于这些管道在厂区外带来的危害，需要合理的设计，建造，定期检查来避免，同时，对于可能由此产生的意外事故，应该做好应急响应方案。

对于厂区内可能产生的事故，应该按照其风险评估中预估到的最严重后果制定相应的应急响应预案，应急响应预案应该符合有针对性，分工明确，任务清晰，可执行等要求，并且正对应急响应预案应该有针对性的演习和定期审核以保证预案的响应速度和可执行性。

但是对于厂外事故，由于设计区域更广，牵扯组织更复杂，人员更多，责任不清晰等因素，制定，执行，演习和审核厂外事故的应急响应预案就非常复杂，以致于成为很多地方不可能完成的任务。这一不可能完成的任务难住了政府，限制了企业，也致使事故真正发生时不能有效作出早期预警处理，最大程度减免损失。

本文试图从信息如何有效分享的角度出发，分析出从信息简化的角度看，如何可以多方合作出厂外事故的应急响应预案，并有效的对其演习和审核以保证有效性。

对于厂外发生的危险化学品运输安全事故，由于其应急处理预案与厂外管道事故类似，但是发生环境更加多变，设计危险化学品数量较少，所以在本文中不做具体讨论，但是在应急响应预案的制定，演习和审核方面的要求和利弊是共通的。

2. 现状分析

2.1 应急响应的制定过程

相比于厂内事故的多样性，厂外事故一般起因比较简单，厂外设备一般不会存在化学反应，且设备内的化学物理性质一般保持恒定，同时厂外设备没有复杂的管路布置，一般为点对点传输的单个管道，从以上分析，厂外可能引发事故的原因大部分为管路内物质泄漏，进一步引发后续的次生事故。尽可能早的发现并正确应对泄漏的应急响应预案将大大降低事故的最终损失。

对于制定一个完备可执行的应急响应预案需要以下步骤：

- 收集相关资料
- 进行风险评估
- 确定最严重后果
- 对最严重后果制定应急响应预案的范围，责任和行动
- 对应急响应预案进行演习
- 对应急响应预案进行审核

2.2 各相关方优劣势分析

由于厂外事故是多方相关的，为了更好理解如何能够在各方协调下制定出一个可行的预案，我们对上面每一个步骤中各个相关方去完成这一步骤的难易程度进行逐步讨论，例如对于第一步“收集相关资料”，公共监管部门（政府/管委会）将更容易完成该工作，而作为管道所有方的企业将是较难完成的，作为其他相关方的其他企业和个人更加难以完成。对每一步进行相同的问题，我们可以得出表 1。

表 1 应急响应预案制定步骤中各相关方的优势分析

应急响应预案制定步骤	政府/管理部门	危险源所属企业
收集相关资料	易	难
进行风险评估	难	易
确定最严重后果	难	易
对最严重后果制定应急响应预案的范围，责任和行动	难	易
对应急响应预案进行演习	易	难
对应急响应预案进行审核	难	难

从上表的分析可以看出，政府和相关管理部门在信息收集，组织演习等需要多方合作方面相对容易，而危险源所属企业更多的掌握危险源性质，可能情形，如何防范等专业知识。而只有将这两种能力相结合才可能制定并确保有执行力的应急响应预案。

2.3 可能的合作方式

以上分析是基于理想情况下，双方都有合作意愿前提下，双方的合作可能，如何能够实现这种能力本身互补的合作，还需要一定的制度安排来实现，本节将具体讨论每一个步骤的合作可能和一些好的合作经验。

2.3.1 信息收集的完整化

完备的工艺管道信息是做好风险评估的关键所在，而这些管道信息不仅仅包含危险工艺管道，而且包括可能对泄漏危险品提供传输，受限空间等危险条件的其他管道。而地下管网由于深埋地下，更难被察觉，所以为了实现长期的信息有效性，对所有地下管网实现注册制是非常必要。由于GPS定位技术的广泛应用也为地下管网的登记，查询，对比提供了更加便利的条件。通过各政府部门通力合作实现各管网使用统一标准登记并叠图使用已经在一些工业区进行了推广。相信随着地下综合管网的登记完成，将为制定出可执行的应急响应预案打下良好的信息基础。

2.3.2 最严重事故情景的确定

在收集到地下综合管网信息之后，可以依照区域内管网的危险源，管道复杂性以及周围涉及人员数对危险区域进行初筛，确定出相对危险区域。政府依照一定规则对相对危险区域进行初筛将确定出可能发生严重事故的重点防范区域，做到有的放矢。

即使是厂外事故，危险源所属企业也是最重要的责任方，同时也对于该危险源的防护，物性，工艺，应急拥有更具体的知识。因此，进行具体的风险评估并确认最严重后果事故情形这一步骤，应该更多的由责任企业做出。而政府/管理部门可以通过信息分享对这一过程提供支持，并且通过要求企业完成应急响应备案，定期更新等方式对这一过程进行监督。

更高质量的风险评估和定量分析对于判别事故等级从而做出更有针对性的应急预案的有巨大的帮助。所以对厂外事故提出更高的风险评估要求是必要的，且由于厂外事故主要以泄漏为主，更加容易完成较为准确的定量分析。同时定量风险评估也为应急响应预案提供更清晰的事故等级临界点，影响范围和响应时间，这都将有效提高应急响应预案的可执行性。

2.3.3 应急响应的制定

由于危险源的不同，工业厂外事故在应急响应方式上将不同于普通非工业事故，在事故应急中应该充分尊重危险源所属企业在危险源方面的防护知识，并充分利用已经在定量风险评估中制定的事故模型，对事故响应做到更加准确的分级，预警和以及利用上下游工艺控制措施设计不同的应急处置方式。

在应急响应预案中要更加综合的考虑厂外事故中所能够利用的应急资源，这些资源不仅仅包括通用的消防，医疗等资源，还应该包括针对特定危险源的特别资源，例如国内外消防应急普遍缺少的化学品防护资源。当出现这类资源短缺时，应该要求由作为事故责任方的危险源所属企业提供，在一定区域内可以考虑多个企业共用这些资源，必要时可以建立起联防机制。

在应急响应预案中，更加应该强调量化的事故等级规定，从而为事故现场处置提供更加直观的判断标准，并对不同的事故等级，制定不同的处置方式，例如疏散范围，响应等级等，明确，简洁的事故处置方案表述将为事故现场处置节省非常多的时间。

即使对于同一危险源，由于外部环境的不同，例如风向，降雨等因素，也可能存在多种最严重事故情形，必要时应针对每一个情形制定不同的处置方案。

2.3.4 应急响应制度的演习

演习作为最常用的确保应急响应制定有效的方式，用以检查应急响应的可执行性与预估是否一致。厂外事故的应急处理所涉及的相关方更多，为了确保有效配合，演习更加必要，但是要进行演习所需要涉及的范围更大，难度也更大。政府/管理机构更应该成为演习的组织方，以确保企业制定的应急预案的可行性。

为了避免影响范围过大，同时又能检验制度的有效性，一些企业采用桌面演习和分解单项演习的方法来达到检验的目的。

桌面演习主要用于检验事故应急预案中的事故分级响应方式和沟通机制。通过预设事故情景，由多方合作来熟悉事故不同等级时不同的处置方法。同时检验事故处置各方的沟通和理解的一致性。由于桌面演习涉及资源较少，影响较小，更加适合进行无预警演习以检验事故相关方对于预案的熟悉程度。

分解单项演习主要正对特殊事故情景中设备人员的组织和使用，尤其是对于需要特殊处置的化学品，应该提前演习以确保处置设备和技能得到有效掌握和沟通。分解单项演习应该更加注重单项任务的完成度和耗时是否符合应急响应预案的要求。

2.3.5 应急响应预案的审核

应急响应制度不仅可以通过演习来检验，也可以通过专业的第三方审核来检查其可执行性。这种审核将不仅关注应急设备和资源，同时也通过对建立应急响应预案各个步骤的审核来确保整个流程的完整性和及时更新。

厂外事故的应急响应预案虽然涉及多个相关方，但是事故责任方主要是危险源所属企业，应急响应预案也由该企业制定，故而审核应该由政府组织以实现监督目的，并邀请专业第三方进行。企业在可以通过自查来对预案进行审核。

3. 结论

由于厂外事故中政府/管理部门和危险源所属企业在掌握信息，危险源处置知识，协调沟通能力方面各有优势。故而本文通过在应急响应预案制定各个步骤中各方的优劣势入手，讨论了在理想状态下，各方合作制定应急预案的方式。

进而，对于制定预案的每个步骤做了详细讨论，分享了如何可能实现这一合作，并分享了成功合作的经验。

尤其针对危险源所属企业在处置知识上的优势，做出了在进行风险评估，评估最严重后果，制定应急预案等步骤中，应该有责任企业承担更多责任的建议。

Electrostatic Hazards in the Chemical Industry

The Nature of the Problem and Practical Measures for its Control

Vahid Ebadat, Ph.D.
Chilworth Technology, Inc. (A DEKRA Company)
113 Campus Drive, Princeton, New Jersey 08540, USA
Email: vahid.ebadat@dekra.com
www.chilworth.com

Keywords: static electricity; electrostatic ignition; static discharge; minimum ignition energy; spark

Abstract

Many processes and operations in a chemical plant involve handling and processing of liquids and powders, most of which could under certain conditions be flammable. Flammable gas, vapor, and dust cloud atmospheres can be ignited if a sufficiently energetic ignition source is present. It is noteworthy that most flammable gas/vapor atmospheres and some dust clouds can ignite by very small energy ignition sources. One potential ignition source that, in many cases could be an inherent part of the process/operation is an electrostatic discharge. Electrostatic discharges result from the generation and accumulation of electrostatic charges. Electrostatic charge generation most commonly occurs whenever two materials, liquids and/or solids, make and then break contact with each other. The accumulation of the electrostatic charge can result in electrostatic discharges. This suggests that there are many occasions during the transfer, handling, processing, and packaging of flammable liquids and powders in a chemical plant where flash fire and explosion hazards may exist.

This paper discusses the practical measures that can be considered to prevent/control electrostatic ignition hazards that are associated with some common liquid and powder handling/processing operations in the chemical industry.

1. Introduction to Static Electricity

Static electricity is the charge that is generated whenever any two contacting surfaces move relative to each other. Static electricity is generated during many common chemical processing operations, such as when liquids flow relative to pipe walls or when powder particles come into contact with the surfaces of the processing and conveying equipment, among other mechanisms.

Contrary to some common belief, static electricity is predictable, quantifiable, and controllable and in fact the predictable and controllable nature of electrostatic charge is used for our benefit in a multitude of applications, including photocopying, air pollution control, and spray coating. There are also many negative instances of static electricity, including electrostatic nuisances, problems, and hazards. The shock that we receive from a doorknob after walking across a carpeted floor is certainly a nuisance. Static electricity can also cause materials handling and product quality problems by causing bulk solids and powders to agglomerate and adhere to the walls of vessels and conveyances. Actions taken to control electrostatic hazards and nuisances may have the collateral benefit of resolving such electrostatic problems. Measures for correcting quality and materials handling problems associated with static electricity are beyond the scope of this paper, however.

Of course, the greatest concern posed by static electricity in an industrial setting is the risk of fire and explosion due to the ignition of flammable atmospheres by electrostatic discharges. Flammable gases, liquids, and powders are used in many chemical processing operations and precautions are required to prevent their inadvertent ignition. One area of precautions involves the exclusion or elimination of potential electrostatic ignition sources from locations where flammable atmospheres may reasonably be expected to exist. It is this area that is the focus of this paper and it is the controllable nature of static electricity that makes such precautions possible and effective.

2. Electrostatic Charge Generation

There are several mechanisms by which electrostatic charge is generated, including: (1) contact electrification; (2) double layer charging; and (3) induction charging.

2.1 Contact Electrification

Contact electrification -- also known as triboelectrification -- occurs when any two solid materials come into contact and then (quickly) separate. There is a rearrangement of electrons at the contacting surfaces where one surface is left with a surplus of electrons and is said to be negatively charged, while the other surface is left with a deficit of electrons and is said to be positively charged. Contact electrification occurs when for example powder particles collide/rub against the conveying chutes during pouring and when the soles of footwear make contact with insulating flooring. The level of charge that is generated by contact electrification is influenced by factors such as the (chemical) nature of the two contacting surfaces, the true contact area, and surface conductivity. The handling and processing of bulk solids and powders typically result in the development of relatively high levels of static charge because many powders are insulating in electrostatic terms and because of the relatively large surface area available for contact electrification.

Contact electrification on conductive materials, can generally be controlled by grounding. For insulating materials, it may be possible to control contact electrification by controlling the speed of contact and sometimes by increasing the relative humidity.

2.2 Double Layer Charging

Electrostatic charge is generated during the handling and processing of liquids by a mechanism known as double layer charging. Double layer charging can occur at liquid/solid and liquid/liquid interfaces. Ions of uniform (like) polarity are adsorbed at the interface. The adsorbed ions attract ions of opposite polarity within the liquid forming a double layer within the liquid near the interface. This layer is thin for conductive liquids and more diffused in relatively less conductive liquids due to reduced ion mobility. As the liquid flows relative to the interface, the oppositely charged ions repel and move apart increasing the potential within the liquid. This effect -- combined with the competing effect of ion recombination at the vessel wall -- results in an unbalancing of charge within the liquid. This is the type of static charging that occurs as liquids flow through pipes.

The principal difference between low and high conductivity liquids is the rate at which ion recombination occurs. Liquids having conductivities less than 100 pico-siemens per meter (pS/m) are generally considered low conductivity (insulating). However, both low and high conductivity liquids can develop static charge during processing. When the rate of double layer electrostatic charging exceeds the rate at which ions recombine at the charge interface, charge accumulates in the liquid. Examples of electrically insulating liquids include non-polar solvents such as toluene, hexane, and heptane. Examples of conductive liquids include alcohols, such as methanol and isopropanol, and ketones, such as acetone. The factors that can aggravate charge generation in liquids include:

- Presence of immiscible solids and liquids.
- Presence of strainers and filters.
- Splashing and spraying; and.
- Stirring and mixing.

2.3 Induction Charging

Induction charging occurs when a conductive object that is electrostatically isolated from ground enters an electric field, such as that emanating from an adjacent charged object. The charged object causes a polarization of the otherwise balanced charges (equal numbers of positive and negative charges) in the conductor. If an isolated conductor sufficiently charged by induction is subsequently exposed to ground, an electrostatic discharge may result. It is generally possible to control such induction charging through the grounding of conductive objects.

3. Charge Accumulation

Generation of electrostatic charge is not in and of itself hazardous. However, if static charge is allowed to accumulate, it can give rise to hazardous electrostatic discharges. Electrostatic discharges occur between objects or surfaces at different potentials (voltages). As an object or surface accumulates charge, the electric field strength above it intensifies. If the electric field strength between the two objects or surfaces exceeds the breakdown strength of the atmosphere between them, an electrostatic discharge will occur. Electrostatic charge can accumulate on liquids, on insulating solids, and on conductive objects, and personnel isolated from ground.

4. Electrostatic Discharges

When the difference in the electrical potential between two objects exceeds the breakdown strength of the atmosphere between the two objects, an electrostatic discharge could occur, provided that at least one of the objects is conductive in electrostatic terms. Such discharges may be sufficiently energetic to ignite proximate flammable atmospheres, including gases, vapors, and dust clouds. The types of electrostatic discharge that may be encountered during materials handling and processing include: (1) spark discharges; (2) brush discharges; (3) corona discharges; (4) cone discharges; and (5) propagating brush discharges.

4.1 Spark Discharges

Spark discharges are the type of electrostatic discharges that occur between two conductors at different electrical potentials. Typically, one of the conductors is electrostatically charged while the other is electrically grounded (zero potential). Examples of spark discharges include discharges from metal piping, flanges, vessels, tools, decking, drum chimes, and other metal plant items that have been inadvertently or otherwise isolated from ground.

Spark discharges are characterized by a distinct energy channel between the conductors. Charge flows rapidly through the channel until virtually all of the available energy on the charged conductor is dissipated. Because charge flows readily in a conductor, virtually all of the available charge may be transferred during the discharge. The energy of a spark discharge is dependent upon the capacity of the charged conductor to store charge (capacitance) and the potential to which the conductor is charged.

4.2 Brush Discharges

Brush discharges are the type of electrostatic discharges that are produced from a charged insulating surface to a grounded conductor having a relatively large radius of curvature (> 1 mm), i.e., a rounded or blunt electrode. For example, polyethylene drum liners can become electrostatically charged during the pouring of bulk solids and powders. Electrostatic charge accumulated on such liners can give rise to brush-type electrostatic discharges to personnel or to the manway through which the solid is being poured.

Since charge mobility on insulating surfaces is impeded by the resistivity of the material, only charge from a limited surface area participates in a brush discharge. Also, charge cannot flow to form one distinct discharge channel and instead a network of smaller channels is formed. Brush discharges are known empirically to exhibit effective energies of as much as about 4 millijoules (mJ). Consequently, brush discharges are capable of igniting many flammable gas and vapor atmospheres. Brush discharges should also be theoretically capable of igniting explosible dust clouds having an MIE of 4 mJ or less. However, the ignition of an explosible dust cloud by brush-type electrostatic discharges has not been reported.

4.3 Corona Discharges

Corona discharges arise when a pointed conductor, i.e., a conductor having a sufficiently small radius of curvature (< 1 mm), is raised to a high electrical potential (voltage) or is inserted into a strong electric field when grounded. In either case, the field strength at the tip of the pointed conductor is very strong and the atmosphere within the vicinity of the tip is ionized. Charged ions are then accelerated either towards or away from the pointed electrode depending upon their polarity.

Corona discharges can pose an ignition hazard only to the most sensitive flammable gas atmospheres, e.g., hydrogen and oxygen-enriched vapor/gas atmospheres.

4.4 Cone (Bulking Brush) Discharges

Electrostatic discharges that may occur on the surface of highly-charged high resistivity (insulating) bulk solids and powders as they accumulate (bulk) in vessels or containers are known as cone discharges. As high resistivity bulk solids or powders accumulate in a vessel, the electric field on the surface of the material intensifies — a phenomenon known as charge compaction. The strong electric field causes ionization of the air just above the surface of the material, giving rise to cone discharges from the powder heap to the vessel walls.

While the effective energy of cone discharges – also referred to as bulking brush discharges – is dependent upon the volume resistivity, particle size, and mass charge density of the bulking solid and the diameter of the vessel in which it is accumulating, cone discharges are known empirically to exhibit maximum effective energies on the order of 10 – 25 millijoules (mJ) and thus are capable of igniting flammable atmospheres, including some dust clouds evolved during vessel filling.

4.5 Propagating Brush Discharges

Propagating brush discharges are highly-energetic electrostatic discharges capable of igniting many flammable atmospheres, including gases, vapors, and dust clouds. These discharges -- exhibiting effective energies as high as 1,000 mJ or higher -- derive their energy from the formation of a double layer charge on opposite sides of an insulating surface such as a fabric, film, or coating backed by a grounded conductive surface. Additionally, insulating (plastic) pipes and hoses used for pneumatic conveying of particulate solids or transfer of low conductivity liquids as well as some insulating containers could also give rise to propagating brush discharges.

5. Flammable Atmospheres

Perhaps the greatest hazard posed by an electrostatic discharge is as a potential ignition source for flammable atmospheres. Three elements are required for a fire or deflagration to occur: (1) a fuel; (2) an oxidant, and (3) a sufficiently energetic ignition source. These three elements are commonly referred to as the fire triangle. The first two of these elements -- the fuel and oxidant -- when in an appropriate ratio are referred to as a flammable atmosphere.

Flammable atmospheres are atmospheres that are capable of being ignited and supporting combustion and flame propagation. The ignition of a flammable atmosphere in an enclosed vessel or area can create an explosion. Fuels commonly associated with industrial fire and explosion incidents include hydrocarbon gases, organic solvent vapors, alcohols, and dust clouds. The most common oxidant is the oxygen in air.

For an electrostatic discharge to ignite a flammable atmosphere, the discharge energy must be greater than the Minimum Ignition Energy (MIE) of the flammable atmosphere. Minimum Ignition Energy (MIE) is defined as the smallest electrostatic discharge energy that is capable of igniting the flammable atmosphere at its most easily ignitable concentration. The MIE of a flammable atmosphere can be compared with the effective energy of potential electrostatic discharge in order to determine the extent of required precautions by identifying which type of discharge may pose a threat. The MIE of most hydrocarbon gases and organic vapors ranges from 0.1 - 1.0 millijoules (mJ), while the MIE of most dispersed dusts ranges from < 3 - 1000+ mJ. While some literature data for MIEs of gases, vapors, and dusts are available, it is generally recommended that such data be developed on a process specific basis since slight variations in composition or physical and chemical characteristics can have significant effects. For example, the MIEs of powders and dusts decrease markedly as particle size and moisture content decrease.

6. Controlling Electrostatic Hazards Associated with Specific Equipment and Operations

This section discusses some of the measures that could be considered in order to reduce the risk of flash fires and explosions due to electrostatic discharges from some common operations in chemical industry.

6.1 Metal Tanks, Vessels, and Reactors

As described above, electrostatic charge generation is expected during processing of liquids and bulk solids. Such charge may be retained when liquids and solids are introduced to reactors, tanks, and other vessels. Charge retained on liquids and solids can pose an electrostatic hazard when flammable vapors or dust clouds are evolved within the vessel. Potential electrostatic ignition sources include: (1) electrostatic discharges from the surface of charged liquids and solids; (2) spark-type electrostatic discharges from isolated conductors; and (3) spark-type electrostatic discharges from isolated personnel.

The following measures are offered to control the potential electrostatic ignition hazard associated with metal reactors, tanks, and other vessels when flammable gases, liquids, vapors, or explosible dusts may be processed, present, or evolved:

1. All metal vessels and ancillary piping and equipment must be suitably grounded. Where the bonding/grounding system is all metal, resistance in continuous ground paths typically is less than 10 ohms. The suitability of grounding should be checked regularly by direct measurement.

2. Avoid splashing when flammable liquids are transferred to a vessel in order to avoid the formation of electrostatically-charged sprays and mists and also minimize the turbulence in

the bottom of the vessel. This is typically accomplished through the use of a grounded metal fill pipe (dip leg) or by introducing liquids through a valve located at or close to the bottom of the vessel. A fill pipe should extend sufficiently close to the bottom of the vessel such that it is almost immediately submerged beneath the liquid upon filling.

3. For an uncontaminated single-phase liquid having a conductivity of 100pS/m or less, the flow velocity of liquid entering a vessel should be limited to <1m/s until the inlet or end of the fill pipe has been submerged beneath the liquid. The fill rate can then be increased up to 7m/s.

   For multiphase or contaminated liquids having a conductivity <100 pS/m, the liquid flow velocity should not exceed 1 m/s during the entire fill cycle.

4. Cleaning of vessels should be performed in a manner that minimizes the generation of electrostatically charged flammable sprays and mists. The use of non-flammable cleaning agents should be considered. If spraying of flammable cleaning agents is required, it should be done using fixed spray balls or nozzles with the vessel closed and inerted. The cleaning of vessels by spraying flammable liquids through an open manway is strongly discouraged.

5. The interior of vessels should be visually inspected and any tools, fasteners, cans, and other tramp metal should be removed. Such items may float or otherwise become isolated from ground and electrostatically charged as liquids are introduced to the vessel. Electrostatically charged tramp metal can give rise to spark-type electrostatic discharges sufficiently energetic to ignite flammable vapor evolved in the vessel ullage (headspace). An inspection for tramp metal should always be performed before filling a vessel.

6. The presence of liquid filters upstream can greatly exacerbate electrostatic charge development in liquids during vessel filling. Suggestions to control electrostatic hazards associated with liquid filters are presented in a subsequent section.

7. To minimize the potential fire and explosion risk posed by electrostatic cone discharges during filling of vessels with explosible powders, inerting and/or explosion protection for the vessel should be considered when the volume resistivity of the powder being processed is greater than 10^9 ohm-meters, the MIE of the dust cloud is 25 mJ and the powder is being introduced to a dry vessel, i.e., no liquid in the vessel.

6.2 Metal Vessels with Internal Linings or Coatings

Many operations involving high or low pH liquids necessitate the use of vessels and piping having chemical-resistant linings or coatings such as glass or polytetrafluoroethylene (PTFE). Such materials can be highly insulating in electrostatic terms. Consequently, these materials may present an electrostatic hazard, particularly when present in vessels receiving and containing flammable liquids or explosible solids.

Electrically-insulating linings and coatings may become electrostatically charged when charged liquids or solids are introduced to the vessel. Such linings and coatings may also inhibit the relaxation of static charge from liquids and solids. Linings and coatings may also become charged during mixing or agitation of the vessel contents.

Potential electrostatic ignition sources include: (1) electrostatic discharges from the surfaces of charged liquids or solids; (2) spark discharges from proximate isolated conductors; (3) propagating brush discharges through electrically-insulating liners and coatings; (4) brush discharges from the liner or coating, such as during vessel cleaning. Flammable atmospheres may include vapors, dust clouds, hybrid atmospheres, and mists generated during vessel cleaning.

Many of the suggestions presented for unlined metal vessels remain applicable for metal vessels having internal linings or coatings. The following additional suggestions are offered to control the potential electrostatic ignition hazard associated with metal vessels having internal lining or coatings:

1. The outer metal shell of a vessel having an internal lining or coating must be suitably grounded. All ancillary metal items and equipment should be similarly grounded with resistance-to-ground less than ten (10) ohms.

2. It is suggested that a static dissipative lining/coating with a volume resistivity of less than 10^9 ohm-meters ($\Omega \cdot m$) or surface resistivity less than 10^{11} ohm per square be used.

3. While control of filling and agitation rates and raising the conductivity of the liquid may reduce the probability of propagating brush discharges in metal vessels lined with electrically-insulating materials, these measures cannot eliminate the probability entirely. Therefore, for metal vessels having an electrically-insulating lining or coating (volume resistivity $>10^9$ ohm-meter), safety should be based on inert gas blanketing (exclusion of oxygen).However, it should be noted that although inert gas blanketing will reduce electrostatic ignition risk, the possibility of creation of pinholes through the lining or coating due to propagating brush discharges will still remain. The possibility of pinhole creation will be reduced by the provision of a suitable route to ground for electrostatic charges on the vessel contents.A pathway to electrical ground may typically be provided by: (a) a tantalum patch below the liquid level; (b) a conductive bottom discharge valve; or (c) a grounded dip leg.

4. Cleaning of vessels having electrically-insulating internal linings or coatings should be performed in a manner which: (a) minimizes static charging of the internal lining or coating; and (b) minimizes the generation of flammable and electrostatically-charged sprays and mists. It is suggested to consider the use of a non-flammable cleaning agent. If spraying of flammable cleaning agents is required, it should be done with fixed nozzles or spray balls and the vessel closed and inerted. The spraying of flammable liquids through an open manway is strongly discouraged.

6.3 Vessel Gauging and Sampling

In chemical processing operations, it is sometimes necessary for quality or process control purposes to take a simple measure of the quantity of liquid or solid product in a vessel or to obtain a sample of such product. These activities are referred to as gauging and sampling respectively. Gauging and sampling typically involve the insertion of a gauging or sampling device into the vessel through a manway. Often, gauging devices are simply rods with graduation marks, while sampling devices are rods with small cups on the end.

Gauging and sampling can present an electrostatic hazard when: (1) the liquid or solid contents of a vessel have retained a hazardous level of static charge; (2) when the gauging or sampling device itself gives rise to a discharge; or (3) when the personnel conducting the gauging or sampling operation give rise to a discharge.

The following suggestions are offered to control the potential electrostatic ignition hazard associated with gauging and sampling operations when flammable gases, liquids/vapors, or explosible dusts may be processed, present, or evolved:

6.3.1 General

1. All personnel performing gauging and sampling in or around flammable atmospheres having a minimum ignition energy (MIE) of 25millijoules or less should be suitably grounded in order to prevent the accumulation of electrostatic charge and thereby minimize the probability of ignition from electrostatic discharges from their bodies.

   Grounding of personnel may be accomplished through the use of grounding wrist straps or the combination of static dissipative footwear and static dissipative flooring. Regardless of the method used, personnel should be grounded such that their resistance-to-ground is less than 1×10^8 ohms (100 MΩ).

2. Devices used to gauge or sample vessels containing flammable liquids or explosible powders or dusts should be constructed of metal, static dissipative plastics, or natural materials such as wood. Such devices should be grounded during gauging or sampling operations in order to prevent the accumulation of electrostatic charge and thereby minimize the probability of ignition from spark-type electrostatic discharges.

3. Gauging and sampling devices should be grounded such that their resistance-to-ground is less than ten (10) ohms. Static dissipative devices should be constructed from materials having a volume resistivity less than 10^9 ohm-meters. Wooden devices should be unfinished, i.e., uncoated, and grounded.

6.3.2 Liquids

4. For pure and single-phase liquids having conductivities less than 100 picosiemens per meter (pS/m), no sampling or gauging should be performed during operations where electrostatic charge may be generated. Such operations include vessel loading, mixing, agitation, or settling of solids and immiscible materials.

After the cessation of such operations, gauging or sampling may be performed after a period of 30 minutes to allow for the relaxation of electrostatic charge from the vessel contents, assuming that a suitable route to ground has been provided. If a shorter waiting period is desired, increasing the conductivity of the liquid through the addition of an antistatic additive or conductive liquid should be considered.

5. For liquids having conductivities of 10,000pS/m or greater, gauging and sampling may be performed at any convenient time as long as a suitable route to ground has been provided to allow for relaxation of charge from the vessel contents.

6.3.3 Solids

6. Sampling should be performed only after a suitable waiting period, not necessarily for purposes of charge relaxation but to allow for settling of dispersed powder or dust particles. This waiting period should reflect the particular characteristics of the powder or dust in question.

6.4 Plastic Items and Implements

In any chemical facility, various plastic items and implements such as tools, funnels, and scoops are used during materials handling operations. Items made from plastics and other electrically-insulating materials can accumulate electrostatic charge and give rise to brush discharges. These discharges are known empirically to have an effective energy of as much as four (4) millijoules (mJ) and are sufficiently energetic to ignite flammable gas and vapor atmospheres.

Consequently, when flammable atmospheres – including gases and/or vapors may be evolved during materials handling operations – the exclusion of items and implements made from plastics is generally recommended. The use of such items and implements may be allowable, however, when it can be demonstrated that the level of electrostatic charge will be insufficient to produce discharges capable of igniting the prevailing flammable atmosphere.

Alternatively, when the probability of ignition from brush discharges cannot be discounted, the use of items made from metal and static dissipative or conductive plastics should be used. Such items should be electrically grounded during use. Static dissipative items and implements should be made from materials having a volume resistivity of less than 10^9 ohm-meters.

6.5 Piping and Hoses

As liquids and solids flow through pipes and hoses, electrostatic charge can be generated. Liquids typically become charged due to double-layer charging, while solids become charged by contact electrification. Piping and hoses made from electrically-insulating materials can also become charged due to external charging, such as when personnel rub or brush against such piping, or due to contact with charged mists or sprays. Static charge can accumulate on non-conductive liquids and on powders (regardless of whether they and conductive or insulating) being conveyed, insulating pipes and hoses, metal piping and components isolated from ground, electrically-insulating lining or coating materials used in pipes, and nearby personnel if isolated from ground.

Potential electrostatic ignition sources include: (1) spark discharges from isolated conductors, such as metal piping, flanges, connectors, valves, and clamps; (2) brush discharges from charged insulating materials, such as electrically-insulating piping, hoses, and lining or coating materials; (3) propagating brush discharges from the inside surfaces of insulating hoses and pipes and electrically-insulating linings or coatings; and (4) discharges from electrically-insulating liquids and/or powders in the receiving vessel. Flammable atmospheres may include flammable gases, vapors, and dusts within pipes and hoses and within the receiving vessels/containers.

The following suggestions are offered to control the potential electrostatic ignition hazard associated with piping and hoses when flammable gases, liquids, vapors, or explosible dusts may be processed, present, or evolved:

1. All metal piping and hoses should be suitably grounded such that their resistance-to-ground is less than ten (10) ohms. If metal piping is isolated from electrical ground due to electrically insulating materials used in flange connections, e.g., gaskets, washers, or bushings, then the use of bonding cables over the flanges may be considered. The grounding of metal piping should be checked periodically by measurement, and particularly after maintenance operations that involve the opening or removal of a pipe section, i.e., breaking pipe.

2. Static dissipative and conductive hoses and piping elements should be made from materials having a volume resistivity of less than 10^9 ohm-meters and be electrically grounded during use.

3. Some piping and hoses constructed from electrically-insulating materials may have some associated metal elements, such as flanges, connectors, fasteners, clamps, sheaths, and reinforcing spirals and meshes. Such metal elements should be grounded such that their resistance-to-ground is less than ten (10).

4. The use of piping and hoses constructed from electrically-insulating materials or metal pipes with an insulating lining or coating on the inside is generally not recommended for conveying any powder or liquids having conductivities of 100 picosiemens per meter (pS/m) or less due primarily to the risk of ignition from brush and propagating brush discharges. Instead, electrically grounded conductive or static dissipative piping must be used.

6.6 Pneumatic Conveying

Pneumatic conveying is used for the convenient transfer of bulk solids and powders and involves using a flowing gas, typically air, to move the powder particles from one vessel to another. Because of the high flow rate of solids through the system, and the consequent contact between the particles and the pipe walls and between the particles themselves, pneumatic conveying is characterized by the development of high levels of electrostatic charge. Potential electrostatic ignition sources include: (1) spark discharges from isolated conductors; (2) cone discharges in the filter receiver or collection hopper; and (3) brush discharges from insulating filter media.

Flammable atmospheres may include dust clouds within the main transfer pipeline, cyclone, filter receiver, or collection hopper, and flammable vapors from vessels into which solids are being introduced.

The following suggestions are offered to control the potential fire and explosion hazard associated with pneumatic conveying operations when flammable gases, liquids/vapors, or explosible dusts may be processed, present, or evolved:

1. All metal elements of pneumatic conveying systems should be electrically grounded such that the resistance-to-ground is less than ten (10) ohms in order to prevent the accumulation of electrostatic charge and thereby minimize the probability of ignition from spark discharges. Metal elements that should be grounded include: (a) source and receiving vessels; (b) cyclones; (c) filter receivers; (d) filter cages in filter receivers; (e) transfer piping and hoses; and (f) all piping elements, including valves, couplings, fittings, and flanges. Grounding should be checked regularly by measurement.

2. The use of piping and other components made from ordinary plastics and other electrically-insulating materials is generally not recommended in pneumatic conveying systems where flammable atmospheres may exist due to the potential risk of ignition from brush and propagating brush discharges. Instead, piping and other components made from metal and static dissipative or conductive materials should be used. Such piping and components should be electrically grounded.

3. Wands used with pneumatic conveying systems in which explosible powders are processed should be made from metal or another suitable static dissipative or conductive material and electrically grounded.

4. When a flammable vapor atmosphere may be present in the filter receiver, the use of static dissipative filter media should be considered. Static dissipative filter media should be fabricated from materials having a surface resistivity of less than 1×10^{11} ohms per square and be electrically grounded during its service.

5. If the volume resistivity of the powder is 10^9 ohm-meters or greater and the minimum ignition energy (MIE) of the dust evolved from the powder is 25 millijoules (mJ) or less, explosion protection or inert gas blanketing measures should be considered for receiving vessels and collection hoppers into which pneumatically conveyed powders are transferred in order to minimize the risk of ignition from cone discharges.

6. Due to the nature of filter receivers, it is very difficult to base safety on the exclusion or elimination of potential ignition sources. Electrostatic charge will inevitably be generated due to the flow of solids through the system. Also, hot embers or tramp metal may occasionally be sucked into the system and present a potential ignition source. Consequently, safety in dust filters is usually based on explosion protection, primarily explosion relief vents. Explosion pressure and products from filter receivers should be vented to a safe area, i.e., away from personnel thoroughfare and other vessels and equipment. When explosion

pressure and products cannot be vented to a safe area, or the filter receiver is located too far away from an external wall, explosion suppression or quench tubes should be used.

6.7 Operators' Footwear

The human body is an electrical conductor. Footwear or flooring that are insulating in electrostatic terms can cause the body to be isolated from ground and susceptible to accumulation of electrostatic charge. Charge accumulated on personnel can give rise to spark discharges upon exposure to conductors at a lower potential, e.g., ground. Such discharges can be sufficiently energetic to ignite flammable atmospheres, including gases, vapors, and dust clouds with a minimum ignition energy (MIE) less than 25mJ.

The following suggestions are offered to control the potential electrostatic ignition hazard posed by personnel in or around flammable gases, liquids, vapors, or explosible dusts:

1. When personnel may be exposed to flammable gases, vapors, or dusts having a minimum ignition energy (MIE) less than 25 millijoules (mJ), they should be electrically grounded such that their resistance-to-ground is less than 10^8 ohms. This will prevent the accumulation of electrostatic charge on their bodies and therefore minimizes the probability of ignition from spark discharge.

2. Grounding of personnel can be accomplished through the use of personnel grounding straps, which are typically attached to the wrist of personnel, or through the combination of static dissipative footwear for personnel and the specification of static dissipative or conductive flooring for operating areas. Since grounding wrist straps can restrict the range of motion of personnel and pose a trip hazard, the combination of static dissipative footwear and flooring is generally preferred.

3. Suitable static dissipative footwear and flooring should be made from materials having a volume resistivity less than 10^8 ohm-meters.

4. Where materials of low ignition energy, such as explosives and propellants, are handled it is generally recommended that a resistance-to-ground through the footwear and the floor of less than 10^6 ohms is maintained. Conductive footwear should not be used where a possibility for electrocution by line voltages exists.

5. Gloves worn by operators should be static dissipative or conductive with a volume resistivity less than 10^9 ohm-meters. The requirements for the use of static dissipative gloves are applicable to situations where operators use conductive tools and implements in their gloved hands. Alternatively, the tools and implements can be grounded separately using grounding cables and clamps, if practical.

7. Fire and Explosion Prevention and Protection

If the formation of a flammable atmosphere cannot be prevented or all ignition sources cannot be reasonably excluded or eliminated, then the risk of a fire or explosion will persist and measures must be taken to protect against their consequences. One method of preventing a flammable atmosphere from forming is inert gas blanketing or purging. This involves the introduction of an inert gas to a vessel or system to reduce the relative oxygen concentration to a level below which a fire or explosion cannot occur, i.e., the limiting oxygen concentration (LOC).

Fire and explosion protection measures include containment, relief venting, and suppression. Containment is the design of vessels and equipment so that they are sufficiently robust to be capable of withstanding and containing the maximum explosion pressure. Relief venting is the installation of special panel(s) in vessels and equipment that are designed to open at a pressure lower than the design pressure of the vessel and equipment. This protects the vessel and equipment from overpressure and relieves explosion pressure, resulting fireball and products to a safe area. Suppression is the rapid introduction of an inerting agent to the incipient explosion. This prevents the course of the explosion from reaching its maximum pressure and thereby protects the vessel and equipment from overpressure. These measures are often coupled with isolation, which is the use of equipment or methods to prevent the propagation or communication of fires and explosions from one unit to another.

8. Concluding Remarks

Electrostatic charge generation most commonly occurs whenever two materials, liquids and/or solids, make and then break contact with each other. The accumulation of the electrostatic charge can result in electrostatic discharges. This suggests that there are many occasions during the transfer, handling, processing, and packaging of flammable liquids and powders in a chemical plant where flash fire and explosion hazards exist. The assessment of electrostatic ignition hazards normally involves:

- Determining the magnitude and polarity of the charges that are generated;
- Investigating the ability of the processing equipment and materials to retain and accumulate the static charge;
- Determining the type and igniting power of the discharges that could result from electrostatic charge generation and accumulation; and
- Determining the sensitivity of the flammable atmosphere to ignition by the resulting discharges (minimum ignition energy)

If the energy of the electrostatic discharge is greater than the minimum ignition energy of the flammable atmosphere a fire or explosion hazard will exist and measures should be taken to avoid the conditions under which such discharge could occur. If the exclusion or elimination of electrostatic ignition sources is not practical, other measures such as avoidance of flammable atmospheres or explosion protection must be considered. Finally, whatever preventative and protective measures are considered (hardware or procedural), they have to be effectively implemented, installed, and maintained. Clearly, training of staff to understand electrostatic hazards is also important.

9. Useful References

You may refer to the following references for more detail on the topics that have been discussed in this paper:

[1] NFPA 77, Recommended Practice on Static Electricity, 2014 Edition, National Fire Protection Association, 1 Batterymarch Park, Quincy, MA 02169-7471.
[2 Vahid Ebadat, et al., Reducing Electrostatic Hazards Associated with Chemical Processing Operations, Chapter 36, Handbook of Chemical Health and Safety, Oxford University Press, 2001.
[3] Britton, L. G., "Avoiding Static Ignition Hazards in Chemical Operations", (1999).

2nd CCPS CHINA CONFERENCE ON PROCESS SAFETY

A Survey Study on Training Needs and Effectiveness for Responding Hazardous Material Incident in the Central Region of Taiwan

Jyun-Hong Shen
Graduate School of Engineering Science and Technology, Yunlin University of Science and Technology (YunTech)
123 University Road, Section 3, Douliou, Yunlin 64002, Taiwan, R.O.C.
cgh1210@gmail.com

Chla-Chieh Shen & Jao-Jia Horng
Department of Safety, Health, and Environmental Engineering, Yunlin University of Science and Technology (YunTech)
123 University Road, Section 3, Douliou, Yunlin 64002, Taiwan, R.O.C.
shencc.tw@gmail.com, horngjj@gmail.com

Prepared for Presentation at
American Institute of Chemical Engineers
2nd CCPS China Conference on Process Safety
Qingdao, China
August 28 – 29, 2014

UNPUBLISHED

A Survey Study on Training Needs and Effectiveness for Responding Hazardous Material Incident in the Central Region of Taiwan

Jyun-Hong Shen
Graduate School of Engineering Science and Technology, Yunlin University of Science and Technology (YunTech)
123 University Road, Section 3, Douliou, Yunlin 64002, Taiwan, R.O.C.
cgh1210@gmail.com

Chla-Chieh Shen & Jao-Jia Horng
Department of Safety, Health, and Environmental Engineering, Yunlin University of Science and Technology (YunTech)

Keywords: HAZMAT, response personnel, training course, questionnaire surveys.

Abstract

In responding to complex hazardous material (HAZMAT) incidents, the specialized training on HAZMAT response personnel is an essential part of prevention strategies. The successful training courses not only consider the needs and interests of trainees but also practical needs. This study used questionnaire surveys to investigate five dimensions of response system needs: needs, objectives, course contents and effectiveness on those response employees in chemical industries in the central region of Taiwan. Among 340 surveys with effective return rate of 64%, our analysis indicated that the needs were the highest for technical (95%) and operational (92%) levels, followed by commander, awareness, and liaison officer levels. Five dimensions on needs, objectives, course contents and effectiveness were not different by different types of companies and personal backgrounds. However, those who had encountered HAZMAT incident showed strong agreements. However, the training objectives showed some differences by genders, the course contents by job types and previous training, and effectiveness by education levels. Since our study was for the region of central Taiwan, the results should be applicable for planning the future needs on training for HAZMAT response.

1. Introduction

The dangerous chemicals (DCs) generally refer as dangerous goods, hazardous materials (HAZMAT), hazardous chemicals, toxic chemicals and so on. Those chemicals may cause life, environmental consequences and property losses when leaking or firing by accident or human errors. As Taiwan's economy developed and the domestic industry prospered, the numbers of DCs have increased to more than 64,000 [1] and the amount of DC increased steadily. Many DCs have been categorized as toxic industrial materials (TIM) and chemicals of interest, thereby increasing the risk of operation. Once those DCs causing accidents, they might create hazards along with economic importance, social impacts, media attention, environmental risks, or others.

Therefore, how to enforcing prevention measures and emergency responses have become a crucial issue for the government and industries to lessen the possible effects or impacts.

The central region of Taiwan not only an transportation passage but also important industrial bases, especially the Mailiao Industrial Zone and the Central Taiwan Science Park. DCs are frequently transported crossing as well as the production of raw, intermediate and final products to distribute throughout Taiwan or to export out. Moreover, those operation of DCs would compound with dense populations near transportation routes or production sites and further breaching the security of industrial bases by the possible DCs or HAZMAT.

Many previous studies have discussed the security assessment and procedures in order to prevent deliberate attacks and possible damages or impacts, including threat analysis, vulnerability analysis, security countermeasures, mitigation and emergency response [2-4]. However, there are few studies on training needs and efficiency assessment for adults [5-7] and those possible responding to HAZMAT incidents. It is needed to know in-depth about the specialized components for training on response personnel in order to respond complex HAZMAT incidents [8-10] effectively. Moreover, for successful training courses it is needed to consider the needs and interests of trainees but also to practical applications [5-8]. For these reasons, we investigated the personnel who were responsible for operation or responding to HAZMAT in chemical industries in the central Taiwan by questionnaire surveys. It is hope that our results could provide a reference for the government and enterprises in their future planning on training.

2. Methodology

2.1 questionnaire design

Our subjects was to studied on training needs and effectiveness for those who would operate or respond to HAZMAT in chemical related industries in the central region of Taiwan. Our survey questionnaires contained six main portions including, basic information and five dimensions of system needs, training needs, objectives, course contents and effectiveness (Fig. 1).

Using the Likert Scale, we compared the score distribution for each dimension which briefly described as follows:

(1) Basic information: Items included industry type, company size, job types, work experience, gender, education level, job function and experience (management or participation) regarding to HAZMAT response, and previous experience of training courses or workshop.
(2) System needs: Items included levels of response personnel do and need in HAZMAT response (Q1), or possible needs in environmental protection, fire fighting, industrial hygiene and plant operation (Q2-Q5)?
(3) Training needs: Items included necessity of HAZMAT training (Q6), training level (Q7), re-training every year? (Q8), joint training with governmental agencies and enterprises (Q9), differences in training between management and response personnel (Q10).
(4) Objectives: Items included response capability (Q11), practical operation (Q12), training objectives for different response levels (Q13), knowledge of overall response process (Q14), training objectives for different units (Q15).
(5) Course contents: Items included course contents for different units (Q16), training contents

for different response levels (Q17), overall response process (Q18), contents with case studies and exercises (Q19), response theory (Q20).

(6) Effectiveness: Items included test needs after class (Q21), teaching assessment (Q22), evaluation by grouping testing (Q23), government-leading course planning and assessment (Q24), self- assessment by enterprise needs (Q25).

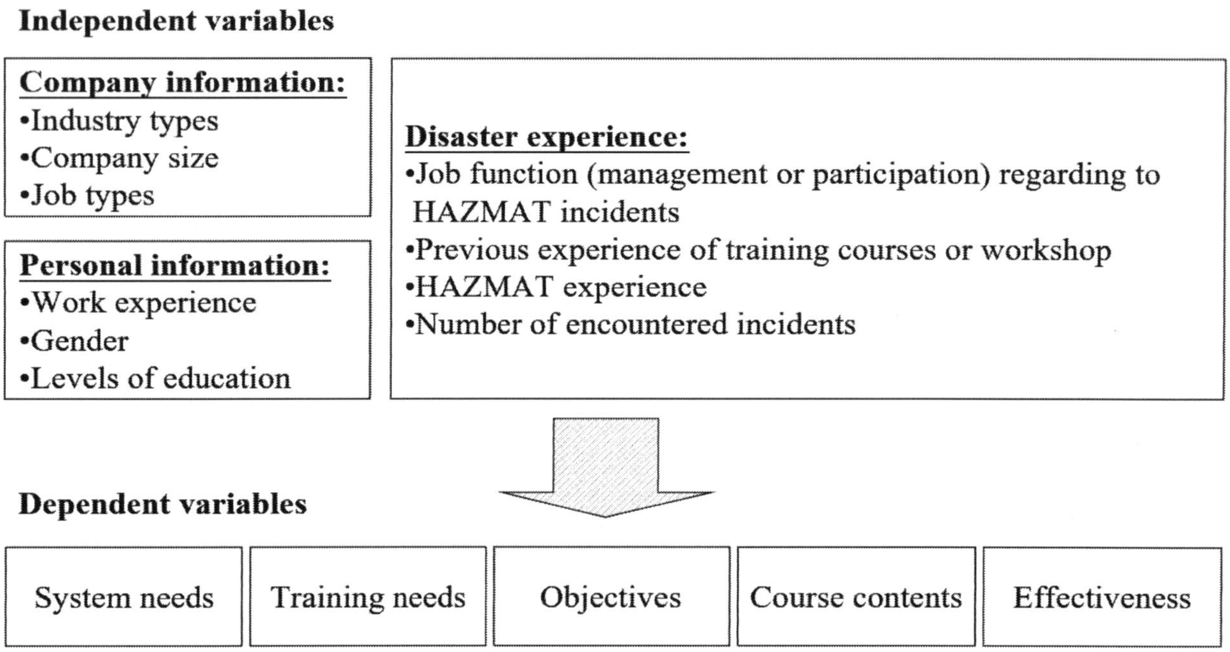

Fig. 1: Questionnaire contents

2.2 questionnaire analysis

The questionnaires were surveyed, verified, and then registered for further data analysis. First, the questionnaires would be checked on its verifications and then input the data in MS Excel. And, we employed SPSS 20.0 for the data analysis and hypothesis testing:

(1) Descriptive statistics: Our statistical analysis was to determine the importance of respondents to those questionnaires as variables and to conduct analysis of means and others. We categorized all questions into five dimensions and assessed the overall awareness of respondents towards HAZMAT training needs and effectiveness. Higher means indicated greater awareness and lower means indicated less awareness.

(2) Independent samples t-test and one-way analysis of variance (ANOVA): We also determined certain significant differences among respondents by t-test and ANOVA for significant differences. In this paper, only significant differences would be shown, while unshown portions were no significant.

2.3 limitations of the study

This study was subjected to several limitations from data collection and the perceptions of subjects. Firstly, we focused only on chemical industries in the central region of Taiwan. Secondly, we strived to be concise in questionnaire design and were unable to obtain complete information. Finally, the differences in attitude and truthfulness of respondents on those questionnaires could not be eliminated and may therefore cause differences as results.

3. Results and Discussion

3.1 basic information analysis

The subjects of this questionnaire were personnel in chemical industries. Most of respondents were male (80.2 %) with university/graduate school (65.9%). Industry types were mostly traditional (24.4%) and chemical (23.9%), and following by high-tech industry, petrochemical, academic institutions, government agencies and other departments. Companies or organization sizes were more than a hundred people (52%). The respondents were mostly working-level staffs (77%) as technicians, engineers, environmental safety and health personnel, low-level supervisor, operator and contractors. In addition, there was only 17.5% of respondents had actual experience of HAZMAT response. Table 1 presented the relevant data.

3.2 descriptive statistics for system needs

The system needs for response levels were multiple choices with importance ranking of technical (95%), operational (92%), commander (83%), awareness (82%) and liaison officer (75%) as shown in Fig. 2a. More than 70% of respondents considered all levels of response personnel were importance in the HAZMAT response system with technical level as highest, followed by commander level for environmental units (Fig. 2a). The technical and operational levels were thought to highly important for fire-fighting units and technical and awareness levels were major for industrial hygiene units (Fig. 2b). The major levels were operational, awareness and technical for enterprises (Fig. 2c). In addition, we found that the needs for liaison officer were the lowest for all. It might be because most domestic HAZMAT events were relatively small and did not need the function of liaison officers.

3.3 significant differences for other dimensions

3.3.1 training needs

Table 2 showed that training needs did not appear to have significant differences for industry type and size, job type and work experience, gender and education level. However, there were significant differences at Q6 “Necessity of HAZMAT training” and Q8 “Re-training every year” for those who had encountered HAZMAT incidents showing strong agreements. From our results, it seemed that those who have the experience of encountering HAZMAT incident in an agreement to HAZMAT training. And, those agreements increased as they seeing more incidents.

Table 1. Relevant data of respondents

Variables	Sub-items	Percentage
Industry types	Traditional	24.4%
	Chemical	23.9%
	High-tech	9.6%
	Transportation	1.3%
	Petrochemical,	7.8%
	Academic institutions	3.6%
	Hygiene units	1.3%
	Environmental units	6.4%
	Fire-fighting units	3.2%
	Government agencies	13.3%
	Others	4.6%
Company size	10 people	9.2%
	10-29 people	19.8%
	30-49 people	11.1%
	50-99 people	7.8%
	100-299 people	20.7%
	300 people and upward	31.3%

Variables	Sub-items	Percentage
Job types	Technicians	11.1%
	First superintendent	5.5%
	Engineers	8.3%
	Safety and health personnel	35.9%
	Department superintendent	7.8%
	Manager	3.7%
	Ordinary staffs	16.6%
	Section chief	6%
	Others	5.1%
Work experience	1 year	6.5%
	1-5 years	31.8%
	6-10 years	18.9%
	11-15 years	19.3%
	16-20 years	9.7%
	21 years and upward	13.8%
Gender	Male	80.2%
	Female	19.8%

Variables	Sub-items	Percentage
Education level	Graduate school	22.1%
	University	43.8%
	Vocational school	24.9%
	Senior high school	9.21%
	Junior high school	0%
Management or participation regarding to HAZMAT response	Yes	87.1%
	No	12.9%
Previous experience of training courses or workshop	Yes	81.6%
	No	18.4%
HAZMAT Experience	Yes	17.5%
	No	82.5%
Number of encountered incidents	0	82.5%
	1	8.3%
	2	4.1%
	3 and upward	5.1%

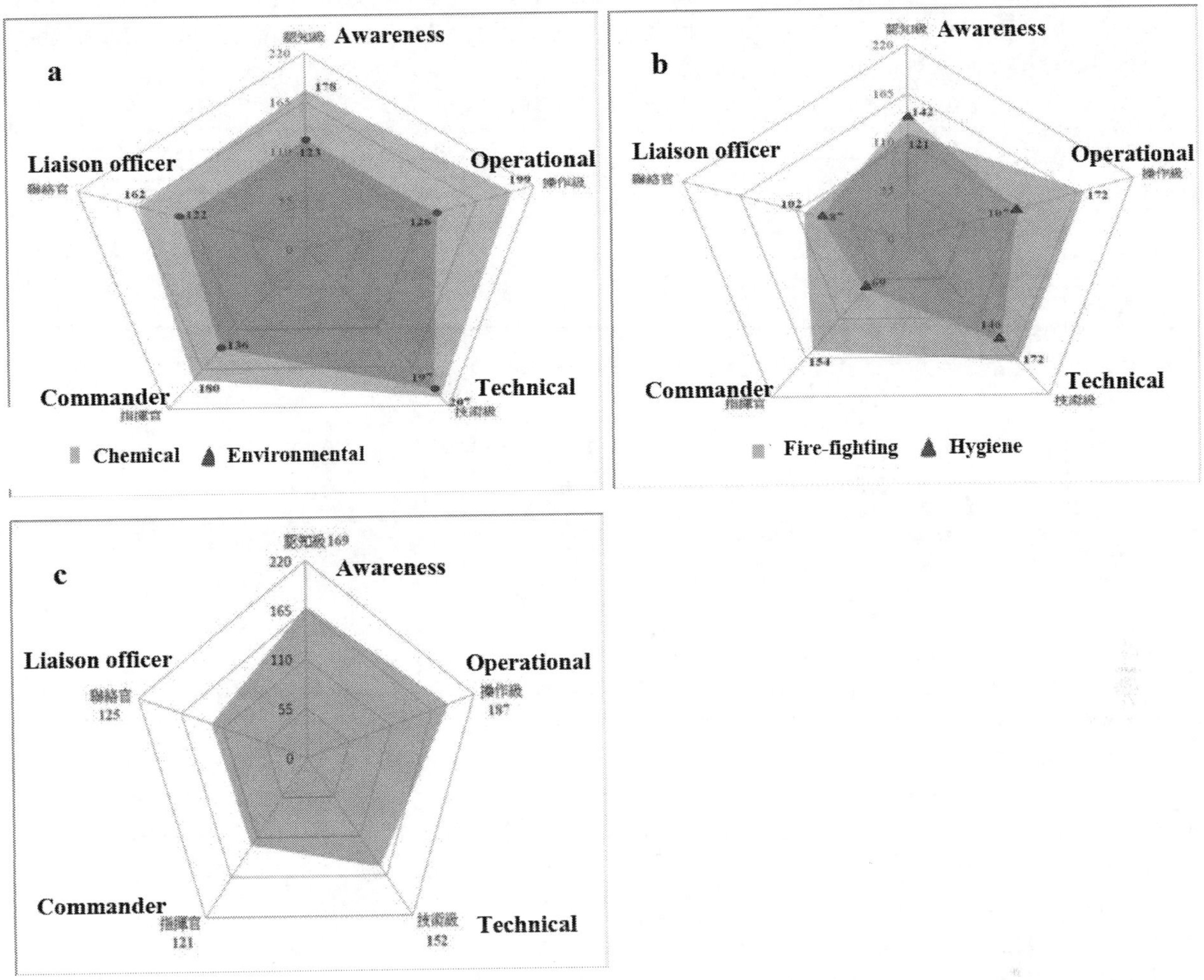

Fig. 3: System Needs of Response Personnel for Different Units
(a: Chemical and environmental units; b: Fire-fighting and hygiene units; c: Enterprises)

Table 2. Differences on Personal Background in Training Needs

Variables	Items	Q6	Q7	Q8	Q9	Q10
HAZMAT experience	(1) Yes	(1)>(2)*	*ns*	*ns*	*ns*	*ns*
	(2) No					
Number of encountered incidents	(1) 0	*ns*	*ns*	(4)>(1)*	*ns*	*ns*
	(2) 1					
	(3) 2					
	(4) 3 and up					

ns p>.05 **p*<.05

3.3.2 objectives

For dimension on training objectives, there were some significant differences on respondents' background of gender and response experience as shown in **Table 3**. At Q13 "Different training objectives with different levels for response personnel ", respondents who had encountered

HAZMAT incident showing higher agreements than those had no experience. Similar results were seen at Q14 "Understanding overall response process" while female respondents had higher agreements than male.

Table 3. Differences on Personal Background for Training Objectives

Variables	Items	Q11	Q12	Q13	Q14	Q15
Gender	(1) Male (2) Female	*ns*	*ns*	*ns*	**(2)>(1)***	*ns*
HAZMAT experience	(1) Yes (2) No	*ns*	*ns*	**(1)>(2)***	**(1)>(2)***	*ns*
Number of encountered incidents	(1) 0 (2) 1 (3) 2 (4) 3 and up	*ns*	*ns*	**(4)>(1)***	*ns*	*ns*

ns $p>.05$ *$p<.05$

3.3.3 Course Contents

For dimension on course contents of "Whether they manage or assist in relief operations", "Whether they attend the related training course or workshop", and "Experience of HAZMAT incidents" showed no significant differences (**Table 4**). For Q19 "Contents including case studies and exercises", respondents who have assisted in HAZMAT activities, attended the training courses, had experience of HAZMAT incidents all agreed highly that case studies and exercises should be blended into training courses as necessary parts. At Q20 "Contents including response theory", those respondents who had experience on HAZMAT incidents showed higher agreements that response theory should be an essential part.

Table 4. Significant Differences on Personal Background for Course Contents

Variables	Sub-items	Q16	Q17	Q18	Q19	Q20
Assisting in relief operations	(1) Yes (2) No	*ns*	*ns*	*ns*	**(1)>(2)***	*ns*
Attending related training course	(1) Yes (2) No	*ns*	*ns*	*ns*	**(1)>(2)***	*ns*
HAZMAT experience	(1) Yes (2) No	*ns*	*ns*	*ns*	**(1)>(2)***	**(1)>(2)***
Number of encountered incidents	(1) 0 (2) 1 (3) 2 (4) 3 and up	*ns*	*ns*	*ns*	**(4)>(1)***	*ns*

ns $p>.05$ *$p<.05$

3.3.4 Effectiveness

The dimension on training effectiveness, some differences were observed for education levels on Q21 and experience of HAZMAT incidents on Q22, Q23 and Q25 (**Table 5**). It was known that respondents had higher education levels and experience of HAZMAT incidents wanted the training with practical operation. Thus, responders who were familiar with the technical aspects in response operation would ask more evaluation of training effectiveness. In addition, teaching assessment, group testing and enterprise' self-assessment needs were recognized by those encountered HAZMAT. It seemed that there was an expectation of raising training effectiveness. Since it has not developed official teaching and assessment methods yet, this study also found that responders were neutral on government-lead planning, training practices and measures.

Table 5. Significant Differences on Personal Background for Effectiveness

Variables	Sub-items	Q21	Q22	Q23	Q24	Q25
Levels of education	(1) Graduate school (2) University (3) Vocational school (4) Senior high school	**(1)>(3)***	*ns*	*ns*	*ns*	*ns*
Experience of HAZMAT incidents	(1) Yes (2) No	**(1)>(2)***	**(1)>(2)****	**(1)>(2)****	*ns*	**(1)>(2)***
Number of encountered incidents	(1) 0 (2) 1 (3) 2 (4) 3 and up	*ns*	**(4)>(1)** (4)>(3)***	*ns*	*ns*	**(4)>(1)***

ns $p>.05$ *$p<.05$ **$p<.01$

4. Conclusion and Suggestions

With complication of HAZMAT incidents, the specialized training for HAZMAT response have become an essential part of prevention strategies. In order to improve the present training courses and planning for future needs, this study surveyed by questionnaire on training needs for the operator and responders of chemical industries in the central region of Taiwan. Our results showed some conclusions for those training needs as the follow.

(1) Most of respondents were male (80.2%) with university/graduate school (65.9%) in relative large company(52%) in traditional (24.4%) and chemical (23.9%), and following by high-tech industry, petrochemical, academic institutions, government agencies and others. The respondents were mostly working-level staffs (77%) as technicians, engineers, environmental safety and health personnel, low-level supervisor, operator and contractors with17.5% actual experience of HAZMAT response.

(2) More than 70% of respondents considered all levels (chemical, environmental, fire-fighting, hygiene, etc.) of response personnel were importance in the HAZMAT response system with technical level as highest, followed by operational and commander levels. The needs for

liaison officer were not significant probably because most domestic HAZMAT events were relatively small and did not need it.

(3) The training needs appeared no significant differences for industry type and size, job type and work experience, gender and education level. However, those who had encountered HAZMAT incidents showed many strong needs on "Necessity of HAZMAT training,", "Re-training every year,""Different training objectives with different levels for response personnel ", "Contents including case studies and exercises", and"Contents including response theory".

(4) The dimension on training effectiveness, it was known that respondents had higher education levels and experience of HAZMAT incidents wanted the training with practical operation. Thus, responders who were familiar with the technical aspects in response operation would ask more evaluation of training effectiveness. Since it has not developed official teaching and assessment methods yet, this study also found that responders were neutral on government-lead planning, training practices and measures.

With these findings, we would apply those results toward our class planning for training HAZMAT personnel for chemical industry and governmental agencies. However, we also need to develop the hand-on or table-top exercises that fit into the training classes in order to fulfill the effectiveness of practicality of training.

5. References

[1] Kuo, C. Y., "Security vulnerability analysis of chemical transport in the central region of Taiwan," Thesis of Master's Degree in Chinese, Graduate School of Engineering Science and Technology, YunTech, Taiwan, ROC, 2010

[2] Gupta, J. P., "A course on inherently safer design," *Journal of Loss Prevention in the Process Industries*, Volume 13, Pages 63–66, 2000

[3] Jaeger, C. D., "Chemical Facility Vulnerability Assessment Project," *Journal of Hazardous Materials*, Volume 104, Pages 207–213, 2003

[4] Bajpai, S. & Gupta J. P., "Site security for chemical process industry," *Journal of Loss Prevention in the Process Industries*, Volume 18, NO. 4, Pages 301–309, 2005

[5] Roberson, D. N. & Merriam, S. B., "The Self-directed Learning Process of Older, Rural Adults," *Adult Education Quarterly*, Volume 55, NO. 4, Pages 269–287, 2005

[6] Romiszowski, A. J., "The Selection and Use of Instructional Media: for Improved Classroom Teaching and for Interactive, Individualized Instruction (2nd ed.)," NewYork : Nichols Pub, 1998

[7] Schuler, R. S., "Strategic Human Resources Management Linking the People with the Strategic Needs of the Business," *Organizational Dynamics*, Volume 21, Pages 18-32, 1992

[8] "California Specialized Training Institute," http://www.calema.ca.gov/CSTI/Pages/Landing%20Page.aspx (accessed Feb. 10, 2013)

[9] "Engineering Extension Service", Texas A & M University, http://www.teex.org/esti/ (accessed Feb. 10, 2013)

[10] "NFPA 472: Standard for Professional Competence of Responders to Hazardous Materials Incidents, 2002 Edition (NFPA)," http://www.nfpa.org/catalog/product.asp?pid=47202&title=Code-472-2002-Professional-Competence-of-Responders-to Ha & category_name = & target_pid = 47202 & source_pid = 47213 & link_type = edition_change (accessed Feb. 11, 2013)

2nd CCPS CHINA CONFERENCE ON PROCESS SAFETY™

Interactions between Indigenous and Immobilized Microorganisms in Bioremediation of Petro-contaminated Soil

Yutong Han
Graduate student, Department of Environmental and Safety Engineering, China University of Petroleum (East China), No. 66, Changjiang West Road,Qingdao Economic & Technical Development Zone, Qingdao 266580, China
Tel:15666220213 Email:hyt8983@163.com

Xiuxia Zhang, Xuejing Bai, Haijie Wu, Shoujuan Zhang, Han Zhang

Zhang Xiuxia, professor, Department of Environmental and Safety Engineering, China University of Petroleum (East China). Xuejing Bai, Haijie Wu, Shoujuan Zhang, Han Zhang, Graduate student, Department of Environmental and Safety Engineering, China University of Petroleum (East China),
Email: zhxiuxia@upc.edu.cn,

Prepared for Presentation at
American Institute of Chemical Engineers
2nd CCPS China Conference on Process Safety
Qingdao, China
August 28 – 29, 2014

UNPUBLISHED

Interactions between Indigenous and Immobilized Microorganisms in Bioremediation of Petro-contaminated Soil

Yutong Han
Graduate student, Department of Environmental and Safety Engineering, China University of Petroleum (East China), No. 66, Changjiang West Road,Qingdao Economic & Technical Development Zone, Qingdao 266580, China
Tel:15666220213 Email:hyt8983@163.com

Xiuxia Zhang, Xuejing Bai, Haijie Wu, Shoujuan Zhang, Han Zhang

Zhang Xiuxia, professor, Department of Environmental and Safety Engineering, China University of Petroleum (East China). Xuejing Bai, Haijie Wu, Shoujuan Zhang, Han Zhang, Graduate student, Department of Environmental and Safety Engineering, China University of Petroleum (East China),
Email: zhxiuxia@upc.edu.cn,

Keywords: immobilized microorganism; indigenous microorganism; petro-contaminated soil; bioremediation

Abstract

In this study, straw SY (leaves) was used as a carrier to adsorb and immobilize high-efficiency petroleum-degrading bacteria. The immobilized microorganisms were used to repair petro-contaminated soil. The interactions between indigenous and immobilized microorganisms were evaluated. The results show that the immobilized microorganisms and the indigenous microorganisms were allowed to interact with each other under the suitable conditions. And the degradation rate of petroleum hydrocarbons during the 35 days of the experiment, using the immobilized and indigenous microorganisms together, could reach 20.78%. It reveals the indigenous microorganism and immobilized microorganism have synergistic effect, and improves the bioremediation efficiency of petro-contaminated soil.

1. Introduction

An increasingly serious environmental problem may be caused by oil pipelines spill. Research suggests that the use of bioremediation technology to repair petro-contaminated soil has excellent economic and social benefits (Acevedo, et al., 2010; Guwy, et al., 1999); however, the use of this technology entails many difficulties and challenges (Lu, et al., 2010), including the toxicity and hydrophobicity of organic pollutants and the complexity of soil structures (Viteri & Schmidt, 1996). The counts of microbes capable of digesting petroleum hydrocarbons increase rapidly when the soil is contaminated by petroleum, demonstrating that petroleum hydrocarbons can stimulate or induce the growth and reproduction of hydrocarbon-digesting microorganisms (Das & Mukherjee, 2007; Wolicka, et al., 2009). Research has shown that the use of indigenous microorganisms is an effective way to repair groundwater and soil contaminated by petroleum, although this method may be time-consuming, and results may be hindered by competition between the indigenous microorganisms and other dominant bacteria (Gomes et al., 2013).

Research has also suggested that immobilized microorganisms may be particularly effective at restoring soils in complex environments (El-Naas, et al., 2009; Su, et al., 2006; Su D, et al., 2006; Wang Xin, et al., 2006; Zhang Xiuxia, et al., 2008a; Zhang Xiuxia, et al., 2008b); therefore, it would be advantageous to develop approaches to increase petroleum degradation efficiency. To address this, Liang reported the use of porous materials to immobilize the microorganisms, which resulted in a 12-fold increase in the concentration of microorganisms and a 3-fold increase in dehydrogenase activity (Liang, et al., 2009). Additionally, Cunningham embedded petroleum-degrading strains in polyethylene to immobilize the microorganisms, thus creating a biological composting technology that was used to efficiently degrade diesel fuel–contaminated soil over a 32-day period (Cunningham, et al., 2004).

In this study, straw SY was used as a carrier to adsorb and immobilize high-efficiency petro-degrading bacteria. The immobilized microorganisms were used to repair the petro-contaminated soil in a pot to simulate the in-situ remediation. In order to research the interactions between the indigenous and immobilized microorganisms during the repair of petro-contaminated soil, three experimental conditions were evaluated: using the immobilized microorganism only, using the indigenous microorganism only, and using both simultaneously, the degradation rate of petroleum hydrocarbons, soil enzyme activity, soil microbe counts, and the diversity of the soil microbe community were determined. It will reveal the relationship between indigenous microorganism and immobilized microorganism, provide theory support for the immobilized microorganism repair oil contaminated soil.

2. Material and Methods

2.1 Soil Samples

Soil samples, polluted by petroleum, were collected from petroleum-contaminated areas in the city of Karemary, Xinjiang, China. Plant residues and sandstones were removed from the soil samples using 3-mm sieve. Soil samples were allowed to dry naturally for 7 days.

2.2 Media

(1) The beef-protein medium (pH = 7.0-7.2) comprised 5.0 g/L beef extract, 10.0 g/L peptone, and 5.0 g/L sodium chloride in 1000 ml distilled water. The medium was sterilized at 121°C for 20 min.

(2) The inorganic salt solution (pH=7.5) comprised 5.0 g/L NaCl, 1.0 g/L $(NH_4)_2SO_4$, 0.25 g/L $MgSO_4 \cdot 7H_2O$, 2.0 g/L $NaNO_3$, 4.0 g/L KH_2PO_4, and 10.0 g/L $K_2HPO_4 \cdot 3H_2O$ in 1000 ml deionized water. The medium was sterilized at 121°C for 20 min.

(3) The petroleum-decreasing medium (pH=7.0–7.2) comprised 50 ml inorganic salt solution and 0.5 g petroleum. The medium was sterilized at 121°C for 20 min.

(4) The selective medium comprised 5.0 g Na_2SO_4, 2.7 g NH_4Cl, 0.35 g KH_2PO_4, 0.27 g K_2HPO_4, 0.1 g $MgCl_2$, 0.1 g $CaCl_2$, 1.0 g yeast extract, 2.5 g $NaHCO_3$, and 0.5 g ammonium ferrous sulfate. The medium was sterilized at 121°C for 20 min.

(5) The G1 medium comprised (pH = 7.4-7.6) 20 g/L soluble starch, 0.5 g/L NaCl, 0.5 g/L $K_2HPO_4 \cdot 3H_2O$, 1.0 g/L KNO_3, 0.5 g/L $MgSO_4 \cdot 7H_2O$, 0.1 g/L $FeSO_4 \cdot 7H_2O$, 15-25 g/L agar gel, 1000 ml water. The medium was sterilized at 121°C for 20 min.

(6) The Martin medium comprised 1.0 g/L KH_2PO_4, 0.5 g/L $MgSO_4 \cdot 7H_2O$, 5.0 g/L peptone, 10 g/L glucose, 15-20 g/L agar gel, and 3.3 ml 1% rose Bengal in 1000 ml medium solution. Under aseptic conditions, 0.3 ml of 1% streptomycin was also added per 100 ml medium solution prior to any experiments. The medium was sterilized at 121°C for 20 min.

2.3 Determination of physical and chemical properties of soils

Soil organic materials were identified using the potassium bichromate titrimetric-dilution heat method (Bao,S.-d, 2000). Total nitrogen was determined by semimicro-Kjeldahl determination (Bao,S.-d, 2000). Available phosphorus was determined using the $NaHCO_3$ extracting molybdenum-antimony anti-spectrophotometric method (Cheng, L.-m, et al, 2009).

2.4 Determination of total petroleum hydrocarbons in soil

Put 0.001 g soil sample into a 50 ml centrifuge tube. And the sample was mixed with 10 ml petroleum ether, then, the centrifuge tube was placed in ultrasonic extraction for 15 min. Ultrasonic intensity was maintained at 360 W, and temperature was maintained below 40°C. Samples were then centrifuged for 10 min at 7500 rpm. The supernatant was extracted, collected in a erlenmeyer flask, and dehydrated using anhydrous sodium sulfate. The supernatant should be extracted three times. The erlenmeyer flask and funnel were washed twice with 5 ml supernatant extract, and the extracted supernatant was then transferred to a volumetric flask. The extracted supernatant was maintained at a constant volume by addition of petroleum ether, with a boiling range of 60-90°C. The concentration was diluted to the range of measurement, and the absorbance was determined using an ultraviolet spectrophotometer at λ (max) =225 nm. The total concentration of petroleum hydrocarbons, united by mg/L, was searched in the petroleum standard curve through the absorbance (Wang, Ru.-G, et al, 2010).

2.5 Determination of the degradation rate of soil petroleum hydrocarbons

The soil remediation conditions were as follows: the microorganism inoculation quantity was 15% (Volume than quality, v/w), surfactant content was equal to 0.1% (v/w), C/N/P ratio was adjusted to 100:10:1(quality than quality, w/w), and the area of damaged soil was plowed once daily for 35 days. Three separate conditions were evaluated in pots to simulate the in-situ remediation: addition of the immobilized microorganism to sterilized soil samples; addition of the indigenous microorganism to unsterilized soil samples; and addition of the immobilized microorganism to unsterilized soil samples. The degradation rate of petroleum hydrocarbons were measured every 7 days during the repair process.

2.6 Determination of soil enzyme activity

We evaluated variations in urinary enzyme activity, polyphenol oxidase activity, catalase activity, and dehydrogenase activity in petroleum-contaminated soil every 7 days. The determination refered to the standard methods (Zhou,Li.-K, et al, 1980). The urinary enzyme activity was determined by sodium phenolate colorimetric analysis. The polyphenol oxidase activity was determined by pyrogallol colorimetric analysis. The catalase activity was determined by permanganate titration. The dehydrogenase activity was determined with the TTC reduction method.

2.7 Quantification of soil microbes

Soil microbe quantity was determined by colony counting every 7 days. The proportions of bacterium, actinomycetes and fungi were determined with the CFU (the abbreviation of colony-forming unit method. Soil samples (1g) were mixed with 9 ml of aseptic water in sterilized tubes. Tubes were oscillated for 10 min to disperse the soil evenly and form a petroleum sludge suspension. The petroleum sludge suspension was processed by gradient dilution, and the appropriate dilutions were used to coat the plates. The coating densities were 10-4-10-6, 10-4-10-6, and 10-2-10-3, for bacteria, actinomycetes and fungi, respectively. The counting media were beef-protein medium, G1 medium, and Martin medium for bacterium, actinomycetes and fungi, respectively.

2.8 Determination of soil microbe diversity

Bacterial DNA was extracted from soil and examined by PCR for 16Sr DNA V3 and DGGE (the abbreviation of denaturing gradient gel electrophoresis).

3. Results and discussion

3.1 Determination of the physical and chemical properties of the soil

In order to investigate the optimum conditions for the biodegradation of petro-contaminated soil, the soil physicochemical properties were showed in Table 1.

Table 1. Soil physicochemical properties

Moisture content (%)	Organic matter (g/kg)	Total nitrogen (g/kg)	Available phosphorus (mg/kg)	Total petroleum hydrocarbon (%)
22.66	116.76	18	12	4.02

3.2. Petroleum hydrocarbon degradation rate during remediation

Fig.1 shows the petroleum degradation rates observed.

Fig. 1 Petroleum degradation rates

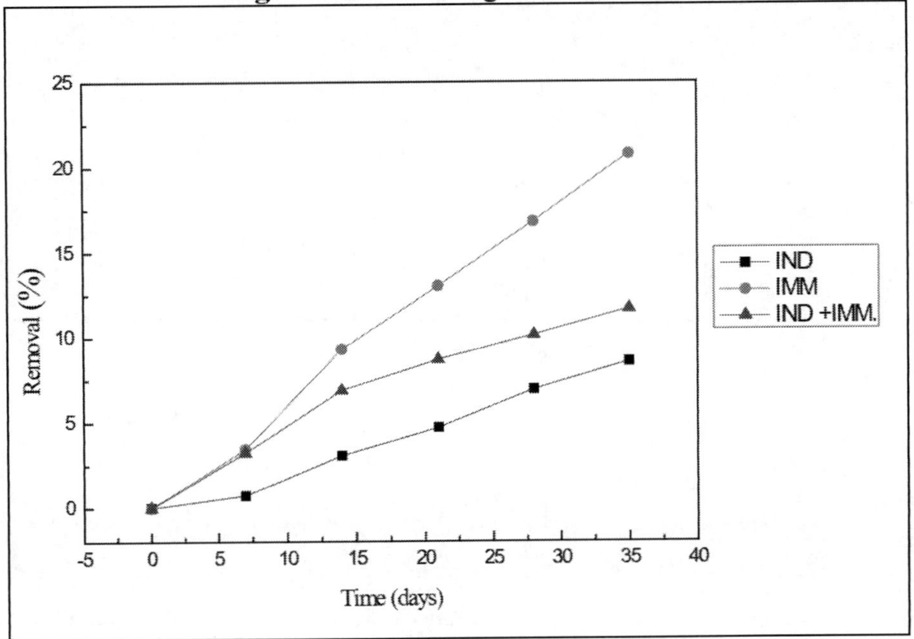

IND is the abbreviation of the indigenous microorganisms, IMM is the abbreviation of the immobilized microorganisms, IND+IMM is the abbreviation of the indigenous microorganisms + the immobilized microorganisms.

It was inferred that efficient bacterial degradation improved the efficiency of the repair process. The degradation rate of petroleum hydrocarbons during the 35 days of the experiment, using the immobilized and indigenous microorganisms together, reached 20.78%. This was significantly higher than that achieved with either the immobilized or indigenous microorganisms alone, although the immobilized microorganisms produced a greater repair effect than the indigenous microorganisms.

3.3. Changes in soil enzyme activity during the repair process

3.3.1. Activity of soil urinary enzymes

Soil urease activity initially increased and then subsequently decreased with time during the repair process (Fig. 2).

Fig. 2 Changes in soil urease activity

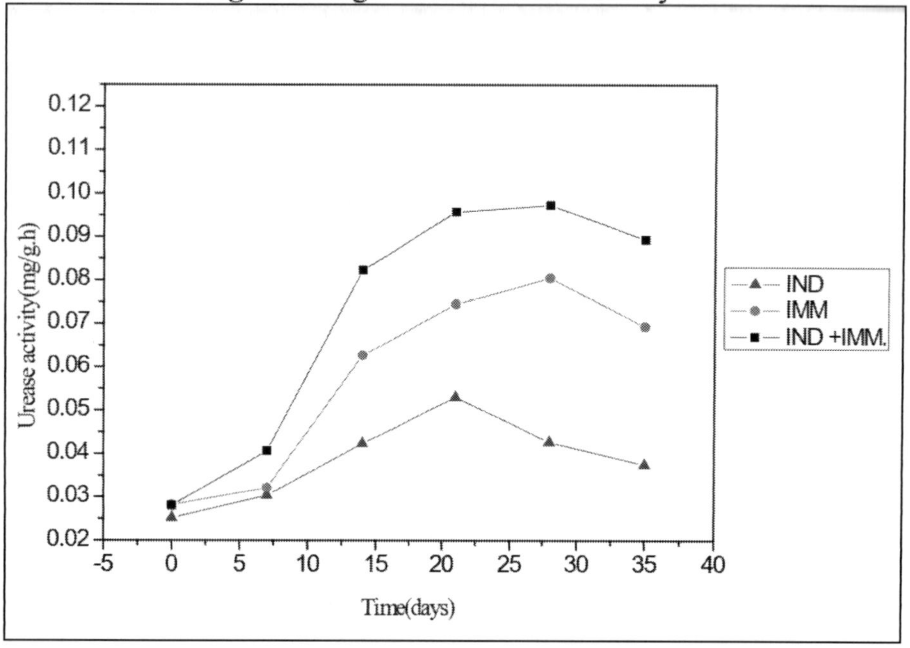

Urease activity in soil treated only with indigenous microorganisms decreased after 20 days, while the urease activity in soils treated with immobilized microorganisms and with indigenous microorganisms plus immobilized microorganisms began to decrease after 28 days. Urease is an important functional enzyme in organic nitrogen decomposition reactions. Organic nitrogen is hydrolyzed to urea and then converted to nitrogen, which can be absorbed and used by microorganisms directly. At the beginning of the experiments, soil urease activity increased because of the nitrogen source added to the soil, which provided sufficient nutrition for the microorganisms. Over time, soil urease activity decreased with decrease in the nitrogen content in the soil; therefore, it was inferred that urease was very sensitive to changes in the nutrition content in the environment.

3.3.2. Activity of soil polyphenol oxidase

Soil polyphenol oxidase activity was observed to increase with time during the repair process (Fig. 3).

Fig. 3 Changes in soil polyphenol oxidase activity

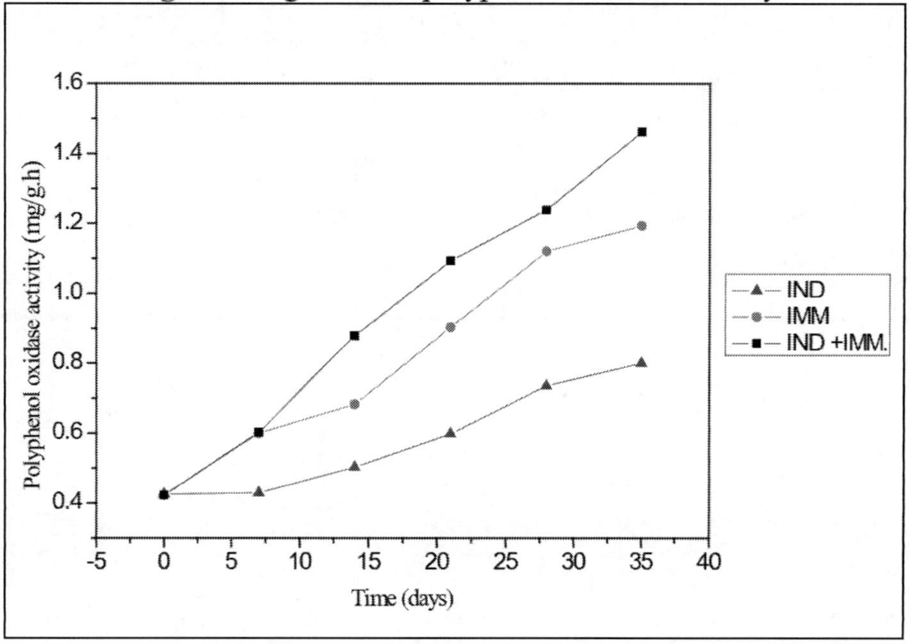

The main function of polyphenol oxidase is to catalyze the oxidation of benzodiazepines to benzene quinone. It can also oxidize a single phenol mono-oxygenase substrate. Thus, polyphenol oxidase is an important functional enzyme in petroleum hydrocarbon oxidation decomposition. The results of this study indicated that polyphenol oxidase activity increased on phenol stimulation; therefore, soil polyphenol oxidase activity increased gradually during the repair process, consistent with results reported in previous studies(D. Kumar jha, et al., 1992; Zhang Lili, et al., 2003).

3.3.3. Activity of soil catalase

Soil catalase activity increased in all three experiments during the repair process (Fig.4).

Fig. 4 Changes in soil catalase activity

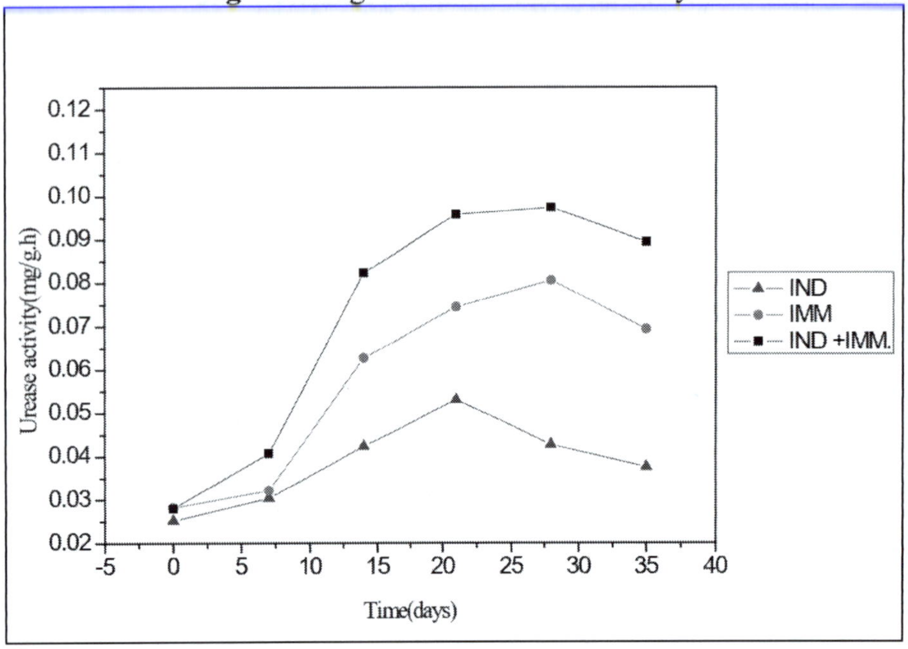

Catalases (H_2O_2 : H_2O_2-oxidoreductase, EC 1.11.1.6.) use a two-electron transfer mechanism to split hydrogen peroxide into molecular oxygen and water. Catalases scavenge H_2O_2 and protect cells from damage caused by reactive oxygen species (Guwy, et al., 1999). A certain concentration of hydrogen peroxide will be produced during microbial respiration and during decomposition of organic matter in soil; therefore, microbes secrete an enzyme to remove hydrogen peroxide in order to prevent oxidation of their cell membranes.

3.3.4. Activity of soil dehydrogenase

Soil dehydrogenase activity increased with time during the repair process (Fig. 5).

Fig. 5 Changes in soil dehydrogenase activity

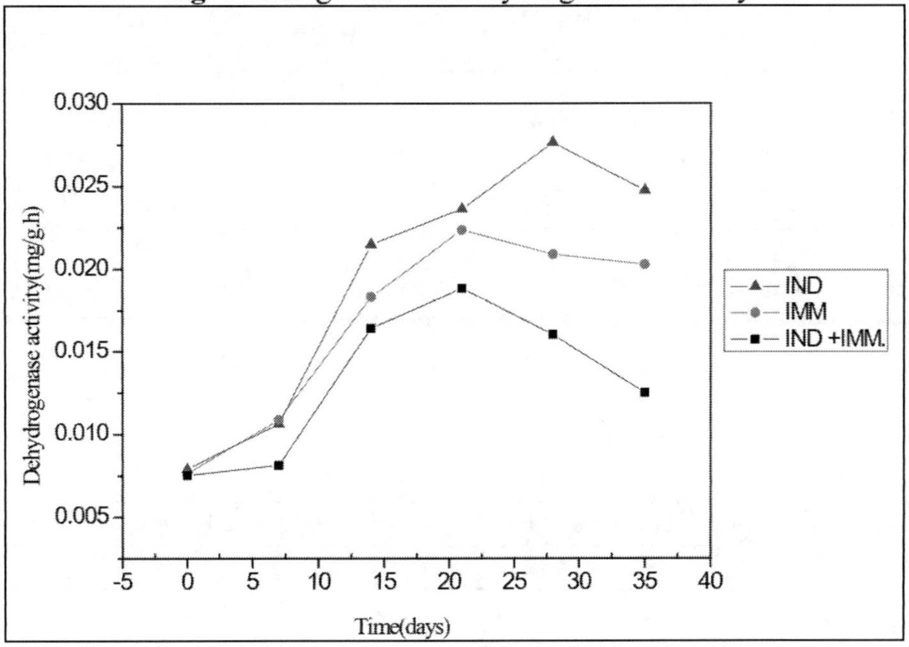

Dehydrogenase is critical at the beginning of bioremediation, as it can activate the hydrogen atoms in organic matter, which can increase the transfer rate of hydrogen atoms and the oxidation decomposition rate of petroleum hydrocarbons. Therefore, repair efficiency of contaminated soil can be approximated by soil dehydrogenase activity.

In general, soil enzyme activities were observed to increase with time during the repair process (Figs. 2-5) for all three experimental conditions, with the highest activity produced by combining the indigenous microbes and immobilized microbes, followed by that produced with the immobilized microorganisms alone. These results suggest that the indigenous and immobilized microorganisms were able to interact with each other. Additionally, the enzyme activity correlated with the concentration of petroleum hydrocarbons in the soil; therefore, the conditions of the repair system could be determined based on changes in soil enzyme activity. These experimental results indicated that the indigenous and immobilized microorganisms could not only coexist, but when combined, were able to improve remediation under suitable conditions.

3.4. Changes in soil microbe counts

Soil microbe quantity is the key factor in determining the effects of soil remediation. Petroleum hydrocarbons in soil can be metabolically degraded only when sufficient microbial cells are available. In an oil-polluted environment, petroleum hydrocarbons can stimulate or induce hydrocarbon microbial growth and reproduction. In this experiment, soil samples were collected once every 7 days during the repair process to measure the changes in the counts of bacteria, fungi and actinomycetes, as summarized in Tables 2, 3, and 4, respectively.

Table 2 Changes in bacterial counts during the repair process

Time (days)	IND (×CFU·g-1)	IMM (×CFU·g-1)	IND+IMM (×CFU·g-1)
0	3.2 × 106	5.4 × 106	7.1 × 106
7	2.7 × 106	4.7 × 106	6.8 × 106
14	8.5 × 106	1.2 × 107	9.7 × 107
21	1.0 × 107	2.8 × 108	9.2 × 108
28	2.2 × 107	2.4 × 109	1.1 × 1010
35	6.2 × 107	2.9 × 109	1.9 × 1010

IND is the abbreviation of indigenous microorganisms, IMM is the abbreviation of immobilized microorganisms, IND+IMM is the abbreviation of indigenous microorganisms and immobilized microorganisms.

Table 3 Changes in fungi counts during the repair process

Time (days)	IND (×CFU·g-1)	IMM (×CFU·g-1)	IND+IMM (×CFU·g-1)
0	not detected	not detected	not detected
7	not detected	2.0 × 104	1.1 × 104
14	2.0 × 104	2.7 × 105	1.9 × 105
21	3.2 × 105	4.1 × 105	9.3 × 106
28	6.8 × 105	2.1 × 106	8.7 × 106
35	6.7 × 105	6.1 × 106	9.8 × 106

Table 4 Changes in actinomycetes counts during the repair process

Time (days)	IND (×CFU·g-1)	IMM (×CFU·g-1)	IND+IMM (×CFU·g-1)
0	1.6 × 104	2.7 × 104	2.6 × 104
7	1.3 × 104	3.2 × 104	3.5 × 104
14	2.7 × 105	3.2 × 105	2.8 × 106
21	1.6 × 105	7.3 × 105	8.8 × 105
28	1.8 × 105	9.1 × 105	1.5 × 106
35	2.6 × 105	7.9 × 105	2.5 × 106

The counts of all three microorganisms increased with time during the 35 days of repair. The soil fungal count was relatively lower than the bacterial count during the repair process, while actinomycetes were the least abundant of the three. During the early stages of the repair process, bacteria were the first microorganisms to exhibit logarithmic growth, followed by actinomycetes and fungi. This finding suggested that the existence of petroleum hydrocarbon pollutants during the initial stages of the repair process stimulated the growth of actinomycetes and bacteria. Subsequently, as soil properties improved during remediation, the fungal count increased.

The greatest microbial count with the richest species diversity was observed when indigenous and immobilized microorganisms were combined. Because immobilized microorganisms have higher cell density, they remained active for a longer time, were more resistant to external adverse factors, and promoted the growth of immobilized microorganisms. Combining straw with the immobilized microorganisms improved soil perviousness while simultaneously promoting the growth of bacteria and fungi. Additionally, straw contains cellulose, hemicellulose, and lignin, which can be transformed into high-quality carbon sources capable of promoting the growth of microorganisms, including hydrocarbon-degrading bacteria.

In the first 7 days of the experiments, the separate counts for the indigenous and immobilized microorganisms were greater than the counts for the combination of indigenous and immobilized microorganisms. This was likely due to competition between the exogenous immobilized microorganisms and the indigenous microorganisms in the contaminated soil. However, during the repair process, the rate of growth for the combination of indigenous and immobilized microorganisms exceeded the sum of the growth rates observed for indigenous and immobilized microorganisms individually. This was because the exogenous immobilized microorganisms adapted to the new environment gradually, thus allowing the indigenous and immobilized microorganisms to develop a cooperative relationship and promote the other's performance.

Researches (Cerqueira, et al., 2011; Karamalidis, et al., 2010) have suggested that the presence of multiple microorganisms can increase the bioremediation process of oily sludge, much better than a single microorganism.

The results of these different studies led us to conclude that the presence of indigenous microorganisms in the contaminated soil, combined with the addition of exogenous immobilized microorganisms, could increase microbial quantity and microbe species diversity, thus improving the petroleum hydrocarbon degradation rate.

3.5. Diversity of the soil microbe community

Bacterial DNA was extracted from soil and examined by PCR for 16Sr DNA V3 and DGGE. The microbial communities and diversity changes in contaminated soil samples were studied, as were the competition and synergy between the indigenous and immobilized microorganisms, during the repair process.

Based on the analysis of our experiments, we selected representative soil samples at day 0 (immediately following addition of the immobilized microorganism), at day 14 (during the competition process between indigenous and immobilized microorganisms), and at day 35 (following stabilization of the community structure). DNA extracted from these three soil samples was examined by PCR and DGGE (Fig.6).

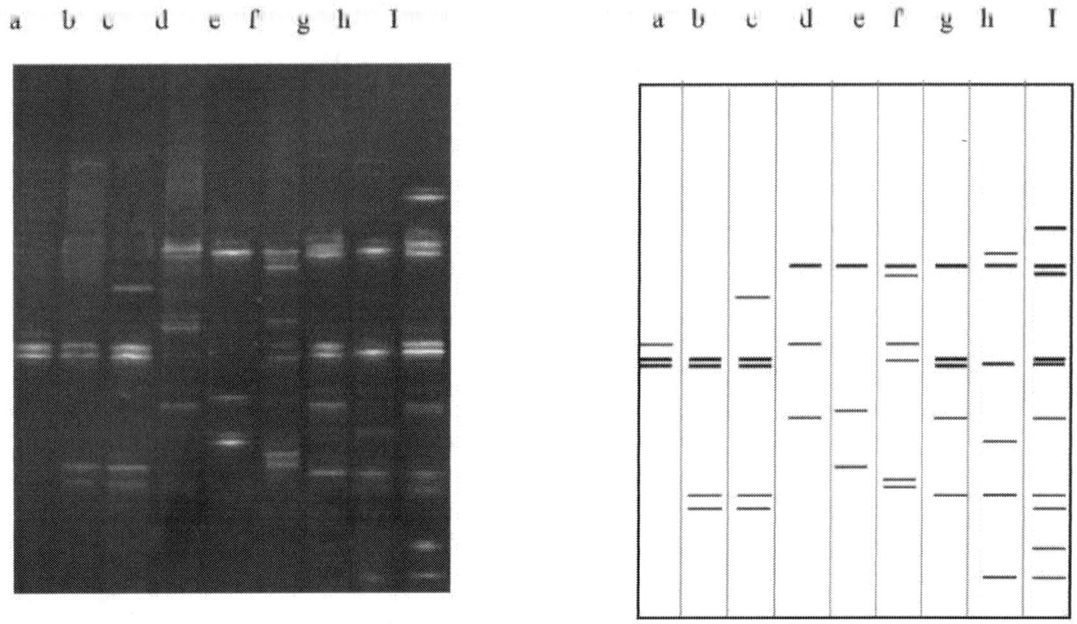

Fig. 6 DGGE spectrum diagram of bacterial community of different training soil for:

Line-a: represented soil sample repaired by the indigenous microorganism for 0 day.

Line-b: represented soil sample repaired by the indigenous microorganism for 14 days.

Line-c: represented soil sample repaired by the indigenous microorganism for 35 days.

Line-d: represented soil sample repaired by the immobilized microorganism for 0 day.

Line-e: represented soil sample repaired by the immobilized microorganism for 14 days.

Line-f: represented soil sample repaired by the immobilized microorganism for 35 days.

Line-g: represented soil sample repaired by the indigenous microorganism and the immobilized microorganism for 0 day.

Line-h: represented soil sample repaired by the indigenous microorganism and the immobilized microorganism for 14 days.

Line-i: represented soil sample repaired by the indigenous microorganism and the immobilized microorganism for 35 days.

On the basis of the results shown in Fig. 6, it was inferred that microbial diversity was low in the contaminated soil prior to remediation efforts, which could be due to the availability of specific flora capable of degrading petroleum hydrocarbons, thereby leading to the disappearance of other types of bacteria through a process of prolonged natural selection.

In our current study, we observed an increasing trend toward microbial diversities under all experimental conditions, thereby indicating that soil microbial diversity was restored after bioremediation. For soil with indigenous microorganisms only, analysis of three samples exhibited two bands representing the dominant flora of the indigenous microorganisms, while for the soil with immobilized microorganisms only, analysis of three samples exhibited a single band representing the high-efficiency petroleum-degrading bacteria. The presence of additional flora in the three samples may be due to contamination of the immobilized microorganisms during preparation, or the introduction of other bacteria from air or water into the soil during the remediation process. The diversity of the microbial community structure in soil of the indigenous and immobilized microorganisms at day 0 was approximately the same as the sum of the diversity of the microbial community structure of indigenous microorganisms at day 0 and the immobilized microorganisms at day 0. The diversity of the microbial community structure in soil had also changed after 14 days, which suggested a competition relationship between indigenous and immobilized microorganisms. Additionally, the diversity of the microbial community structure in soil increased significantly during the 35 days of observation, thus demonstrating that the immobilized microorganisms had adapted to the environment and that interaction between the immobilized and indigenous microorganisms produced a synergistic effect.

4. Conclusions

This study demonstrated that the degradation rate of petroleum hydrocarbons, the activity of soil enzymes (i.e., urinary enzyme, polyphenol oxidase, catalase, and dehydrogenase activities), soil microbe counts (bacterium, actinomycetes, and fungi), and the diversity of the soil microbe community increased over time and under optimal degradation conditions and that these changes correlated with the concentrations of petroleum hydrocarbons in the soil. The best restoration effects were observed in soil treated with a combination of immobilized and indigenous microorganisms. Therefore, we concluded that the indigenous and immobilized microorganisms not only coexisted but apparently interacted with each other synergistically to enhance remediation.

5. Acknowledgements

The authors would like to thank the China National Petroleum Corporation (CNPC) Innovation Foundation (2009D-5006-07-01) and the Fundamental Research Funds for the Central Universities (11CX05011A, CX-1219 and CX2013035) ,Science and Technology Development Guidance Program of Qingdao(KJZD-12-65-jch) for the financial support provided to conduct this study.

6. References

[1] Acevedo, F., Pizzul, L., Castillo, M.d., Gonz ├ílez, M.E., Cea, M., Gianfreda, L., Diez, M.C. 2010. Degradation of polycyclic aromatic hydrocarbons by free and nanoclay-immobilized manganese peroxidase from Anthracophyllum discolor. Chemosphere, 80(3), 271-278.

[2] Bao, S.-d. 2000.Analysis of agricultural soil. China Agticultural Press, 978-7-109-06644-1.

[3] Bao, S.-d. 2000.Analysis of agricultural soil. China Agticultural Press,39-45.

[4] Cerqueira, V.S., Hollenbach, E.B., Maboni, F., Vainstein, M.H., Camargo, F.í.A.O., Peralba, M.d.C.R., Bento, F.í.M. 2011. Biodegradation potential of oily sludge by pure and mixed bacterial cultures. Bioresource Technology, 102(23), 11003-11010.

[5] Cheng, L.-m., Chen,L., Liu, J.-l., Liu, S.-f., Liu, T.-h. 2009. Precautions of the NaHCO3 extracting molybdenum-antimony anti-spectrophotometric method. Anhui Agriculture, 2(3),205.

[6] Cunningham, C.J., Ivshina, I.B., Lozinsky, V.I., Kuyukina, M.S., Philp, J.C. 2004. Bioremediation of diesel-contaminated soil by microorganisms immobilised in polyvinyl alcohol. International Biodeterioration & Biodegradation, 54(2-3), 167-174.

[7] Das, K., Mukherjee, A.K. 2007. Crude petroleum-oil biodegradation efficiency of Bacillus subtilis and Pseudomonas aeruginosa strains isolated from a petroleum-oil contaminated soil from North-East India. Bioresource Technology, 98(7), 1339-1345.

[8] El-Naas, M.H., Al-Muhtaseb, S.A., Makhlouf, S. 2009. Biodegradation of phenol by Pseudomonas putida immobilized in polyvinyl alcohol (PVA) gel. Journal of Hazardous Materials, 164(2-3), 720-725.

[9] Gomes, H.I., Dias-Ferreira, C., Ribeiro, A.B. 2013. Overview of in situ and ex situ remediation technologies for PCB-contaminated soils and sediments and obstacles for full-scale application. Science of The Total Environment, 445-446(0), 237-260.

[10] Guwy, A.J., Martin, S.R., Hawkes, F.R., Hawkes, D.L. 1999. Catalase activity measurements in suspended aerobic biomass and soil samples. Enzyme and Microbial Technology, 25(8-9), 669-676.

[11] Jha, D.K., Sharma, G.D., Mishra, R.R. 1992. Soil microbial population numbers and enzyme activities in relation to altitude and forest degradation. Soil Biology and Biochemistry, 24(8), 761-767.

[12] Karamalidis, A.K., Evangelou, A.C., Karabika, E., Koukkou, A.I., Drainas, C., Voudrias, E.A. 2010. Laboratory scale bioremediation of petroleum-contaminated soil by indigenous microorganisms and added Pseudomonas aeruginosa strain Spet. Bioresource Technology, 101(16), 6545-6552.

[13] Liang, Y., Zhang, X., Dai, D., Li, G. 2009. Porous biocarrier-enhanced biodegradation of crude oil contaminated soil. International Biodeterioration & Biodegradation, 63(1), 80-87.

[14] Lu, M., Zhang, Z., Qiao, W., Wei, X., Guan, Y., Ma, Q., Guan, Y. 2010. Remediation of petroleum-contaminated soil after composting by sequential treatment with Fenton-like oxidation and biodegradation. Bioresource Technology, 101(7), 2106-2113.

[15] Su, D., Li, P.-j., Frank, S., Xiong, X.-z. 2006. Biodegradation of benzo[a]pyrene in soil by Mucor sp. SF06 and Bacillus sp. SB02 co-immobilized on vermiculite. Journal of Environmental Sciences, 18(6), 1204-1209.

[16] Su, D., Li, P.-j., Ju, J.-l, Xu, H.-x. 2006. Remediation of polycyclic aromatic hydrocarbons in non-fluid medium with immobilized microorganism technique. Applied Ecology, 17(08), 1530-1534.

[17] Viteri, S.E., Schmidt, E.L. 1996. Ecology of indigenous soil rhizobia: selective response of Bradyrhizobium japonicum to a soybean meal. Applied Soil Ecology, 3(3), 187-195.

[18] Wang, Ru.-G., Wang, Min., Niu, Xiao.-W., Tang, Jing.-C. 2010.Determination of Total Petroleum Hydrocarbons Content in Soil by Ultrasonic-Soxhlet Extraction-Gravimetric Analysis. Chinese Journal of Analytical Chemistry,38(3),417-420.

[19] Wang, X., Li, P.-j, Song, S.-z , et al. 2006. Immobilization of introduced bacteria and degradation of pyrene and benzo (a) pyrene in soil by immobilized bacteria. Applied Ecology, 17(11), 2226-2228.

[20] Wolicka, D., Suszek, A., Borkowski, A., Bielecka, A. 2009. Application of aerobic microorganisms in bioremediation in situ of soil contaminated by petroleum products. Bioresource Technology, 100(13), 3221-3227.

[21] Zhang, L.-l, Chen, L.-j, Liu, G.-f, Wu, Z.-j. 2003. Advance in enzymological remediation of polluted soils. Applied Ecology ,14(12), 2342-2346.

[22] Zhang, X.-x, Fang, M.-m, Zhao, C.-c, Wu, W.-l, Shan, B.-l, Zhang. J.-j. 2008a. The capability of removing quinoline of immobilized bacteria Q5. Chinese Journal of Environmental Engineering, 2(09), 1265-1268.

[23] Zhang, X.-x, Geng, C.-x, Fang, M.-m, Wu, W.-l, Shan, B.-l. 2008b. Bioremediation of Petroleum Contaminated Soil by Immobilized Microorganisms. Acta Petrolei Sinica (Petroleum Processing Section), 24(04), 409-414.

[24] Zhou, L.-k, Zhang, Z.-m. 1980. Determination of soil enzyme activity. Chinese Journal of Soil Science,5, 37-38+49.

The Study of Three-Dimensional Fire Risk Analysis Technique and Its Application on a Semiconductor Cleanroom

Abstract: The purpose of this research is to establish an integrated method that can assess risk of fire disaster within a three-dimensional space. The research employed Fire Dynamics Simulator (FDS) to explore the consequences of personnel injury under fire in a semiconductor cleanroom. Later, the three-dimensional fire risk analysis technique (3D-FRAT) software programmed by the C# language was used to assess the individual risk of the personnel within the cleanroom laboratory. This study designed and simulated an isopropanol (IPA) burning event within a chemical station. The research not only assessed the impact of hazard factors on the equipment and personnel of a semiconductor plant, but also calculated the maximum physical effect, the death probability, and the individual risk of three types of hazardous factors (thermal radiation of fire, carbon monoxide, smoke). Finally, the research output the dynamic images, iso-surface plots, and transection plots via the graphical software to implement the consequence and risk analysis of the simulation results. This paper proposes a new method to calculate the thermal radiation dose of fire and thoroughly discussed the relation of the unharmed lower limit and the cumulative thermal radiation dose. It was found that the thermal radiation of fire individual risk of the cleanroom laboratory is between 2.74×10-7 (person/year) to 1.68×10-5 (person/year); the carbon monoxide individual risk of the cleanroom laboratory is between 1.52×10-23 (person/year) to 3.99×10-6 (person/year); and the smoke individual risk of the cleanroom laboratory is between 2.74×10-7 (person/year) to 1.68×10-5 (person/year), if the laboratory has no protection or all its mitigation measures are failure.
Keywords: Three-Dimensional Fire Risk Analysis Technique (3D-FRAT); Fire Dynamics Simulator (FDS): thermal radiation; unharmed lower limit; cleanroom; risk assessment

1. 前言

高科技產業的蓬勃發展帶動著台灣的經濟脈動，也導致更多複雜性工業程序、危害性物質及化學反應被開發出來，一旦此類化學製程發生失誤將可能導致化學災變，進而造成人命或金錢的巨大損失。目前對於毒性物質洩漏、火災及爆炸的危害評估方式，已逐漸由二維延伸至三維，利用計算流體力學（Computational Fluid Dynamics, CFD）軟體可以評估障礙物的影響，故在模擬的準確度上也相對提高。藉由 CFD 後果模擬數據，再搭配機率單位公式計算 3D 風險，進而評估爆炸危害的方式在 2008 年首先由易逸波等人所發表（I, 2008），但在火災風險的量化方面一直未臻完善，本研究嘗試針對 CFD 火災模擬提出較為完整的風險評估方法，期望給予相關人員更全面的參考依據。

2. 研究方法

2.1 半導體廠房設計

本案例模擬目標位於某晶圓廠生產棟 Clean FAB 區之製程實驗室，其長、寬、高為 19 m×9 m×9 m，室內溫度維持在 22°C，無塵室等級為 Class 1,000，其中的機台配置如圖 1 所示，假設濕式清洗台為起火源。模擬目標屬於單層垂直層流型之無塵室，其氣流流動方向由天花板之 FFU 向下吹，並穿過高架地板孔板的孔洞，由高架地板下方經由迴風道循環回天花板上方（如圖 2 所示）。該無塵室的 3D 模型配置如表 1 所示。

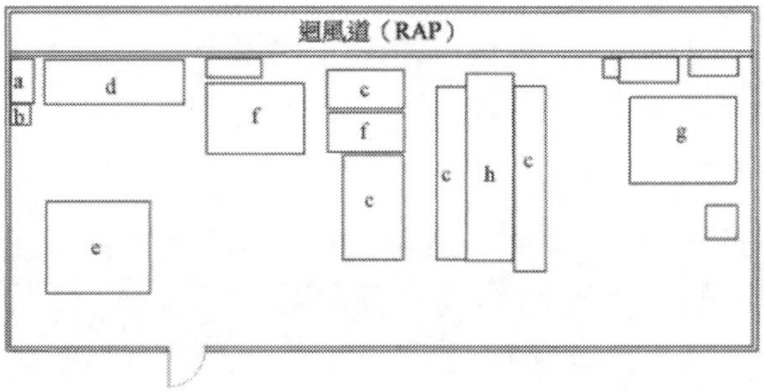

圖 1　模擬區域設備機台配置平面圖
a: 物料放置櫃；b: 備品放置櫃；c: 工作台（4 組）；d: 濕式清洗台；e: 光罩對準儀；f: 半導體作業機台（2 台）；g: 微波電漿化學氣相沉積系統；h: 高溫氧化擴散爐系統（2 台）

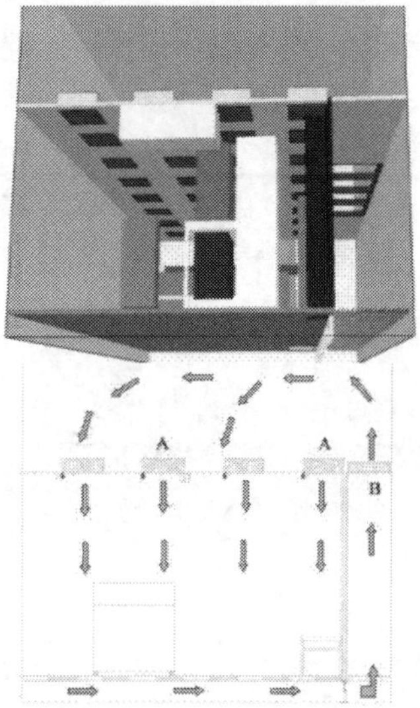

圖 2　模擬區域無塵室循環氣流示意圖，其中 A：風扇過濾機組（FFU）、B：管盤（Dry Coil）

表 1　無塵室區域條件

設備名稱	尺寸（L×W×H）	數量	說明
風扇過濾機組	1.2 m × 0.9 m × 0.3 m	32 台	覆蓋率 25%、出口風速 0.45 m/s
管盤	2.0 m × 1.0 m × 0.1 m	4 片	表面風速 2.5 m/s
高架地板-孔板	0.6 m × 0.6 m × 0.08 m	73 片	覆蓋率 20%、開孔率 17%
高架地板高度	0.6 m	—	—
無塵室淨空間高度	4.6 m	—	—

2.2 FDS 軟體之火災後果模擬方法

FDS是由美國NIST所公開發表的一套模擬火場用CFD軟體，可在Windows及Linux作業系統環境下使用，本研究假設目標區內之濕式清洗台為起火源（火源大小為 0.36 m^2），模擬條件如表 2 所示，總模擬時間為 800 秒，其中前 200 秒是要讓半導體廠中的循環迴風穩定，故火災發生時間是在第 201 開始。模擬中所使用的火源為異丙醇（IPA），並假設濕式清洗台內還有其他化學品丙酮及乙醇。在廠房中的濕式清洗台內層及相關管線為PVC材質。其餘工作台與機台的材質皆為不銹鋼製。

表 2　FDS 模擬條件基本設定

參數	設定值	參數	設定值
模型邊界尺寸	19 m × 9 m × 9 m	環境溫度	22°C

網格大小	0.1 m × 0.1 m × 0.1 m	起火源燃燒物	IPA
總網格數	1,539,000 格	火災發生時間	模擬時間第 200 秒後
熱釋放率	1,000 kW/m^2	總模擬時間	800 秒

2.3 火災風險量化方法

火災造成人員、建築物或其他財產損失的最主要原因，為燃燒時產生的火焰、熱輻射及濃煙。人員在室外空地受到火焰的灼傷程度，隨暴露於熱輻射的強度、時間與火源的距離而定。機率單位（probit）公式係由D.J. Finney所提出，而由Eisenberg等人（1975）大量應用於工業安全衛生之各種損傷估計，見式 1，其中Y=probit值，I為熱輻射通量（kW/m^2），t為有效暴露時間（sec），V為熱輻射劑量 [$t\,I^{4/3}\times10^{-4}$]。

$$Y = -14.9 + 2.56ln(tI^{4/3} \times 10^{-4}) \quad (1)$$

在評估熱輻射通量對於人體致死的過程時，通常會使用到Mudan（1984）所建立的熱輻射通量與致死程度的關係圖（如圖 3），其中不同顏色代表不同之致死百分比。吾人以圖 4 之火災溫度曲線來做說明不同方法間的差異，其中I_0為環境溫度所對應之熱輻射通量，I_1則代表「熱輻射無害下限值」，其定義為當所處環境之熱輻射通量低於此界限時，就算暴露時間再長也不會對人體造成影響。

方法一（見圖 3 藍線）採用舊的計算方式（I，2008），時間間隔採 FDS 內定的一秒鐘，隨著熱輻射通量增加，藍線逐漸往右行進，當到達或超過 A 點時，即為 100%致死。此種計算方式累積了環境溫度以上的無害熱輻射劑量，且將不同熱輻射通量所累積之劑量視為等效，故整體而言有高估現象。

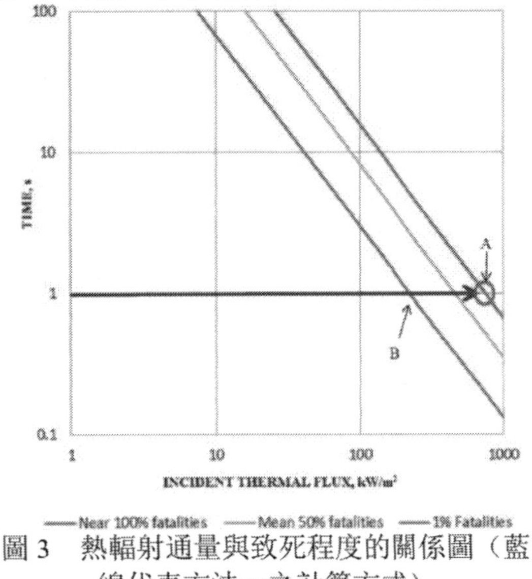

圖 3　熱輻射通量與致死程度的關係圖（藍線代表方法一之計算方式）

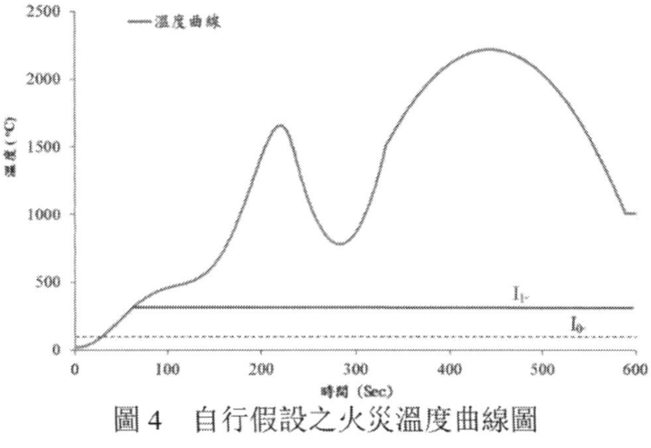

圖 4　自行假設之火災溫度曲線圖

方法二採用新的計算方式，其觀念為人體對不同熱輻射通量的耐受時間不同、致死機率也不同，必須分開計算後再比較何者造成的傷害最大。如圖 5 所示，當溫度超過I_1之熱輻射無害下限值後，此時每一種不同強度的熱輻射通量（I_1~I_5）皆會對人體造成傷害，其個別所轉換之熱輻射通量會隨著時間開始進行累加而成為不同效應之熱輻射劑量。圖 6 中I_1~I_5為圖 5 不同溫度下所對應之熱輻射通量，其所對應的熱輻射劑量有自己個別的時間變數，各自跑各自的時間；若計算過程中，有任何一個熱輻射劑量點的數值超過圖 6 中的紅線（100%致死）時，則判定該網格之人員致死機率為 100%。

圖 7 為I_1~I_5按照上述方式進行累加之結果，在累加過程中每一條線之斜率均維持固定（熱輻射通量愈大者其斜率愈大），若某一時段下網格內之溫度經換算後低於其所對應之熱輻射通量閾值時，累加步驟會暫停，造成劑量累加線出現中斷的情形。上述步驟計算完畢後，將每個時間下所對應之各個熱輻射劑量點取其最大值，即可得到為最後之結果。圖 8 為將圖 4 火災溫度曲線經前述兩種方法轉換成熱輻射劑量曲線的比較圖，此圖之方法二中熱輻射無害下限I_1採用 7.53 kW/m^2，小於I_1的數值皆予以忽略。方法二中若A、B之間，作為判定用的熱輻射通量點取得愈多愈密，理論上會越精準。本文因限於篇幅，其驗證在此予以省略。

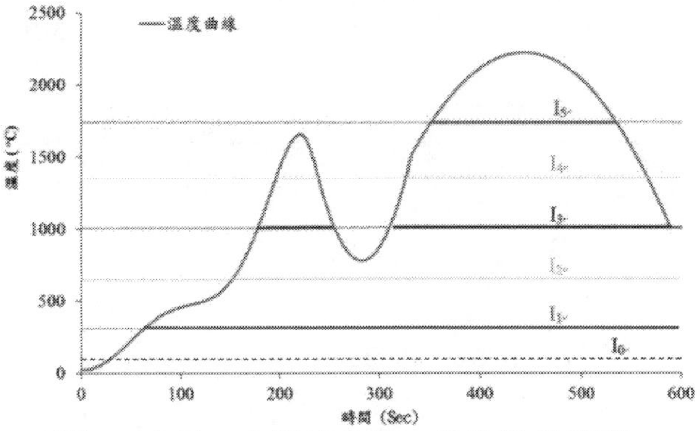

圖 5　方法二在溫度曲線上的取點示意圖

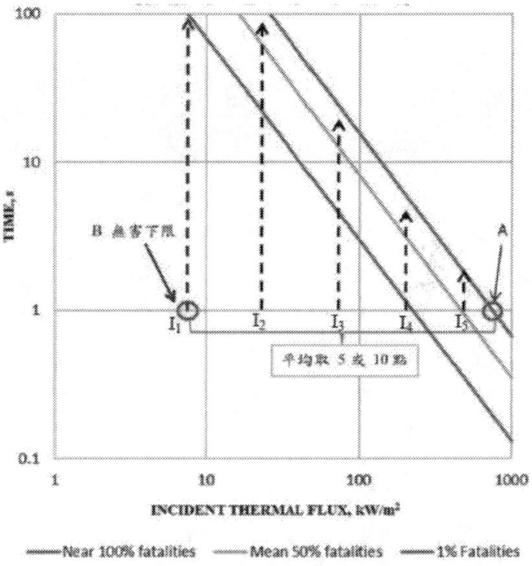

圖 6　方法二熱輻射取點及累加示意圖

圖 7　五個平均間隔點累加示意圖

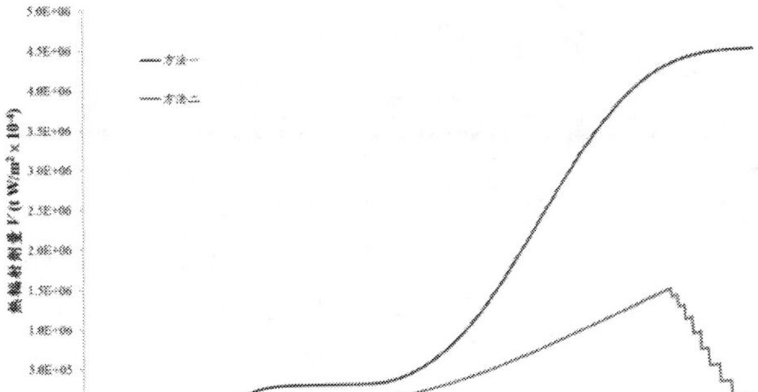

圖 8　由火災溫度曲線經兩種方法轉換成熱輻射劑量曲線圖

在火災效應中的CO危害評估採用Witlox（1999）提供之值（見式 2），其中$Y_{x,y,z}$=（x,y,z）座標下之probit值；C_{co}為三維空間中每一個網格的CO濃度值，t為暴露時間（min）。煙密度因未有相關文獻的實驗數據，故自行假設推算出其效應模式，主要透過FDS模擬之能見度來推估煙密度值，假設其為線性關係，危害下限訂為 10 公尺、上限訂為 0 公尺，藉由煙密度危害上下限解聯立方程式求出K_1及K_2，得到煙密度效應模式（見式 3）。

$$Y_{x,y,z=} \ -37.95 + 3.7\ \ln\left(\sum C_{co}{}^{1}{}_{x,y,z}\, t\right) \tag{2}$$

$$Y_{x,y,z} = -19.43 + 4.49\ \ln\left(\sum C_{soot}{}^{1}{}_{x,y,z}\, t\right) \tag{3}$$

其中C_{soot}=煙粒子濃度（ppm）。

2.4 火災風險效應分析

火災包含高溫熱輻射、CO及濃煙的風險探討，其個人風險計算方式可修改Considine模式，而以式 4 進行計算。其中$IR_{x,y,z}$為特定位置（x, y, z）之個人風險；n為所列舉之危害事件總數；F_i為危害事件i之發生頻率；$P_{i,i}$為點火源機率；$P_{wind,i}$為風向機率；P_Z為人員在座標（x, y, z）出現機率；$P_{D,i}$為危害後果案例i之致死機率。危害事件之發生頻率則由邱淑靜（2008）之論文資料計算而得，危害致死機率則由前述之CFD軟體產生之後果計算而得。表 3 為操作人員出沒機率一覽表。

$$IR_{x,y,z} = \sum_{i=1}^{n} F_i P_{i,i} P_Z P_{wind,i} P_{D,i} \tag{4}$$

表 3　假設之人員出沒機率

位置	工作型態	人員數	出現機率	人員數×出現機率	人口密度**（人/網格）
地板至 1.5 m 處	上班（每天 8 hr）	3	0.25	0.75	4.71×10^{-5}
	小維修（每周 2 hr）	2	0.01	0.02	
1.6 m 至天花板	小維修（每月 8 hr）	2	0.01	0.02	7.44×10^{-7}

**人口密度=人員數×出現機率÷人員會出沒的總網格數

3. 結果與討論

3.1 火災後果分析與風險評估

(1) FDS 火災後果分析

本研究模擬了無塵室的循環迴風系統（如圖 9），為了使風場穩定，本研究在 201 秒後才進行濕式清洗台起火之模擬（見圖 10），當 201 秒時 IPA 著火，隨後點燃旁邊的丙酮及甲醇；煙在清洗台內部的空間進行累積，濃煙因為循環迴風的關係，會先向下沉降再往橫向移動，380 秒時濃煙充滿整個模擬空間直至 800 秒。

圖 9　FDS 風場向量圖（X= 0.5 m, t=20 sec）

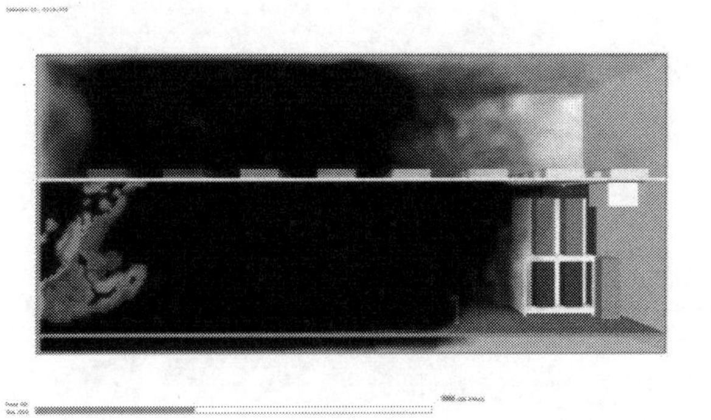

圖 10　FDS 火災模擬現場火及煙隨時間變化 XZ 平面圖（t=320 sec）

圖 11 之紅圈處為火災偵測點，位處無塵室出入口，吾人將該處四個偵測點數據取出來探討，由圖 12 可發現當火災發生後 80 秒左右，靠近出入口位置約 1.5 公尺以上的高度即到達 600℃，而在 4.5 公尺處更可達到 900℃；若未吸入高溫蒸氣，人體僅能在 140℃下忍受一段時間，若在這段時間內人員能通過出入口，即有很大機會能逃生。

圖 11　FDS 火場測點配置圖（黃色為偵煙器）

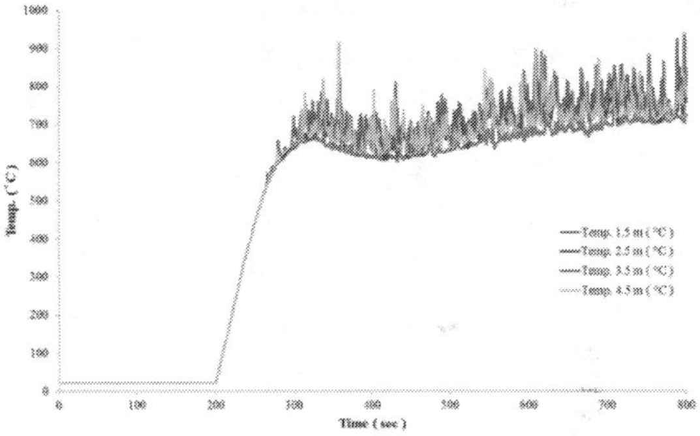

圖 12　無塵室入口測點溫度變化趨勢

圖 13 為出入口處距地板 1.5 公尺高之煙密度變化趨勢，由於無塵室出入口離火源較近，加上循環迴風關係，會使濃煙下降時間提早。由圖 14 可發現，濃煙在此空間中的橫向擴散速度會隨著水平距離增加而變慢（煙在 smoke_1、2、3、4 之間的流速分別為 0.16 m/s、0.09 m/s、0.02 m/s），其原因為每一個出風口的風柱皆類似防煙垂壁的效果，會降低濃煙的橫向流速，若出入口在無塵室右側，則可提高約 200 秒的逃生時間，故建議需有兩個出入口。

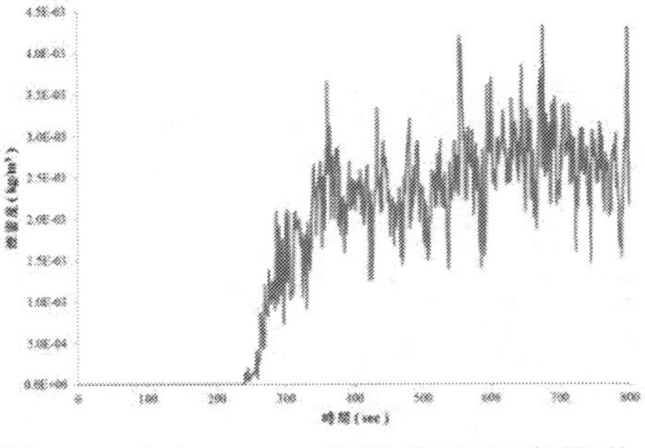

圖 13　出入口 1.5 公尺處煙密度變化趨勢

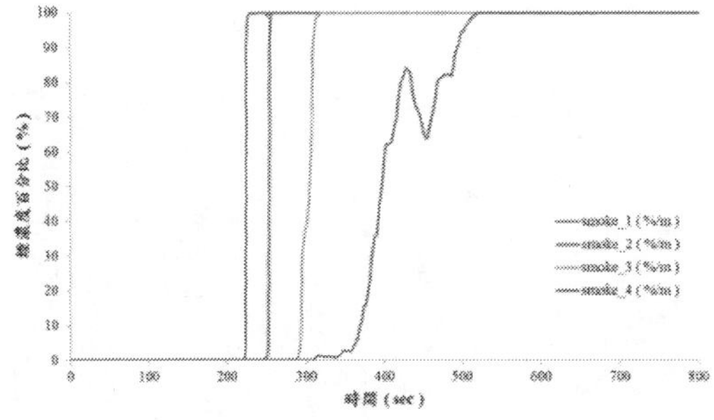

圖 14　偵測點之濃煙百分比隨時間變化圖

(2) 熱輻射危害風險評估

本研究藉由 FDS 的模擬結果取得三維物理量，再利用自行撰寫之 3D-FRAT 軟體進行數據剖析（見圖 15(a)），即可得到流場溫度之最大物理量及致死機率，再搭配火災發生頻率及人員分佈資料即可進行各種危害的風險繪圖輸出（見圖 15(b)）。

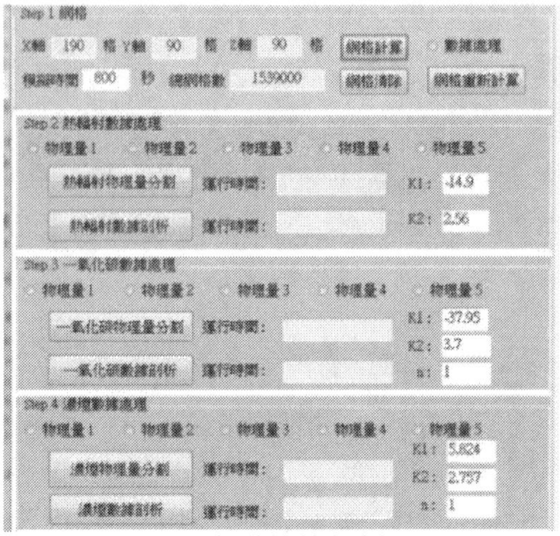

(a) 火災數據剖析

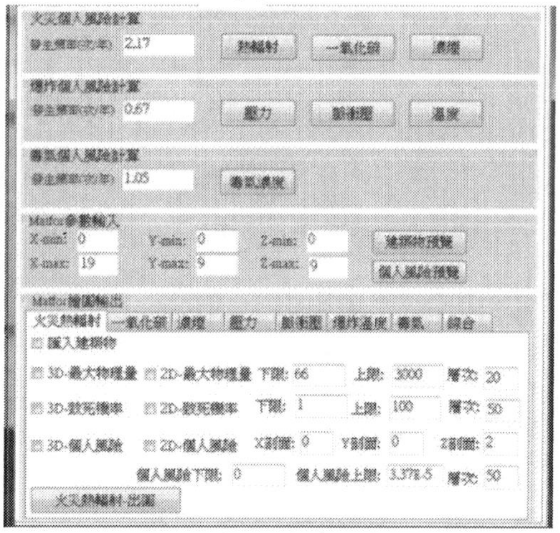

(b) 火災風險計算

圖 15　3D-FRAT 火災熱輻射數據分析及風險計算介面

圖 16 (a) 為火災溫度的最大物理量等曲面圖，本模擬最大溫度 2,820 K發生於火源上方，其他未受到火焰影響區域的高溫係濃煙所造成，藉由 3D-FRAT最大溫度物理量可評估設備是否會受到影響。致死機率主要是探討人員受到熱輻射影響的致死率，由圖 16 (b) 得知，火災發生後 10 分鐘，整個區域的致死機率皆會達到 100%。個人風險所代表的意義為每年目標廠區因為火災熱輻射導致死亡的人數，由圖 16 (c)、(d) 得知在離地 1.5 公尺處之個人風險皆為 1.68×10^{-5}（人/年），在離地 4 公尺處之個人風險為 2.74×10^{-7}（人/年）。

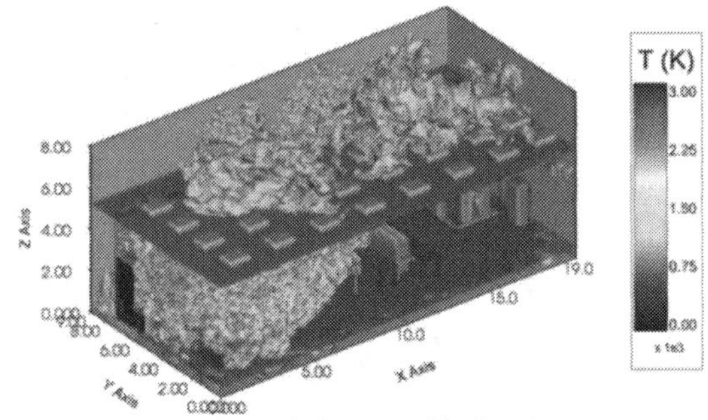

(a) 溫度最大物理量等曲面

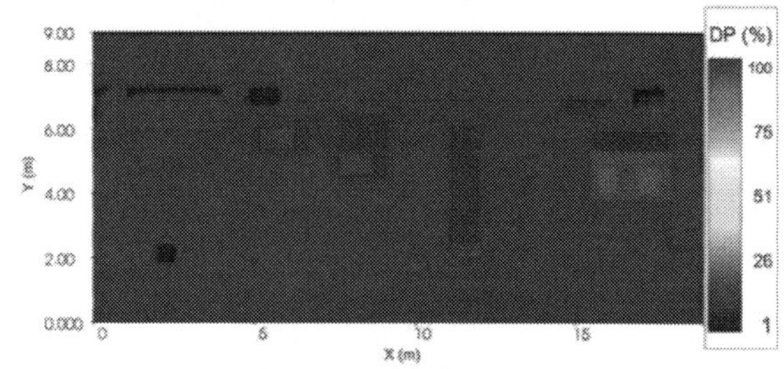

(b) 致死機率剖面

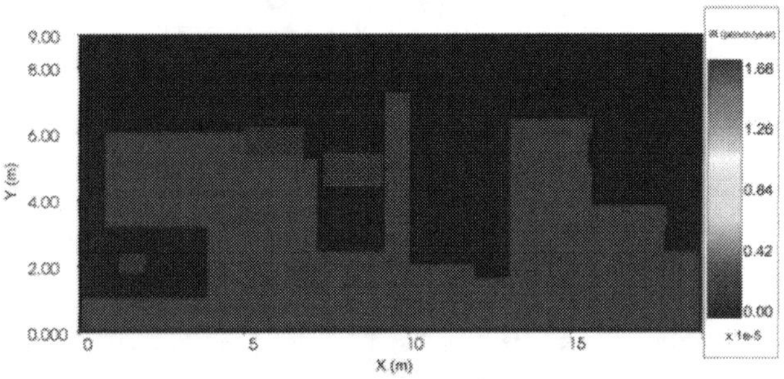

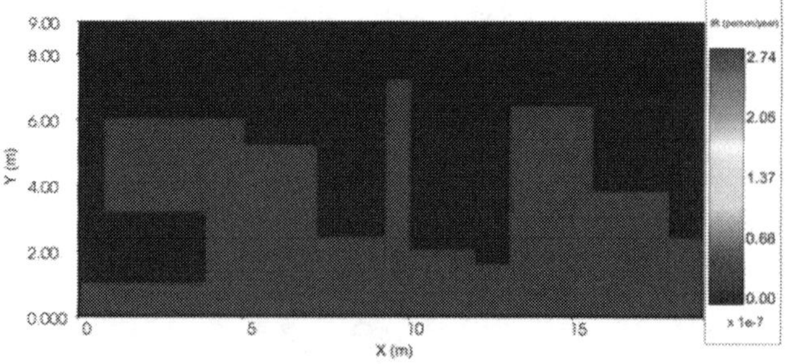

(c) Z=1.5 m 之個人風險　　　　　　　　　　(d) Z=4 m 之個人風險

圖 16　火災熱輻射風險分析圖

(3) CO 危害風險評估

圖 17 (a) 為CO濃度剖面圖，可以看到CO的擴散由於受到循環迴風的牽制，侷限在火源區附近、出入口（約X=5 m處）及迴風通道中。其致死機率等曲面如圖 17 (b) 所示，致死區域主要分布在火源周圍。而CO的個人風險值介於 1.5×10^{-23}~3.99×10^{-6}（人/年）間（見圖 17 (c)、(d)），其最大值在離地 1.5 公尺處為 3.99×10^{-6}（人/年），在離地 4 公尺處則為 2.58×10^{-7}（人/年）。

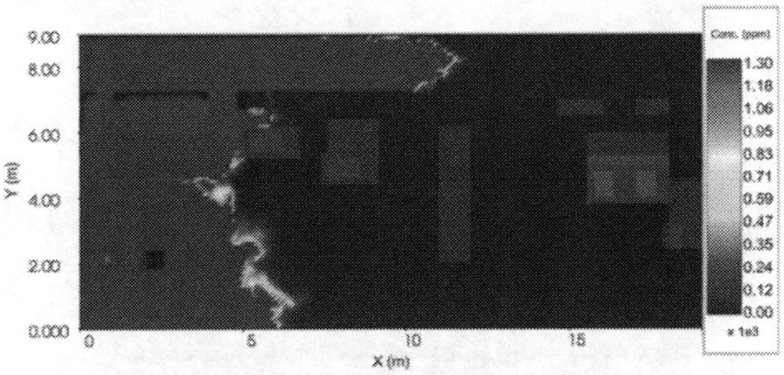

(a) CO 濃度最大物理量剖面圖（Z=1.5 m）

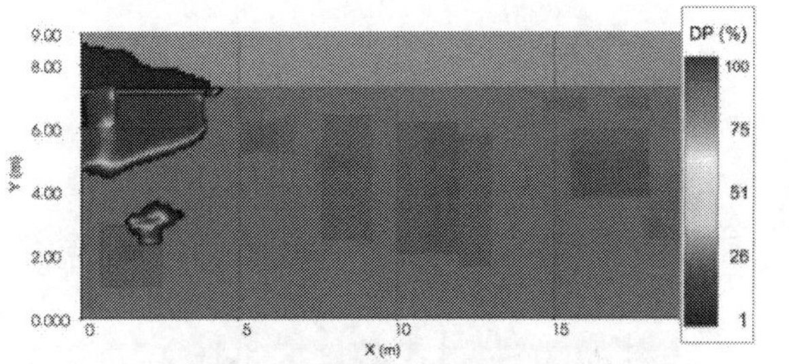

(b) 致死機率圖（Z= 1.5 m）

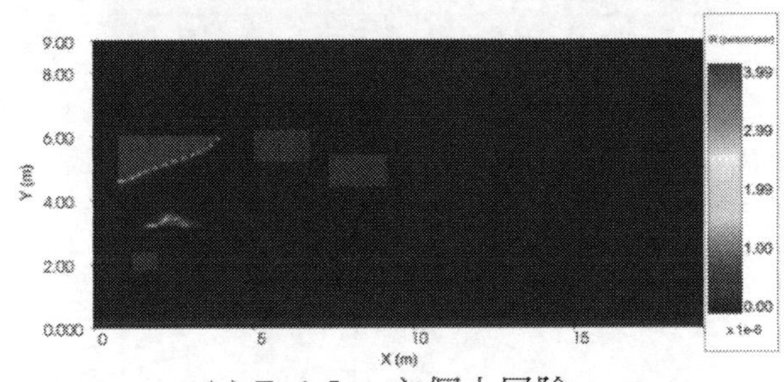

(c) Z=1.5 m 之個人風險

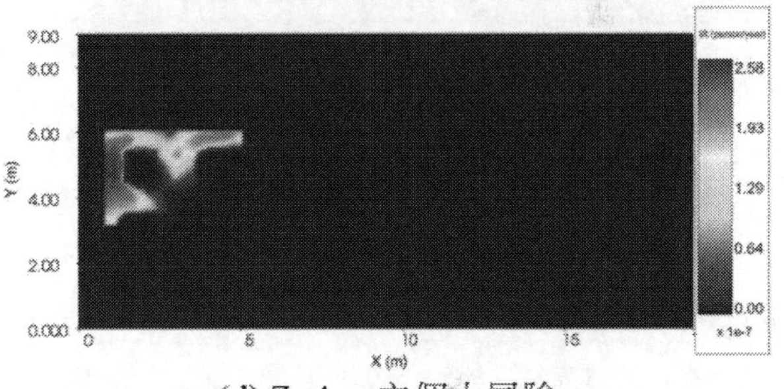

(d) Z=4 m 之個人風險

圖 17　CO 風險分析圖

(4) 濃煙危害風險評估

濃煙的最大物理量如圖 18 (a)所示，其最大值 11,857 ppm為圖中深紅色位置，離火源越遠其濃度越低，但在牆壁角落及機台四周會因蓄積而造成高濃度。在致死機率部分（見圖 18 (b)），由 1.5 公尺高之濃煙致死機率切面顯示，在 800 秒時除了右側迴風道少數區域外，無塵室中幾乎全部到達 100 %致死範圍。而由圖 18 (c)、(d)得知，濃煙個人風險值在離地 1.5 公尺處為 1.68×10^{-5}（人/年），而離地 4 公尺處則為 2.74×10^{-7}（人/年）。

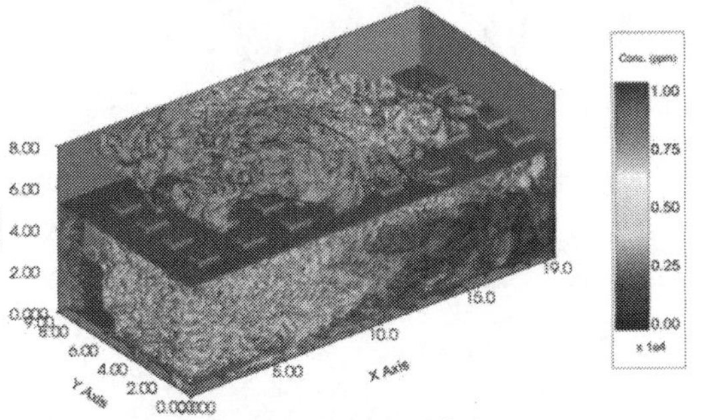

(a) 濃煙最大物理量等曲面

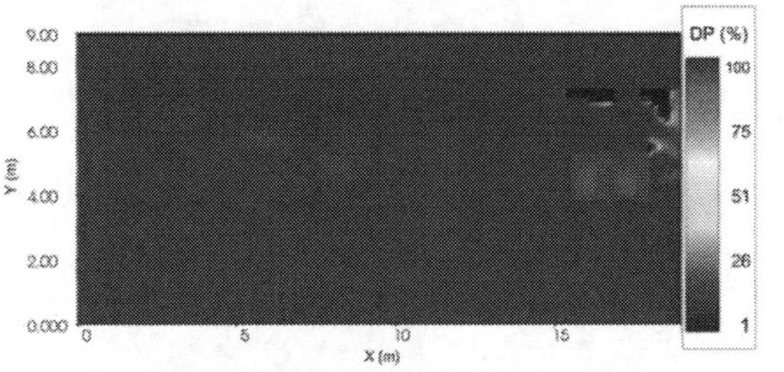

(b) 致死機率剖面（Z= 1.5 m）

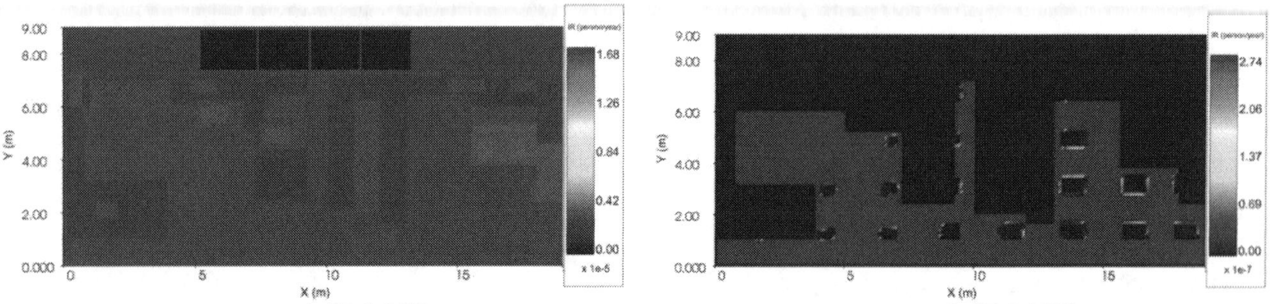

(c) Z=1.5 m 之個人風險　　(d) Z=4 m 之個人風險

圖 18　濃煙個人風險圖

4. 結論

本研究嘗試建立一套評估廠區火災危害之風險評估方法，能展現熱輻射、CO 及濃煙等危害之風險分析程序。在研究過程中獲致以下經驗：

(1) 火災熱輻射若利用方法一，將熱輻射劑量隨時間累積並全部加總，會有高估的現象，若使用方法二訂定無害下限，並在無害下限與 100 %致死曲線間取固定間隔進行熱輻射劑量之估算應較為合宜。上述間隔取得越密其結果會更精準。

(2) 無塵室之循環迴風對於煙流流動有很大的影響，在後果分析時需先讓風場穩定再進行模擬。本研究發現循環迴風系統有類似防煙垂壁之功效，未來可利用在性能式設計中。

(3) 本研究初步嘗試評估濃煙危害，未來若有較多之濃煙毒性效應資料，則可對此部分修正。而火災熱輻射劑量無害下限之訂定亦尚未有一定標準，未來需尋求更合適之規範。

5. 參考文獻

[1] Mudan, K. S., 1984, "Thermal radiation hazards from hydrocarbon pool fires", Progress in Energy and Combustion Science.

[2] Eisenberg, N.A. *et. al.*, 1975, "Vulnerability Model: A Simulation System for Assessing Damage Resulting from Marfne Spills", NTIS, AD-A015-245.

[3] Witlox, H.W.M., Holt, A., 1999, "A Unified Model for Jet, Heavy, and assive dispersion Including Droplet Rainout and Re-Evaporation", International Conference and Workshop on Modeling the Consequence of Accidental Releases of Hazardous Materials, pp.315–344.

[4] I, Y.-P., Cheng, T.-L., 2008, "The Development of a 3D Risk Analysis Method", Journal of Hazardous Materials, Vol. 153, Iss. 1–2, 600–608.

[5] 邱淑靜, 2008, 台灣半導體相關產業重大災害原因探討與分析, 長榮大學健康科學學院職業安全與衛生學研究所, 碩士論文

The Development of an Integrated Quantitative Risk Analysis System and Its Application on a Complex Disaster with Toxic, Fire and Explosion Disasters

Abstract: The purpose of this research is to establish an integrated method that can analyze risk of toxic and chemical disaster within a three-dimensional space. This research employed Fire Dynamics Simulator (FDS) and Flame Acceleration Simulator (FLACS), both belong to the Computational Fluid Dynamics (CFD) fire and explosion simulation software, to explore the consequences of personnel injury under fire, explosion and toxic gas release within a semiconductor cleanroom. Later, an Integrated Quantitative Risk Analysis System (IQRAS) programmed by the C# language was used to analyze the individual risk of the personnel. This research designed and simulated the following cases: the isopropanol (IPA) burning, the hydrogen explosion event, and the ammonia dispersion event. The research not only analyzed the impact of hazard factors on the equipment and personnel of a cleanroom, but also calculated the maximum physical effect, the death probability, and the individual risk of seven types of hazardous factors (thermal radiation of fire, carbon monoxide, smoke, pressure, impulse pressure, thermal radiation of explosion, and toxic gas dispersion). Finally, the research output the dynamic images, iso-surface plots, and transection plots via the graphical software to implement the consequence and risk analysis of the simulation results. This research successfully integrates two grid systems of different CFD software and fully demonstrates the implementing procedures of the integrated 3D risk analysis techniques. It was found that the integrated individual risk of the cleanroom is between 1.71×10-9 (person/year) to 3.21×10-5 (person/year) if the process has no protection or all its mitigation measures are failure. In the future, as long as the simulation case is implemented via the CFD software such as FDS or FLACS, its risk value can be calculated using the method developed in this study. The results of this research are expected to be able to give policy makers and researchers a better reference basis for risk analysis.
Keywords: Integrated Quantitative Risk Analysis System (IQRAS); Fire Dynamics Simulator (FDS); Flame Acceleration Simulator (FLACS); fire and explosion; toxic gas; risk analysis

1. 前言

當人類生活水準不斷地提昇，在經濟穩定且生活無憂的情況下，即會開始注重自身之生命財產安全。為了滿足全球日益增加的人口需求，伴隨著更多的危害性化學物質被開發出來，一旦此類製程發生失誤，將可能引發重大的人命財產損失。而化學災變的發生通常是由物質外洩開始，目前對於毒性物質洩漏、火災及爆炸的評估方式，已逐漸由二維延伸至三維計算流體力學（CFD）的電腦模擬，目前仿間有不少計算流體力學軟體，可用來模擬特定危害之後果，但是某些事故有可能屬於複合性災害，其相關風險並非單一種軟體所能分析，故整合不同軟體之特長以便用在類似災害之風險分析實有其必要。故本研究希望能提出一種全面型的廠區事故風險分析方法，以便提供決策者作為預防策略擬定時之參考。

2. 研究方法

2.1 半導體廠房設計

本案例模擬目標位於某晶圓廠生產棟 Clean FAB 區，其大小為 19 m×9 m×9 m，室內溫度維持在 22°C，無塵等級為 Class 1,000，屬於單層垂直層流型之無塵室，氣流流動方向由天花板之 FFU 向下吹，並穿過高架地板孔板的孔洞，由高架地板下方經由迴風道循環回天花板上方（如圖 1 所示）。

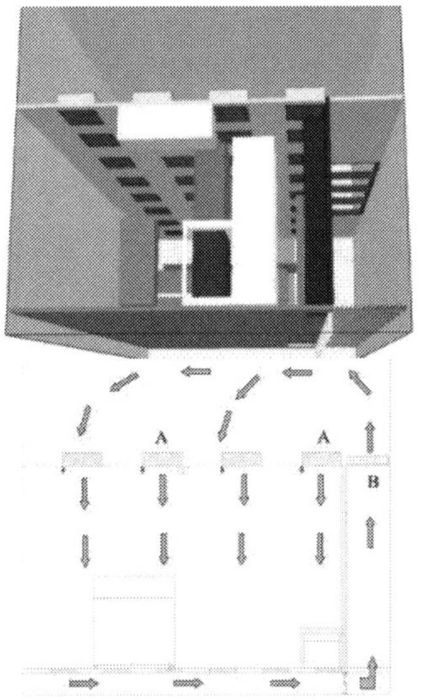

圖 1　模擬區域無塵室循環氣流示意圖。
其中 A：風扇過濾機組、B：管盤

2.2 FDS 火災後果模擬

FDS 是由美國 NIST 所公開發表的一套模擬火場用 CFD 軟體。本研究假設目標區之濕式清洗台起火，火源為異丙醇（IPA），模擬條件如表 1 所示，總模擬時間為 800 秒，其中前 200 秒為了讓循環迴風穩定，故火災由第 201 秒開始。假設廠房中的濕式清洗台內還有丙酮及乙醇，清洗台內層及相關管線為 PVC 材質，其餘工作台與機台材質皆為不銹鋼製。

表 1　FDS 模擬條件

`	設定值	參數	設定值
模型邊界	19 m×9 m×9 m	環境溫度	22°C
網格大小	0.1 m×0.1 m×0.1 m	起火源物質	IPA
網格總數	1,539,000 格	起火時間	第 201 秒
熱釋放率	1,000 kW/m^2	總模擬時間	800 秒

2.3 火災效應評估

本研究採用 FDS 的輸出結果檔（.q file）來進行火災熱輻射、毒性氣體及濃煙等危害之效應評估（見表 2），處理之數據為各種三維危害物理量與時間關係的四維陣列，運算數據相當龐大，利用自行研發之程式來擷取 FDS 輸出數據，再該根據表 2 之效應模式計算出 probit 值，最後利用式 1 轉換為人員致死機率，以便用來計算個人風險。

表 2　火災效應模式一覽表

效應	單位	效應模式
熱輻射	kW/m^2	$Y_{x,y,z} = -14.9 + 2.56\ln\left(\frac{t_e I_{x,y,z}^{4/3}}{10^4}\right)$
毒性氣體	ppm	$Y_{x,y,z} = -37.95 + 3.7\ln\left(\sum C_{co}{}_{x,y,z}^{1} \cdot t\right)$
濃煙	ppm	$Y_{x,y,z} = -13.22 + 2.76\ln\left(\sum C_{soot}{}_{x,y,z}^{1} \cdot t\right)$

$$P = 50\left[1 + \frac{Y-5}{|Y-5|} erf\left(\frac{|Y-5|}{\sqrt{2}}\right)\right] \quad (1)$$

2.4 FLACS 爆炸及毒氣擴散後果模擬

FLACS 為一套 CFD 模擬軟體，可針對危害性氣雲洩漏所引起毒氣擴散或火災爆炸進行動態後果模擬，本研究假設連接機台的氫氣管線發生洩漏爆炸，見表 3。在進行毒氣擴散模擬時，先給 300 秒讓氣流穩定達到層流特性，在 301 秒時才開始氨氣洩漏擴散模擬，見表 4。

表 3　FLACS 氣體爆炸模擬條件

參數	設定值	參數	設定值
模型邊界	19 m×9 m×9 m	洩漏氣體	氫氣
網格大小	0.2 m×0.2 m×0.2 m	氣體濃度	100 %
總網格數	192,375 格	環境溫度	22℃
洩漏形式	氣雲	點火時間	0.1 秒
氣雲體積	18 m^3	總模擬時間	5 秒

表 4　FLACS 毒氣洩漏模擬條件

參數	設定值	參數	設定值
洩漏形式	氣雲擴散	空調風速	0.45 m/s
氣雲體積	60 m^3	空調形式	循環迴風
洩漏氣體	氨氣	點火時間	無
氣體濃度	100 %	氣流穩定時間	300 秒
環境溫度	22℃	總模擬時間	900 秒

2.5 爆炸及毒氣擴散效應評估

利用 FLACS 模擬之*.r3file 檔進行爆炸過壓、脈衝壓、熱輻射及毒氣擴散效應評估，以了解人員損傷程度，見表 5。用式 1 將 probit 值轉換為人員致死率，以便估算個人風險。

表 5　爆炸及毒氣擴散效應模式一覽表

效應	單位	效應模式
過壓	barg	$Y_{x,y,z} = -77.1 + 6.91\ln\left(P_{x,y,z}\right)$
脈衝壓	Pa·s	$Y_{x,y,z} = -46.1 + 4.82\ln\left(J_{x,y,z}\right)$
毒性氣體	ppm	$Y_{x,y,z} = -35.9 + 1.85\ln\left(\sum C_{NH_3}{}^{2}_{x,y,z} \cdot t\right)$

2.6 量化風險整合

各種危害之風險整合如圖 2，本階段將採用前述之人員致死率，配合大氣條件、點火機率、事件頻率等參數來計算個人風險，並將其結果輸出至 3D 風險等曲面圖中。

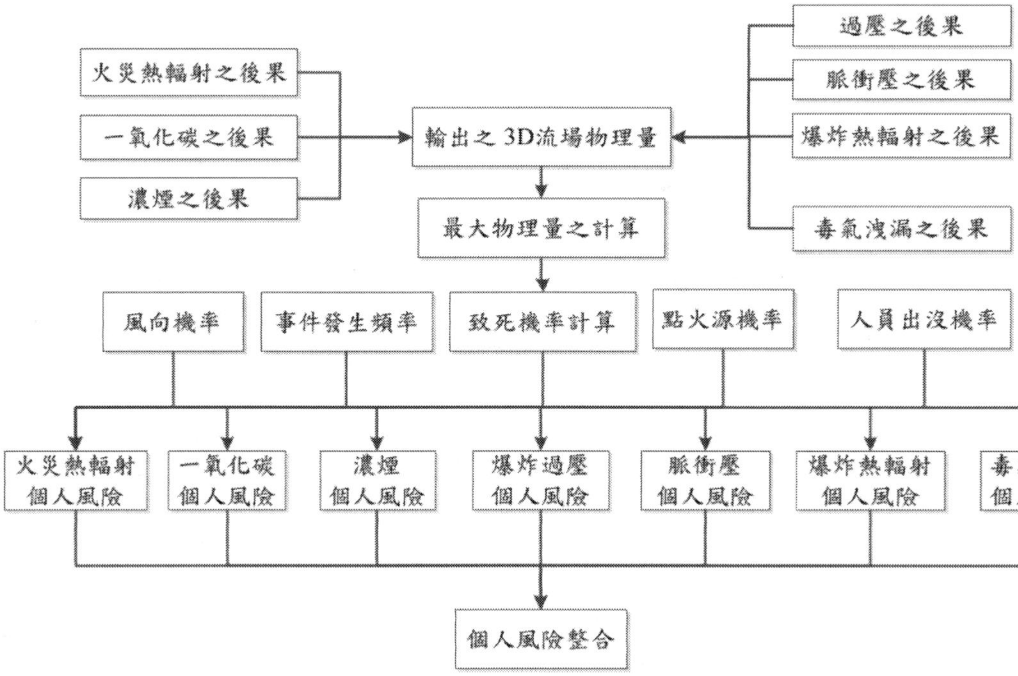

圖 2　危害風險整合流程

要找出無塵室中發生火災、爆炸及毒氣洩漏造成目標區內人員的損傷，需先行計算出各個危害之致死率。個人風險可依照公式 2 進行計算，其中$IR_{x,y,z}$為特定位置（x, y, z）之個人風險；n為所列舉之危害事件總數；F_i為危害事件i之發生頻率；$P_{i,i}$為點火源機率；$P_{wind,i}$為風向機率；P_Z為人員在座標（x, y, z）出現機率；$P_{D,i}$為危害後果案例i之致死機率。式 2 中風向機率設為 1（在室內）；事件發生頻率可由表 6（邱淑靜，2008）計算而得；點火機率設為 0.33；人員出沒機率則見表 7。

$$IR_{x,y,z} = \sum_{i=1}^{n} F_i P_{i,i} P_Z P_{wind,i} P_{D,i} \tag{2}$$

表 6　台灣半導體產業事故統計表（1995-2012）

事故類型	發生次數	百分比	頻率（次/年）
火災	39	55%	2.17
毒性物質外洩	19	27%	1.05
爆炸	12	17%	0.67
供應系統中斷	—	—	—
其他		1%	
總計	70	100%	

表 7　目標區人員出沒機率

位置	工作型態	人員數	出現機率	人員數×出現機率	人口密度*（人/網格）
地板至 2 m 處	上班（每天 8 hr）	3	0.25	0.75	4.24×10^{-5}
	小維修（每周 2 hr）	2	0.01	0.02	
	大維修（每月 8 hr）	2	0.01	0.02	
2 m 至天花板	大維修（每月 8 hr）	2	0.01	0.02	5.09×10^{-7}

*人口密度=人員數×出現機率÷人員會出沒的總網格數

在風險整合部份，FLACS 與 FDS 的網格系統存在著顯著差異，FLACS 的數據代表網格中心點，而 FDS 則是分布在格點上，故本研究將 FDS 三維空間中每八個數據合併為一筆數據，以便與 FLACS 進行對應，見圖 3。

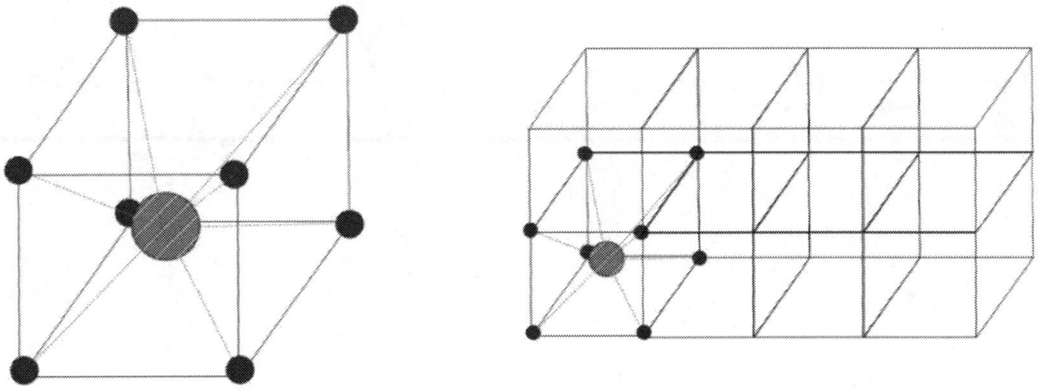

圖 3　FDS 數據八合一示意

3. 結果與討論

3.1 FDS 火災後果分析與風險分析

(1) 火災後果分析

無塵室的循環迴風模擬如圖 4，在 201 秒風場穩定後進行濕式清洗台起火模擬（見圖 5），IPA 著火後會點燃旁邊的丙酮及甲醇，濃煙在清洗台內部空間累積，循環迴風會使煙先向下再橫向移動，380 秒濃煙充滿整個空間直至 800 秒。

圖 4　FDS 風場向量圖（t=20 sec）

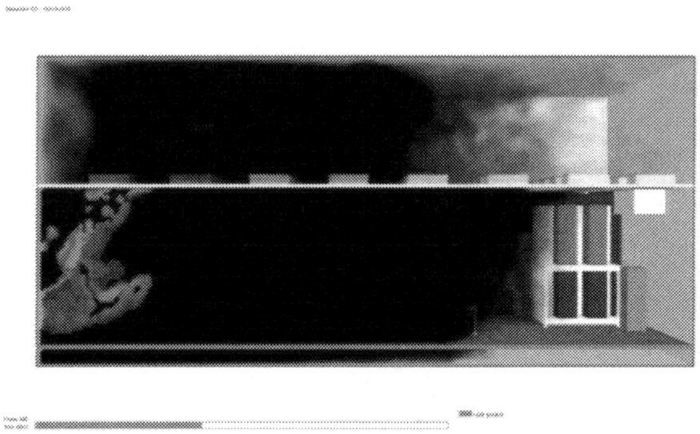

圖 5　FDS 火煙變化剖面圖（t=320 sec）

(2) 火災熱輻射危害風險分析

本研究藉由 FDS 的模擬結果取得三維物理量，再利用自行撰寫之 IQRAS 軟體（含 3D-FRAT 功能）進行數據剖析（見圖 6(a)），即可得到流場溫度之最大物理量及致死機率，再搭配火災發生頻率及人員分佈資料即可進行各種火災危害的風險繪圖輸出（見圖 6(b)）。

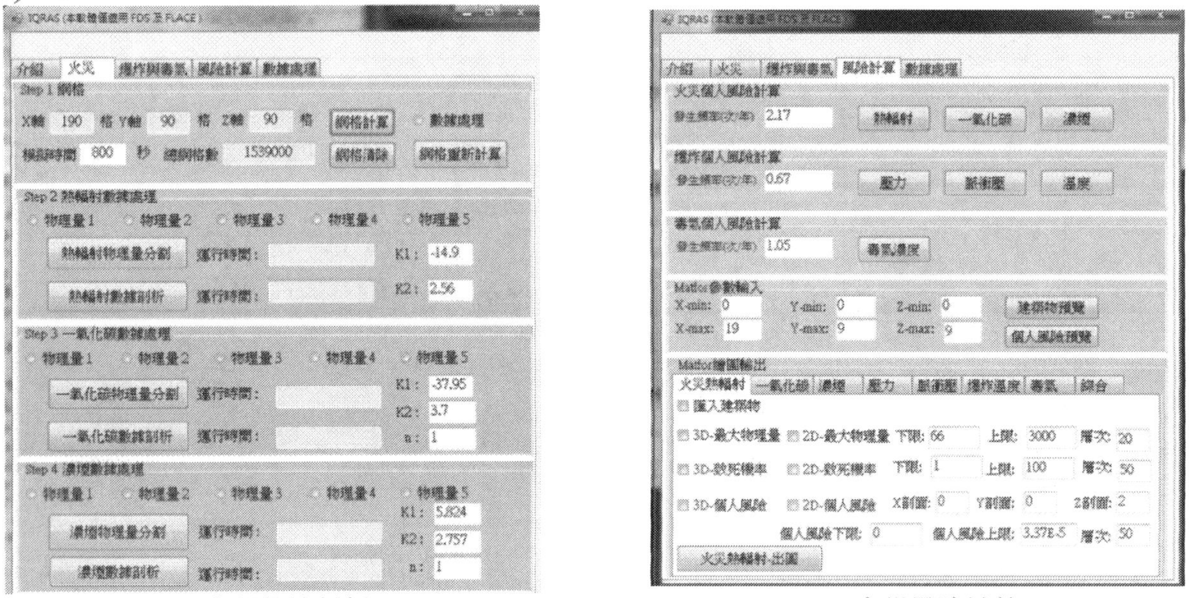

(a) 火災數據剖析　　(b) 火災風險計算

圖 6　IQRAS 介面

個人風險代表每年無塵實驗室因為火災熱輻射導致死亡的人數，見圖 7，火災熱輻射個人風險在離地高度 2 m處為 7.59×10^{-6}（人/年），在離地高度 4.6 m處之風險為 9.09×10^{-8}（人/年）。

(3) CO 危害風險分析

CO濃度每分鐘要達到 59,070 ppm才會到達 1%致死機率，如圖 8 所示，本研究的CO致死區域主要分布在火源周圍。而在人員可能出沒的區域，CO的個人風險值介於 1.5×10^{-23}~7.59×10^{-6}（人/年）間，離地 2 m之風險最大值為 7.59×10^{-6}（人/年），而離地 4.6 m之風險最大值為 9.09×10^{-8}（人/年）。

(4) 濃煙危害風險分析

除了右側的迴風道以外，濃煙在 2 m高無塵室中幾乎全部到達 100%致死。而由圖 9 的個人風險可知，在離地 2 m處，其濃煙致死之風險為 7.59×10^{-6}（人/年），而離地高度 4.6 m處之風險為 9.09×10^{-8}（人/年）。

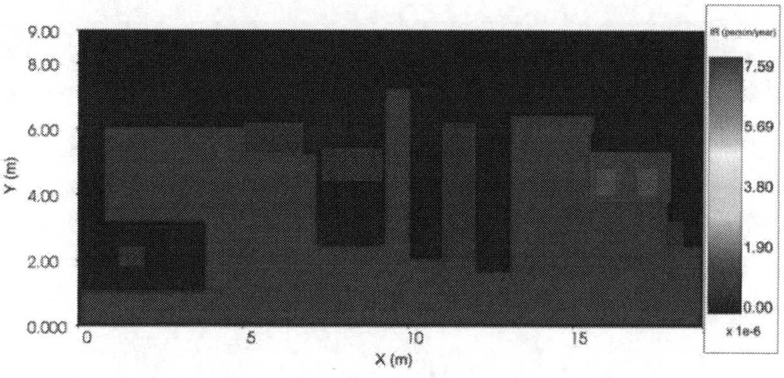

圖 7　火災熱輻射個人風險

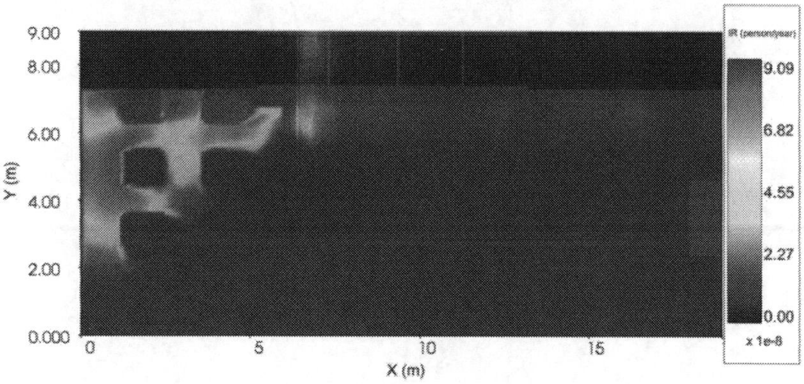

圖 8　CO 個人風險

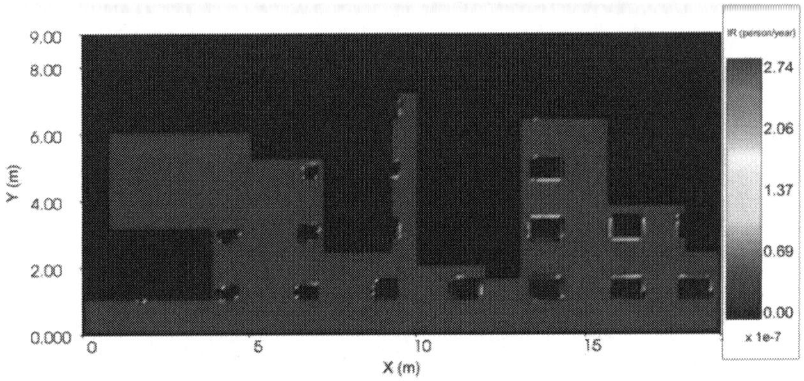

圖 9　濃煙個人風險

3.2 FLACS 爆炸後果分析與風險分析

(1) 爆炸後果分析

爆炸過壓之結果如圖 10 所示，壓力的最大值 0.15 barg 發生在剛開始爆炸的瞬間，出現在潔淨室內部與迴風道間之牆壁邊，此壓力對於沒有強化之混凝土結構會造成破壞，故在此點上可能造成牆壁的毀損。其餘空間的壓力約在 0.03~0.08 barg 之間，其壓力可能會造成窗戶玻璃破裂或邊框毀損扭曲。圖 11 為爆炸測點配置圖，其中測點 M2、M19 及 M24 之位置分別為出入口、右側牆壁及左側牆壁，由圖 12 中可以看到出入口測點的壓力最大值為 0.0065 barg，其壓力約僅能造成小型窗戶的破損，而左側牆壁的最大值約 0.005 barg、右側為 0.004 barg，均未達到牆壁結構損壞之下限值。

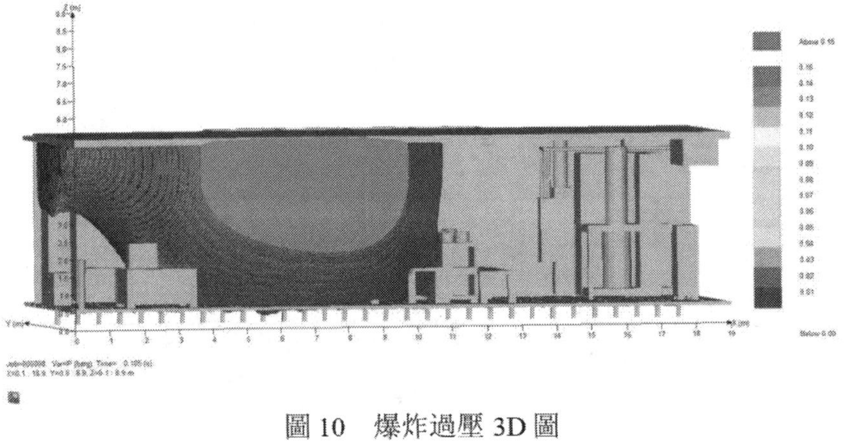

圖 10　爆炸過壓 3D 圖

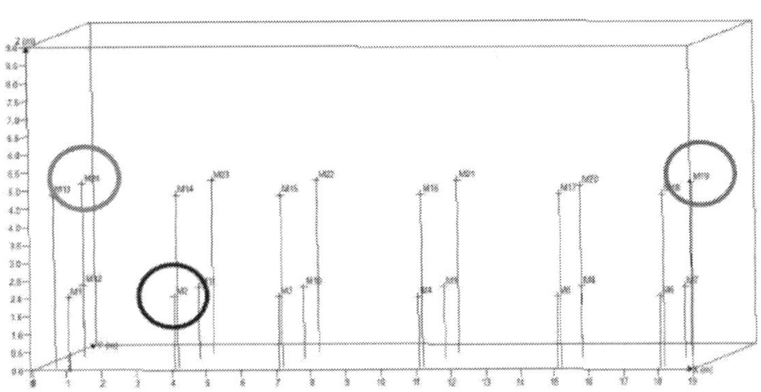

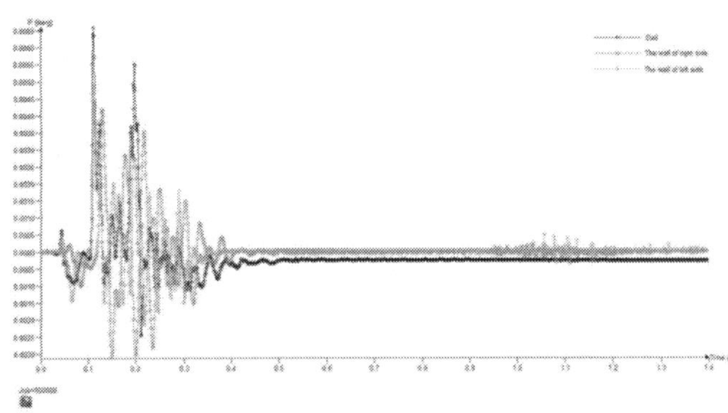

圖 11　FLACS 爆炸測點配置圖　　　　圖 12　FLACS 測點之過壓趨勢圖

爆炸脈衝的 3D 圖如圖 13 所示，脈衝壓為壓力與時間的積分，本例中最大值為 85 (Pa·s)故基本上無害。圖 14 為脈衝壓趨勢圖，由圖中可見靠近出入口之最大的脈衝壓約 10 (Pa·s)，而左邊牆壁及右邊牆壁則都低於 5 (Pa·s)。

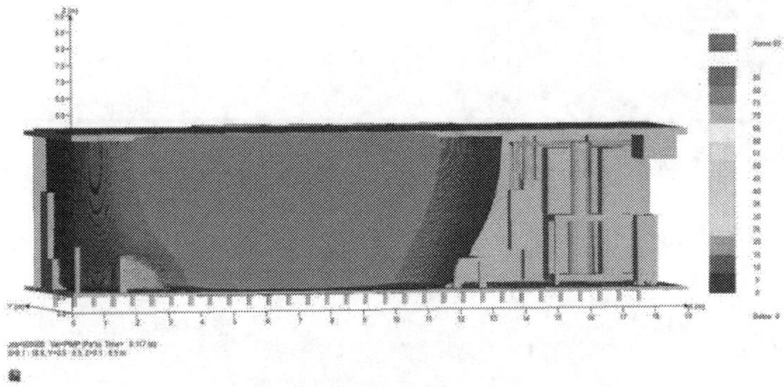

圖 13　爆炸脈衝壓 3D 圖

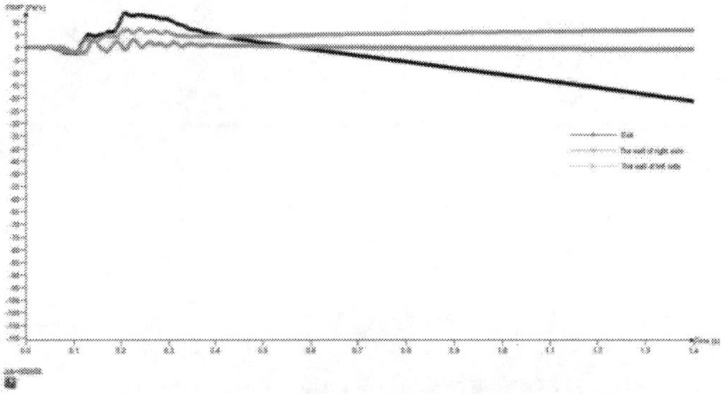

圖 14　FLACS 測點之脈衝壓趨勢圖

圖 15 為爆炸後產生的溫度 3D 圖，爆炸開始後，高溫氣流會迅速向四周圍擴張，在 0.2 秒後即產生高達 2,400 K 之溫度，可能造成設備損壞；直至 5 秒模擬結束，空間 2.5 m 高以上的區域仍有 1,200 K 以上的高溫蓄積，並呈現較高區域溫度愈高的現象，而物件周圍的溫度會較其他地方為低。圖 16 為前述三測點所量到之高溫趨勢圖，其中在左側牆壁及出入口處，其最高溫度達到 2,400 K，而右側牆壁則到了約 1.8 秒後到達了 2,200 K，左右兩側的測點在高 4.5 公尺處，而出入口的側點在高 2.1 公尺處，故可發現爆炸後高溫氣流會在高處逗留，而較低的位置溫度則會有逐漸下降的趨勢。

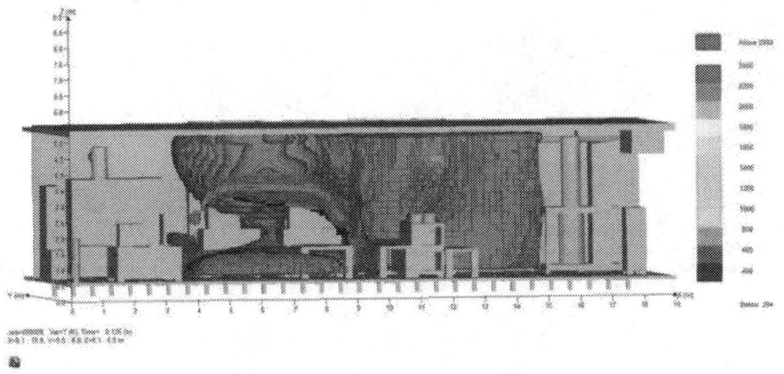

圖 15　爆炸溫度 3D 圖

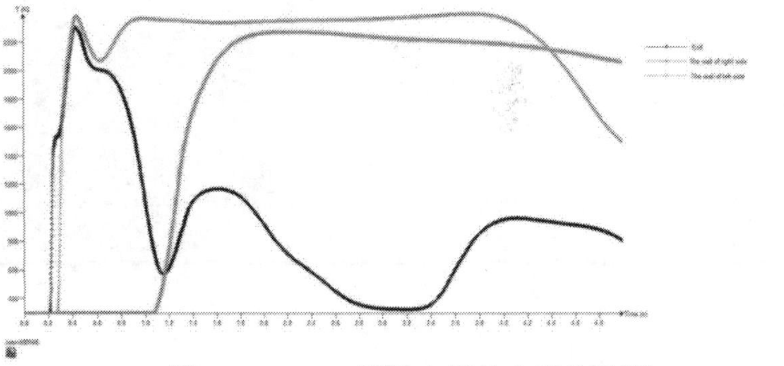

圖 16　FLACS 測點之爆炸高溫趨勢圖

(2) 過壓危害風險分析

爆炸過壓最大值為 0.0363 barg，可能造成大小窗戶破裂，但要造成 1%人員致死則需超過 1.05 barg，本案例在過壓危害部分其致死機率為 0，故其個人風險也為 0。

(3) 脈衝壓危害風險分析

脈衝壓最大值僅 88.4 (Pa·s)，基本上對人體及設備皆無傷害性，若要造成 1%人員致死需超過 20150 (Pa·s)，因脈衝壓未達到上述標準，故其致死機率與個人風險皆為 0。

(4) 爆炸熱輻射危害風險分析

圖 17 為爆炸熱輻射之個人風險，其範圍為 5.76×10^{-22}~9.37×10^{-6}（人/年），在人員可能出沒離地 2 m切面處，除了爆炸點周圍 5m處均為 9.37×10^{-6}（人/年）外，其餘個人風險

值約在 2.34×10^{-6}~7.03×10^{-6}（人/年）間，在離地 4.6 公尺處因僅有維修人員出沒，其個人風險值為 1.13×10^{-7}（人/年）。

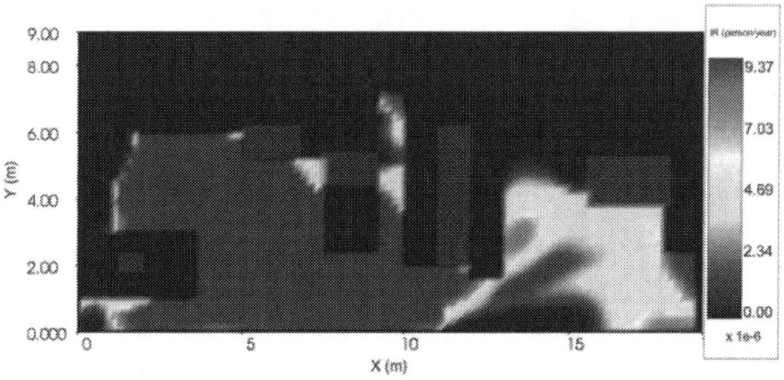

圖 17　爆炸熱輻射個人風險切面

3.3 FLACS 毒氣擴散後果分析與風險分析

(1) 毒氣擴散後果分析

由 FLACS 模擬結果中得知氨氣一開始向四周圍擴散，在橫向擴散時受到循環迴風氣流的影響會開始下降，但由於氨氣的比重較空氣輕，故仍會飄在空中。直至模擬結束，氨氣氣雲幾乎充滿整個空間，但上層的濃度較高危害也較大。在 304 秒時氣雲濃度約達 80 %以上，而周圍濃度約 15%，最終時上層濃度約 30%。

(2) 毒氣擴散危害風險分析

氨氣在離地 2 公尺處之個人風險值皆為 0，在 4.6 公尺處之個人風險為 1.15×10^{-21}~1.13×10^{-7}（人/年），其原因為高濃度氨氣會聚積在空中，對於維修人員可能造成風險，見圖 18。

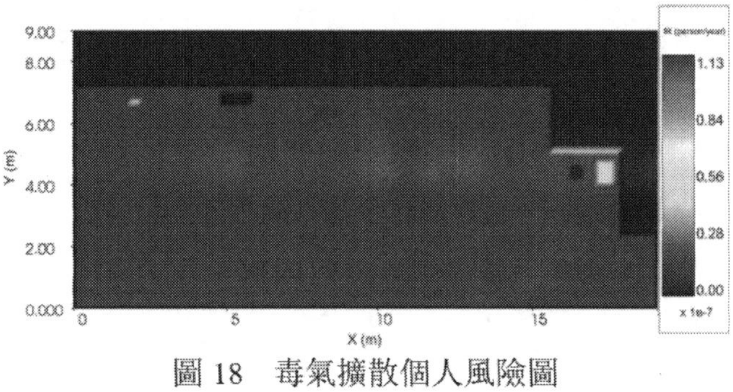

圖 18　毒氣擴散個人風險圖

3.4 整合風險分析

此階段係利用 IQRAS 的風險整合介面將前述七種危害之個人風險進行整合，見圖 19。FDS 部分需先用圖 19(a)之模組將網格縮減至與 FLACS 相同，再利用圖 19(b)中之整合風險模組將七個相同網格矩陣之個人風險進行加總，其結果可用 Matfor 繪圖展現。

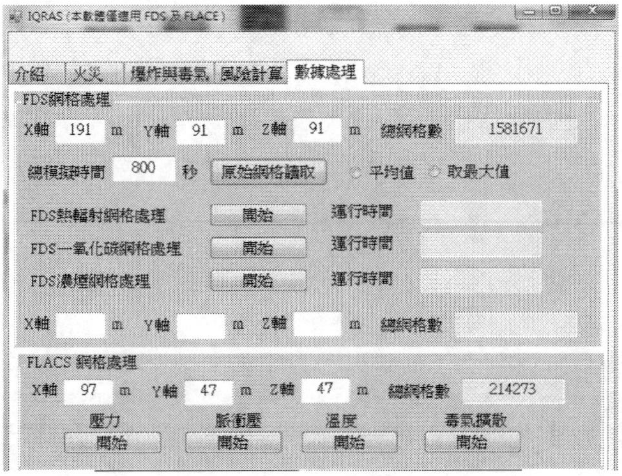

(a) 網格處理

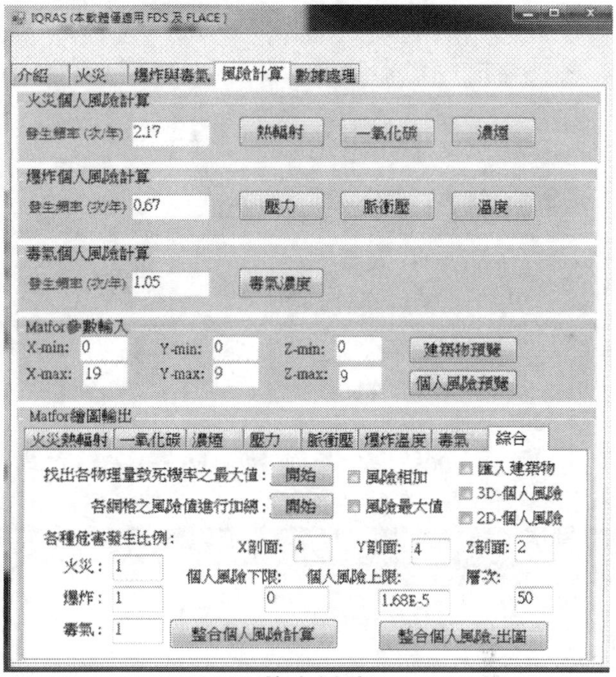

(b) 整合風險

圖 19　風險整合介面

圖 20 顯示最終個人風險值介於 1.71×10^{-9}~3.21×10^{-5}（人/年）之間，一般操作人員活動範圍離地板最高約 2 m，其風險如圖 20(a)所示，在火源附近最大值為 3.21×10^{-5}（人/年），橘色部分約 2.41×10^{-5}（人/年），而靠地板處（綠色）約 1.61×10^{-5}（人/年）。維修人員之活動區域可達離地高度 4.6 m處，其個人風險如圖 20(b)所示，其中紅色區域之個人風險值為 2.73×10^{-7}（人/年）的部分，而天花板FFU之綠色區域個人風險值約為 1.36×10^{-7}（人/年）。

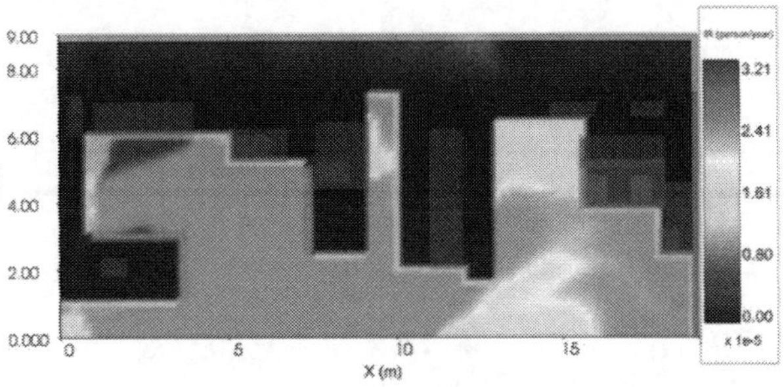

(a) Z= 2 m

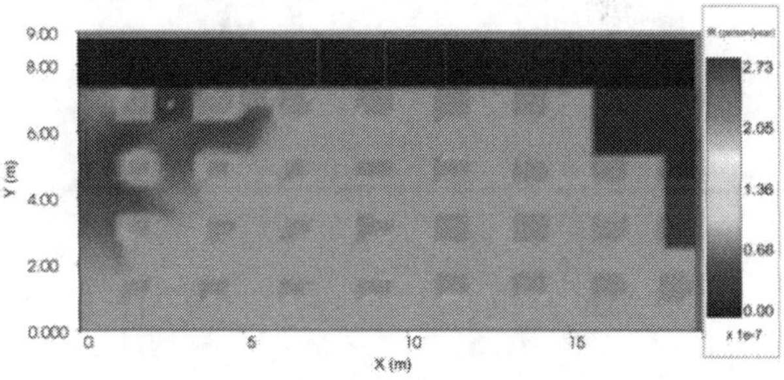

(b) Z= 4.6 m

圖 20　潔淨室各種危害之個人風險相加圖

4. 結論

本研究建立了一套能夠完整評估廠區化學物質危害之風險分析方法，能展現火災、爆炸及毒氣擴散等危害之風險整合程序。由模擬得知氫氣爆炸所產生之高溫多存在空間上層及牆壁邊，而氨氣擴散時會在物件附近滯留，上述現象用二維模擬軟體均無法評估，故以三維模擬軟體較為合適。藉由自行撰寫之IQRAS軟體將FDS及FLACS等CFD軟體模擬出來之數據進行剖析，最終計算其個人風險再加以彙整輸出，得到在沒有任何防護措施的狀況下，本研究所設定之無塵室其毒化災個人風險值介於 1.71×10^{-9}~3.21×10^{-5}（人/年）之間。未來只要採用FDS或FLACS軟體，皆可利用IQRAS軟體來預測目標廠區內部人員之風險。

5. 參考文獻

[1] Eisenberg, N.A., et al., 1975, "Vulnerability Model: A Simulation System for Assessing Damage Resulting from Marfne Spills", NTIS, AD-A015-245.
[2] I, Y.-P.; Cheng, T.-L., 2008, "The Development of a 3D Risk Analysis Method", Journal of Hazardous Materials, Vol. 153, Iss. 1–2, 600–608.
[3] 林逸青，2010，以FDS軟體驗證12吋半導體廠房消防系統設計之研究，國立雲林科技大學環境與安全衛生工程系，碩士論文.
[4] 邱淑靜，2008，台灣半導體相關產業重大災害原因探討與分析，長榮大學健康科學學院職業安全與衛生學研究所，碩士論文.
[5] 邱奕龍、吳世忝，2005，半導體廠房內可燃性氣體之火災爆炸模擬，國立雲林科技大學環境與安全衛生工程系實務專題.
[6] 黃奕豪，2009，三維火災風險分析技術之開發及其在公共場所事故模擬之應用，國立雲林科技大學環境與安全衛生工程所，碩士論文.

2nd CCPS CHINA CONFERENCE ON PROCESS SAFETY

Extinguishing efficiency of liquid nitrogen against trichlorosilane fires.

Liao Kuangyu
+886-5-5342601-4450
No.123, Sec. 3, Daxue Rd., Douliu City, Yunlin County 640, Taiwan (R.O.C.)
liaoky@gmail.com

Horng Jaojia, Liou Kehun, Liao Yuanchun, Shen JyunHong
+886-5-5342601-4450
No.123, Sec. 3, Daxue Rd., Douliu City, Yunlin County 640, Taiwan (R.O.C.)
horngjj@gmail.com, soul0801@gmail.com, steveycliao@gmail.com, cgh1210@gmail.com

Prepared for Presentation at
American Institute of Chemical Engineers
2nd CCPS China Conference on Process Safety
Qingdao, China
August 28 – 29, 2014

UNPUBLISHED

Extinguishing efficiency of liquid nitrogen against trichlorosilane fires.

Liao Kuangyu
+886-5-5342601-4450
No.123, Sec. 3, Daxue Rd., Douliu City, Yunlin County 640, Taiwan (R.O.C.)
liaoky@gmail.com

Horng Jaojia, Liou Kehun, Liao Yuanchun, Shen JyunHong
+886-5-5342601-4450
No.123, Sec. 3, Daxue Rd., Douliu City, Yunlin County 640, Taiwan (R.O.C.)
horngjj@gmail.com, soul0801@gmail.com, steveycliao@gmail.com, cgh1210@gmail.com

Keywords: Trichlorosilane (TCS), Extinguishing, Liquid nitrogen

Abstract

TCS (Trichlorosilane, SiHCl3) and other Chlorosilane were used abundantly in semiconductor and solar energy industries as the principal resource to generate ultrapure silicon in production. They are mostly acutely toxic, water-reactive and corrosive substances. While in the fires, TCS might form Silicon dioxide, Chlorine and Hydrogen gases and were difficult to extinguish. There were numerous fires due to TCS's flammability, such as overturned truck in Michigan, US, 1989 and leaking in solar manufacturer in Pingtung, Taiwan, 2011. Many concerns aroused among the relevant agencies on how to put out fires and how to reduce hazards.

Our study showed liquid nitrogen could yield extinguishing ability with its cooling and suffocation effects. Vaporizing liquid nitrogen drew the heat from combustion reaction and reduced the flammable vapor to quench flames. The key parameter was liquid nitrogen application in large amount could effectively extinguish TCS fires. Cooled TCS showed unchanged properties and still had potential to flame.

1. Introduction

Along with the popularity of energy saving and carbon reduction globally, there are many alternatives and demands on national energy policies being developed including the Solar photovoltaic (PV) industry. The demands for PV have been grown explosively in the past decade after year 2000. Among them, China occupied about 45% of production for global PV [1][2] . Since Trichlorosilane (TCS) was one of the principal precursors generating silicon in semiconductor, solar energy and chemical industries. And, the processing quantity of TCS has increased dramatically in recent years. Not only TCS is a poison, water-sensitive and corrosive substance but also its derivatives (SiO_2 、Cl_2 、H_2) are harmful to human respiratory tract[3][13][14], especially as TCS and the derivatives involving in fire incidents. As a fire fighting became rather complicated in an event in 2008, Hubei, China when Yichannan's manufacture plant leaking TCS that caused 18 people injured[15]. In 2011, in Pintung, Taiwan, one solar energy company leak TCS in an accident. In this case, spillage dikes were established for the storage of outfalls from extinguishing agents and foams to avoid those unwanted side effects. There were many difficulties involving on extinguishing TCS fires including (1) TCS being water-reactive, (2) water derivatives forming caustic smoked, (3) possible re-ignition due to some undefined reactions among extinguisher agents and TCS, and (4) inefficient extinguisher agents of powder, CO_2, and Halon. It was known that medium expansion foams could effective put out TCS fire [4][5][16] but with explosive derivative, H_2, threatening the surroundings. Therefore, due to TCS's hazardous properties, the design of spillage dike was a necessity for TCS storage.

Since liquid nitrogen (LN_2) is an inert, frozen liquid that yield (1) no reaction with others, (2) displacement of oxygen during its gasification, (3) absorption massive heat [9] for thermal decomposition during its gasification. These advantages of LN_2 are contributed to its properties of low boiling point (-195.8 °C), high latent heat of vaporization (199.1 kJ/kg) and high gas density (1.2506 kg/m^3). Many literatures showed that LN_2 is also an effective suppressant in extinguishing fires caused by metal sodium, water-sensitive substance, chemical pool, and fuel as well as forest fires in fire prevention and/or suppression system [6][7][8].

The objective of this study is to investigate (1) the relations of spillage dike volume and surface area to TCS burning rate per unit area (kg/ m^{2}s), (2) effectiveness and mechanism of LN_2 for the TCS fire suppression, and (3) LN_2 re-ignition possibility.

2. experimental

Since our study would investigate the gasification of liquid nitrogen while suppressing TCS fire. Several possible effects include the absorption of TCS decomposition and reaction heat, the expansion of gas vaporization to expel oxygen from above the fuel. Our preliminary experiments examined the burning rate of TCS and the quantification of the effectiveness of LN_2 suppressing TCS fire and the possibility of re-ignition. Our study conducted outdoor in a firefighting training field where outdoor temperature, humidity, wind speed and direction were all measured and considered.

The experimental schematic was shown in Fig.1. Iron pans used in small (40cm length, 40cm width, 10cm deep), medium (80cm length, 80cm width, 10cm deep), and large (120cm length, 120cm width, 10cm deep) sizes to contain TCS. Electronic scales, thermocouple data loggers, thermal infrared imager, cameras and timers were all used in the experiment. The purpose of exhauster was to divert derivatives and protect the apparatus.

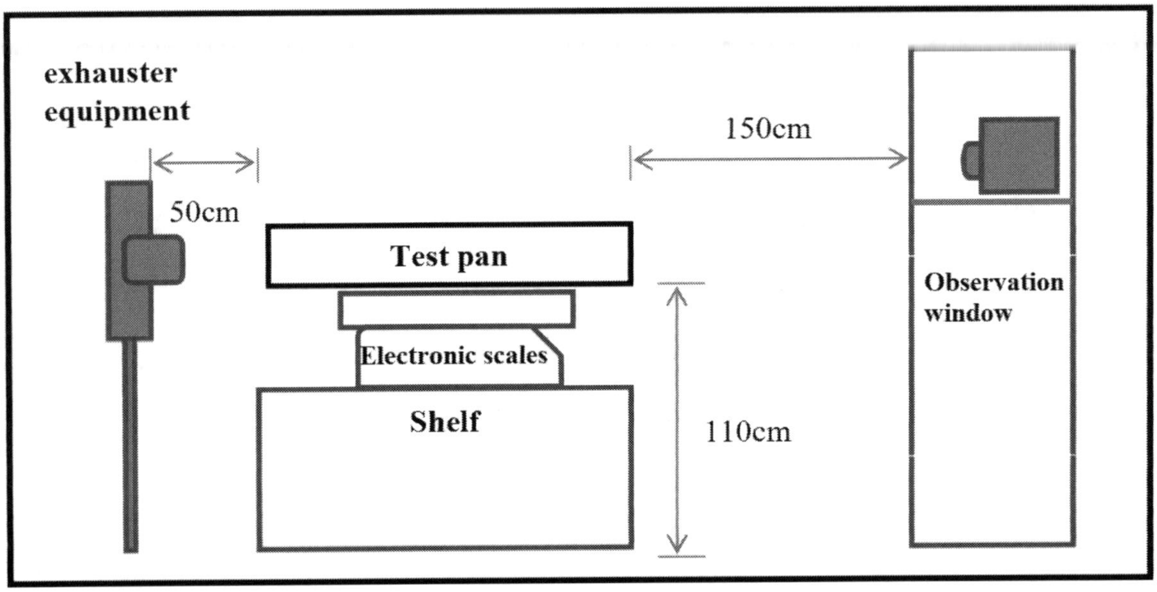

Fig.1. Experimental setup.

2.1 TCS burning rate

The experiments conducted of TCS burning with two distinct configurations [17]: (1) fixed surface area and (2) fixed height by iron pans. The experiments with fixed surface area were designed to observe the burning rates per unit area (kg/ m^2s). Hence, the 2kg, 3kg, 4kg, 8kg of TCS were poured into small pan (area of 40cm * 40cm) and then ignited to observe the burning rates. On the other hand, the experiments with fixed height were designed to observe the relationship between the surface area and burning rate per unit area (kg/ m^2s). The TCS was poured pans with 1cm height in each pan and then TCS was ignited. Several observations of burning phenomenon such as temperature, temperature changes and fuel weight changes were recorded in order to estimate combustion and/or reaction rates.

2.2 Effectiveness of LN2 for TCS fire suppression

Those experiments were to quantify the effectiveness of LN_2 suppressing TCS fires [7]. The 2kg, 3kg and 4kg of TCS were poured into small iron pans and ignited. The fires were allowed to burn for 90 seconds to reach their high temperatures and then suppressing by 200ml of LN_2. The effectiveness of fire suppression was then observed by the measurements of thermal infrared imager (Ching Hsing Computer-Tech Ltd. TVS-200EX).

2.3 Re-ignition of TCS

Some experiments were designed to test the re-ignition of TCS after the applications of LN_2. First, 200ml of TCS was poured into the 500ml beaker and ignited for a minute. Then, LN_2 was added in 100ml per addition in order to observe the temperature changes of flame and other phenomenon. In addition, the remains of TCS were ignited to test for their property changes.

3. Results

Our field tests were done on a training field in an industrial district of Kaoshiung City, Taiwan with daytime temperature at about 30℃, slight wind and sunny in April of 2014.

3.1 TCS burning rate

The observation of TCS burning happened mainly on the combustible vapor above the liquid surface accompanying with heat generation and continuing fuming (fig.2a-2d). The extensive white fume was reported to be the TCS byproducts of HCl and SiO_2 releasing from the flame (fig.2e-2f, fig.3b). Using thermal imager to measure the temperature distribution while TCS burning, we observed highest temperature in flame above 510℃ (detection limit of thermal imager) in white color (fig. 3a).

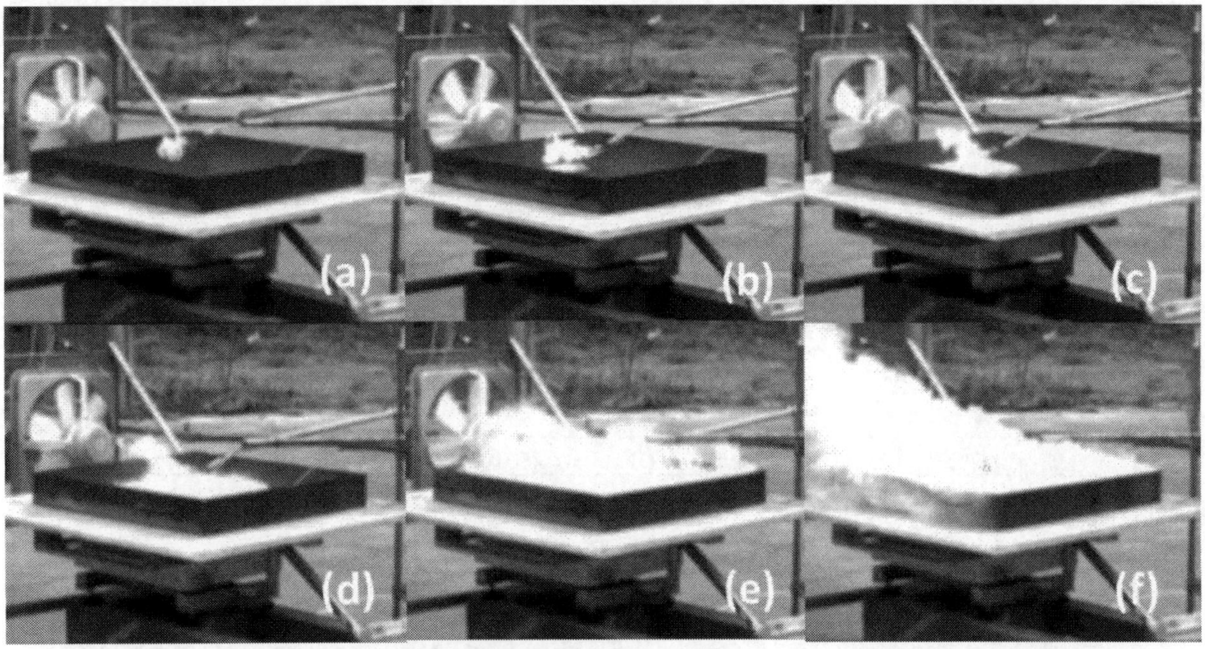

Fig.2. TCS ignition

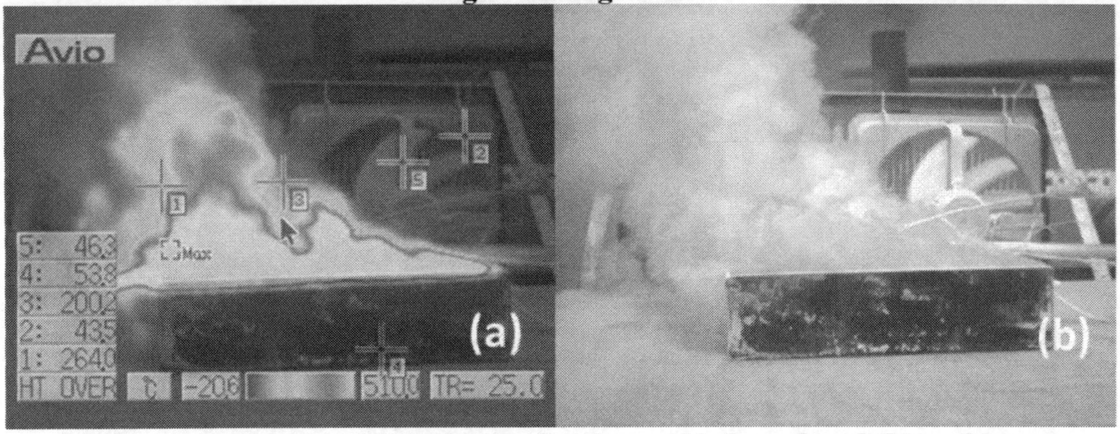

Fig.3. (a) temperature distribution by thermal infrared and (b) real image

(1) Fixed surface area:

Our tests showed a linear relationship between burning time and loss weight of TCS (fig.4 showed the linear slopes). The burning rates of 2kg, 3kg, 4kg, 8kg were 0.0060(kg/s), 0.0069(kg/s), 0.0081(kg/s), and 0.009(kg/s), respectively. The linear correlation between burning rate per unit area (kg/ m^2s) and TCS fuel weight (fig.4) showed a direct proportion which might be related to the evaporation rates of combustible TCS. As the amount of TCS increased, it seemed that the TCS burning rate per unit area (kg/ m^2s) approached a fixed value (fig.5). Thus, under a limited burning surface circumstance the TCS burning would yield an evaporation quantity approached to fixed value regarding as stable burning rate [9].

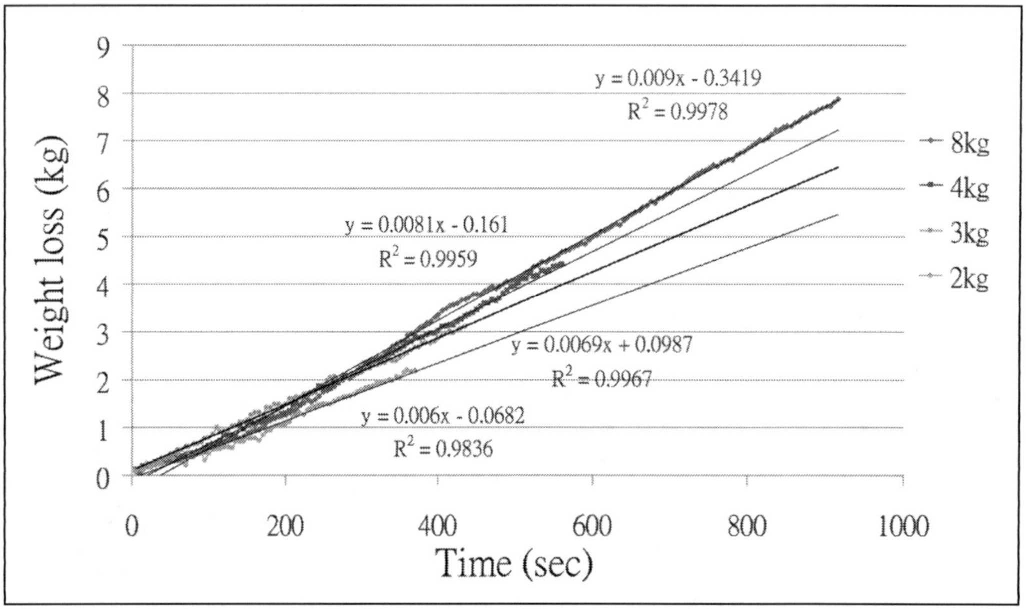

Fig.4. TCS burning rate(fixed burning area 40cm*40cm)

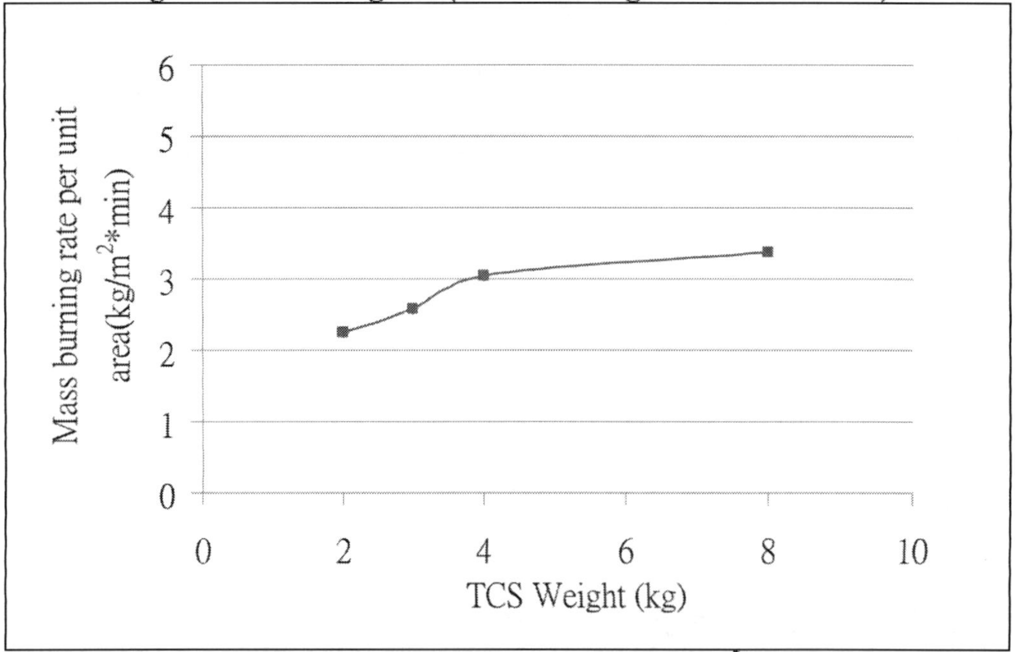

Fig.5. Relationships between burning rate per unit area (kg/ m^2s) and TCS weight (small pan: 40cm*40cm)

(2) Fixed height:

The burning rates (kg/s) of small, medium, and large iron pans were 0.0060(kg/s), 0.0222(kg/s), and 0.0353(kg/s). The linear ratios were 1 : 3.7 : 5.9. (fig.6). Due to the high burning rates and the losses, the TCS could not cover the entire surfaces for medium and large pans after 250 and 300 seconds (fig.7). Hence, the burning rates of medium and large pans were then reduced. If we considered the initial 250sec for TCS burning in small, medium, and large pans, the mass burning rate (kg/s) were 0.0055(kg/s), 0.0255(kg/s), and 0.0571(kg/s), respectively. The burning rates were at ratios of 1 : 4.6 : 10.4—similar to their area ratios of 1 : 4 : 9. Thus, we could conclude that TCS burning rates would be directly proportional to their area at fixed height in those pans. Fig.9 showed that estimated mass burning rates per unit area was 2.41 kg/m^2-min. Therefore, our findings for TCS burning rates implied that as we design the spillage dike area for TCS should be at smaller surface area to reduce burning rates. However, we still need a better way to break the TCS burning conditions while using LN_2.

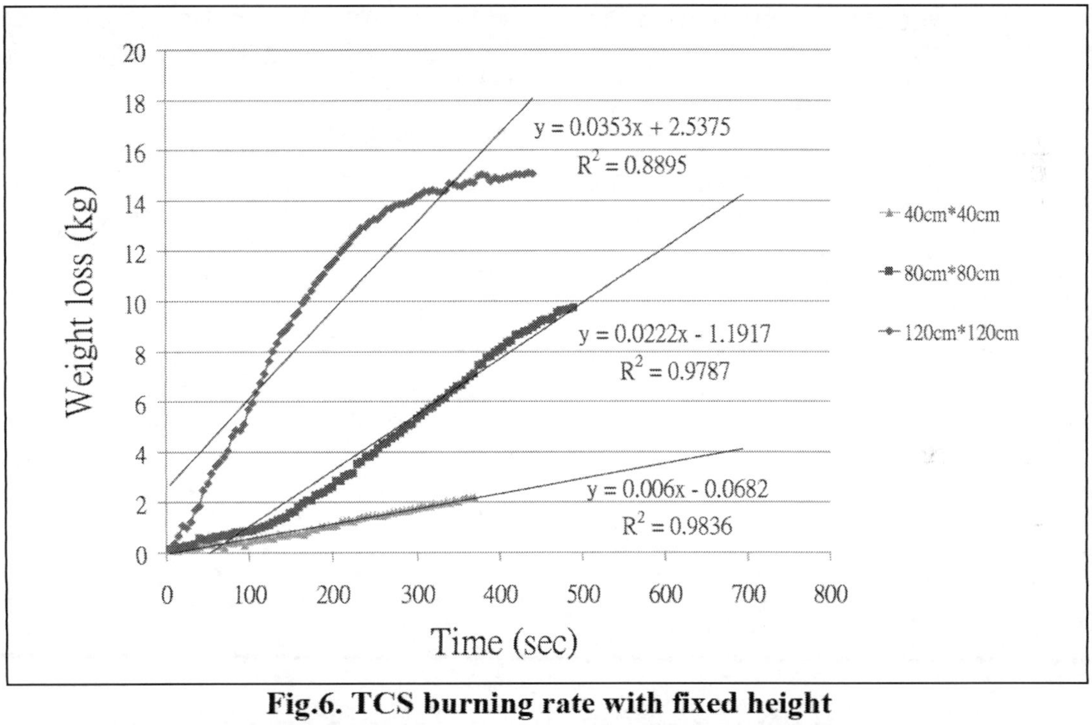

Fig.6. TCS burning rate with fixed height

Fig.7. TCS burning at large pans (a) 30sec, entire area, (b) 300sec, 50% area, (c) 350sec, about 25% area.

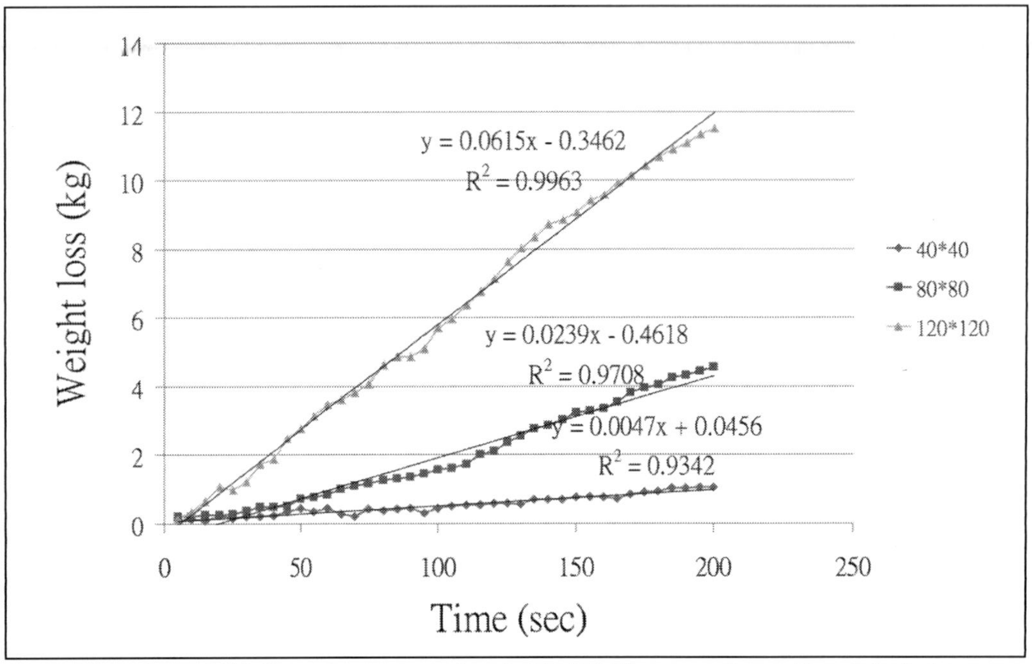

Fig. 8、TCS burning rates for initial 250 seconds at fixed height

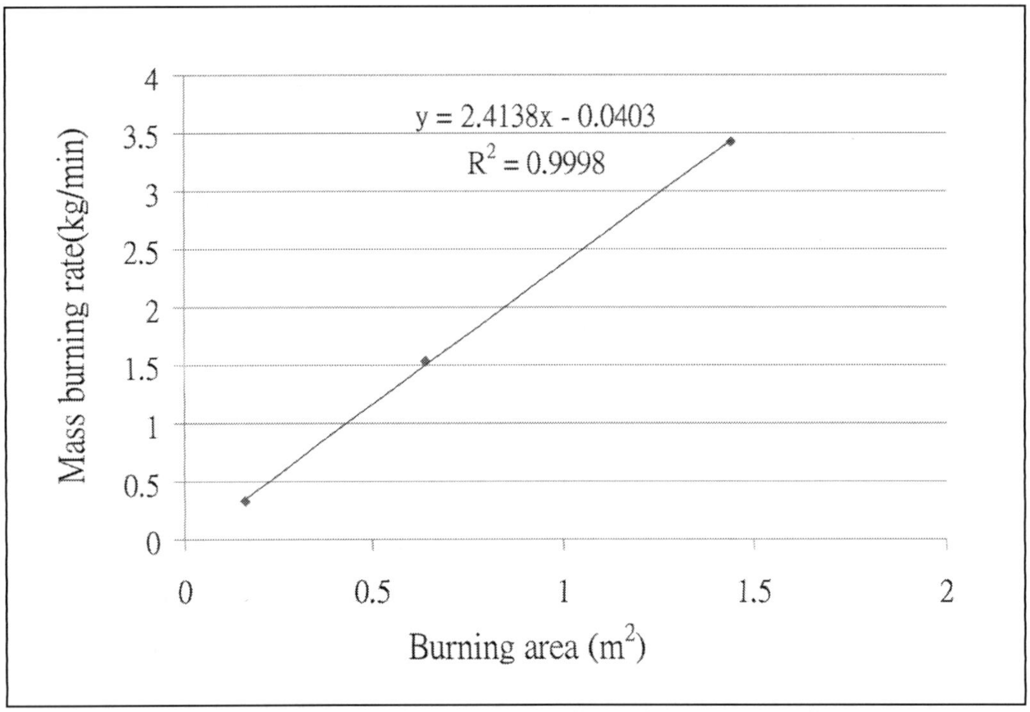

Fig. 9、A linear relationship for burning rates and areas

3.2 Effectiveness of LN2 Suppressing TCS fire

Since our preliminary findings on TCS burning rate at 0.006 kg/s, we should look into the thermodynamic properties of TCS. It has enthalpy of combustion 456.0 kJ/mol[10] and LN_2 has heat of vaporization 199.1 kJ/kg. Thus, we estimated that with 125ml/s of LN_2 could absorb 2kg TCS by the enthalpy of combustion. Considering that LN_2 evaporation would not be 100% effective, our experiments used 200ml of LN_2 for our initial test (fig.10) and failed to suppressing fire. Therefore, a 500ml of LN_2 was used as suppressant and effective extinction of fire started by 2kg TCS as showed in fig.11. Table 1 showed the amount of LN_2 needed to effectively stop TCS combustion at different TCS burning weight. Since LN_2 evaporation would absorb the enthalpy of TCS combustion until the reaction stop, we hypothesized that the amount of LN_2 used at the fixed surface should relate to the weight of TCS remains. The evaporation of LN_2 need to absorb the enthalpy of remaining TCS to below its autoignition point (104C). We observed this cooling by the stop of the TCS combustion reaction. Our results showed a linear regression of TCS weight and effective LN_2 quantity (fig.12).

Effective LN_2 volume (ml) = 177 + 0.287*TCS

The effective volume of LN_2 177ml was larger than the theoretical 125ml to absorb the enthalpy of burning TCS, This effect might be the extra LN_2 needs for ineffective cooling while only the surface contact of LN_2 TCS without proper mixing. The experiments also showed that TCS was cooled to 23℃ in Fig.13d. However, is this suppressing of TCS combustion changing its property? Could this cooled TCS be re-ignited? We need to investigate further.

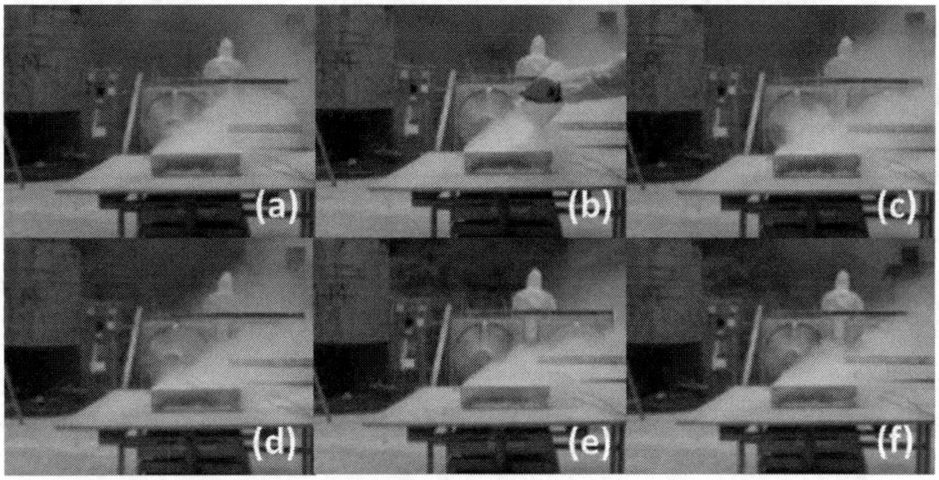

Fig.10. TCS-2kg, LN_2 (200ml) extinguishing series, (a) initial combustion, (b)LN_2 pouring in, (c) 2sec, (d) 4sec, (e)10sec, (f) 30sec.

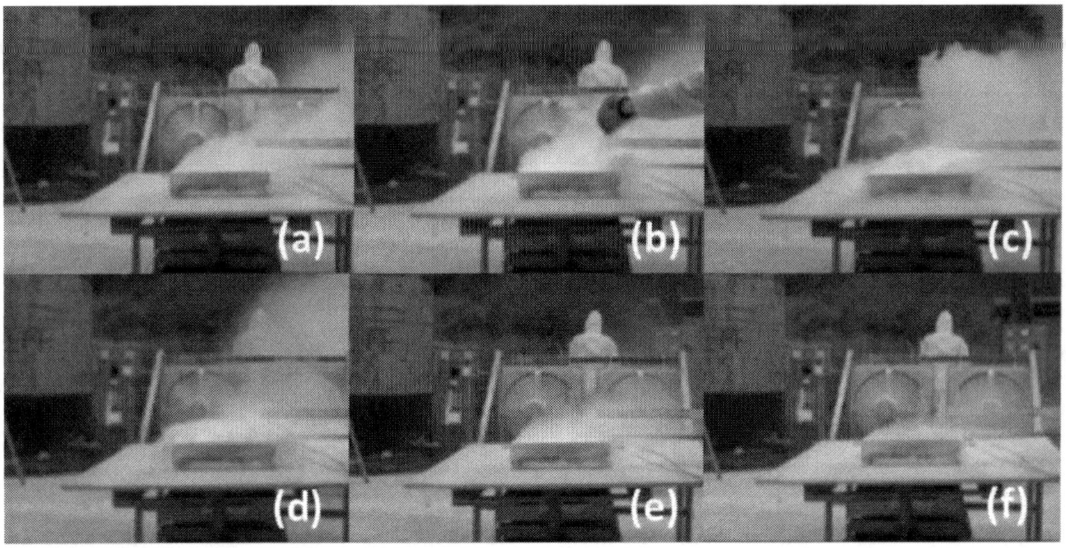

Fig.11.TCS-2kg LN_2 (500ml) extinguishing series, (a) initial combustion, (b)LN_2 pouring in, (c) 2sec, (d) 4sec, (e) 10sec, (f) 30sec, combustion stop

Table 1. The effectiveness and volumes of using LN_2 to extinguish TCS fires

TCS Weight	LN_2 volume (Weight)	Extinguishing Effects
2 kg	200 ml (0.162 kg)	N
	300 ml (0.242 kg)	N
	400 ml (0.323 kg)	N
	500 ml (0.404 kg)	**Y**
3 kg	500 ml (0.404 kg)	N
	600 ml (0.485 kg)	N
	700 ml (0.566 kg)	**Y**
4 kg	700 ml (0.566 kg)	N
	800 ml (0.646 kg)	N
	900 ml (0.727 kg)	**Y**

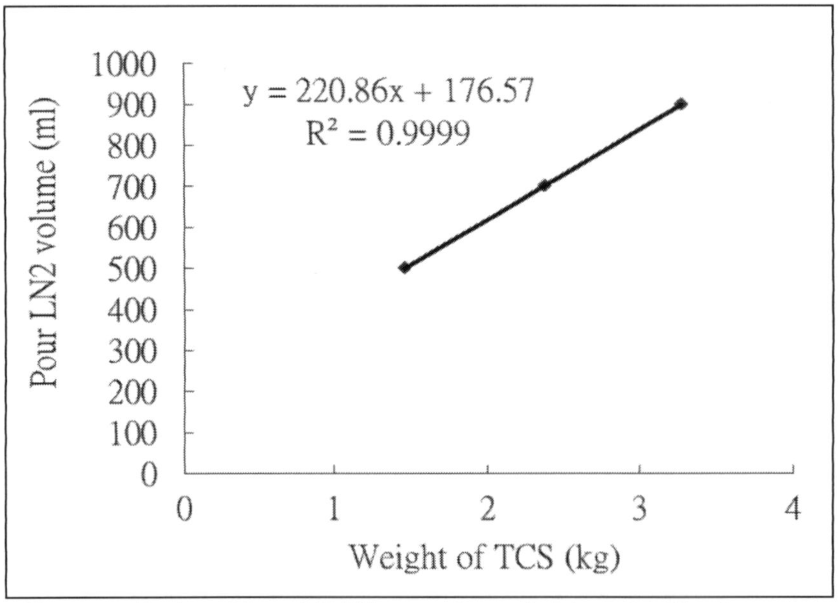

Fig.12. The linear relationship of TCS remain weight and effective amount of LN_2

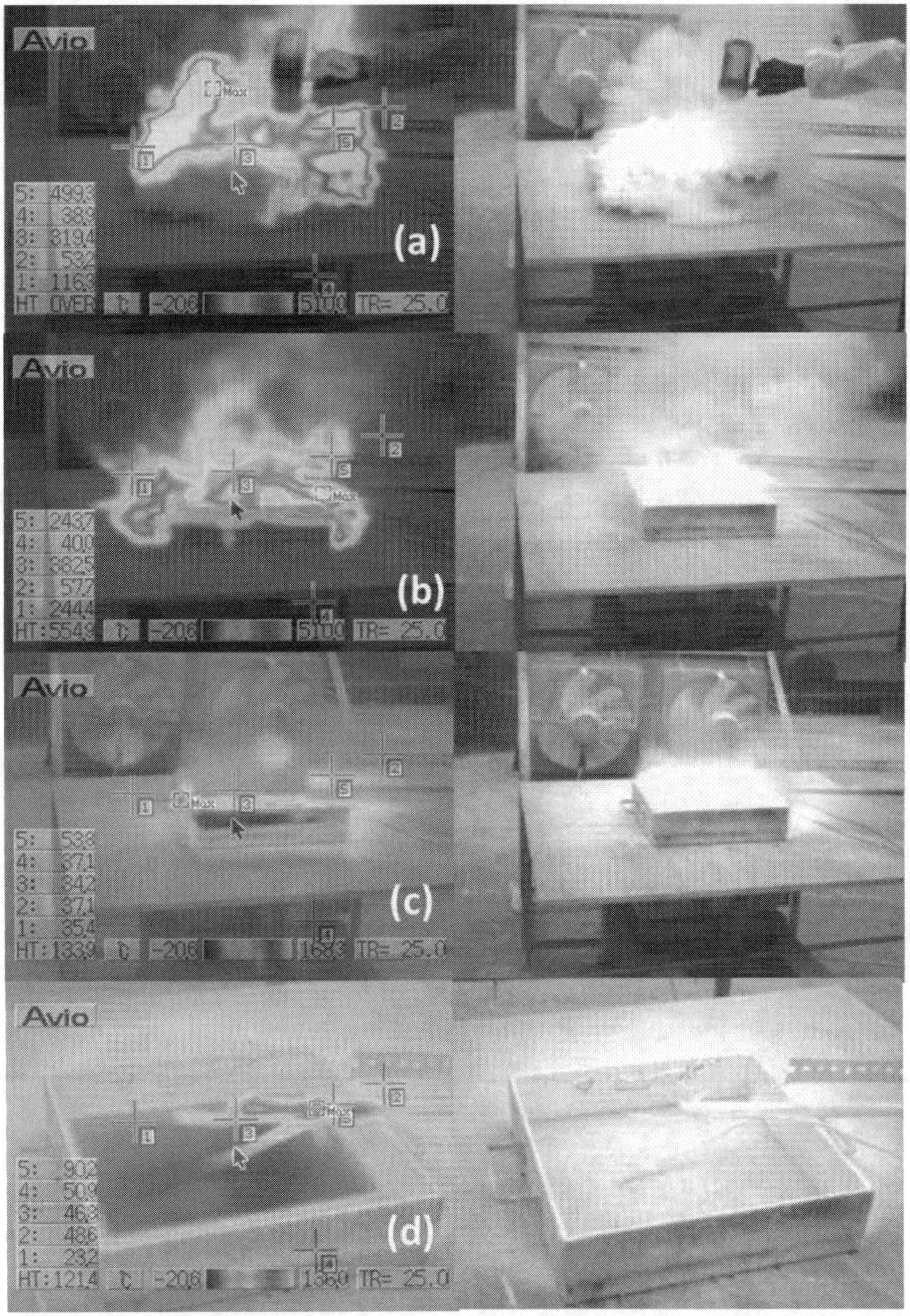

Fig.13. (TCS 2kg, LN_2 500ml)thermal infrared image, (a) as LN_2 pouring in and flame extending outward, (b)temperature dropped, (c)LN_2 gasified constantly, (d)fire extinguished with TCS liquid temperature around 23.2 ℃

3.3 TCS remains re-ignition

As LN_2 pouring into the burning TCS, LN_2 continue to evaporate and cooling the TCS. After 30sec, the TCS surface showed some frozen layers. As LN_2 completely evaporated after 90sec, we found the frozen layer was about 1.5cm thick. Since LN_2 has specific gravity (H_2O=1) of 0.808 and boiling point of -195 ℃ and TCS has specific gravity (H_2O=1) of 1.3 and freezing point of -128 ℃[11] [12]. We assumed that LN_2 cooling had absorbed the heat and the TCS had frozen as show in fig.15. We then processed the re-heating.

The overall data of this freezing and re-heating was showed in Fig.16. Our results indicated that as LN_2 pouring in for rapid cooling of TCS liquid in the first 30 seconds. Then, TCS started cooling and reduced heat conduction in 30s-90s as TCS liquid temperature at 12-15℃. When LN_2 evaporated completely after 90sec, the solidified TCS started to melt and liquid temperature gradually increased. After 240 seconds, TCS reached lowest temperature of -32℃. We then reheated the TCS at the rate of 2.75℃/min until field temperature of 30℃ in 24 min. We then try to re-ignite the TCS.

We found that the remaining TCS could be re-ignited and produced white fume (HCl and SiO_2). We then concluded that the frozen and thawed TCS did not change its chemical properties and LN_2 freezing process could suppress the TCS fire and was merely a physical phenomenon. Thus, the suppressing TCS still need to be processed or treated to avoid further problems.

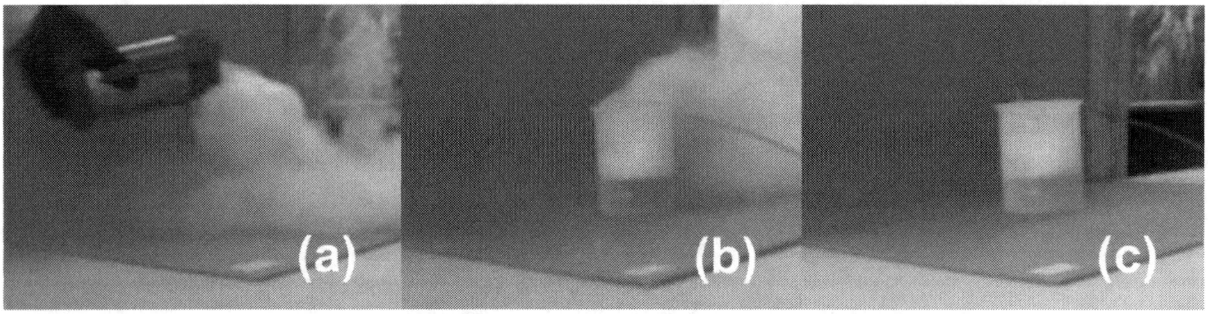

Fig.14. LN_2 pour into TCS, (a)beginning, (b)after 30sec as LN_2 gasifying, (c) after 90sec, frozen.

Fig.15. TCS top layer frozen.

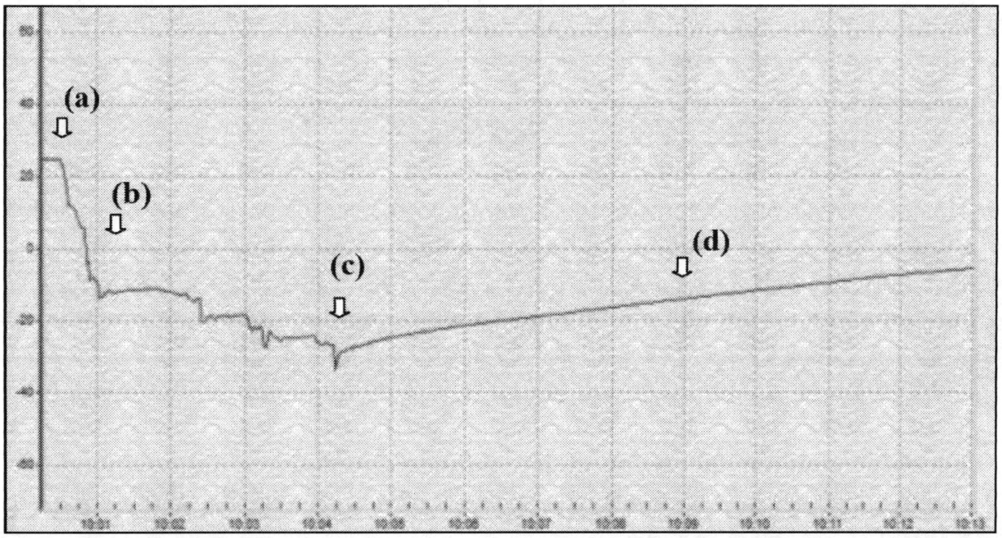

Fig.16. Thermocouple data log on temperature as LN2 pouring into TCS, (a)extinguishing start, (b) 30sec as rapid cooling of TCS liquid, (c) after 240 seconds as TCS reached lowest temperature of -32℃, (d) reheated the TCS at the rate of 2.75℃/min

Fig.17. TCS re-ignition tests

4. Conclusions & suggestions

Were and several conclusions could be drawn from this study.

- The mass burning rate of TCS was estimated to be 2.41 kg/m^2min in our field tests with unobvious TCS flame over 510℃.
- Out tests showed that burning rate suggestion was affected by surface area than fuel weight. Our study suggests spillage dike should be designed to keep the maximum volume but narrowing the surface area in order to decelerate the burning rate and most importantly to reduce the explosive derivative proliferation rate which mainly cause extensive disaster.
- We concluded that LN_2 was an effective suppressant in TCS fire. When burning area was 40cm*40cm pool fire, the effective amount of LN_2 to put out TCS fire with volume = 177+0.287*TCS volume (ml) a regression value from our study. LN_2 could suppress TCS fire by its physical cooling. After fire extinction, the thawed TCS could still react with water vapor, therefore, TCS combustible and hazardous derivative properties still existed.

Suggestions

- TCS spillage and derivative are quite hazardous. In addition to in compliance with national safety regulations, ventilation and water equipment were recommended. Emergency response processes should take disaster area estimation, and personnel or nearby residents evacuation into first consideration.
- In TCS leakage or fire accident, contamination or burning area limitation can decerlerate hazardous evaporation and derivative rates. While handling thawed TCS, any hot work and environmental temperature must be paid attention to in order to avoid re-ignition.
- Using LN_2 on fire fighting, personnels must take necessary precautions against suffocation and frostbite

5. References

[1] Arnulf Jäger-Waldau (2013). Chapter IIC-2 – The Photovoltaic Industry. Solar Cells (Second Edition), Pages565-583.

[2] Peter Woditsch, Wolfgang Koch (2002). Solar grade silicon feedstock supply for PV industry. Original Research Article Solar Energy Materials and Solar Cells, Volume 72, Issues 1–4, Pages 11–26.

[3] Centers for Disease Control and Prevention, International Chemical Safety Cards. (Available from: URL:http://www.cdc.gov/niosh/ipcsnswh/nswh0591.html)

[4] Higgins, J. T., W. E. Hubbs, R. A. Kayser, P. Kremer, S. W. McMahan, L. Patrick, M. Strong (ed.) , 1998. Manual on Chlorosilane Emergency Response Guidelines, ASTM MNL 33.

[5] Liao Kuangyu ,Horng Jaojia, Liou Kehun, Liao Yuanchun, Chen Jianhong (2013). Investigating the efficiency of different extinguisher means on Trichlorosilane fires. CCPS, First Asia-Pacific Conference on Process Safety.

[6] Yiannis Levendis1, Ali Ergut1, Michael Delichatsios (2010). Cryogenic extinguishment of liquid pool fires. Process Safety Progress, Volume 29, Issue 1, pages 79–86.

[7] Eunkwang An, Peter B. Sunderland , Daniel P. Lathrop (2013). Suppression of sodium fires with liquid nitrogen. Fire Safety Journal 58, Pages 204–207.

[8] Yiannis, Michael (2013).University of Ulster. Cryogenic Suppression of Liquid Pool Fires and Wooden Crib Fires.(Available from： http://www.see.ed.ac.uk/FIRESEAT/files11/FS11-Presentation6-Delichatsios.pdf)

[9] American Bureau of Shipping (2014). Guidance notes on fire-fighting systems.

[10] Mores, T. F. (1984). Fundamental Studies of Laser Interaction in Materials Preparation: New Aspects of Chemical Vapor Deposition, Trichlorosilane, Literature Survey and Combustion Experiments. DTIC.

[11] SIGMA-ALDRICH, Inc. (2012). Trichlorosilane Materials safety Data sheet.

[12] Praxair, Inc.(2009). Trichlorosilane Materials safety Data sheet.

[13] Jhy-Charm Soo, Siou-Rong Li, Jenq-Renn Chen, Cheng-Ping Chang, Yu-Fang Ho, Trong-Neng Wu, Perng-Jy Tsai1 (2011). Acid Gas, Acid Aerosol and Chlorine Emissions from Trichlorosilane Burning Processes. Aerosol and Air Quality Research, 11: 323–330.

[14] Institute of Occupational Safety and Health (2009). Evaluation of the hazardous characteristics of trichlorosilane.

[15] Xin hua net, Hubei Yichang CSG company spill occurred trichlorosilane, December 10, 2009.(Available from: URL:http://news.qq.com/a/20081210/002881.htm)

[16] Chen Cheng-Jen, Tsai Peng-Chi, Li Chia-Wei (2008). Trichlorosilane of Emergency Operating Procedures and Implementation of Field Testing Program, Southern Taiwan Science Park Administration.

New Technology for Toxic and Flammable Gas Detection

Barua, R., Senscient
Xu, P.,

Development of gas fields with high levels of H2S has been on a dramatic increase in the last several years. Until now, industry norms for H2S detection have been limited to single-point detectors placed strategically throughout facilities. Senscient has commercialized technology that introduces an advancement in H2S detection by using tuned lasers for monitoring fugitive H2S across an open path. This "open path" detection compliments single-point detection to significantly improve the early warning of fugitive H2S, resulting in improved safety and decreased risk. This technology is ideally utilized to create a detection barrier around the perimeter of a plant, process unit, or storage area.

An open-path H2S detector has been developed using Enhanced Laser Diode Spectroscopy (ELDS). ELDS is an advanced, highly robust form of Tuneable Diode Laser Absorption Spectroscopy (TDLAS) that was specifically developed for use in the most demanding safety applications. ELDS H2S detectors are capable of detecting even small fugitive leaks of H2S with absolute dependability and zero false alarms. This technology is the quickest, most sensitive and most reliable means of detecting H2S leaks or releases that has ever been available.

ELDS open-path H2S detectors utilize laser diodes that operate at wavelengths specially selected for the target gas, to generate a beam that is sent from a transmitter along a path to a receiver. Any target gas present in the beam causes a change in the received signal (called a harmonic fingerprint) which is specific to the target gas and proportional to the quantity of gas in the beam path.

The bottom line is that open-path laser technology provides an additional layer of protection and vastly improved early warning for an industry that is seeking better ways to manage the high levels of risk associated with highly toxic H2S gas. 211 systems have been delivered to China for Phase 1 of the Chuandongbei field development, a joint venture between Chevron and CNPC.

A Call to Action Next Steps for Vision 20/20

Scott Berger
Jack McCavit
Louisa Nara
Center for Chemical Process Safety
AIChE
120 Wall Street, New York, NY, 10005 USA
scotb@aiche.org

Cheryl Grounds
VP Process Safety, Central SO&R, BP
501 Westlake Park Blvd.
Houston, TX 77079
cheryl.grounds@BP.com

Keywords: process safety; vision; culture; competency; learning; standards

Abstract

Process safety is important to our individual health, to our company's success, and to our region's business economy. Companies that handle toxic, flammable, and explosive materials understand the importance of process safety and risk management. So do regulators. So should the community and other stakeholders. Individually these stakeholders have been working for years to improve process safety performance. Accidents continue to occur. It's time to leverage our resources, knowledge and skills to all strive for a common goal of great process safety performance. CCPS's Vision 20/20 describes that vision. It calls for effort from individual companies and effort that goes well beyond any individual company, indeed beyond the industries.The purpose of this paper is to share CCPS's Vision20/20 with the 10th Global Congress on Process Safety and to begin the process of engaging the broader community in the vision and the steps to achieve it.

1. Introduction

In the quest to improve process safety performance, companies have typically centered their efforts on compliance with regulations such as in the US OSHA Process Safety Management, the UK COMAH (1), or North Sea Safety Case (2). The CCPS *Guidelines for Risk Based Process Safety* (3) (RBPS) describes many of the elements that are found in such regulations and company management systems as well as other key features needed to manage process safety.

Despite the attention paid by individuals, companies, industry, and government, accidents continue to occur. Yet there are many good ideas, approaches and insights to be shared - *if we just worked together toward the common vision of improved process safety performance.*

In 2011, the Center for Chemical Process Safety (CCPS) began an effort to approach the challenge differently in a project called Vision 20/20. In this project, participants worked hard to envision the process safety future and ask what will be the characteristics of companies with great process safety performance in the future. What will distinguish the companies with fewer process safety incidents from those with more process safety incidents? How can we work together in companies and with all our stakeholders to further process safety performance?

Envisioning the future, five tenets were identified that companies with great process safety performance will possess. As the tenets were developed, it was recognized that there are four societal themes that, if achieved, would also have a significant positive impact on process safety performance. The five company tenets and four societal themes were introduced at the Global Congress on Process Safety in 2013. Since their introduction, efforts have been directed towards communicating the tenets and themes and identifying and developing tools to support the concepts. The company tenets and societal themes are detailed in this paper. Although this paper is about a future vision, it is written in the present tense recognizing that many aspects of the vision currently exist - just not fully and globally.

This is now our call to action. It is time for individuals, companies, regulators, and academia to embrace the tenets and themes so we can leverage the collective process safety strength built over the last decades. It is time to deeply understand the company tenets and societal themes and evaluate introspectively and collectively what more we can do to further process safety performance.

2. Internal - From a Company Perspective

The five tenets identified for achieving great process safety are:

- Committed Culture
- Vibrant Management Systems
- Disciplined Adherence to Standards
- Intentional Competency Development
- Enhanced Application and Sharing of Lessons Learned

The core principle of fervor for process safety unites and supports adherence to these tenets. It is not sufficient that each company understand the tenet. Additionally, each company should be enthusiastic, focused, and energetic in their process safety efforts. Each tenet is defined and discussed below.

2.1 Committed Culture

Three important aspects of committed process safety culture are emphasized in Vision 20/20. First is felt leadership from senior executives. Felt leadership means more than a periodic mention of process safety in speeches and town hall meetings. It means that the executives are personally involved in process safety activities. Employees know the executives care about process safety because of what they see executives doing in addition to what they hear executives saying. Executive leadership results in a tangible commitment and employee engagement at all levels of the organization. All levels of management and supervision stress the importance of process safety.

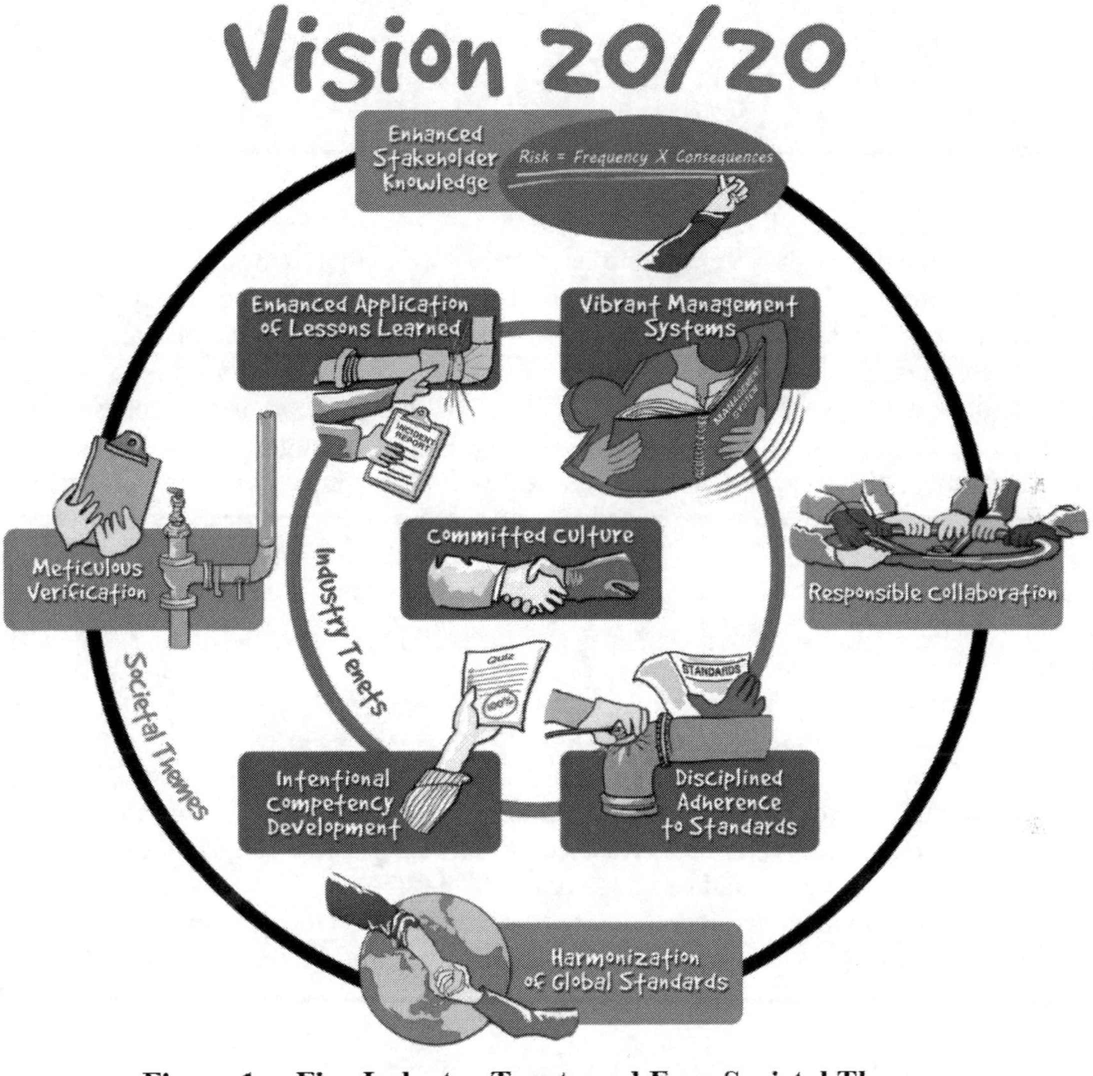

Figure 1. Five Industry Tenets and Four Societal Themes

The second important aspect of process safety culture is operational discipline (4), the performance of all tasks correctly every time. Often operational discipline is considered to be an operator and mechanic issue, but in fact it is a leadership issue too. Sometimes leaders believe they do not have time to verify that work in the field is in conformance with practices and procedures. Sometimes leaders condone poor operational discipline by failing to take action when policies are not followed. In a committed culture leaders know what is happening in the field and respond as needed to ensure that tasks are performed correctly every time. Operators and mechanics know what is expected of them and are doing their best to perform work correctly every time. Operational discipline also means that leaders adhere to policies that apply to them.

The third aspect of a committed process safety culture emphasized in Vision 20/20 is maintaining a sense of vulnerability. Because major process safety incidents occur infrequently, it is natural for employees at all levels to begin to believe unconsciously that their management systems and the implementation of the management systems are so strong that incidents do not happen to them. In a Vision 20/20 committed culture, employees at all levels are aware of the hazards of their processes, knowledgeable of the systems that protect against the hazards, and recognize that incidents can result if the systems are not well executed every time. They have an ever-present sense of vulnerability and respond to early warning signs.

2.2 Vibrant Management Systems

Vision 20/20 recognizes that all employees should have a clear understanding of the expectations of senior management and those expectations should be documented and shared. In a vibrant management system there is a structure that clearly defines and documents expectations for all systems that affect process safety. These systems are implemented with operational discipline throughout the organization. The management systems promote inherently safer design principles and the principles of RBPS. The word vibrant is used to describe the management system because the system needs to be flexible and adaptable to meet ever changing needs and it needs to work well with facilities of all sizes in the organization. Companies with great process safety performance do not impose the most complex systems on their facilities; instead, they require systems that are fit-for-purpose. RBPS provides many alternatives for implementation of process safety element management systems and is a useful tool when companies are developing or refining their fit-for-purpose systems.

2.3 Disciplined Adherence to Standards

Disciplined adherence to standards starts by having clear understanding of what standards apply to various equipment and processes. Most companies have standards for new equipment or hire contractors that have the necessary standards. Too often there is cost and schedule stress with the approach outlined in company standards and shortcuts may be taken. With disciplined adherence to standards, project managers either follow the endorsed standards or use a pre-existing management system to gain approval from executives to modify the standards for a specific application.

The necessity to follow standards affects several disciplines in addition to project engineering. Process design engineers, operations, maintenance, and purchasing also need to be aware of the requirements imposed by the standards. Mechanical, process, instrumentation, engineering, operation, and maintenance all work together as a system. Everyone should play their part in the system for the system to perform as desired. For example, purchasing agents always buy pre-approved valves from pre-approved vendors and do not substitute different valves even if they are more cost effective.

Application of standards to existing equipment is especially important for disciplined adherence to standards in Vision 20/20. Most companies apply standards to new equipment and may or may not apply fit-for-service approaches for existing equipment. There are examples where, after a furnace explodes, the company establishes minimum expectations for furnaces. After a compressor fails catastrophically, minimum expectations for compressors are established. Companies with great process safety performance establish minimum standards for existing equipment before incidents occur and have systems in place to ensure that the existing equipment meets those standards. This does not imply that existing equipment must be the same as new equipment, but decisions are documented in standards that define minimum expectations for existing equipment. To be clear, an existing furnace may not need to have instrumentation that is the same as a new furnace, but a company knows the minimum requirements for all furnaces and a system is in place to periodically ensure that existing furnaces meet those minimum expectations.

It is recognized that standards do not exist for all types of equipment and engineering practices. Risk decision-making is used when applicable standards do not exist.

2.4 Intentional Competency Development

Intentional competency development is ensuring that all employees that affect process safety are fully capable of meeting the technical and cultural requirements of their jobs. Intentional competency development is critical to great process safety performance because no matter how good the culture is, how good the management system is, or how well the company attempts to adhere to standards, highly competent employees are necessary to implement those systems. Competency development needs to be a corporate priority; it requires thought and organizational commitment. It is more than just a fill-in the box exercise during the annual appraisal cycle. The need for competency may even impact required job tenure. Individuals should not be rotated out of jobs before they can contribute effectively.

The need for competency is strong at all levels of the organization, not just new employees. Executives should have basic understanding of process safety and risk management. Corporations should consider how to improve even their strongest and most experienced employees including their technical experts. Companies with great process safety performance consider many forms of competency development including continuing education, seminars and symposia in their field, mentoring, job rotation, and participation on industry committees.

It is a given that companies provide sufficient numbers of personnel to perform the work that must be done.

2.5 Enhanced Application and Sharing of Lessons Learned

Enhanced application and sharing of lessons learned is being fully aware of near misses and incidents and responding in a manner that prevents those incidents from occurring again in the same equipment and in similar equipment throughout the company. There is a culture in which employees are driven to learn from many sources, including benchmarking, near misses and incidents, and jobs done well.

Sorting through near misses, incidents, and other sources of lessons and deciding which ones to emphasize is challenging. Companies with great process safety performance have systems in place to document and respond to these learning opportunities. They do not just share incidents, they respond to the incidents, take action, and learn from the incidents.

2.6 Summary of the Five Tenets

These five tenets, combined with fervor for process safety excellence, make a powerful framework for success. Process safety results will be dramatically improved if a company has:

- A committed culture in which the executives are personally involved, managers drive excellent execution every day, and all employees maintain a sense of vulnerability;
- Vibrant management systems ingrained;
- Disciplined adherence to standards for new and existing equipment;
- Intentional competency development such that all of their employees have full technical and cultural capability to do their jobs well; and,
- Enhanced application and sharing of lessons learned including an expectation and thirst for learning from several different types of opportunities.

3. External - From an Industrial and Societal Perspective

As the CCPS Vision 20/20 committee developed the five tenets for industry, the committee recognized that issues beyond individual companies also strongly impact process safety results. Therefore, external factors must also be addressed to achieve great process safety performance. Four societal themes were identified that would help industry, and beyond, achieve the desired process safety performance. Successful fulfillment of this vision requires that industry, communities, labor, academia, and government embrace the four societal themes described below.

3.1 Enhanced Stakeholder Knowledge

In Vision 20/20 enhanced stakeholder knowledge includes two opportunities, public risk literacy and fundamental process safety education for chemical engineers.

Risk Literacy – The public, government, and industry decision makers have an enhanced understanding of the nature of risk and how it is managed. Middle or high schools introduce risk management concepts so that all of the population has a better understanding of risk. That early education also provides the background needed for more advanced concepts learned in university. Technical risk management becomes a specific area of university study in engineering and business.

Process Safety Fundamentals – Today many chemical engineering students learn about process safety after university when they are on the job. Some companies do an excellent job of educating their new engineers about process safety fundamentals, but many companies do not. In Vision 20/20, chemical engineers are required to take at least one process safety related course to receive a bachelor degree. All engineers receive education in system safety appropriate for their discipline.

3.2 Responsible Collaboration

There are significant opportunities for organizations that address industrial process safety to work together to have a positive impact on process safety results around the world. There are numerous opportunities where technical, government, labor, community, scientific, academic, and industrial organizations can work together locally, regionally, and internationally. Collaboration amongst these different groups can sometimes be challenging because the groups may have different objectives and approaches. If all the organizations could share Vision 20/20 and on this basis pull together, this will help align systems supporting process safety which would allow more time for managing process safety risk as opposed to managing systems.

Additionally, in Vision 20/20, industry, regulatory and investigatory authorities, labor organizations, communities, technical organizations, and universities work together to effectively remove barriers to reporting of incidents, develop reporting databases, and promote mutual understanding of risks and effective process safety systems.

3.3 Harmonization of Global Standards

Currently a number of organizations produce standards, guidelines, and practices for safe design, operation and maintenance of equipment. This cacophony of standards and practices is often confusing to people who are trying to design and maintain equipment to be in conformance with standards and practices, especially as they work in a global marketplace. Writing standards with the global marketplace in mind, would improve understanding, use, and efficiency in application of the resulting standards. In the future, organizations issuing accepted standards, guidelines, and practices work jointly to align and streamline practices, eliminate redundancy, and cooperatively address emerging issues.

3.4 Meticulous Verification

Companies use various verification, assessment, and auditing techniques to assure that their process safety management systems are working as intended. Generally, these verifications, audits, and assessments are made by individuals internal to the companies being assessed. There is a need for commonly available and competent third-party assessment to aid companies in identifying any shortcomings that may not be recognized by internal assessors. In Vision 20/20, third parties, including public or Non-Governmental Organizations (NGOs), are available to evaluate implementation of company process safety programs to help companies ensure that their process safety systems are robust and functioning as intended.

4. Tools Being Developed

The CCPS Vision 20/20 tenets and themes have been created. Now the challenge is to bring them to life starting with publicity and communication, to prompt people to change behaviors, and to all support a common vision for great process safety performance. A number of tools are being developed to promote the Vision 20/20 concepts. Some of the tools are intended to help companies better understand the intent of the Vision. Others will help companies identify opportunities for improvement and help them find constructive resources.

4.1 One-Page Documents

A description of each of the tenets and themes is being developed in one-page documents, which will be published on the CCPS website and worldwide to over a million readers through CCPS's distribution networks. Each document has a brief description of the tenet or theme followed by sections that further explain them.

- What does it mean? This section contains a description of activities that will take place when the tenet or theme is in place. This further illustrates the tenet or theme.
- What is the value? This section addresses the importance of the tenet or theme and how it impacts performance.
- What can I do? This section personalizes the tenet or theme by suggesting activities for individuals to promote the tenet or theme.
- Illustration – Each tenet or theme has a specific illustration that visually explains the concepts in the tenet or theme.

The one-page documents for the tenets Committed Culture and Disciplined Adherence to Standards are in Appendix I.

4.2 A day in the life

To help illustrate the changes that will occur if Vision 2020 tenets are achieved, "A day in the life of" examples have been written. These paint the picture of the difference in behavior between the current state and the envisioned state. "A day in the life of" a unit manager and an academic are attached in Appendix II.

4.3 Assessment tool

Companies that embrace the tenets in Vision 20/20 may see benefit in a tool to assess their position relative to the vision. The assessment tool should differentiate mediocre performance from great performance for each of the tenets so that companies can recognize and prioritize their biggest opportunities for improvement (5). At the same time, if the tool is to be valuable, it cannot be unnecessarily burdensome. Developing a tool that differentiates performance yet is not burdensome will be challenging.

4.4 Resources to support the tenets

Where a company assessment identifies improvement opportunities, then the next step might be identifying potential resources available to help them improve. The Vision 20/20 project is compiling a list of previously published resources (e.g. documented references, training materials) that address identified gaps. There are many resources that address the five tenets. The list will include resources from bodies supporting process safety from around the world.

5. A call to action

After an incident occurs, even if a perfect incident investigation is completed, perfect root causes are identified, and perfect action items are developed to ensure the incident does not happen again, no risk reduction occurs until the action items are completed. The same is true for Vision 20/20. No risk reduction occurs until broad industrial and societal change in behavior is accomplished.

CCPS has taken the first step; the next step is yours. It is time to get personally involved. It is time to take on board and communicate the five industry tenets and four societal themes. It is time for companies to evaluate their performance relative to the tenets and themes and take action to improve where needed. It is time for organizations that impact process safety to understand the societal themes and evaluate their contribution to process safety performance, seek collaboration, and take appropriate action.

We look forward to working together with you to achieve Vision 20/20.

6. References

[1] Control of Major Accident Hazards Regulations (COMAH), United Kingdom Health & Safety Executive, 1999 and 2005. www.hse.gove.uk/comah/.

[2] Safety Case Regulation for Offshore Oil Drilling, United Kingdom Health and Safety Executive, 2005.http://www.legislation.gov.uk/uksi/2005/3117/contents/made.

[3] Guidelines for Risk Based Process Safety, CCPS, AIChE, New York, 2007, www.wiley.com/WileyCDA/WileyTitle/productCd-0470165693.html.

[4] Conduct of Operations and Operational Discipline: For Improving Process Safety in Industry, CCPS, March 2011, http://www.wiley.com/WileyCDA/WileyTitle/productCd-0470767715.html.

[5] Organisation for Economic Co-operation and Development (OECD), Corporate Governance for Process Safety, Guidance for Senior Leaders in High Hazard Industries, OECD Environment, Health and Safety, Chemical Accidents Programme, June (2012).

[6] The Health and Safety of Great Britain – Be part of the solution, United Kingdom Health and Safety Executive, June 2009, http://www.hse.gov.uk/strategy/strategy09.pdf.

[7] Chemical Engineering Matters, IChemE, February 2013, http://www.icheme.org/media_centre/technical_strategy/ chemical engineering matters.aspx.

7. Acknowledgements

We would like to thank the CCPS Technical Steering Committee and Advisory Board who supported this project and the following committee members who were active in identifying the appropriate tenets and societal themes and creating the presented vision.

COMMITTEE MEMBERS

Cheryl Grounds – BP, Committee Chair	
Joe Allaben – Flint Hills Resources	Dave Jones – Chevron
Steve Arendt – ABS Consulting	Pete Lodal – Eastman Chemical Company
Todd Aukerman – LanXess	Jack McCavit – CCPS Staff Consultant
Scott Berger – CCPS	Louisa Nara – CCPS
Mike Broadribb – BakerRisk	Samantha Scruggs – BP
Jeff Fox – Dow Corning	Karen Tancredi - DuPont
Walt Frank – CCPS Emeritus	

Appendix I

One-Page Document for Committed Culture

VISION 20/20 | CCPS

Center for Chemical Process Safety

Committed Culture

A **Committed Culture** is one where the executives are personally involved, managers and supervisors drive excellent execution every day, and all employees maintain a sense of vigilance and vulnerability.

What does it mean?

All employees and contractors commit to "do it right", but have a plan for when it goes wrong.

- Executives personally and visibly lead process safety.
- Operators and mechanics diligently follow procedures and speak up when they suspect a problem or see an opportunity for improvement.
- Supervisors and managers verify work is done properly, intervene to correct situations, and freely communicate bad news up.

What is the value?

Doing "the right thing, the right way" shift after shift, day in and day out, prevents incidents.

A **Committed Culture** supports both safe AND reliable operations.

What can I do?

☑ Perform all tasks, correctly, every time.

☑ Support your leadership to demonstrate a commitment to process safety.

☑ Recognize that "it could happen here..."

More information is available at http://www.aiche.org/ccps/about/vision-2020

March 2014

One-Page Document for Disciplined Adherence to Standards

VISION 20/20 | CCPS

An AIChE Technology Alliance

Center for Chemical Process Safety

Disciplined Adherence to Standards

Disciplined Adherence to Standards means using recognized design, operations, and maintenance standards. These standards are followed everytime, all the time, and are continually improved.

What does it mean in 20/20?

Companies identify, document, and diligently follow standards for new designs.

Companies also identify, document, and diligently follow a set of standards applicable to existing equipment. These standards for existing equipment set the minimum expectations for design, operations, and maintenance. Companies identify and manage process safety risks arising from gaps against these standards.

As industry standards evolve, companies codify significant new learnings in their identified standards for existing equipment.

What is the value?

Use of standards promotes efficiency, reduces major accident potential, and minimizes opportunities for error in design, operations, and maintenance.

Disciplined Adherence to Standards supports both safe AND reliable operations.

What can I do?

- ☑ Ensure a system exists to effectively use industry and company standards.
- ☑ Make sure that the system applies to existing equipment.
- ☑ As a leader, assure standards are followed and respond accordingly.

More information is available at http://www.aiche.org/ccps/about/vision-2020

March 2014

Appendix II

"A day in the life of …."

Written by Todd Aukerman, LanXess

The Unit Manager 2012	The Unit Manager 2020
Arriving shortly before the overnight shift concludes, Andrew brings up the latest production numbers from the night shift on his monitor. He sees that for once everything seems to be operating smoothly. Small upsets have caused production numbers to be below target so he's happy to see a shift with no problems. Looks like he won't have to visit the control room. He observes, however, that he still needs to act on the incident that occurred at the end of yesterday's day shift. With a few clicks of his mouse, he deftly assigns it to his engineer Cameron. He allows a small sigh of relief…at least that's done. He begins the 0900 unit meeting with the usual review of production and quality numbers. Following this, he asks Cameron if he saw that he'd been assigned the investigation lead. He follows with: "Good, address it quickly." Then going around the table, he asks each of his team if they have any issues; when it's Cameron's turn, he begins to ask those present about potential safety and environmental impacts from the MOCs he has to coordinate. As some of the team members begin to respond, Andrew interjects "Folks, this isn't a safety review meeting. Schedule a separate meeting." At a site management lunch with the CEO, the plant manager and the plant safety manager approach Andrew and ask about yesterday's incident. He observes the CEO eavesdropping on the conversation and makes it clear that it's being investigated and that there were no safety or environmental impacts. That afternoon, he smiles inwardly as the CEO compliments his unit on their worker injury numbers. Andrew's has made sure that his employees always wear PPE and take their time doing maintenance tasks and similar physical activities. Although he stresses over meeting production goals, he does emphasize worker safety.	Zach arrives at his unit at 0630 every morning. His first stop is always the control room to chat with the night shift. Looking at the shift log and chatting with his operators gives him a much better feel for what's really going on; it's also allowed his operators to talk openly with him about any issues. He's glad to hear that everything was smooth last night; his production numbers have been below target; any shift without an upset is an improvement. He also asks the shift supervisor for any insight on yesterday's overpressure incident. Arriving in his office, his first order of business is to assign his unit engineer James as the investigation lead for yesterday's incident. Zach observes that this is just one of several similar incidents; he decides he's going to keep close tabs on this investigation through the on-line incident system. He starts the 0900 unit meeting with a safety moment about using a ladder at his home. After getting his team in the right frame of mind, his first topic is the incident. He's pleased that James already has scheduled the first investigation meeting and he passes along the input from the night shift. Later, he's pleased that James wants to address the MOCs right now in the meeting. Zach knows it won't take too much more time but these changes should smooth out some of the upsets; Zach readily supports the review at that time. During lunch with the CEO, Zach is a bit surprised to hear the CEO dive directly into a discussion on his unit's incidents. Nevertheless, he calmly contributes that he's similarly concerned about the repeat nature of the events, has received some insight from his night shift, and is personally monitoring this investigation. As the CEO tours his unit, Zach stays in the background as she chats with his operators. He's impressed with her knowledge of process safety and her effort to emphasize it, especially to the supervisors. He's sure her efforts will help improve process safety and production.
The Academic 2012	**The Academic 2020**
Julia strides out of the faculty fitness center and heads	Julia strides down the hall towards her first class of the

towards the campus coffee café. Her workout, combined with a good night's sleep, has left her mind clear and refreshed. It's the first day of classes and she'll need to be sharp--as an assistant professor at the university's school of business, one of the top-rated schools in the nation, she'll be facing an MBA class of aggressive "up and comers" who'll be full of questions trying to impress her and their classmates.

An hour later, she coolly welcomes the students to her Integrated Business Analysis class. She reviews the syllabus, spending a few minutes on each of the topics. She then asks for questions on the overall course content. The first question catches her slightly off guard: "Is risk management limited to financial risk?" Huh? Of course, what other kind of risk…Recovering, she replies: "Business risk focuses on financial effects; therefore, our focus is on financial risk." As she fields other questions, something gnaws at the back of her mind.

Serendipity strikes that afternoon when her department head stops by her office and informs her that the chemical engineering department has requested the business school help prepare a lecture on risk management for the process safety element in their design course. He's assigning that task to her. Her momentary annoyance fades as that gnawing in her mind returns with a vengeance…

That evening she spends a couple hours at the monthly meeting of the local chapter of a major outdoor conservancy organization. Being an avid outdoors person, she's passionate about preserving nature's beauty. Her mind wanders a bit as the officers plod through the regular standing business items but snaps back to reality when the president mentions the meeting's main topic:"risk" from a local chemical facility. There's that word again…

new semester. As a senior faculty member in one of the nation's premier business schools, she's been entrusted with one of the most important classes in the MBA program. She knows she'll be peppered with questions but she also knows she'll be broadening their vision of what makes a successful business leader. Walking purposefully into the classroom, she casually announces "Welcome to Business Risk Management: An Integrated Approach." Fifty minutes later, she concludes forcefully with "As future business leaders, you'll be confronted with a myriad of risks. Your job will be to manage that risk---all of it. If you focus just on financial risk, at best you will limit your success, at worst you will fail miserably. Apply the principles learned in this course and you will be prepared to make risk decisions competently; ignore these principles not only at your *own* peril but to the peril of the lives and well-being of your co-workers, neighbors, friends, and family."

After class, she drives to a local high school in a low-income area to deliver the required "Introduction to Risk Concepts" element of secondary education to the sophomore chemistry class. Risk management has become her passion, propelling her academic career to unforeseen heights. She knows that improving risk literacy in society ultimately promotes more effective risk management for society as a whole. She considers it an honor and a duty to be today's "guest teacher."

She spends her evening preparing for an upcoming "verification" of the process safety management systems at a local chemical facility. She represents the local chapter of a major conservancy group on the community "verification team." As she reviews the latest on-line facility data, casually noting an incident just the day before, she ponders the unique connection to her own life: much of her essential outdoor equipment is made with products produced by that facility…

Pipeline Risk Assessment: Third Party Interference

Nguyen, A., PSRG
Weber, R., PSRG LATIN AMERICA LTDA.

According to Wall Street Journal review, there were 1,400 pipeline spills and accidents in the US 2010-2014. Four in every five pipeline accidents are discovered by local residents, not the companies that own the pipelines. Nowadays, third party interference caused 50% accidents in pipeline industry. The causes of the third party interference event are hard to be forecasted and sometimes beyond control of pipeline operator. In order to facilitate risk assessment, it is crucial to classify the threats of third party interference. By grouping the risk of third party interference, the decision makers can make appropriate safety-related actions to prevent the incidents and to mitigate their possible consequences.

Yuxing Li

August 28, 2014

A. Development Background

B. Basic Knowledge

C. Management platform

D. forecasting

中国油、气管网图

Oil and gas pipeline business has achieved rapid development all over China since the tenth five-year plan.

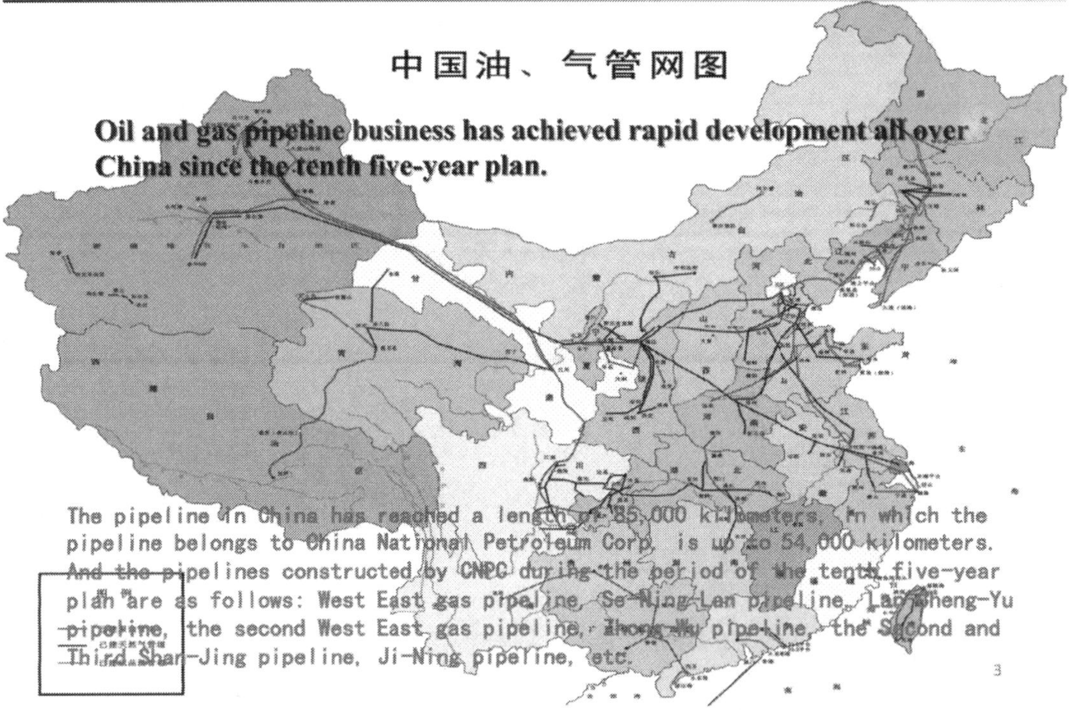

The pipeline in China has reached a length of 85,000 kilometers, in which the pipeline belongs to China National Petroleum Corp. is up to 54,000 kilometers. And the pipelines constructed by CNPC during the period of the tenth five-year plan are as follows: West East gas pipeline, Se-Ning-Lan pipeline, Lan-Cheng-Yu pipeline, the second West East gas pipeline, Zhong-Wu pipeline, the Second and Third Shan-Jing pipeline, Ji-Ning pipeline, etc.

3

1. Pipeline Safety Management - Cannot Be Ignored

- 1971-1976 Oil pipeline rupture accident took place for three times in the Northeast China
- 1992 Rupture accident took place for 14 times during the pressure testing process of Lun-Ku oil pipeline
- 1971-1990 Pipeline failure accident took place for 108 times in the south trunk line of the Sichuan gas pipeline network
- A serious oil pipeline leak and explosion accident took place in Huangdao district last year, during which 62 people died and 136 people were hurt.

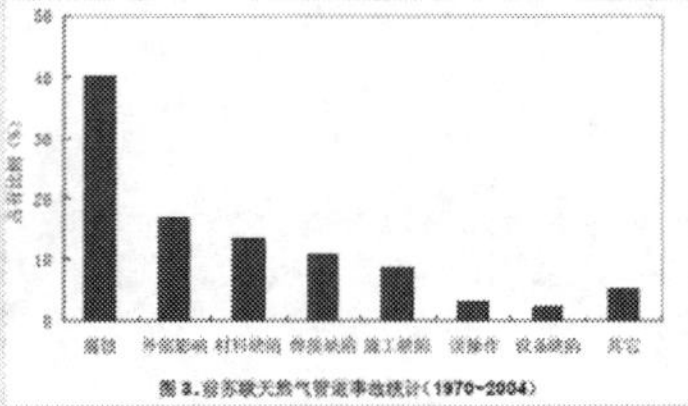

1. Pipeline Safety Management - Cannot Be Ignored

The Iceberg Theory

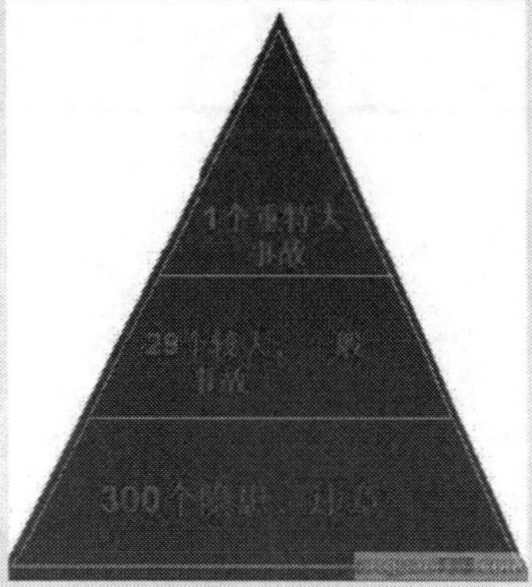

300:29:1

Herinrich's Law

PIPELINE INTEGRITY MANAGEMENT

2. The Development Process of Pipeline Integrity Management

- The safety assessment and pipeline integrity management first appeared in America in 1970s.

In 1987, Amoco started to assess risk of pipeline and tanks by using Kent's scoring method.

Colonial focused on the safety and reliability of the pipeline.

In 2002, the Office of Pipeline Safety in the Department of Transportation issued the management responsibilities of the pipeline operators and checked their integrity management plan and the implementation status.

6

PIPELINE INTEGRITY MANAGEMENT

2. The Development Process of Pipeline Integrity Management

- In 1980s, practical assessment software was explored in Europe, which made the risk assessment more quantitative and intelligent.

In the early 1990s, BG Group established a set of management methods and engineering files.

The implementation of pipeline integrity management helped the SNAM reduce the maintenance cost by one third.

7

2. The Development Process of Pipeline Integrity Management

- In 1990s, Mexico, Canada and some other countries started to explore and imply the pipeline risk management technology.

From 1970s to 1990s, pipeline integrity management (PIM) mainly focused on risk assessment, inspection, maintenance and so on. However, the whole management system was still not well organized until the mid 1990s, during which PIM achieved further development.

During the process of extending the trunk pipeline, NOVA quantified the factors affecting the net present value over the 35 years to make assessment of the pipeline crossing schemes, so that it could select the best one.

From 1980s to the mid-1990s, Enbridge did research on the pipeline integrity and the risk assessment. Thus, the company's PIM process, including planning, implementing, summarizing and improving, was formed. And these four aspects made up the circulation of the whole process.

PIPELINE INTEGRITY MANAGEMENT

- A. Development Background
- B. Basic Knowledge
- C. Management Platform
- D. Forecasting

1. PI and PIM

What is pipeline integrity (PI)?

- Pipelines keep in the safe and reliable service condition.
- Pipelines are integral both physically and functionally. And the pipelines are controlled.
- The pipeline operators has taken measures to prevent accidents. And also, they will keep on doing it.
- Pipeline integrity is closely related to the process of designing, construction, operation, maintenance and management.

10

1. PI and PIM

What is Pipeline Integrity management (PIM)?

According to the constantly changing factors, pipeline company recognize and evaluate the risk factors during the operation. And then, corresponding measures would be taken to control them and improve the pipeline operating state so that the risk standard would be in the acceptable range.

The information about the pipeline integrity state would be acquired by monitoring, detecting, testing and so on. And the information could be used to distinguish the threat and evaluate the applicability. Finally, the pipeline could get continuous improvement and the pipeline accidents could be reduced and prevented. As a result, economically and reasonably, the pipeline could achieve safe operation.

11

2. The Content of PIM

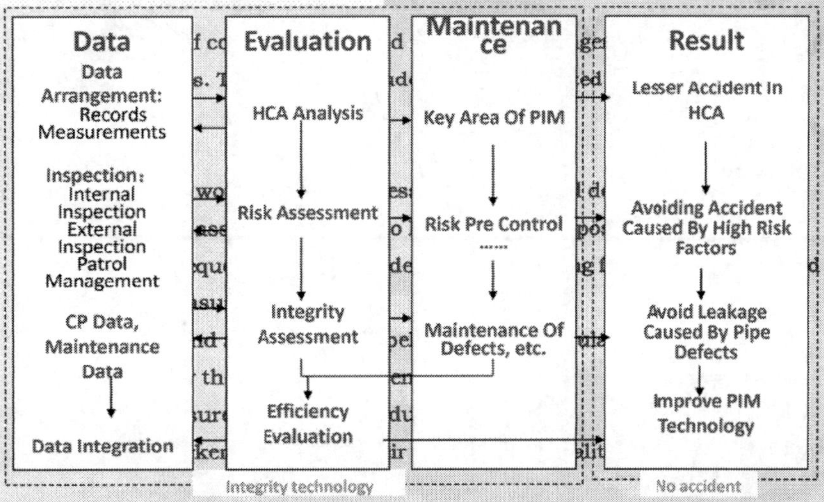

Overall, PIM is a comprehensive method to keep the pipeline system in an integral status both in structure and function. It is to prevent troubles before it happens.

12

3. The Principle of PIM

- The theory and idea of PIM should be kept in mind.
- PIM should be dynamic and flexible according to different pipelines.
- All of the information and data related to PIM should be analyzed and arranged.
- The PIM should be sustained and in circulation.
- All kinds of newly developed technologies should be considered or used in the PIM.
- Corresponding to the PIM, the manned institution and management process should be established.

13

4. The Links of PIM

For now, China National Petroleum Corp. has established the PIM system according to the present situation of Chinese pipe.

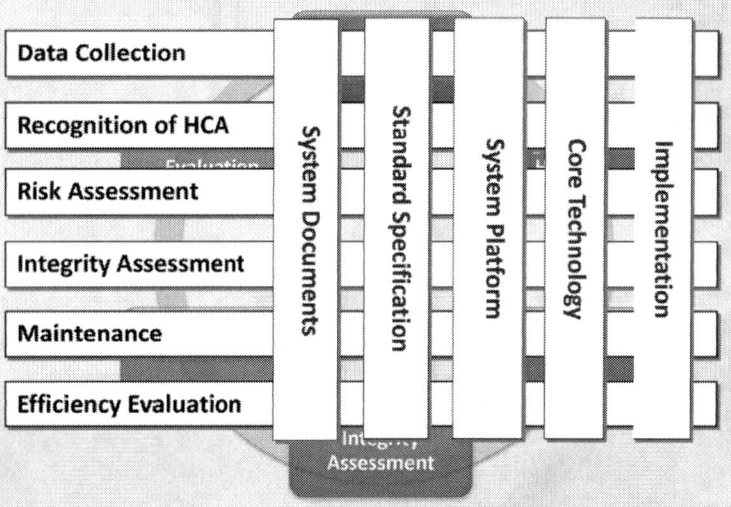

5. Brief Introduction of PIM links

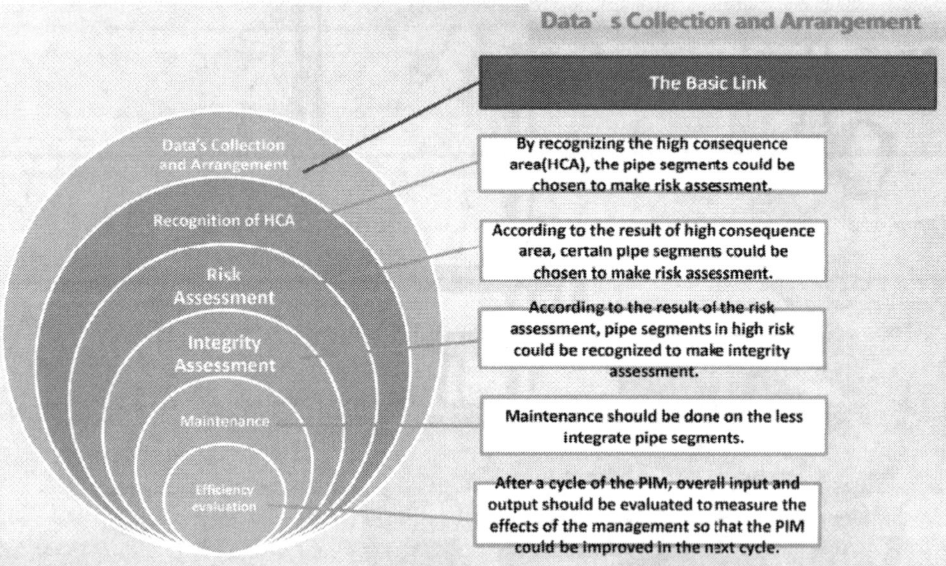

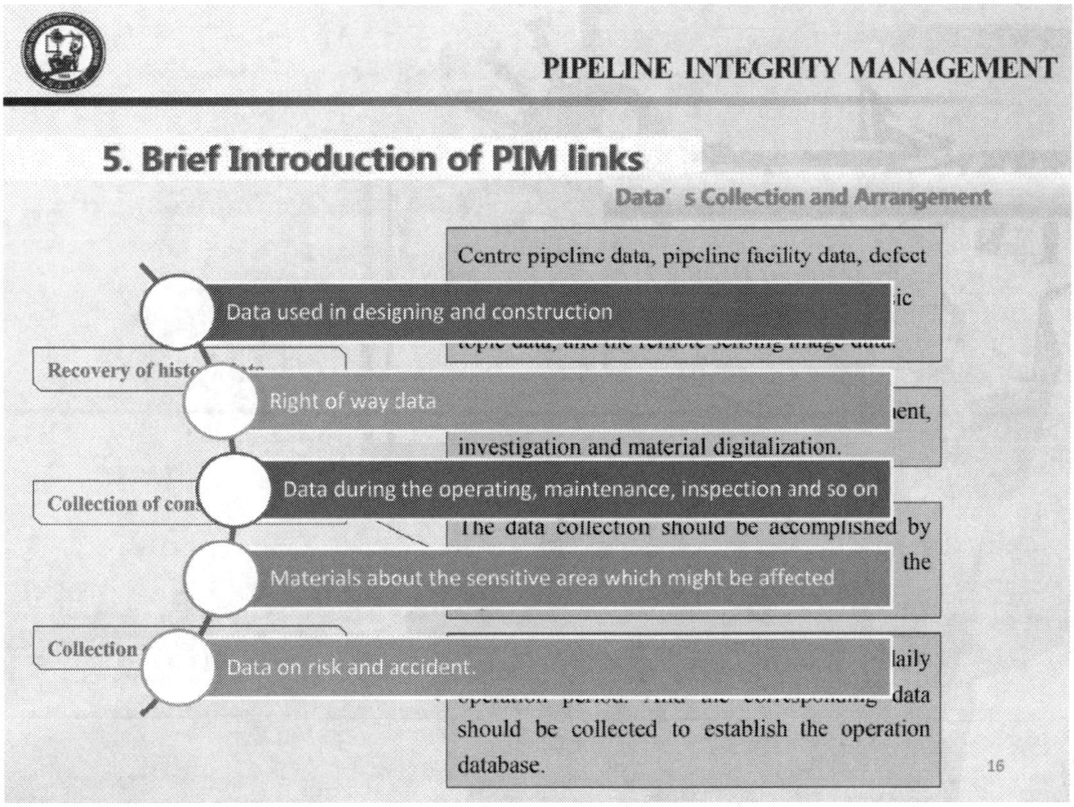
PIPELINE INTEGRITY MANAGEMENT
5. Brief Introduction of PIM links
Data' s Collection and Arrangement
Centre pipeline data, pipeline facility data, defect
Data used in designing and construction
Right of way data
investigation and material digitalization.
Data during the operating, maintenance, inspection and so on
Materials about the sensitive area which might be affected
Data on risk and accident.
should be collected to establish the operation
database.
16

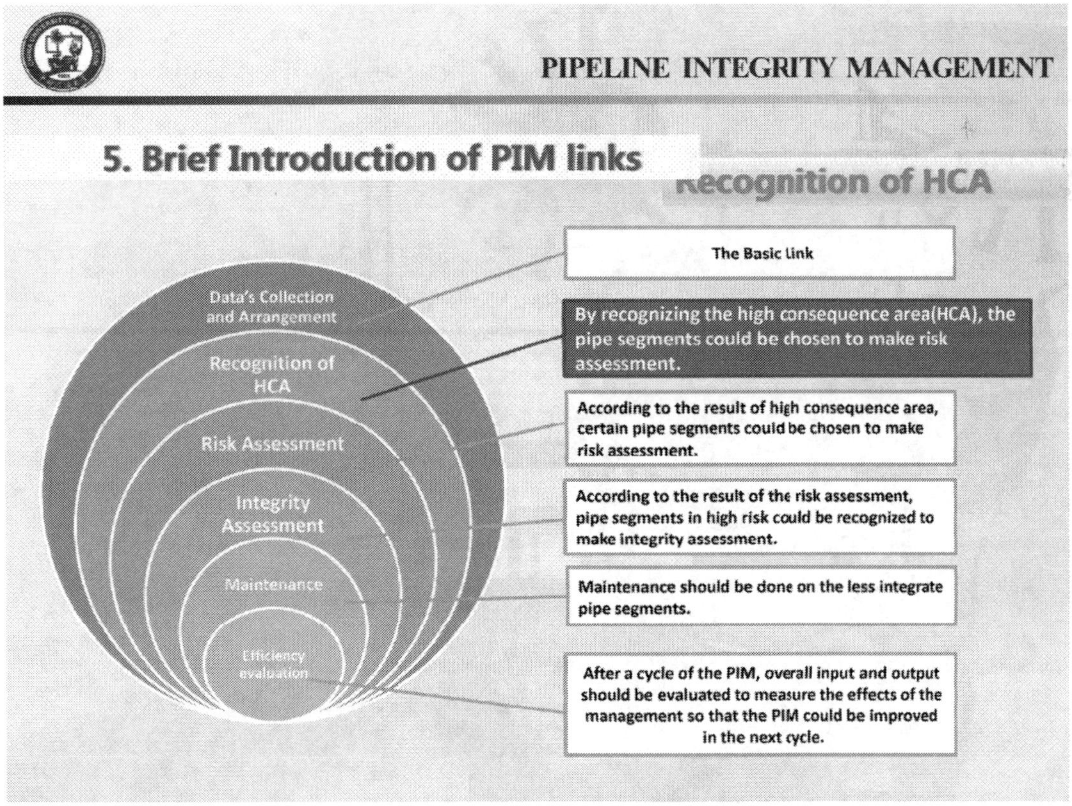
PIPELINE INTEGRITY MANAGEMENT
5. Brief Introduction of PIM links
Recognition of HCA
Data's Collection and Arrangement
Recognition of HCA
Risk Assessment
Integrity Assessment
Maintenance
Efficiency evaluation
The Basic Link
By recognizing the high consequence area(HCA), the pipe segments could be chosen to make risk assessment.
According to the result of high consequence area, certain pipe segments could be chosen to make risk assessment.
According to the result of the risk assessment, pipe segments in high risk could be recognized to make integrity assessment.
Maintenance should be done on the less integrate pipe segments.
After a cycle of the PIM, overall input and output should be evaluated to measure the effects of the management so that the PIM could be improved in the next cycle.

PIPELINE INTEGRITY MANAGEMENT

5. Brief Introduction of PIM links

Recognition of HCA

Recognition of HCA could be accomplished by making an inspection tour or software calculation.

GIS (Geographic Information Systems) technology has been used in the HCA analysis.

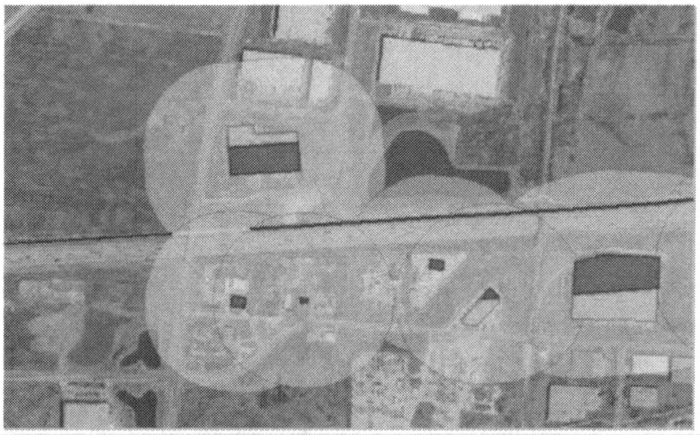

High Definition Image of HCA

18

PIPELINE INTEGRITY MANAGEMENT

5. Brief Introduction of PIM links

Risk Assessment

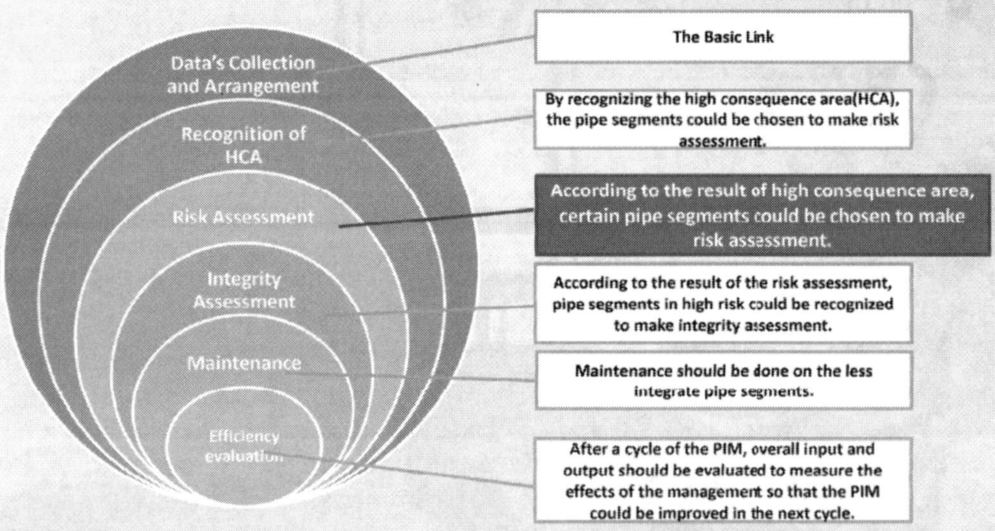

5. Brief Introduction of PIM links

Risk Assessment

Pipeline risk assessment is to identify the risk factors threatening the safe operation of pipelines, to evaluate the possibility and the consequence of the accident and then, to take risk control measures.

Risk assessment is the core link of PIM。

Qualitative Risk Assessment

- Safety Check List
- Preliminary Hazard Analysis
- Hazard and Operability Study
- Fault Tree Analysis
- Accident Tree Analysis
- Risk Matrix

Semi-Quantitative Risk Assessment

- Kent's Pipeline Risk Scoring Method

Quantitative Risk Assessment

- Probabilistic Risk Analysis

PIPELINE INTEGRITY MANAGEMENT

5. Brief Introduction of PIM links

Risk Assessment

Qualitative Risk Assessment——Risk Matrix

		Failure consequence			
Probability of failure		Light	Heavy	Severe	Disastrous
	frequently	Mid	High	High	High
	Occasional	low	Mid	High	High
	Unlikely	Negligible	Low	Mid	High
	Impossible	Negligible	Negligible	Low	Mid

	Level	Description
Probability level	I	Almost never happens
	II	Not so easy to happen
	III	The accident happens occasionally
	IV	The accident happens regularly and will regular occur in future
Consequence level	I	Can be ignored and will not cause system damage or people injuries
	II	It may cause light damage to the system or slight injures to persons
	III	It may cause serious damage to the system or severe injury to persons
	IV	It may lead to system loss or persons' death

PIPELINE INTEGRITY MANAGEMENT

5. Brief Introduction of PIM links

Risk Assessment

Qualitative Risk Assessment——Risk Matrix

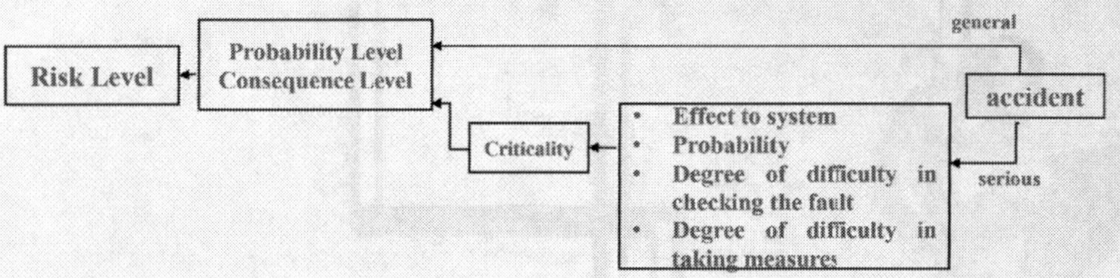

For the accidents with high risk levels, which might lead to death, criticality analysis is necessary. The criticality value can help to get the probability level and the consequence level.

$$C = E \times P(\alpha \times \beta)$$

PIPELINE INTEGRITY MANAGEMENT

5. Brief Introduction of PIM links

Risk Assessment

Semi-Quantitative Risk Assessment
——Kent' s Pipeline Risk Scoring Method

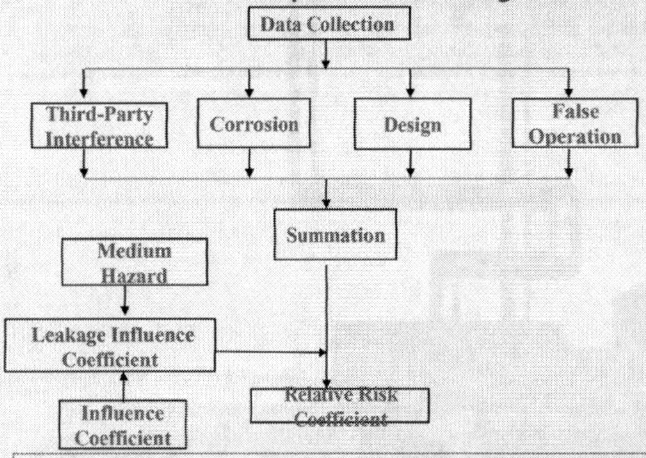

HYPOTHESIS

- Independence
- Worst State
- Relativity
- Subjectivity
- Score Definition
- Homogeneity

$$\text{Relative Coefficient} = \frac{\text{Third-Party Interference} + \text{Corrosion} + \text{Design} + \text{False Operation}}{\text{Leakage Influence Coefficient}}$$

5. Brief Introduction of PIM links

Risk Assessment

Semi-Quantitative Risk Assessment
——Kent' s Pipeline Risk Scoring Method

Relative Coefficient	First Level Index
	Third-party interference Corrosion Design False operation Leakage influence coefficient

First level	Second level index
design	Pipeline safety coefficient System safety coefficient Fatigue effect Water attack potential effect System water pressure test Soil moving The ability of designer

First level	Second level index
Third-parity interference	Minimum embedded depth Activities on the ground Route condition Facility on the ground Patrol frequency Alarm system

First level	Second level index	Third level index	Forth level index
corrosion	Internal corrosion	Product corrosion Internal corrosion protection	
	External corrosion	Atmospheric corrosion	facility Atmosphere type Coating layer
		Corrosion of buried metal	Cathodic protection Coating condition Soil corrosion Other kinds of metal Induced AC Stress corrosion Pipe-ground potential test stake
	Corrosion detection	Inspection tool Inspection frequency	

PIPELINE INTEGRITY MANAGEMENT

5. Brief Introduction of PIM links

Risk Assessment

Semi-Quantitative Risk Assessment
——Kent' s Pipeline Risk Scoring Method

First level	Second level index	Third level index
False operation	False design	Hazard recognition Possibility of MAOP security system Material selection check
	False construction	Testing Material Connection Backfilling transporting Coating layer
	False running	Process flow communication Poison check Safety plan check training Mechanical failure protection
	False maintenance	Documentation programming Maintenance instruction Maintenance frequency Maintenance check Technique of maintenance personnel Responsibility of workers

First level	Second level index	Third level index	Forth level index
Leakage influence coefficient	Medium hazard	Dramatic hazard	Combustibility reactivity toxicity
		Long-term hazard	
	Diffusion coefficient	Liquids or other leaks population density	

5. Brief Introduction of PIM links

Risk Assessment

Quantitative Risk Assessment
——Probabilistic Risk Analysis

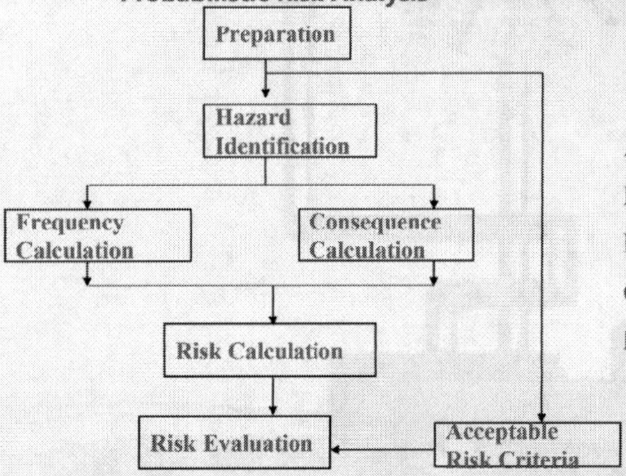

Procedure of Probabilistic Risk Analysis

Probabilistic Risk Analysis mainly consist of Hazard Identification, Risk Analysis, Risk Calculation as well as Risk assessment.

PIPELINE INTEGRITY MANAGEMENT

5. Brief Introduction of PIM links

Risk Assessment

Quantitative Risk Assessment
——Probabilistic Risk Analysis

Risk Value=Failure Probability×Failure Consequence

Failure Probability	Failure Consequence
• Historical Database • Data Modification	• Consequences of Failure Event Tree

Risk Value,

$$R_j = L_{ij} \times (l_j) \times \sum_{k=1}^{3} [(P_{R/F} \cdot Q_{R,k} + P_{L/F} \cdot Q_{L,k})_{ij}]$$

Where,

$$L_i = [\alpha(FR) + (1-\alpha)A + H]$$

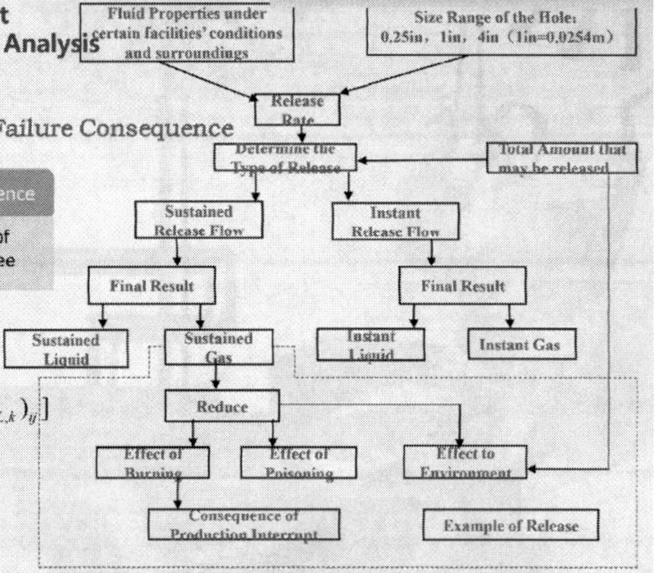

Calculation Chart of Dangerous Fluid Release

PIPELINE INTEGRITY MANAGEMENT

5. Brief Introduction of PIM links

Risk Assessment

Quantitative Risk Assessment
——Probabilistic Risk Analysis

Individual Risk Assessment——this index is to express the influence of casualties. And the value, in fact, is the death probability of a certain point around the pipeline.

- In Britain, one of the common methods is to divide the risk levels.

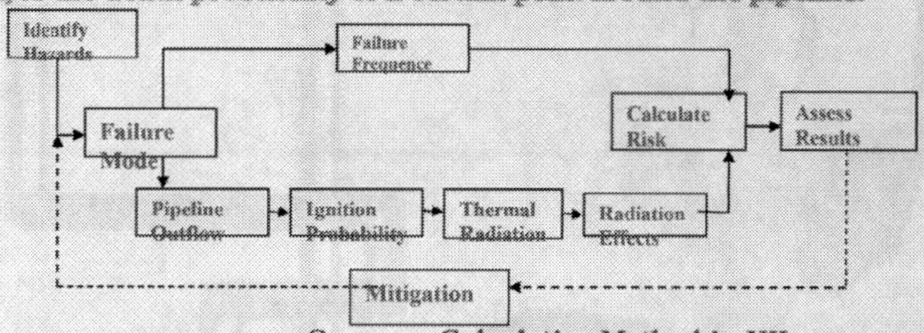

Common Calculation Method in UK

- **The approach used by PEMEX Gas is to grade the risk level by capital cost transferred by comprehensive consume per kilometer per year of operation. (The same method used in China)**

$$ROF\$ = LOF\$ \times COF\$ \begin{cases} LOF\$ = EC\$ + IC\$ + TP\$ + GM\$ + DM\$ + SO\$ \\ COF\$ = PEOD\$ + DAM\$ + IOP\$ \end{cases}$$

28

PIPELINE INTEGRITY MANAGEMENT

Integrity Assessment

5. Brief Introduction of PIM links

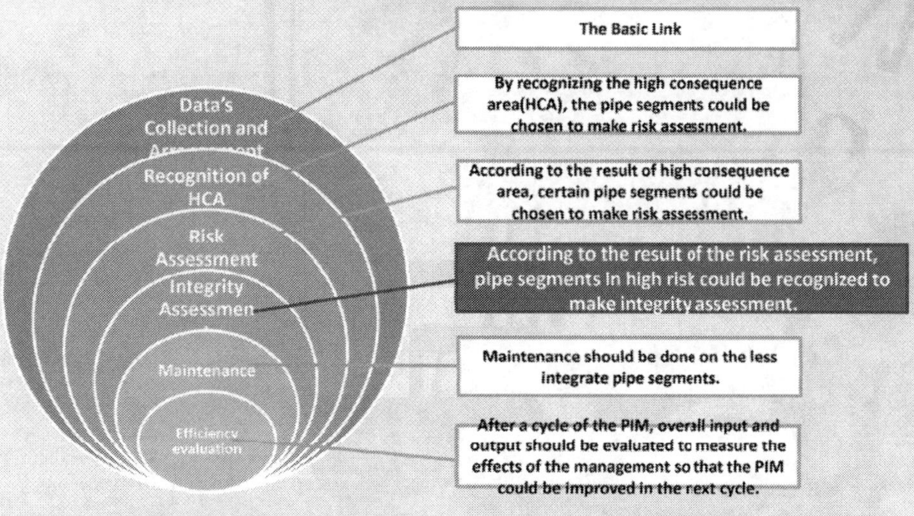

PIPELINE INTEGRITY MANAGEMENT

5. Brief Introduction of PIM links

Integrity Assessment

Integrity assessment is a core step in PIM.

It is to acquire the default information by special technology to evaluate the pipeline integrity. In this way, we could know how safe the pipeline is and what' s the safety level the pipeline can reach at present and in future.

Internal Inspection Method

- To recognize the feature of faults
- The abnormal geometry、metal loss、cracks, etc.

Pressure Testing Method

- to evaluate the PI in certain pressure
- to expose the default that cannot tolerate the pressure

Direct Method

- to evaluate the effect of external corrosion, internal corrosion, and the stress corrosion

PIPELINE INTEGRITY MANAGEMENT

5. Brief Introduction of PIM links

Integrity Assessment

Overall, integrity assessment is based on the result of risk assessment. The high risk segments would be inspected so that the remaining strength and remaining life would be calculated. At last, the maintenance plan would be made.

Remaining Strength → Integrity Assessment (Default Assessment) ← Remaining Life

5. Brief Introduction of PIM links

Integrity Assessment

Corrosion defect remaining life assessment

Remaining life of corrosion default

- Folias semi-empirical formula
- ASME B31G
- RSTRENG 0.85 dL
- RSTRENG effective area
- DNV RP F101

Folias semi-empirical formula $M_s = \dfrac{1-A/A_0}{1-A/(A_0 M_t)}$

Circular fracture stress $\sigma = \sigma_f M_s$

① **ASME B31G-1984**

$M_t = (1+0.4845\lambda^2)^{1/2}$ $\lambda = \dfrac{1.285L}{\sqrt{D_i t}}$

② **RSTRENG 0.85dL**

The flow stress is expressed as σ_s**+68.95MPa**
Area is expressed as $0.85Ld$

③ **RSTRENG effective area**
To evaluate the interactive effects of defects, however, it is not that serviceable

④ **DNV RP F101**

σ_f **is replace by** M_t σ_b **is replaced by** Q

where $Q = \sqrt{1+0.31L^2/(Dt)}$

PIPELINE INTEGRITY MANAGEMENT

5. Brief Introduction of PIM links

Integrity Assessment

Crack defect =remaining strength assessment

Crack defect remaining life

- FAD Chart
- First level evaluation（simplified）
- Second level evaluation（normal）
- Third level evaluation（accurate）

It is mainly based on the theory of partial ligament-plastic damage and crack fracture damage. The main methodology is based on Failure Assessment Diagram (FAD).

S_r **It is to measure the partial ligament-plastic damage.**

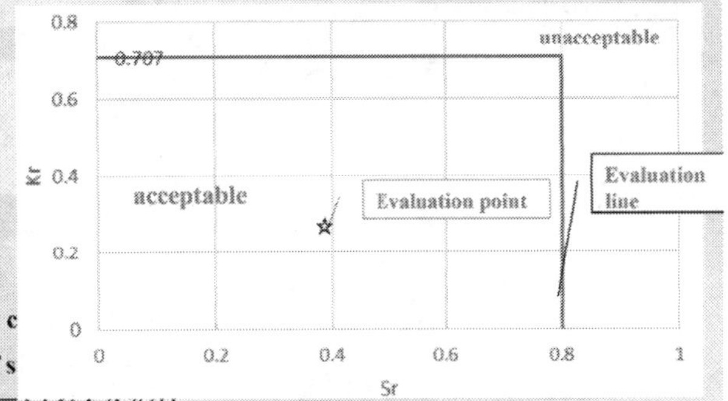

FAD chart of 1A level

For the defect to which led by c

crack-tip regional intensity of s

$K_I = \ldots$

For the circumferential defects，Kastner method is generally used.

Integrity Assessment

5. Brief Introduction of PIM links

Recommended evaluation method of defect under internal pressure load

Remaining strength assessment

Defect type	Criterion	Defect type	Criterion
corrosion	SY/T 6151-2009 SY/T 6477-2000 SY/T 10048-2003 ASME B31G Modified ASME B31G RSTRENG DNV-RP-F101 BS 7910 API RP 579	scratching	BS 7910 API RP 579
sag	—	Manufacturing defect	BS 7910 API RP 579
Ring-shaped welding defect	BS 7910 API RP 579	Straight weld defect	BS 7910 API RP 579
cracking	BS 7910 API RP 579	Environmental cracking	BS 7910 API RP 579
Assembling defect	BS 7910 API RP 579	Defect interaction	BS 7910 API RP 579

PIPELINE INTEGRITY MANAGEMENT

Integrity Assessment

5. Brief Introduction of PIM links

Probabilistic way to predict remaining life

Remaining life assessment

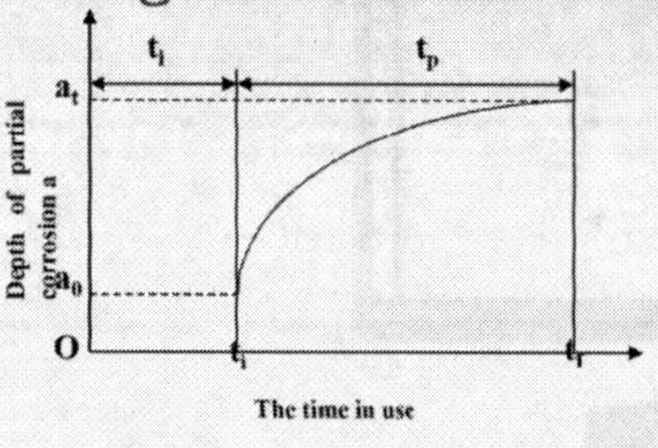

Formula of calculating the pipeline failure pressure

$$p_f = \frac{1.8UTSt}{D}\left(\frac{1-\frac{d_0+V_r(T-T_0)}{t}}{1-\frac{d_0+V_r(T-T_0)}{t}M^{-1}}\right)$$

Limit state function

$$z = p_f - p_{op}$$

When z>0, the pipeline be safe when it runs, which means that the failure frequency is the frequency of z>0.

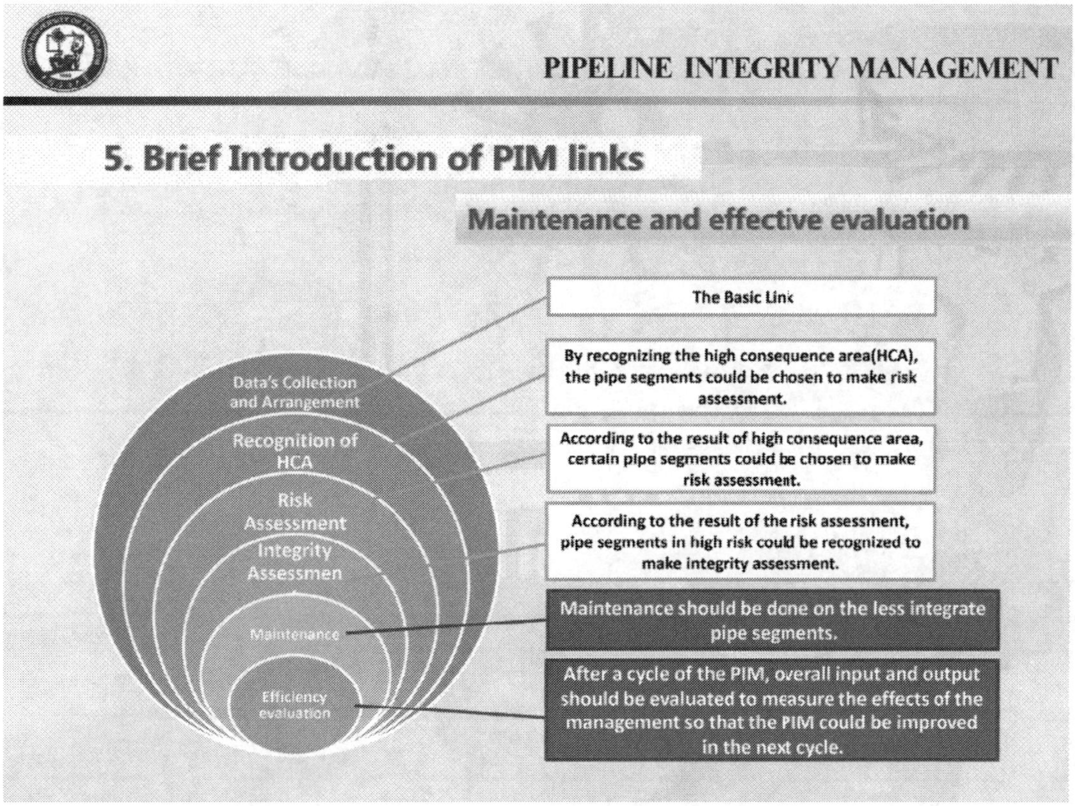
PIPELINE INTEGRITY MANAGEMENT
5. Brief Introduction of PIM links
Maintenance and effective evaluation
Data's Collection and Arrangement
Recognition of HCA
Risk Assessment
Integrity Assessmen
Maintenance
Efficiency evaluation
The Basic Link
By recognizing the high consequence area(HCA), the pipe segments could be chosen to make risk assessment.
According to the result of high consequence area, certain pipe segments could be chosen to make risk assessment.
According to the result of the risk assessment, pipe segments in high risk could be recognized to make integrity assessment.
Maintenance should be done on the less integrate pipe segments.
After a cycle of the PIM, overall input and output should be evaluated to measure the effects of the management so that the PIM could be improved in the next cycle.

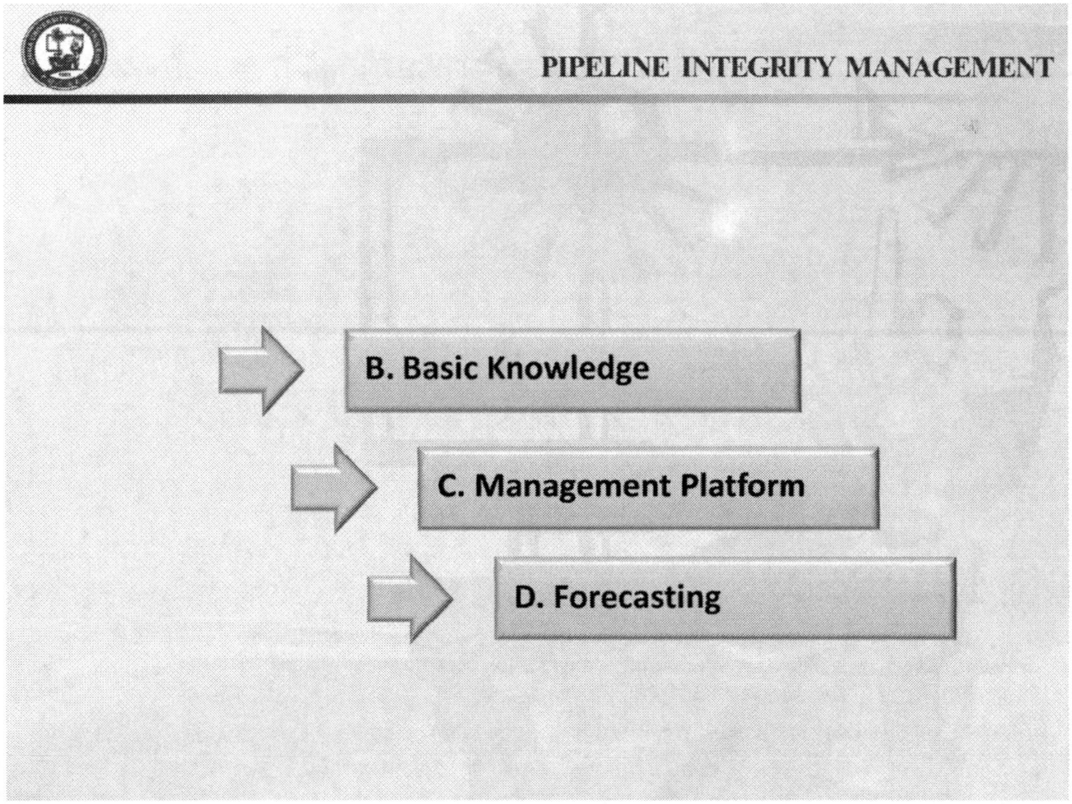
PIPELINE INTEGRITY MANAGEMENT
B. Basic Knowledge
C. Management Platform
D. Forecasting

PIPELINE INTEGRITY MANAGEMENT

The content and achievement

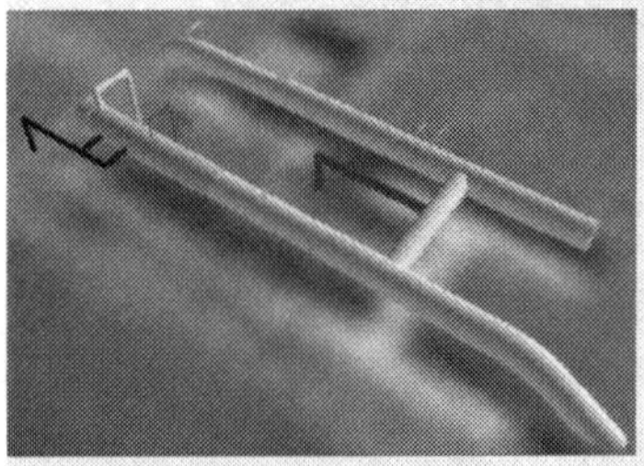

In 2001, PIM was introduced in China

- The pipeline itself
- Geologic hazard and surroundings
- Antiseptic effectiveness
- Station facilities
- UGS and well field

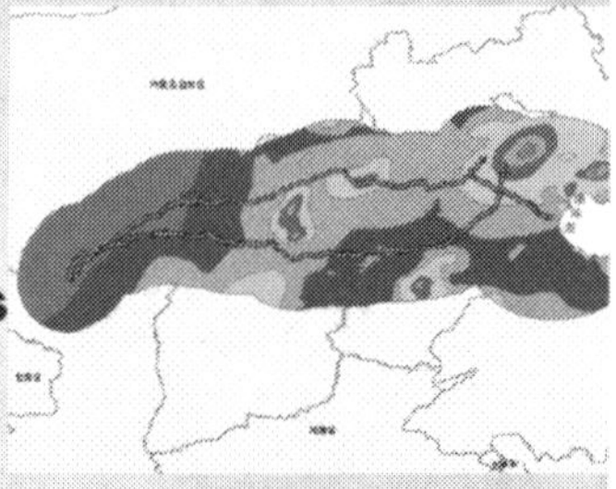

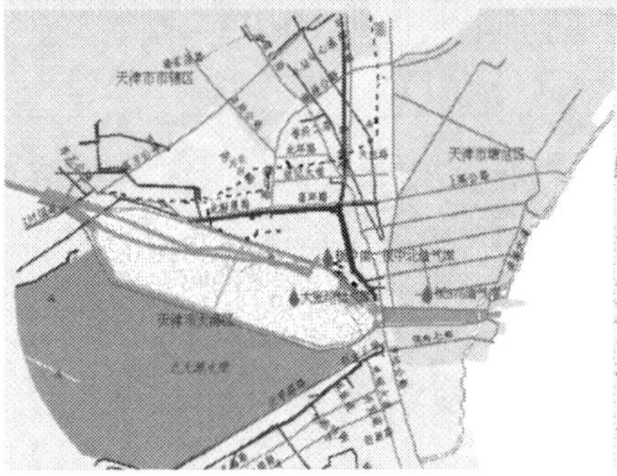

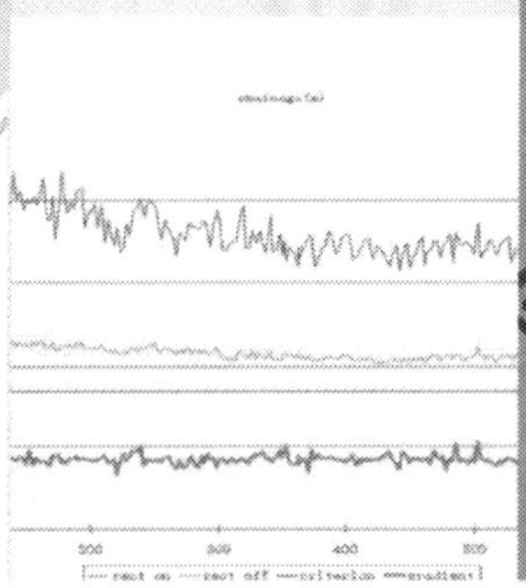

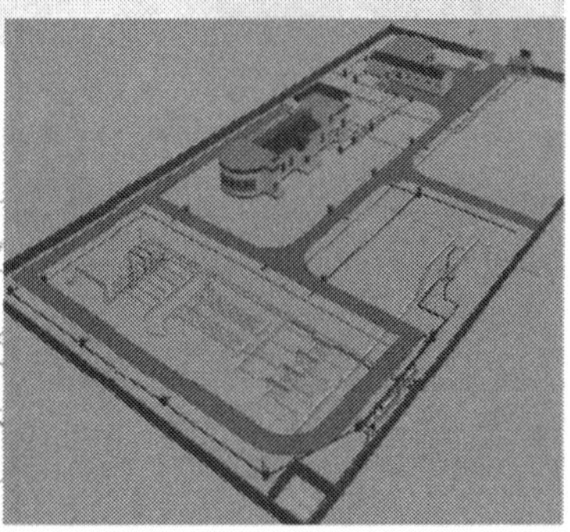

PIPELINE INTEGRITY MANAGEMENT

The content and achievement

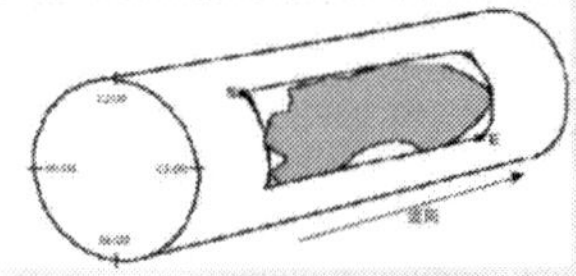

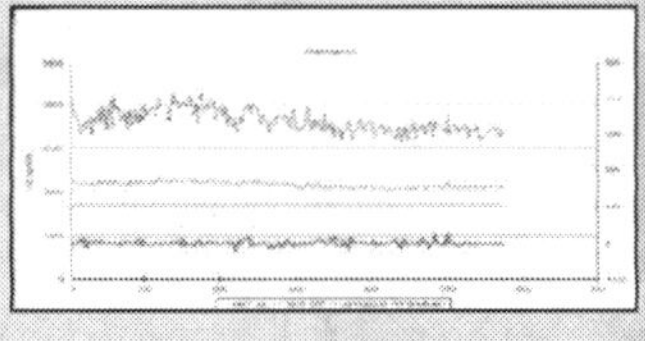

- PIM system has been established and a number of enterprise standards, including the construction period and runtime data collection standards
- Abaqus, API579, DNV and some other risk and integrity assessment technology was introduced. And practical methods were also summarized in Shan-Jing pipeline.
- To evaluate the potential risk, a number of ways were used, such as pipeline detection, external coating testing, ultrasonic corrosion monitoring, HCA analysis and so on.
- Risk assessment is carried out according to the risk detected previously or analyzed.
- The PIM system platform has been built, and a large amount of data has generated to form the database. What's more, risk assessment, integrity assessment can be accomplished by the platform.
- Gradually form a cyclic control system

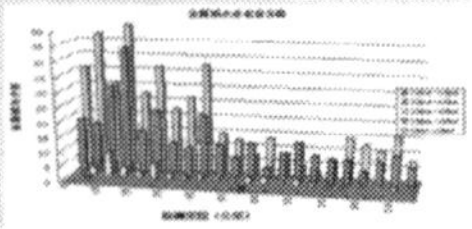

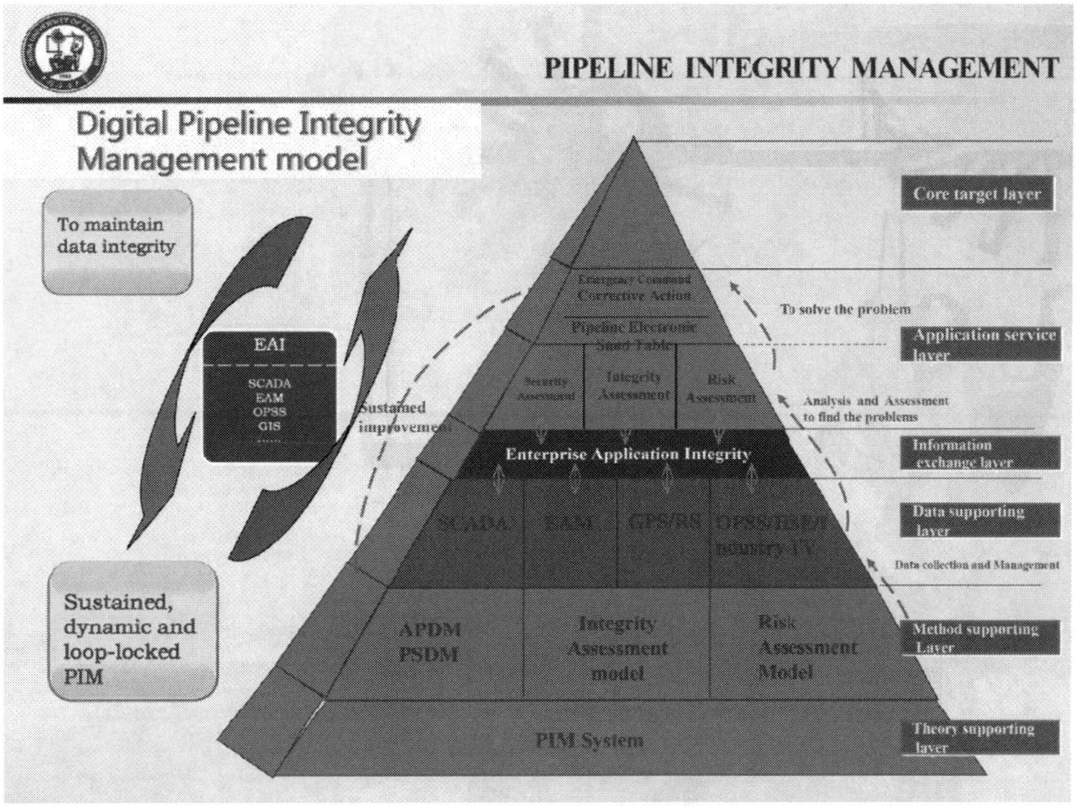
PIPELINE INTEGRITY MANAGEMENT
Digital Pipeline Integrity Management model
To maintain data integrity
EAI
SCADA
EAM
OPSS
GIS
Sustained improvement
Sustained, dynamic and loop-locked PIM
Emergency Command Corrective Action
Pipeline Electronic Sand Table
Security Assessment
Integrity Assessment
Risk Assessment
Enterprise Application Integrity
SCADA
EAM
GPS/RS
OPSS/TBE/Industry TV
APDM PSDM
Integrity Assessment model
Risk Assessment Model
PIM System
Core target layer
To solve the problem
Application service layer
Analysis and Assessment to find the problems
Information exchange layer
Data supporting layer
Data collection and Management
Method supporting Layer
Theory supporting layer

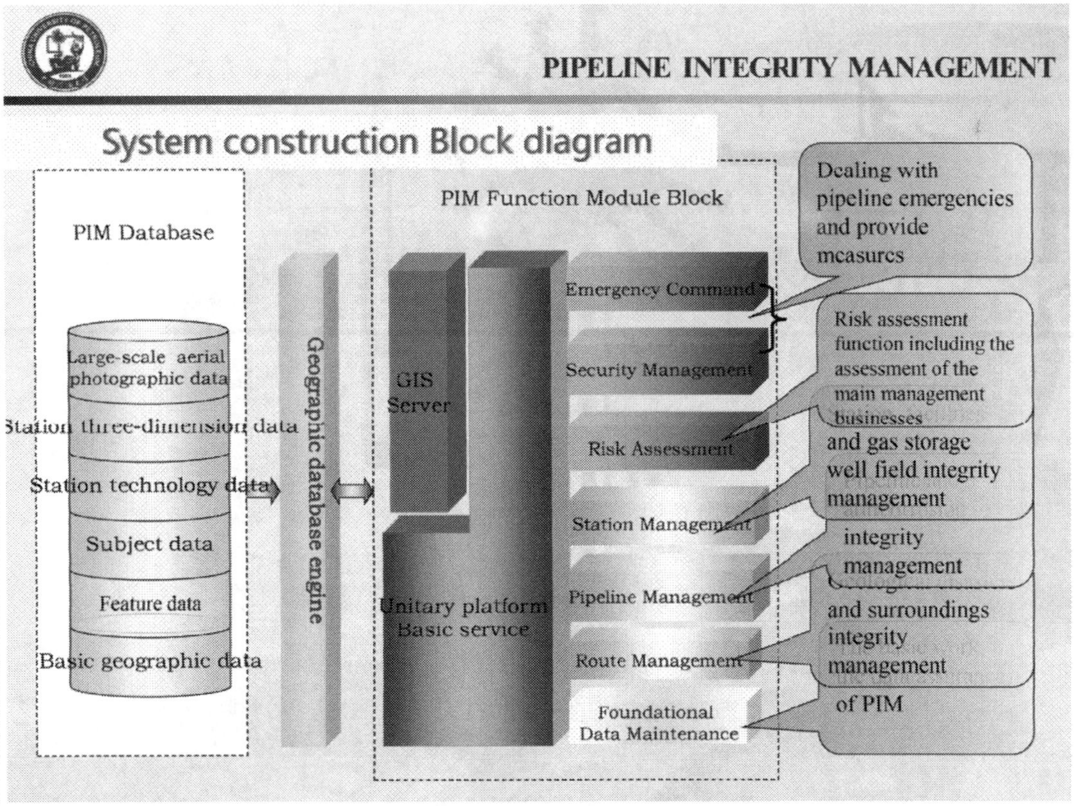
PIPELINE INTEGRITY MANAGEMENT
System construction Block diagram
PIM Database
Large-scale aerial photographic data
Station three-dimension data
Station technology data
Subject data
Feature data
Basic geographic data
Geographic database engine
PIM Function Module Block
GIS Server
Unitary platform Basic service
Emergency Command
Security Management
Risk Assessment
Station Management
Pipeline Management
Route Management
Foundational Data Maintenance
Dealing with pipeline emergencies and provide measures
Risk assessment function including the assessment of the main management businesses
and gas storage well field integrity management
integrity management
and surroundings integrity management
of PIM

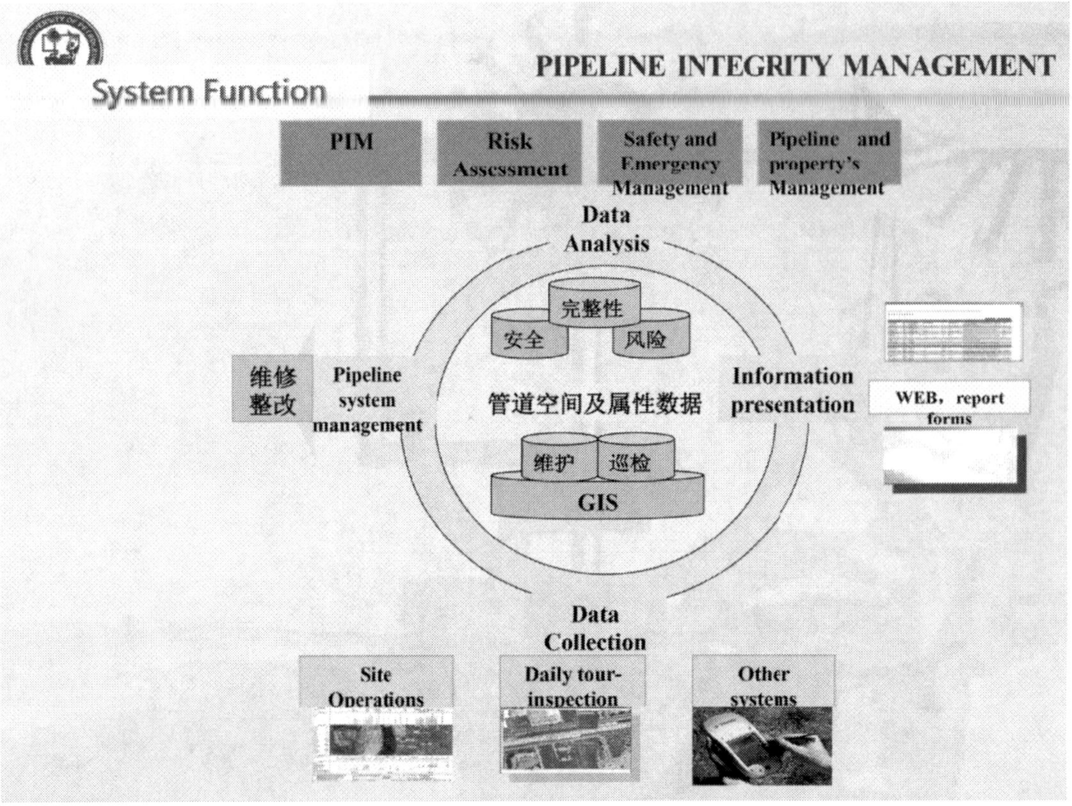
PIPELINE INTEGRITY MANAGEMENT
System Function
PIM
Risk Assessment
Safety and Emergency Management
Pipeline and property's Management
Data Analysis
完整性
安全
风险
维修 整改
Pipeline system management
管道空间及属性数据
Information presentation
WEB，report forms
维护
巡检
GIS
Data Collection
Site Operations
Daily tour-inspection
Other systems

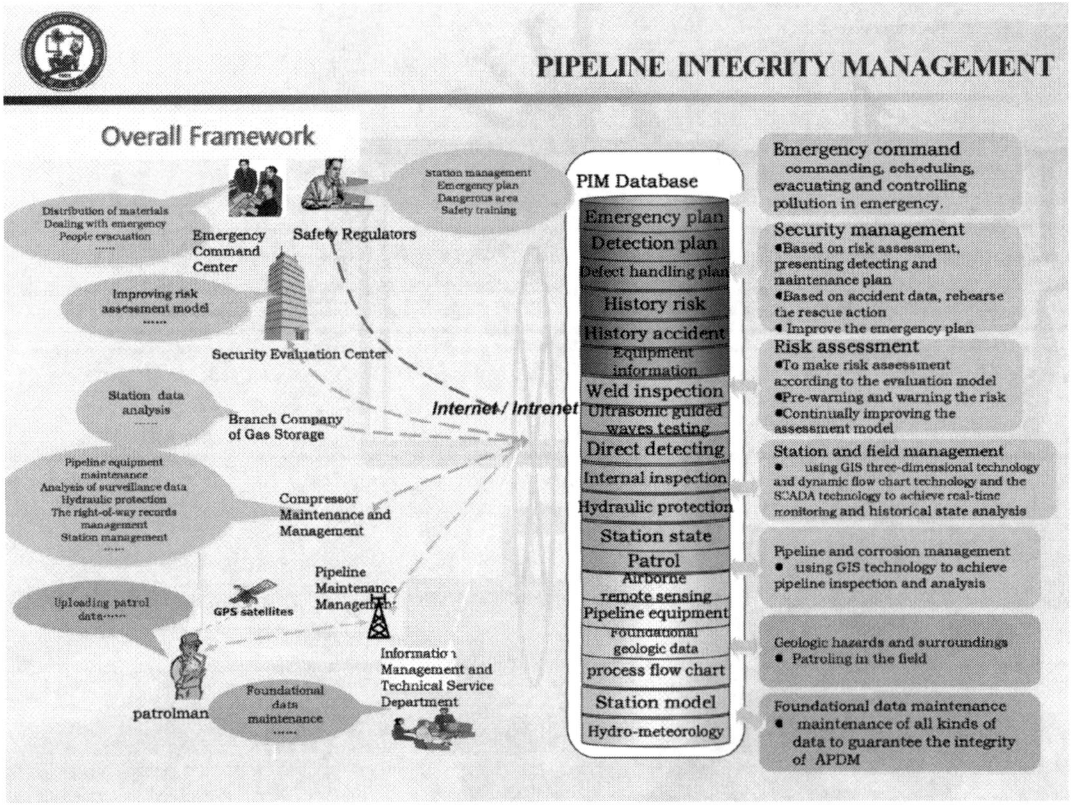
PIPELINE INTEGRITY MANAGEMENT
Overall Framework
Distribution of materials
Dealing with emergency
People evacuation
......
Emergency Command Center
Safety Regulators
Station management
Emergency plan
Dangerous area
Safety training
Improving risk assessment model
......
Security Evaluation Center
Station data analysis
......
Branch Company of Gas Storage
Internet / Intrenet
Pipeline equipment maintenance
Analysis of surveillance data
Hydraulic protection
The right-of-way records management
Station management
......
Compressor Maintenance and Management
Pipeline Maintenance Management
Uploading patrol data......
GPS satellites
Information Management and Technical Service Department
patrolman
Foundational data maintenance
......
PIM Database
Emergency plan
Detection plan
Defect handling plan
History risk
History accident
Equipment information
Weld inspection
Ultrasonic guided waves testing
Direct detecting
Internal inspection
Hydraulic protection
Station state
Patrol
Airborne remote sensing
Pipeline equipment
Foundational geologic data
process flow chart
Station model
Hydro-meteorology
Emergency command
commanding, scheduling, evacuating and controlling pollution in emergency.
Security management
•Based on risk assessment, presenting detecting and maintenance plan
•Based on accident data, rehearse the rescue action
• Improve the emergency plan
Risk assessment
•To make risk assessment according to the evaluation model
•Pre-warning and warning the risk
•Continually improving the assessment model
Station and field management
• using GIS three-dimensional technology and dynamic flow chart technology and the SCADA technology to achieve real-time monitoring and historical state analysis
Pipeline and corrosion management
• using GIS technology to achieve pipeline inspection and analysis
Geologic hazards and surroundings
• Patroling in the field
Foundational data maintenance
• maintenance of all kinds of data to guarantee the integrity of APDM

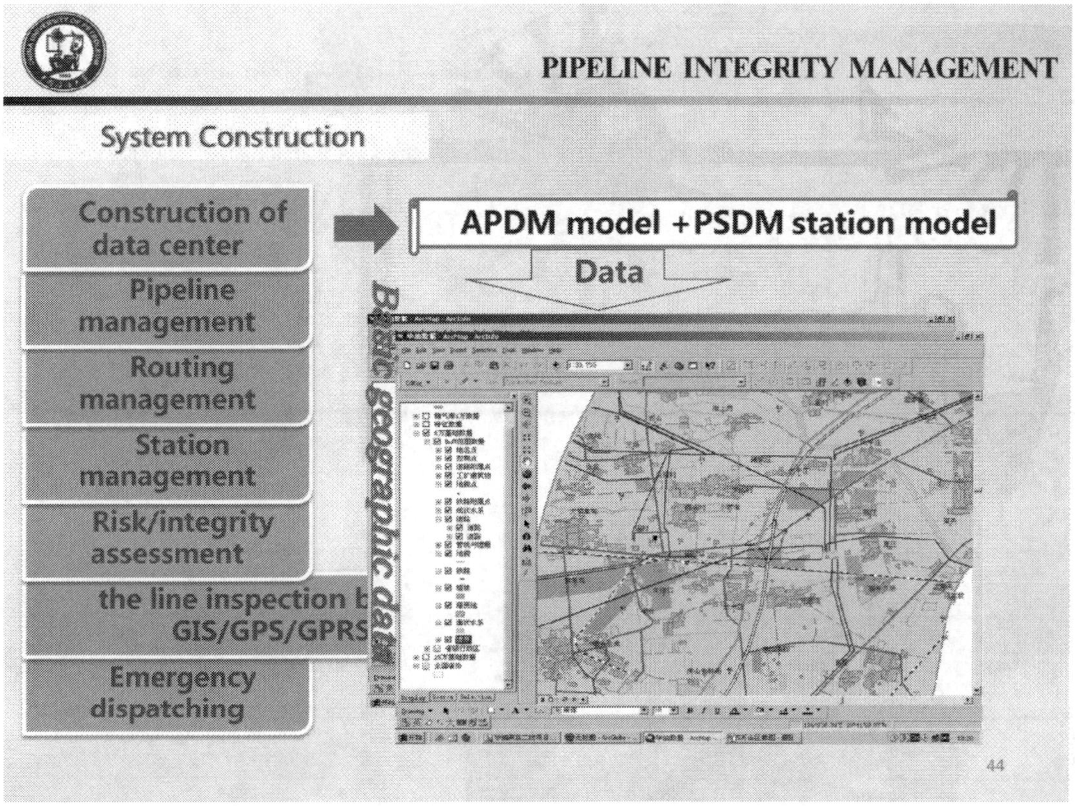
PIPELINE INTEGRITY MANAGEMENT
System Construction
Construction of data center
Pipeline management
Routing management
Station management
Risk/integrity assessment
the line inspection b
GIS/GPS/GPRS
Emergency dispatching
APDM model +PSDM station model
Data
Basic geographic data
44

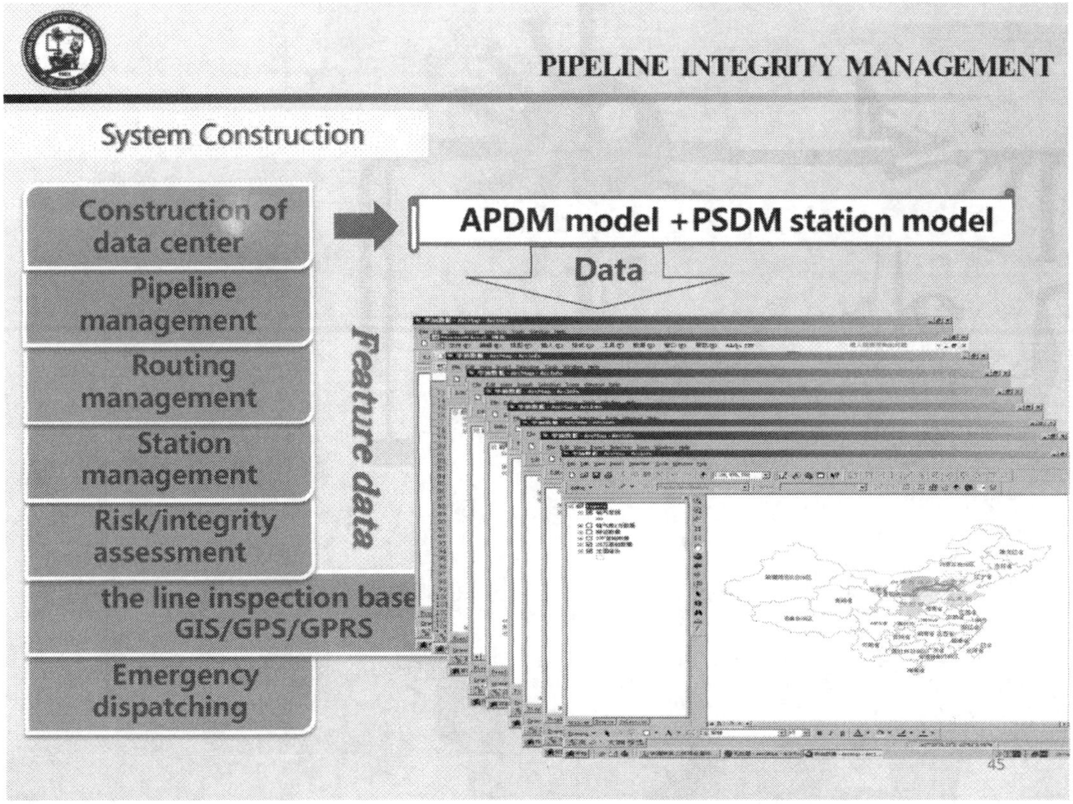
PIPELINE INTEGRITY MANAGEMENT
System Construction
Construction of data center
Pipeline management
Routing management
Station management
Risk/integrity assessment
the line inspection base
GIS/GPS/GPRS
Emergency dispatching
APDM model +PSDM station model
Data
Feature data
45

PIPELINE INTEGRITY MANAGEMENT

System Construction

- Construction of data center
- Pipeline management
- Routing management
- Station management
- Risk/integrity assessment
- the line inspection based GIS/GPS/GPRS
- Emergency dispatching

APDM model +PSDM station model

Data

Subject data

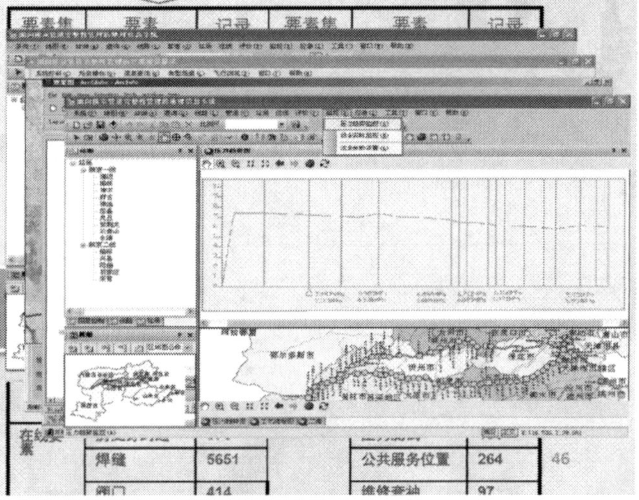

46

PIPELINE INTEGRITY MANAGEMENT

System Construction

- Construction of data center
- Pipeline management
- Routing management
- Station management
- Risk/integrity assessment
- the line inspection based GIS/GPS/GPRS
- Emergency dispatching

APDM model +PSDM station model

Data

Station technology data

Station three-dimension dat

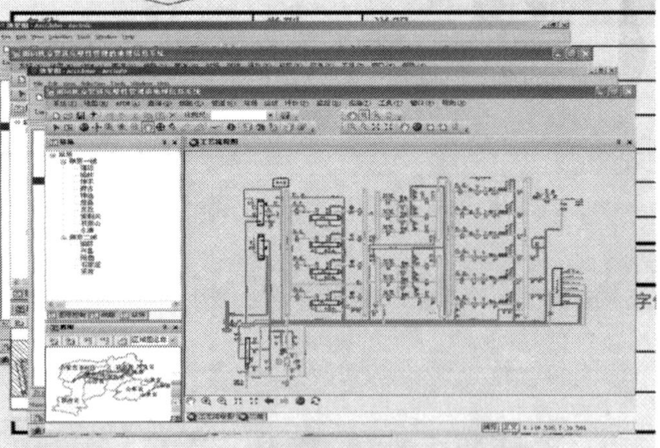

47

PIPELINE INTEGRITY MANAGEMENT

System Construction

- Construction of data center
- Pipeline management
- Routing management
- Station management
- Risk/integrity assessment
- the line inspection based GIS/GPS/GPRS
- Emergency dispatching

→ **APDM model +PSDM station model**

Data

Large-scale aerial photographic data

PIPELINE INTEGRITY MANAGEMENT

System Construction

- Construction of data center
- Pipeline management
- Routing management
- Station management
- Risk/integrity assessment
- the line inspection based on GIS/GPS/GPRS
- Emergency dispatching

→ **Pipeline and corrosion protection**

49

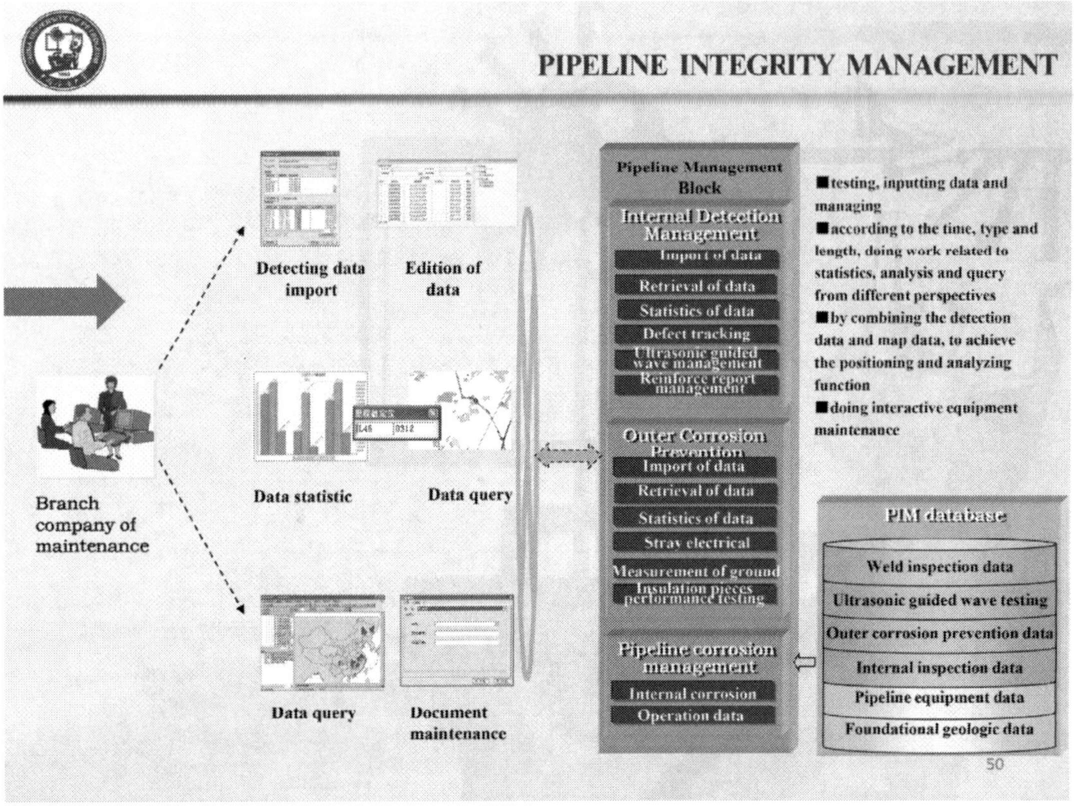
PIPELINE INTEGRITY MANAGEMENT
Detecting data import
Edition of data
Data statistic
Data query
Data query
Document maintenance
Branch company of maintenance
Pipeline Management Block
Internal Detection Management
Import of data
Retrieval of data
Statistics of data
Defect tracking
Ultrasonic guided wave management
Reinforce report management
Outer Corrosion Prevention
Import of data
Retrieval of data
Statistics of data
Stray electrical
Measurement of ground
Insulation pieces performance testing
Pipeline corrosion management
Internal corrosion
Operation data
■testing, inputting data and managing
■according to the time, type and length, doing work related to statistics, analysis and query from different perspectives
■by combining the detection data and map data, to achieve the positioning and analyzing function
■doing interactive equipment maintenance
PIM database
Weld inspection data
Ultrasonic guided wave testing
Outer corrosion prevention data
Internal inspection data
Pipeline equipment data
Foundational geologic data
50

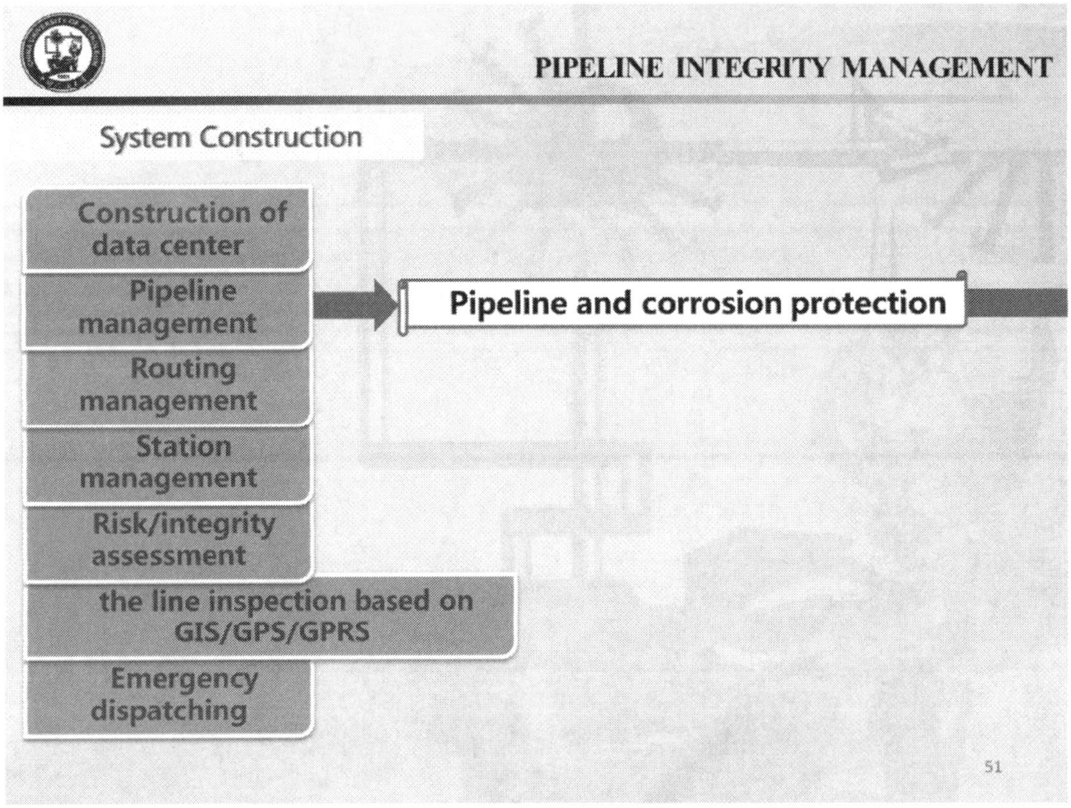
PIPELINE INTEGRITY MANAGEMENT
System Construction
Construction of data center
Pipeline management
Routing management
Station management
Risk/integrity assessment
the line inspection based on GIS/GPS/GPRS
Emergency dispatching
Pipeline and corrosion protection
51

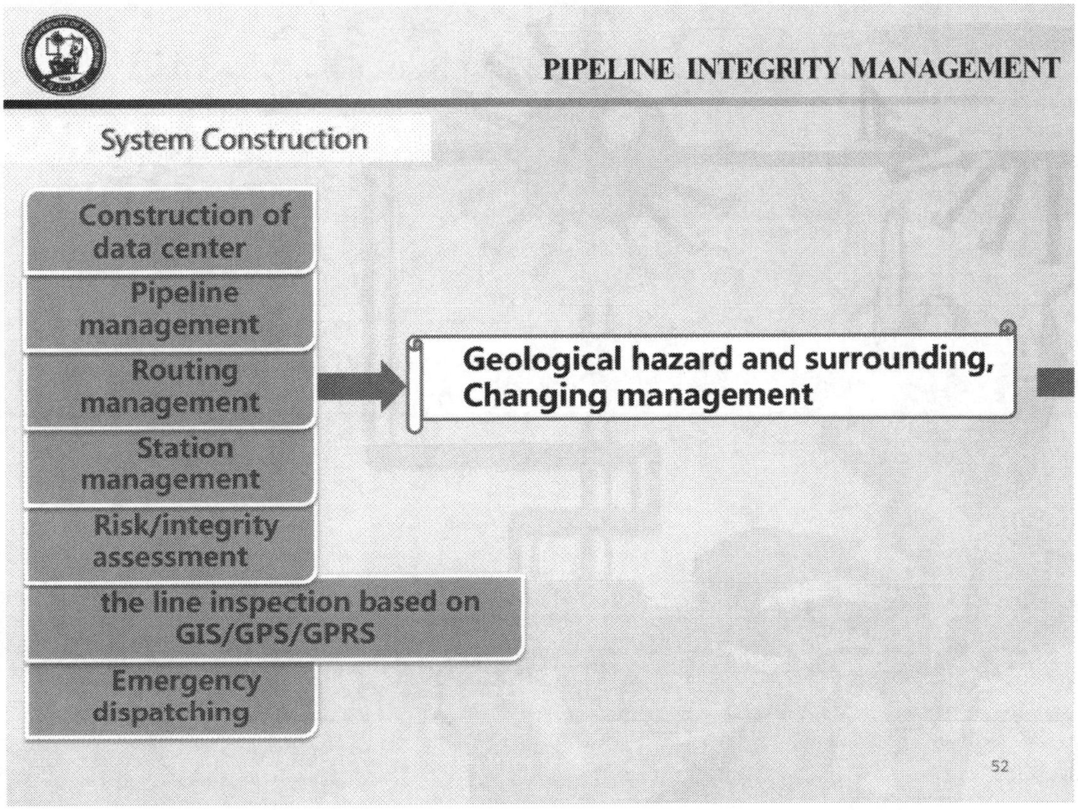
PIPELINE INTEGRITY MANAGEMENT
System Construction
Construction of data center
Pipeline management
Routing management
Station management
Risk/integrity assessment
the line inspection based on GIS/GPS/GPRS
Emergency dispatching
Geological hazard and surrounding, Changing management
52

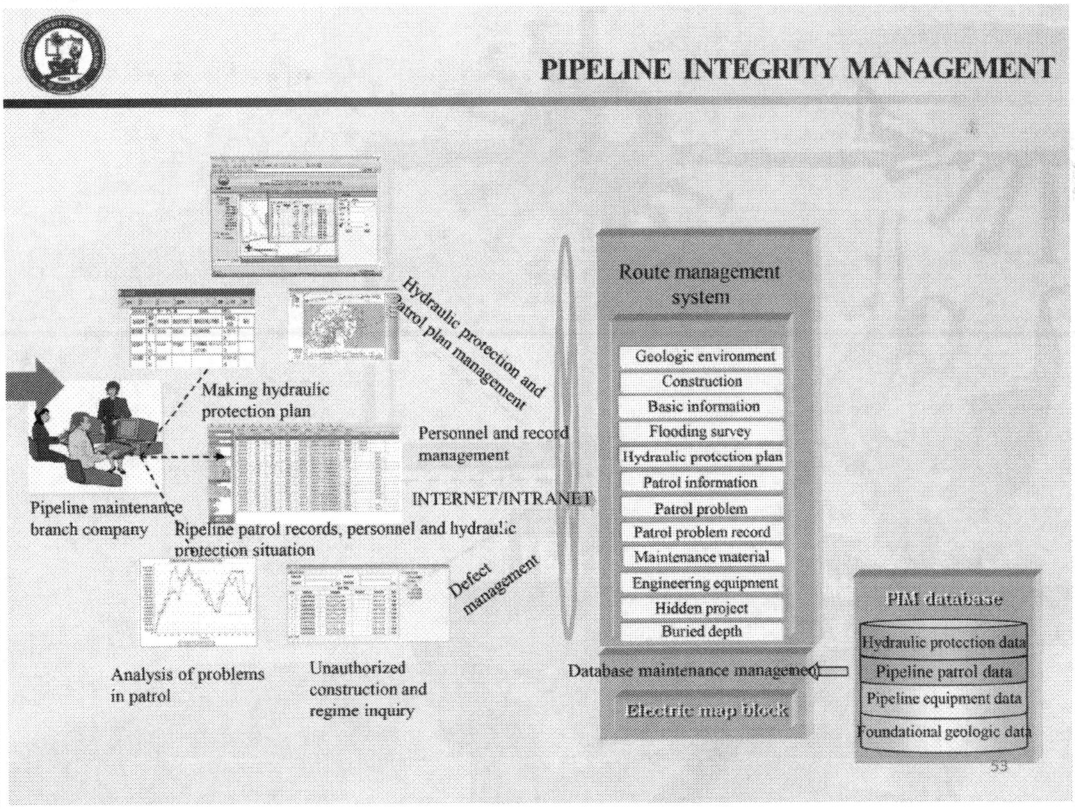
PIPELINE INTEGRITY MANAGEMENT
Hydraulic protection and Patrol plan management
Making hydraulic protection plan
Personnel and record management
INTERNET/INTRANET
Pipeline maintenance branch company
Pipeline patrol records, personnel and hydraulic protection situation
Defect management
Analysis of problems in patrol
Unauthorized construction and regime inquiry
Route management system
Geologic environment
Construction
Basic information
Flooding survey
Hydraulic protection plan
Patrol information
Patrol problem
Patrol problem record
Maintenance material
Engineering equipment
Hidden project
Buried depth
Database maintenance management
Electric map block
PIM database
Hydraulic protection data
Pipeline patrol data
Pipeline equipment data
Foundational geologic data
53

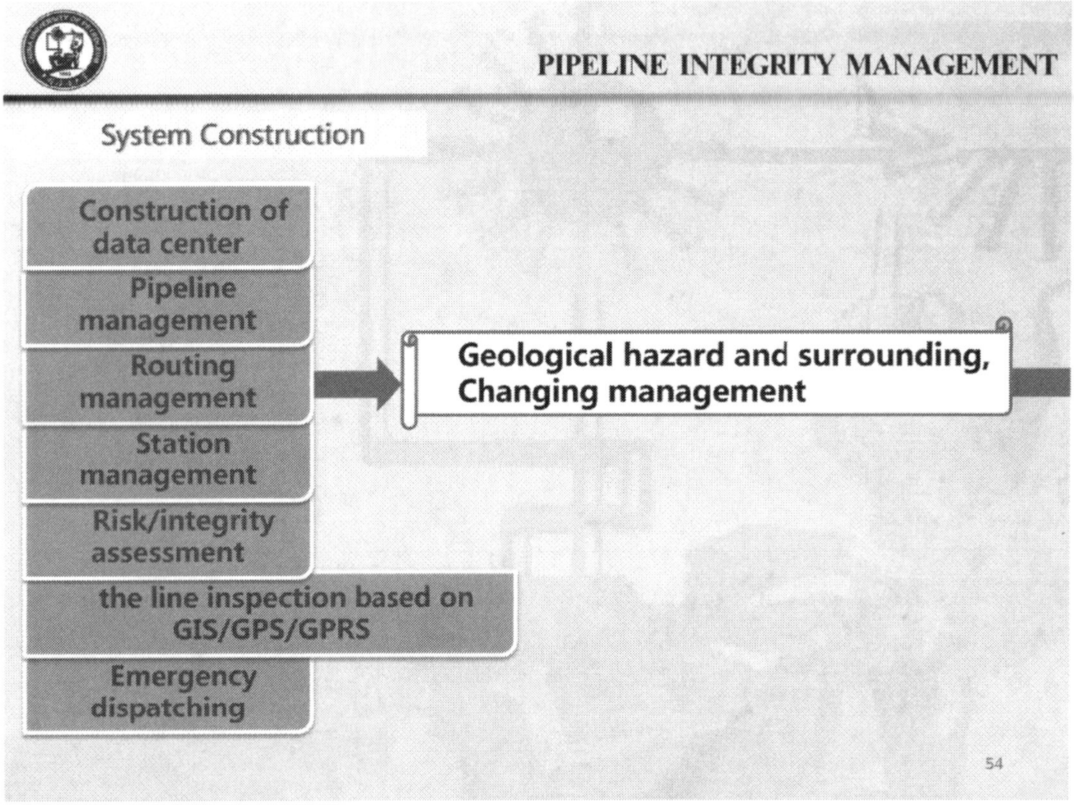
PIPELINE INTEGRITY MANAGEMENT
System Construction
Construction of data center
Pipeline management
Routing management
Station management
Risk/integrity assessment
the line inspection based on GIS/GPS/GPRS
Emergency dispatching
Geological hazard and surrounding, Changing management
54

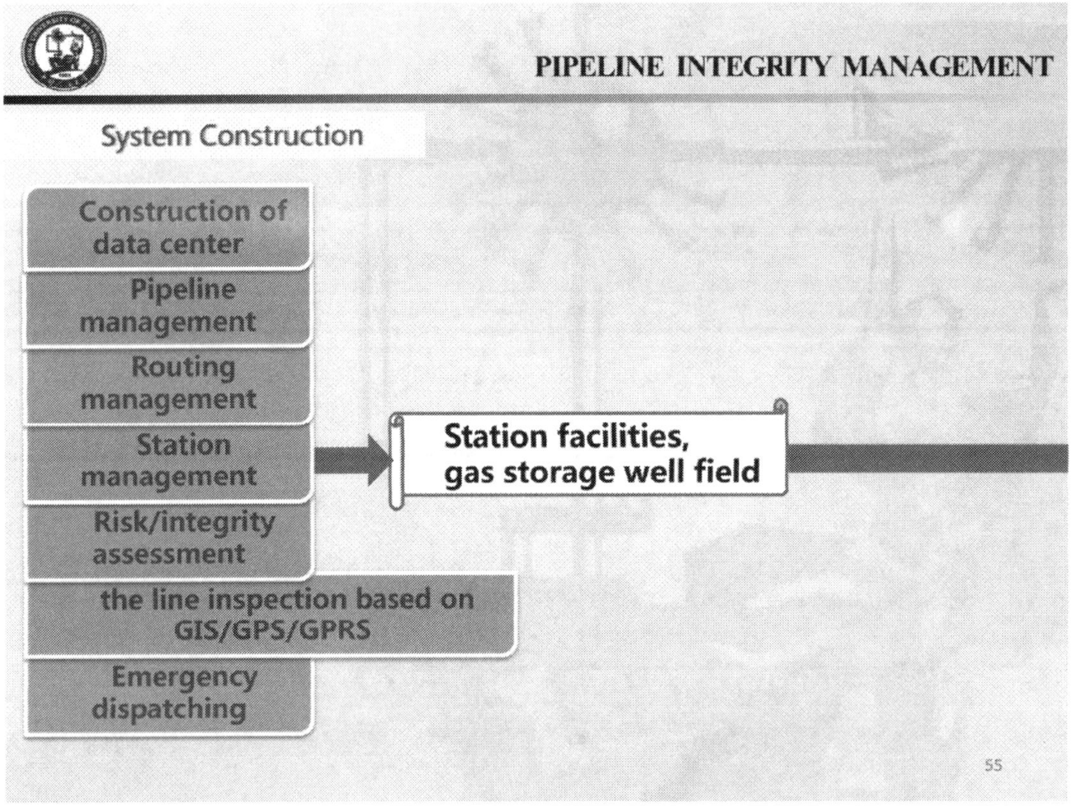
PIPELINE INTEGRITY MANAGEMENT
System Construction
Construction of data center
Pipeline management
Routing management
Station management
Risk/integrity assessment
the line inspection based on GIS/GPS/GPRS
Emergency dispatching
Station facilities, gas storage well field
55

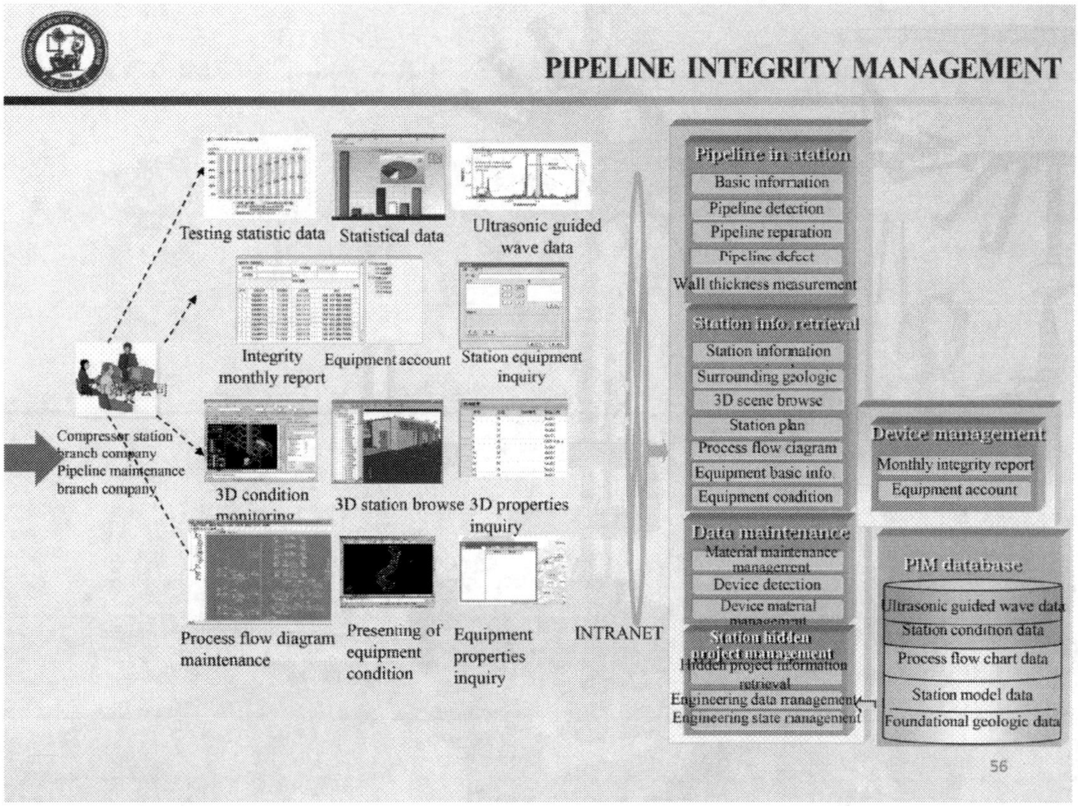
PIPELINE INTEGRITY MANAGEMENT
Testing statistic data
Statistical data
Ultrasonic guided wave data
Integrity monthly report
Equipment account
Station equipment inquiry
Compressor station branch company
Pipeline maintenance branch company
3D condition monitoring
3D station browse
3D properties inquiry
Process flow diagram maintenance
Presenting of equipment condition
Equipment properties inquiry
INTRANET
Pipeline in station
Basic information
Pipeline detection
Pipeline reparation
Pipeline defect
Wall thickness measurement
Station info. retrieval
Station information
Surrounding geologic
3D scene browse
Station plan
Process flow diagram
Equipment basic info
Equipment condition
Data maintenance
Material maintenance management
Device detection
Device material management
Station hidden project management
Hidden project information retrieval
Engineering data management
Engineering state management
Device management
Monthly integrity report
Equipment account
PIM database
Ultrasonic guided wave data
Station condition data
Process flow chart data
Station model data
Foundational geologic data
56

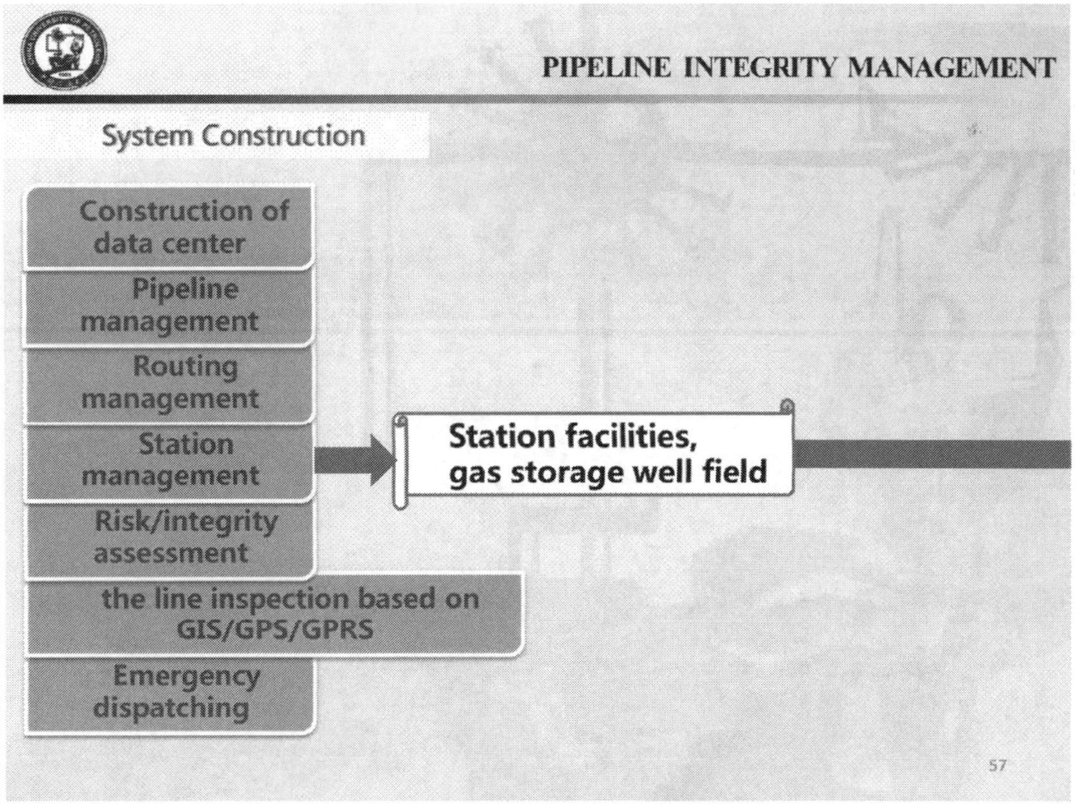
PIPELINE INTEGRITY MANAGEMENT
System Construction
Construction of data center
Pipeline management
Routing management
Station management
Risk/integrity assessment
the line inspection based on GIS/GPS/GPRS
Emergency dispatching
Station facilities, gas storage well field
57

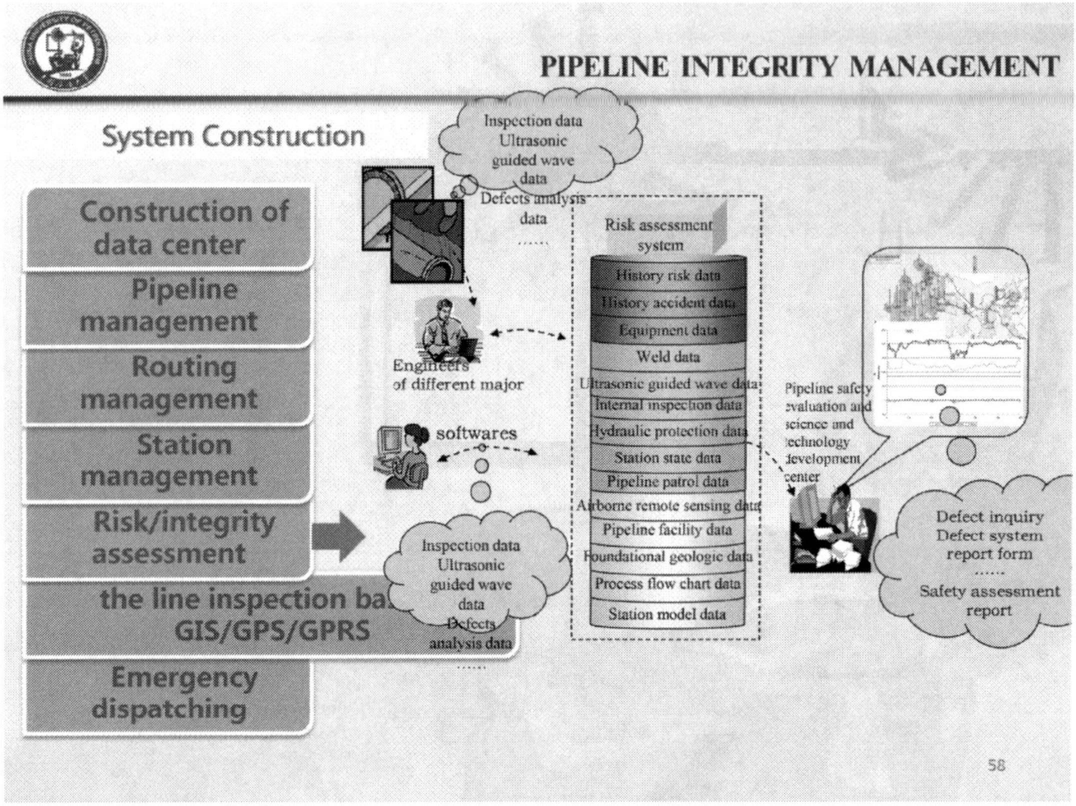

PIPELINE INTEGRITY MANAGEMENT
System Construction
Construction of data center
Pipeline management
Routing management
Station management
Risk/integrity assessment
the line inspection ba... GIS/GPS/GPRS
Emergency dispatching
Inspection data Ultrasonic guided wave data Defects analysis data ……
Risk assessment system
History risk data
History accident data
Equipment data
Weld data
Ultrasonic guided wave data
Internal inspection data
Hydraulic protection data
Station state data
Pipeline patrol data
Airborne remote sensing data
Pipeline facility data
Foundational geologic data
Process flow chart data
Station model data
Engineers of different major
softwares
Inspection data Ultrasonic guided wave data Defects analysis data ……
Pipeline safety evaluation and science and technology development center
Defect inquiry Defect system report form …… Safety assessment report
58

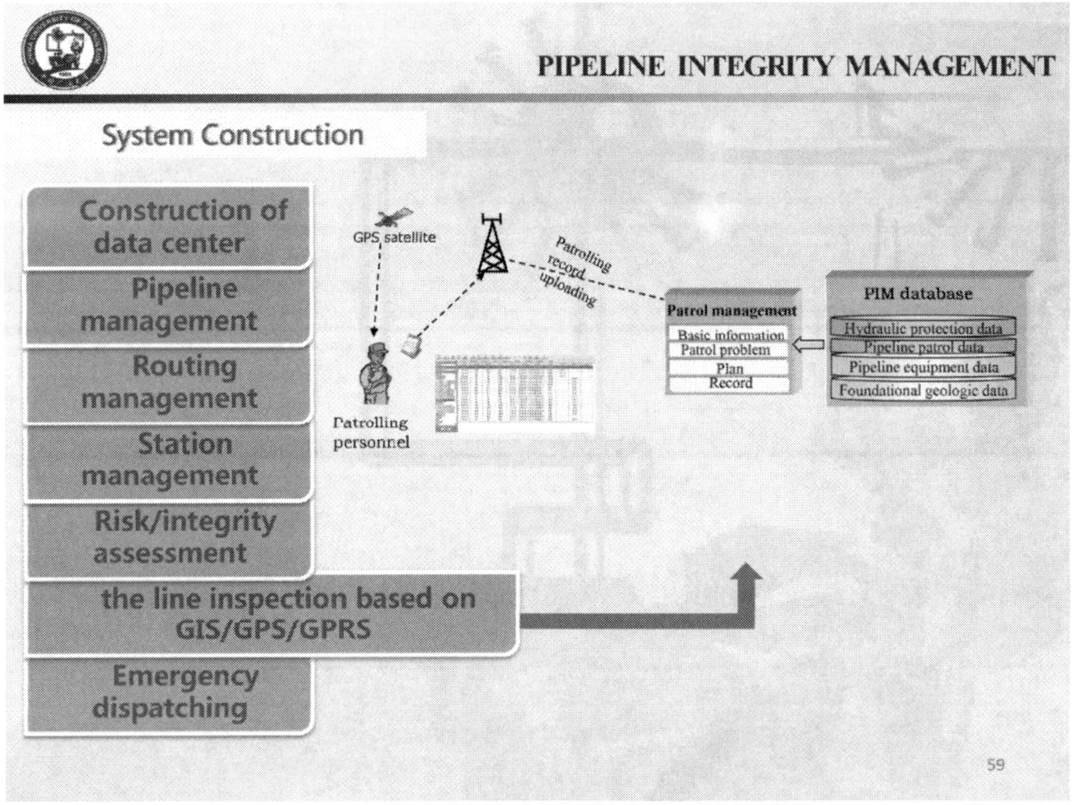

PIPELINE INTEGRITY MANAGEMENT
System Construction
Construction of data center
Pipeline management
Routing management
Station management
Risk/integrity assessment
the line inspection based on GIS/GPS/GPRS
Emergency dispatching
GPS satellite
Patrolling record uploading
Patrolling personnel
Patrol management
Basic information
Patrol problem
Plan
Record
PIM database
Hydraulic protection data
Pipeline patrol data
Pipeline equipment data
Foundational geologic data
59

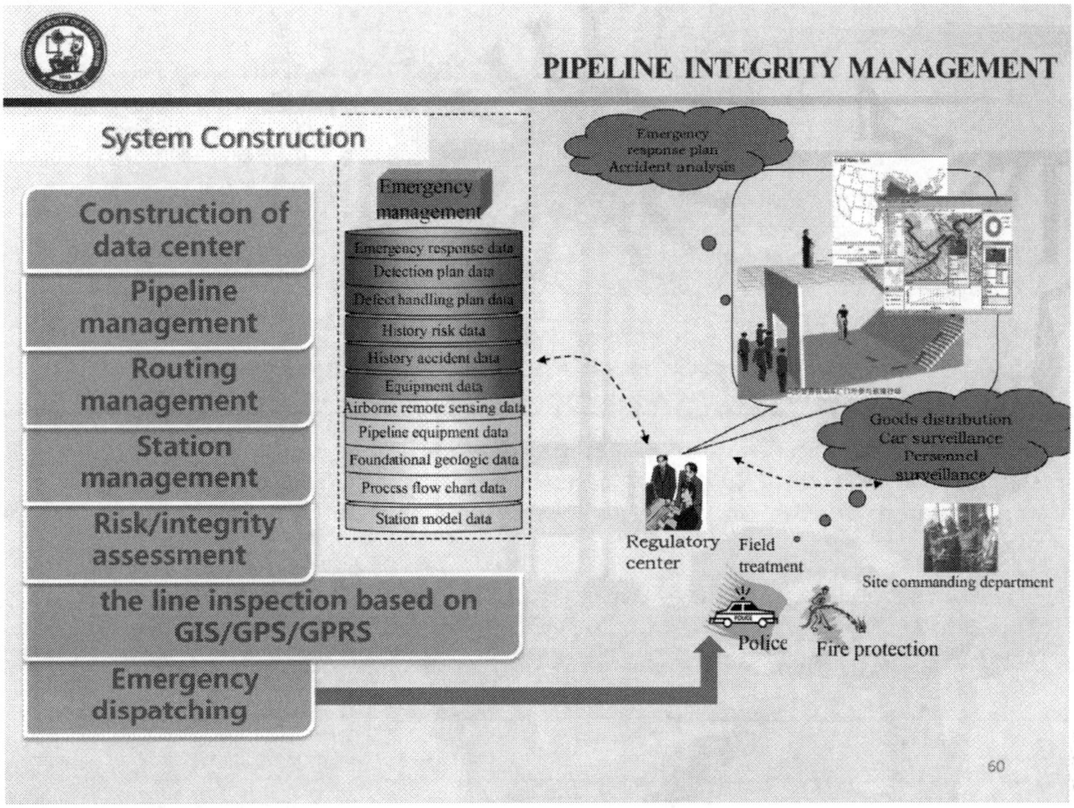
PIPELINE INTEGRITY MANAGEMENT
System Construction
Construction of data center
Pipeline management
Routing management
Station management
Risk/integrity assessment
the line inspection based on GIS/GPS/GPRS
Emergency dispatching
Emergency management
Emergency response data
Detection plan data
Defect handling plan data
History risk data
History accident data
Equipment data
Airborne remote sensing data
Pipeline equipment data
Foundational geologic data
Process flow chart data
Station model data
Emergency response plan
Accident analysis
Goods distribution
Car surveillance
Personnel surveillance
Regulatory center
Field treatment
Site commanding department
Police
Fire protection
60

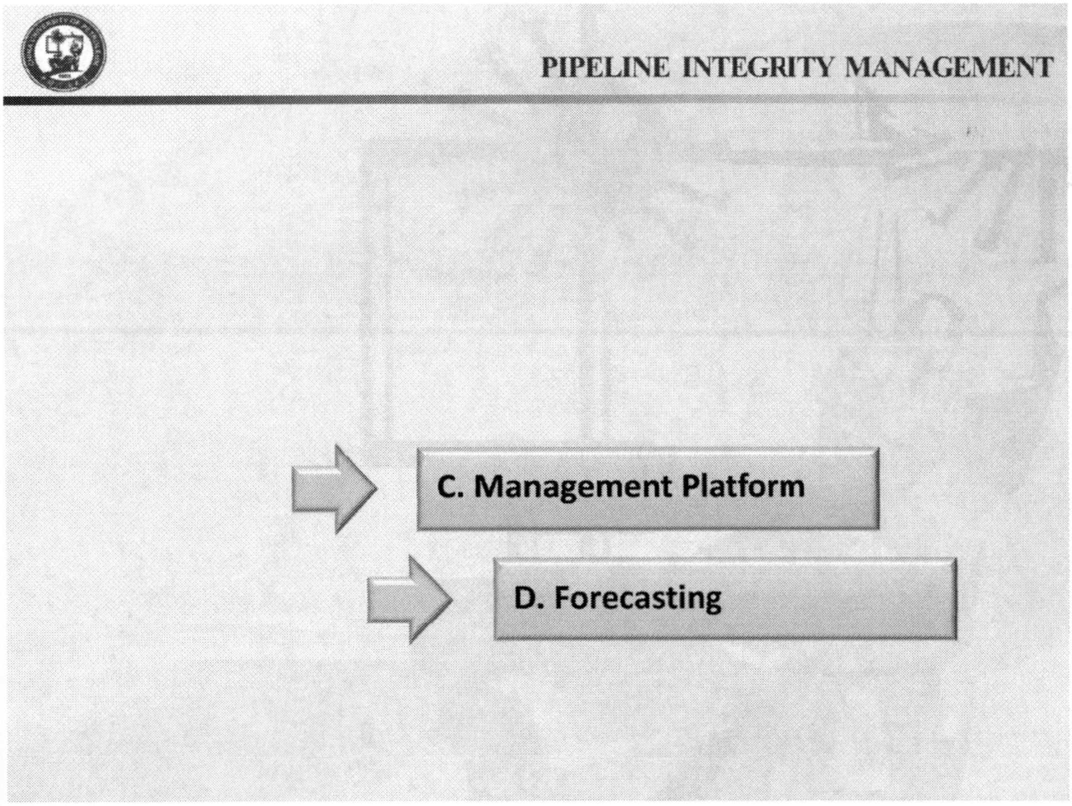
PIPELINE INTEGRITY MANAGEMENT
C. Management Platform
D. Forecasting

Forecast

- The combination of PIM and information technology will gradually lead the decision-making to be intelligent and scientific. Molecular and special construction will be the direction of PIM technology, such as gas storage, LNG/CNG are also important contents in PIM.
- The development of internet of things technology (IOT) demands higher requirements for the popularization and improvement of PIM. The construction of PIM platform will gradually advance to IOT.
- One of the most important management content is to continuously improve and optimize the technologies and related facilities, as well as the whole process. And also, it will be combined with the daily business process。
- One of the challenge we faced with is that the PIM platform could accomplish the function of emergency rescue, emergency decision and also the data integration among main information systems.

62

PIPELINE INTEGRITY MANAGEMENT

Prediction of Development Key Points and Technology Difficulties

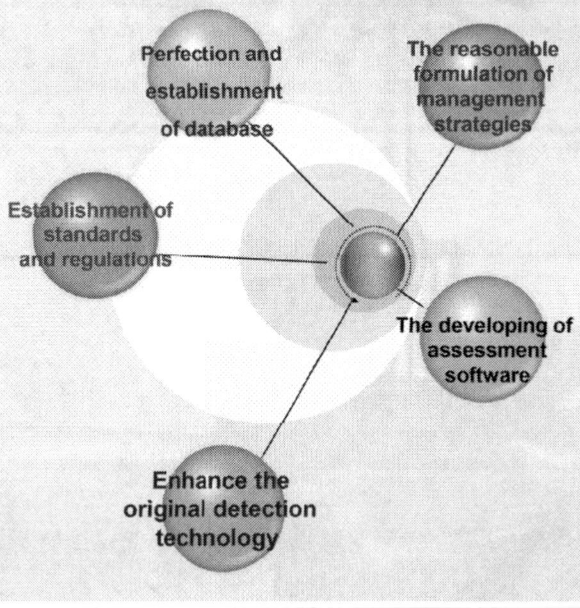

中國石油大學
CHINA UNIVERSITY OF PETROLEUM
Thank You!

2nd CCPS
China Conference
On Process Safety

Some thoughts on LOPA

Qingdao, China
August 28 - 29, 2014

John Zhao

Shegle Risk Consulting Co. Ltd. (Shanghai)

jmzhao@shegle.com

SHEGLe 撷果™

LOPA calculation

$$f_i^C = f_i^I \prod_{j=1}^{J} PFD_j = f_i^I \times PFD_{i1} \times PFD_{i2} \times PFD_{i3} \times \ldots \times PFD_{ij} \qquad (7\text{-}1)$$

where

f_i^C is the frequency for consequence C for initiating event i

f_i^I is the initiating event frequency for initiating event i

PFD_{ij} is the probability of failure on demand of the jth IPL that protects against consequence C for initiating event i.

公式 7-1，CCPS LOPA Concept Book 2001

CCPS CENTER FOR CHEMICAL PROCESS SAFETY

High demand mode? How to calculate?

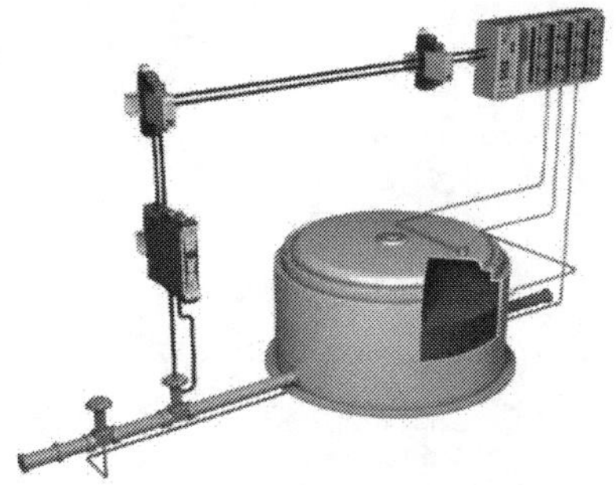

high demand mode - example

- A small tank is filled from a large tank 1400 times per year. A level sensor, logic solver and automatic valve were added to auto start the fill, when there is a high level to stop the fill.

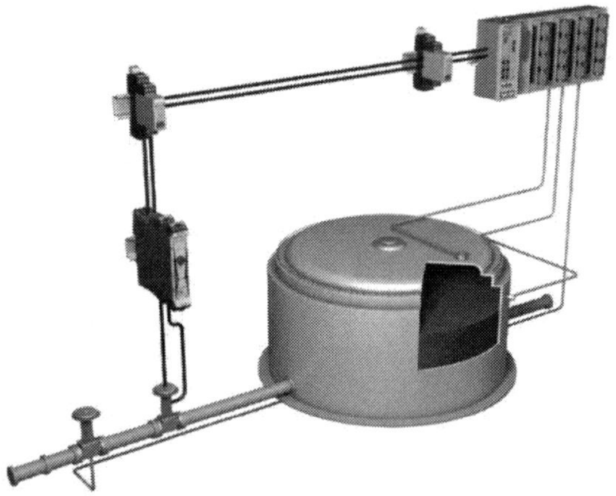

Hazard rate vs. demand rate

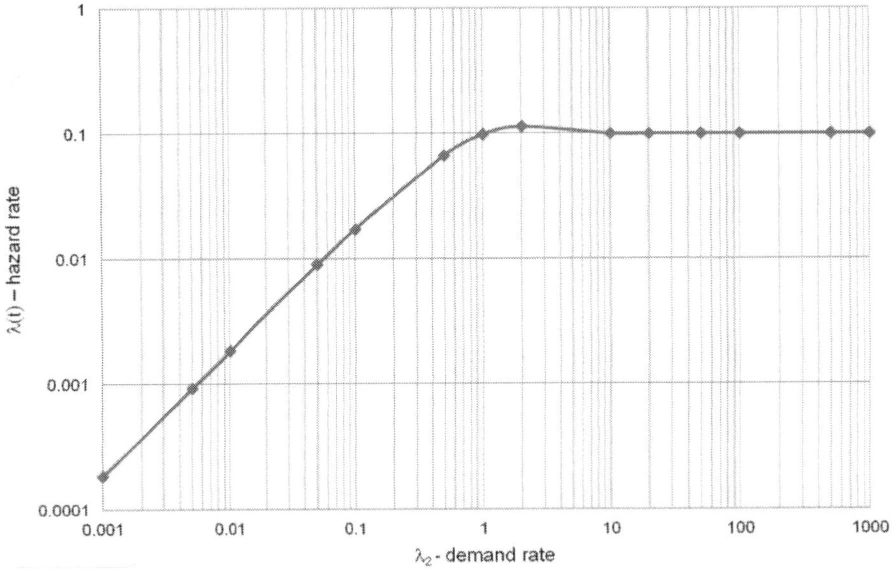

Hazard rate vs. demand rate

$$\lambda(t)=\frac{\lambda_1 e^{-\lambda_1 t}+\lambda_2 e^{-\lambda_2 t}-(\lambda_1+\lambda_2)e^{-(\lambda_1+\lambda_2)t}}{e^{-\lambda_1 t}+e^{-\lambda_2 t}-e^{-(\lambda_1+\lambda_2)t}}$$

where

$\lambda(t)$ is the hazard rate, equals the frequency for consequence C

λ_1 is the failure rate for the IPL1

λ_2 is the demand rate for the initiating event

High demand mode

$$\lambda(t)=\frac{\lambda_1 e^{-\lambda_1 t}+\lambda_2 e^{-\lambda_2 t}-(\lambda_1+\lambda_2)e^{-(\lambda_1+\lambda_2)t}}{e^{-\lambda_1 t}+e^{-\lambda_2 t}-e^{-(\lambda_1+\lambda_2)t}}$$

if

$\lambda_2 \gg \lambda_1$ and $\lambda_2 \geq 1$

then

$$\lambda_1+\lambda_2 \approx \lambda_2$$

$$e^{-\lambda_1 t}+e^{-\lambda_2 t}-e^{-(\lambda_1+\lambda_2)t} \approx e^{-\lambda_1 t}$$

$$\lambda_1 e^{-\lambda_1 t}+\lambda_2 e^{-\lambda_2 t}-(\lambda_1+\lambda_2)e^{-(\lambda_1+\lambda_2)t} \approx \lambda_1 e^{-\lambda_1 t}$$

$$\lambda(t) \approx \frac{\lambda_1 e^{-\lambda_1 t}}{e^{-\lambda_1 t}}=\lambda_1$$

LOPA calculation - high demand

$$f_i^C = f_i^{IPL1} \prod_{j=2}^{J} PFD_j = f_i^{IPL1} \times PFD_{i2} \times PFD_{i3} \times PFD_{i4} \times \ldots \times PFD_{ij}$$

where

f_i^C	is the frequency for consequence C for initiating event i
f_i^{IPL1}	is the failure rate of the first IPL
PFD_{ij}	is the probability of failure on demand of the jth IPL that protects against consequence C for initiating event i.

公式 7-1，CCPS LOPA Concept Book 2001

CCPS CENTER FOR CHEMICAL PROCESS SAFETY

More accurate definition necessary?

LOPA calculation

$$f_i^C = f_i^I \prod_{j=1}^{J} PFD_j = f_i^I \times PFD_{i1} \times PFD_{i2} \times PFD_{i3} \times \ldots \times PFD_{ij} \qquad (7\text{-}1)$$

where

f_i^C or $(f_i^C)_{avg}$?

PFD_{ij} or $PFD_{ij,avg}$?

公式 7-1，CCPS LOPA Concept Book 2001

CCPS CENTER FOR CHEMICAL PROCESS SAFETY

$\lambda(t)$ vs. $\lambda(t)_{avg}$ vs. $\lambda_2 PFD_{1avg}$

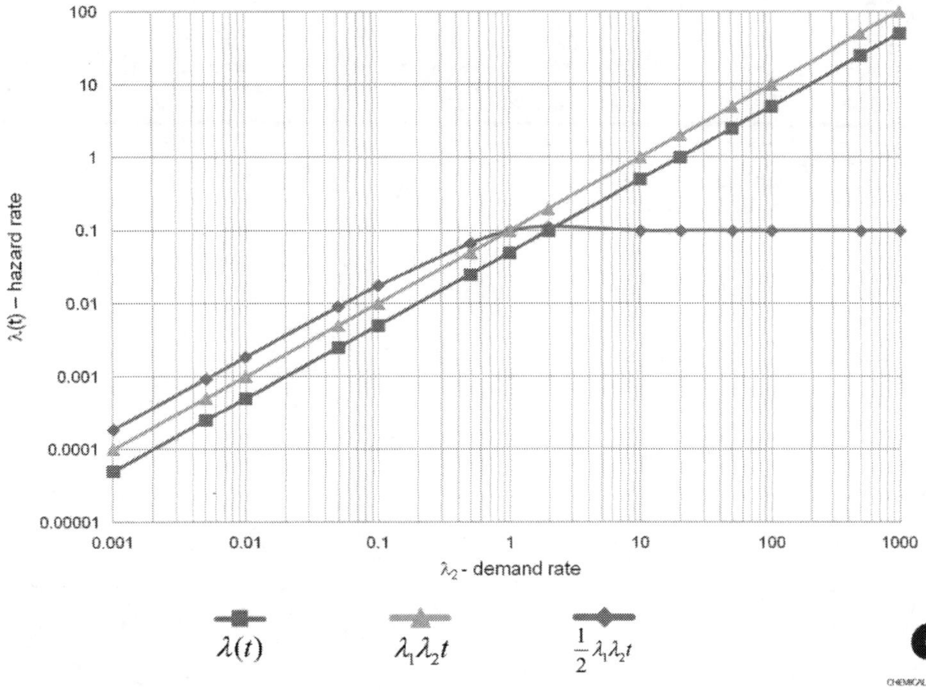

CCPS CENTER FOR CHEMICAL PROCESS SAFETY

1 IPL case

$$\lambda(t) = \frac{\lambda_1 e^{-\lambda_1 t} + \lambda_2 e^{-\lambda_2 t} - (\lambda_1 + \lambda_2) e^{-(\lambda_1 + \lambda_2) t}}{e^{-\lambda_1 t} + e^{-\lambda_2 t} - e^{-(\lambda_1 + \lambda_2) t}}$$

if

$$\boxed{\lambda_1 t, \lambda_2 t \le 0.1}$$

then

$$e^{-\lambda_1 t} \approx 1 - \lambda_1 t \qquad e^{-\lambda_2 t} \approx 1 - \lambda_2 t$$

then

$$\lambda(t) \approx 2\lambda_1 \lambda_2 t$$

since

$$\lambda(t)_{avg} = \frac{1}{t}\int_0^t \lambda(t)dt, \; PFD_1 = \lambda_1 t \;, \; PFD_{1avg} = \frac{\lambda_1 t}{2}$$

then

$$\boxed{\lambda(t)_{avg} = \lambda_1 \lambda_2 t = \lambda_2 \times PFD_1 = 2 \times \lambda_2 \times PFD_{1avg}}$$

CCPS CENTER FOR CHEMICAL PROCESS SAFETY

2 IPLs case

If consider 2 IPLs

$$\lambda(t) = \frac{\lambda_1 e^{-\lambda_1 t} + \lambda_2 e^{-\lambda_2 t} + \lambda_3 e^{-\lambda_3 t} - (\lambda_1 + \lambda_2) e^{-(\lambda_1 + \lambda_2) t} - (\lambda_1 + \lambda_3) e^{-(\lambda_1 + \lambda_3) t} - (\lambda_2 + \lambda_3) e^{-(\lambda_2 + \lambda_3) t} + (\lambda_1 + \lambda_2 + \lambda_3) e^{-(\lambda_1 + \lambda_2 + \lambda_3) t}}{e^{-\lambda_1 t} + e^{-\lambda_2 t} + e^{-\lambda_3 t} - e^{-(\lambda_1 + \lambda_2) t} - e^{-(\lambda_1 + \lambda_3) t} - e^{-(\lambda_2 + \lambda_3) t} + e^{-(\lambda_1 + \lambda_2 + \lambda_3) t}}$$

if

$$\boxed{\lambda_1 t, \lambda_2 t, \lambda_3 t \le 0.1}$$

then

$$e^{-\lambda_1 t} \approx 1 - \lambda_1 t + \frac{\lambda_1^2 t^2}{2!} \quad e^{-\lambda_2 t} \approx 1 - \lambda_2 t + \frac{\lambda_2^2 t^2}{2!} \quad e^{-\lambda_3 t} \approx 1 - \lambda_3 t + \frac{\lambda_3^2 t^2}{2!}$$

$$\lambda(t) \approx 3\lambda_1 \lambda_2 \lambda_3 t^2$$

$$\lambda(t)_{avg} = \frac{1}{t}\int_0^t \lambda(t)dt = \lambda_1 \lambda_2 \lambda_3 t^2 = \lambda_3 \times PFD_2 \times PFD_1$$

$$= 4 \times \lambda_3 \times PFD_{1avg} \times PFD_{2avg}$$

CCPS CENTER FOR CHEMICAL PROCESS SAFETY

Backup slides

High demand mode – IEC61508-4 1997

3.5.12

mode of operation

way in which a safety-related system is intended to be used, with respect to the frequency of demands made upon it, which may be either:

- **low demand mode** – where the frequency of demands for operation made on a safety-related system is no greater than one per year and no greater than twice the proof test frequency; or
- **high demand or continuous mode** – where the frequency of demands for operation made on a safety-related system is greater than one per year or greater than twice the proof check frequency

CCPS CENTER FOR CHEMICAL PROCESS SAFETY

High demand mode – IEC61511-1 2003

3.2.43.1

demand mode safety instrumented function

where a specified action (for example, closing of a valve) is taken in response to process conditions or other demands. In the event of a dangerous failure of the safety instrumented function a potential hazard only occurs in the event of a failure in the process or the BPCS

3.2.43.2

continuous mode safety instrumented function

where in the event of a dangerous failure of the safety instrumented function a potential hazard will occur without further failure unless action is taken to prevent it

NOTE 1 Continuous mode covers those safety instrumented functions which implement continuous control to maintain functional safety.

NOTE 2 In demand mode applications where the demand rate is more frequent than **once per year**, the hazard rate will not be higher than the dangerous failure rate of the safety instrumented function. In such a case, it will normally be appropriate to use the continuous mode criteria.

NOTE 3 The target failure measures for safety instrumented functions operating in demand mode and continuous mode are defined in Tables 3 and 4.

NOTE 4 This term deviates from the definition in IEC 61508-4 to reflect differences in process sector terminology.

CCPS CENTER FOR CHEMICAL PROCESS SAFETY

High demand mode – IEC61508-4 2010

3.5.16

mode of operation

way in which a safety function operates, which may be either

- **Low demand mode**: where the safety function is only performed on demand, in order to transfer the EUC into a specified safe state, and where the frequency of demands is no greater than one per year; or
- **high demand mode**: where the safety function is only performed on demand, in order to transfer the EUC into a specified safe state, and where the frequency of demands is greater than one per year; or
- **Continuous mode**: where the safety function retains the EUC in safe state as part of normal operation;

CCPS CENTER FOR CHEMICAL PROCESS SAFETY

High demand mode – CCPS LOPA concept book

If $f_i^{IPL_{i1}}$ is known:

$$f_i^C = f_i^{IPL_{i1}} \qquad \text{(F-1)}$$

where

f_i^C is the frequency for consequence C for initiating event i

$f_i^{IPL_{i1}}$ Is the failure frequency for the one IPL that protects against consequence C for initiating event i

If $f_i^{IPL_{i1}}$ is not known and PFD_{i1} is known

set $$f_i^C = 2\frac{1}{T_{i1}} \qquad \text{(7-8)}$$

where

T_{i1} IPL_{i1}功能测试的时间间隔（year）

公式F-1，CCPS LOPA Concept Book 2001

CCPS CENTER FOR CHEMICAL PROCESS SAFETY

High demand mode – CCPS LOPA concept book

$$f_i^C = f_i^{IPL_{i1}}(1-e^{-\frac{DT}{2}}) \qquad \text{(F-2)}$$

where

f_i^C 第i个初始事件所导致后果C的频率

$f_i^{IPL_{i1}}$ IPL的失效频率

D 对IPL的的要求频率（1/year），对于只有一个IPL的场景，就是初始事件的频率

T IPL功能测试的时间间隔（year）

公式F-2，CCPS LOPA Concept Book 2001

CCPS CENTER FOR CHEMICAL PROCESS SAFETY

High demand mode – CCPS LOPA concept book

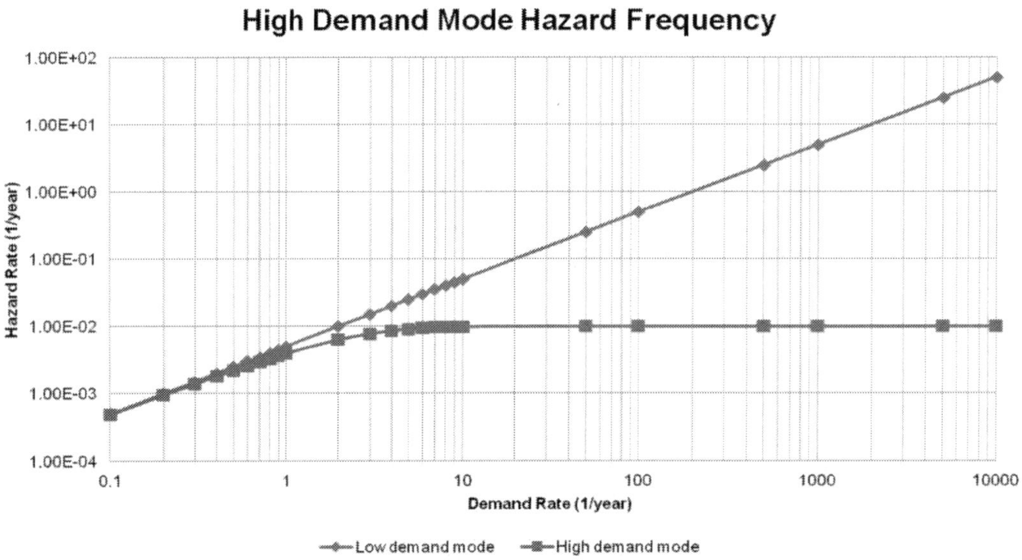

CCPS CENTER FOR CHEMICAL PROCESS SAFETY

Hazard rate vs. demand rate

$$\lambda(t) = \frac{\lambda_1 e^{-\lambda_1 t} + \lambda_2 e^{-\lambda_2 t} - (\lambda_1 + \lambda_2) e^{-(\lambda_1 + \lambda_2)t}}{e^{-\lambda_1 t} + e^{-\lambda_2 t} - e^{-(\lambda_1 + \lambda_2)t}}$$

if

$$\boxed{\lambda_2 \gg \lambda_1 \qquad \text{and} \quad \lambda_2 \geq 1}$$

then

$$\lambda_1 + \lambda_2 \approx \lambda_2$$

$$e^{-\lambda_1 t} + e^{-\lambda_2 t} - e^{-(\lambda_1 + \lambda_2)t} \approx e^{-\lambda_1 t}$$

$$\lambda_1 e^{-\lambda_1 t} + \lambda_2 e^{-\lambda_2 t} - (\lambda_1 + \lambda_2) e^{-(\lambda_1 + \lambda_2)t} \approx \lambda_1 e^{-\lambda_1 t}$$

$$\boxed{\lambda(t) \approx \frac{\lambda_1 e^{-\lambda_1 t}}{e^{-\lambda_1 t}} = \lambda_1}$$

CCPS CENTER FOR CHEMICAL PROCESS SAFETY

Error analysis

Error analysis for numerator

$$A = \frac{e^{-\lambda_2 t} - e^{-(\lambda_1+\lambda_2)t}}{e^{-\lambda_2 t}} = 1 - e^{-\lambda_1 t}$$

$$B = \frac{e^{-\lambda_2 t} - e^{-(\lambda_1+\lambda_2)t}}{e^{-\lambda_1 t}} = \frac{1-e^{-\lambda_1 t}}{e^{-\lambda_1 t}}$$

If $\lambda_1 = 0.1$

λ_2	0.1	0.5	1	2	10	20	50	100
A	9.52%	9.52%	9.52%	9.52%	9.52%	9.52%	9.52%	9.52%
B	9.52%	6.38%	3.87%	1.42%	0.00%	0.00%	0.00%	0.00%

If $\lambda_1 = 0.01$

λ_2	0.01	0.05	0.1	0.5	1	2	10	20	50	100
A	1.00%	1.00%	1.00%	1.00%	1.00%	1.00%	1.00%	1.00%	1.00%	1.00%
B	1.00%	0.96%	0.91%	0.61%	0.37%	0.14%	0.00%	0.00%	0.00%	0.00%

CCPS CENTER FOR CHEMICAL PROCESS SAFETY

Error analysis

Error analysis for denominator

$$C = \frac{\lambda_2 e^{-\lambda_2 t} - (\lambda_1+\lambda_2)e^{-(\lambda_1+\lambda_2)t}}{\lambda_2 e^{-\lambda_2 t}}$$

$$D = \frac{\lambda_2 e^{-\lambda_2 t} - (\lambda_1+\lambda_2)e^{-(\lambda_1+\lambda_2)t}}{\lambda_1 e^{-\lambda_1 t}}$$

If $\lambda_1 = 0.1$

λ_2	0.1	0.5	1	2	10	20	50	100
C	-80.97%	-8.58%	0.47%	4.99%	8.61%	9.06%	9.34%	9.43%
D	-80.97%	-28.76%	1.90%	14.93%	0.04%	0.00%	0.00%	0.00%

If $\lambda_1 = 0.01$

λ_2	0.01	0.05	0.1	0.5	1	2	10	20	50	100
C	-98.01%	-18.81%	-8.91%	-0.99%	0.00%	0.50%	0.90%	0.95%	0.98%	0.99%
D	-98.01%	-90.34%	-81.39%	-30.17%	0.18%	13.67%	0.04%	0.00%	0.00%	0.00%

CCPS CENTER FOR CHEMICAL PROCESS SAFETY

Process incident caused by different fail-safe positions for a control valve under loss of signal and loss of instrument air

J. Windhorst, WEC Inc.

The Fail-Safe State

- The fail-safe position of a control valve (the position it will default to) shall be:
  - Defined as part of the process design.
  - Implemented as part of instrument engineering.
- IEC 61508 (2010): The SF requirements specification shall describe (for all SFs) the manner in which a safe state shall be achieved and maintained.

Heavy Oil Processing Block Diagram

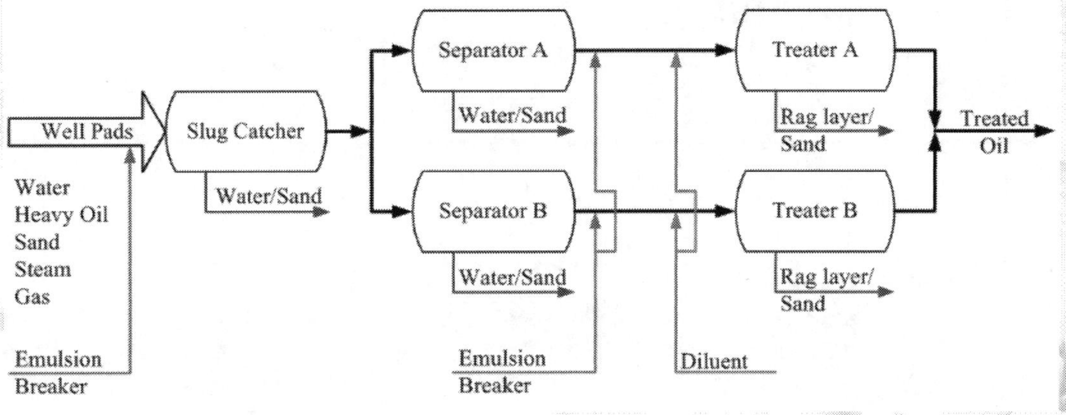

Facility of the S/D Incident (aerial view)

Outline of Heavy Oil Processing Building

Vessel volumes vary from 200 to 500 m^3

Sample taps are inside the building

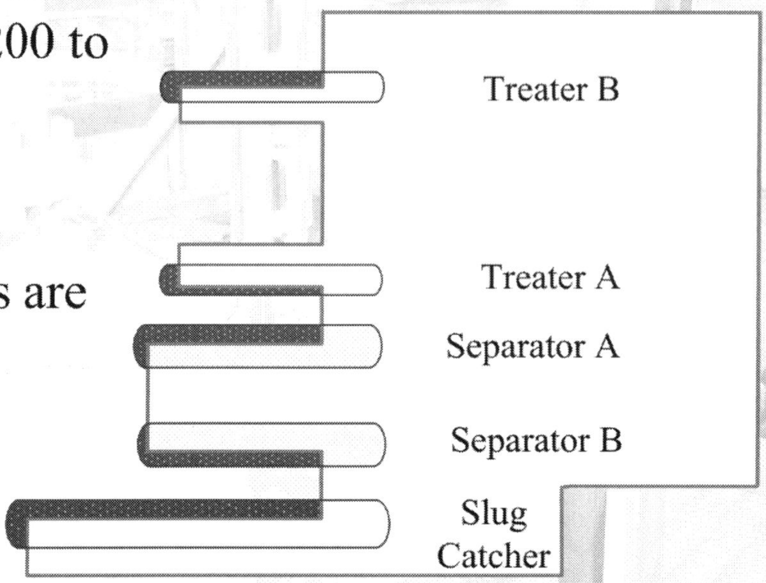

8/29/2014 J Windhorst 5

SEPARATOR B
SEPARATOR A

TREATER B
2V103B

Shut Down Scope

- Vessel Maintenance
  - Slug Catcher: (P_{Design}7.8 bar-g)
  - Separator A & B: (P_{Design}6.9 bar-g)
  - Treater A & B: (P_{Design}5.85 bar-g)
- Flare Stack Repairs
- DCS Upgrade

Normal Shut-Down Execution

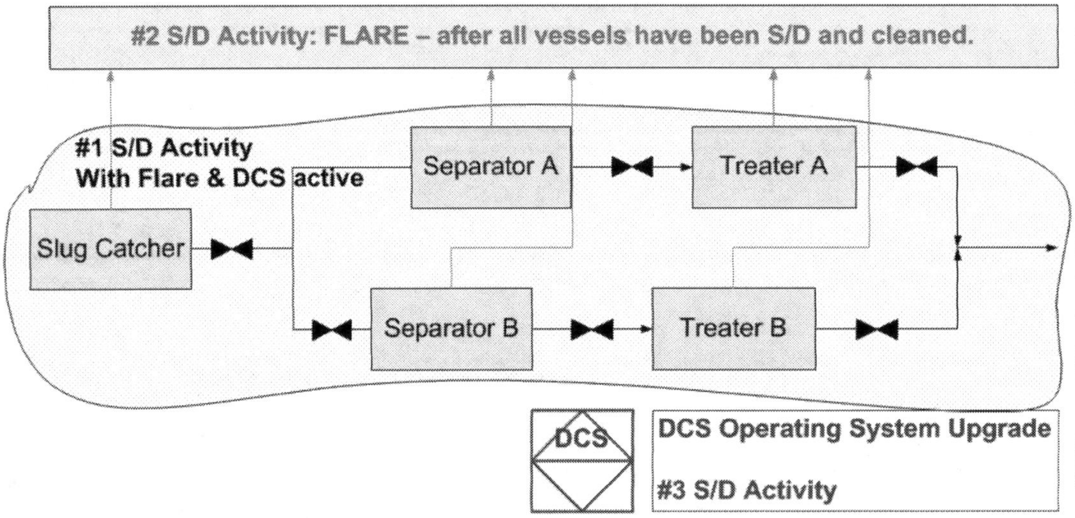

However, "Need for Speed"

Aggressive Shut-Down

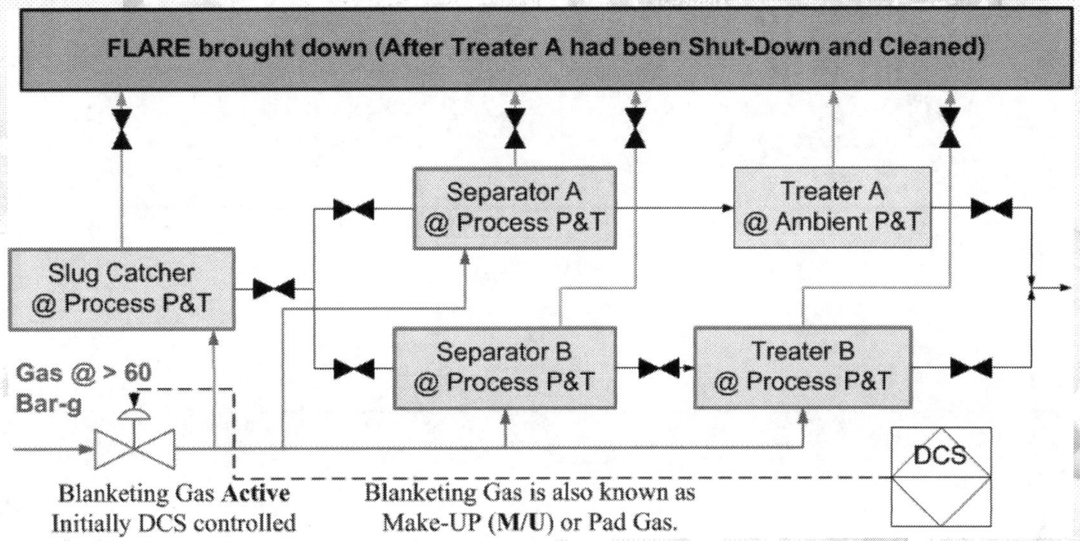

8/29/2014 J Windhorst 11

PHA Issues

- Local vessel venting not possible (→ H2S in the work area).
- Vacuum was not a concern because DCS-controlled M/U gas available.
- DCS S/D was an unknown and a major operations concern; however, a review showed that M/U Gas PCV was FC.

8/29/2014 J Windhorst 12

PHA Issues P&ID

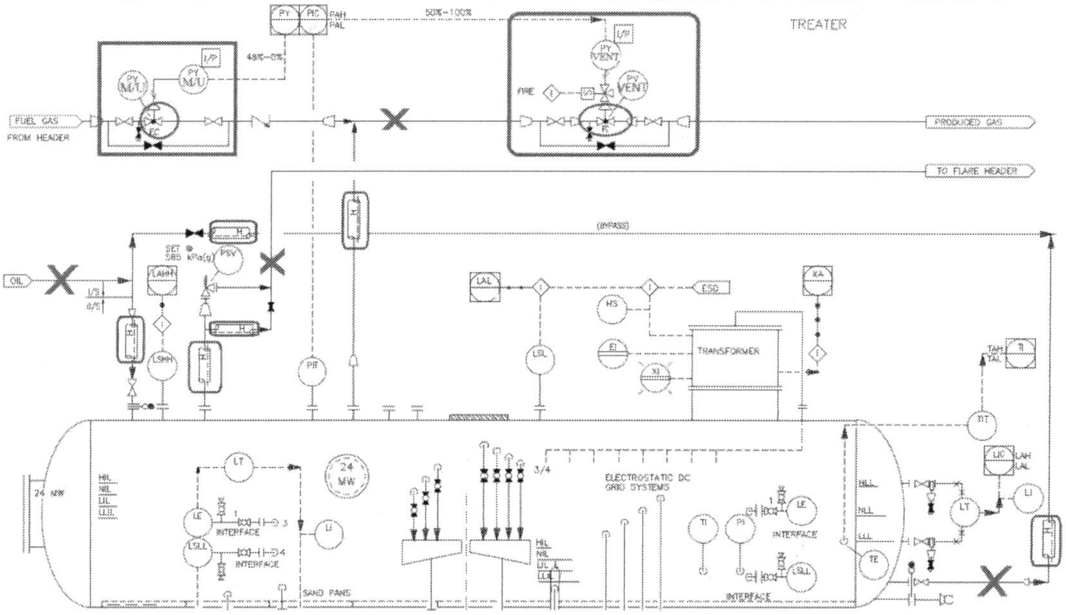

A Vessel System Energy Balance

- Energy leaving the (4) vessel systems, after isolation from other vessels, the flare, and vent gas header, was determined by cooling and leaking flanges (negligible).
- The energy entering the vessel systems was determined by the M/U gas pressure (> 60 bar-g) , the PCV design, and the DCS control system.

Event sequence of the 4 isolated vessels:

- Slow pressure rise;
- Activation of high pressure DCS alarms (not an expected pressure drop!);
- S/D of the DCS (for its upgrade):
  - silenced the PAHs (ISA 18.2);
  - erased DCS memory;
  - Loss of signal drove M/U gas PCVs wide open;
  - No incident process data (..of loss of containment)

8/29/2014 J Windhorst 15

Consequences

- Over-pressurization, some gauges stuck at end of range > 11 bar-g, likely capped by line losses and leaks (vessel); invoking API 579.
- Flammable atmosphere in building (no ignition; near miss w.r.t. a flash fire, vented building explosion and multiple fatalities.
- H_2S in building; exposure of ~ 40 people; duration was mitigated by a quick evacuation.
- M/U gas valves manually closed.

8/29/2014 J Windhorst 16

Treater Damage (Green = No Change)

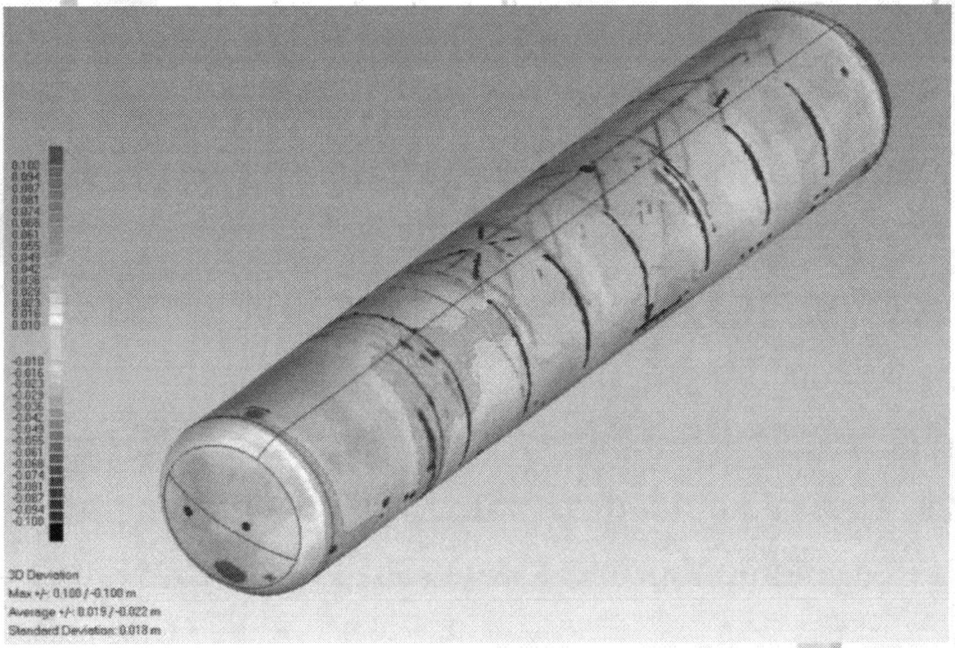

8/29/2014 J Windhorst 17

How did it happen?

Technically

8/29/2014 J Windhorst 18

M/U & VENT PCVs

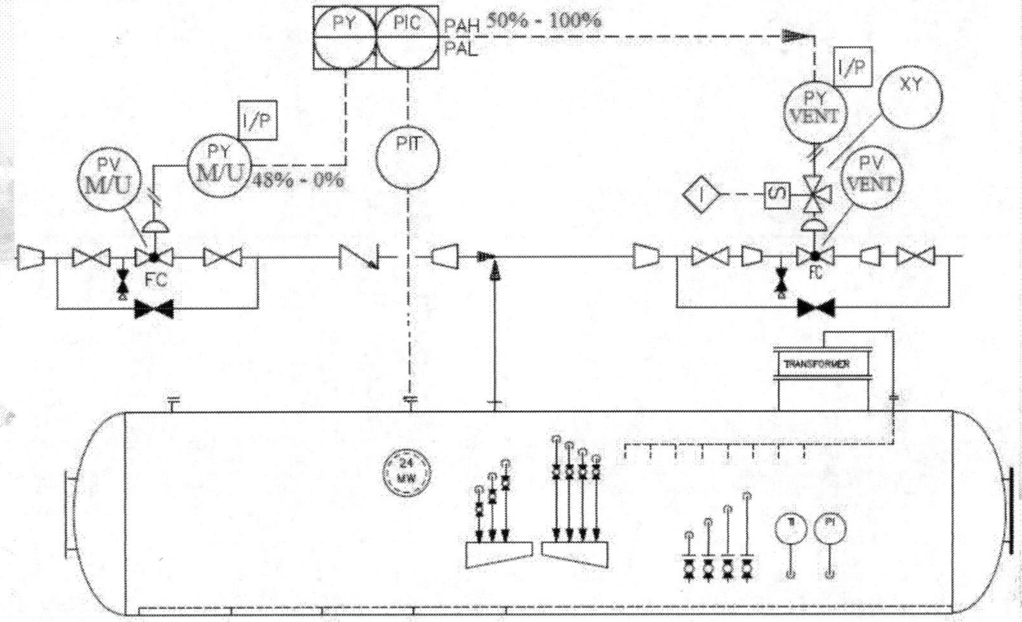

PY
PIC
PAH
PAL
50% - 100%
I/P
PY VENT
XY
PV VENT
PIT
I/P
PY M/U
48% - 0%
PV M/U
FC
S
I
FC
TRANSFORMER
24 MW

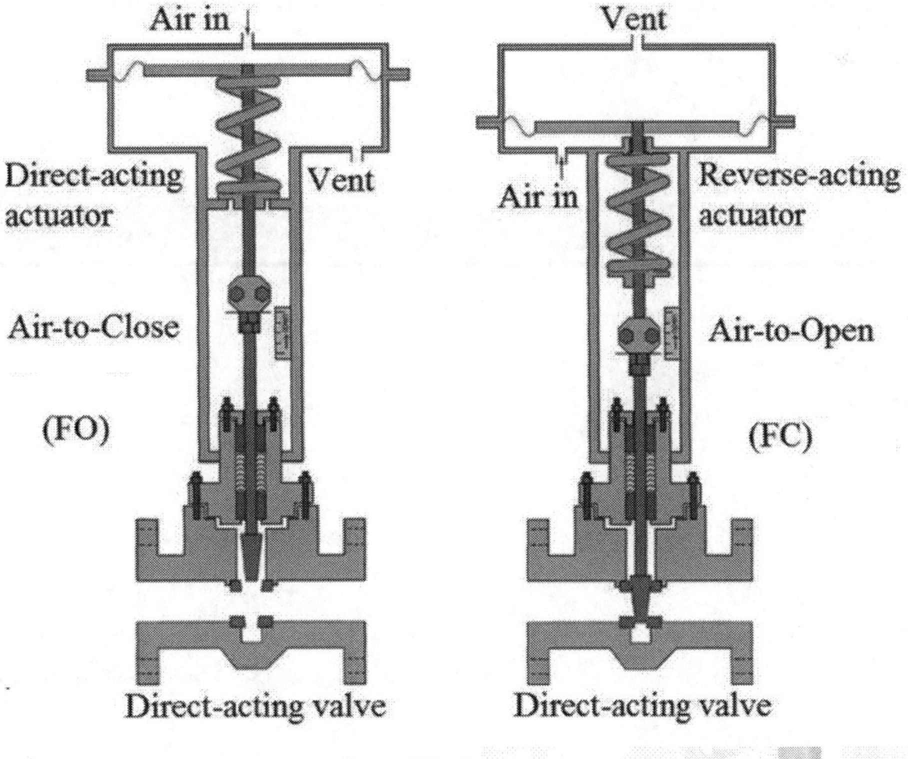

Air in
Direct-acting actuator
Vent
Air-to-Close
(FO)
Direct-acting valve
Vent
Air in
Reverse-acting actuator
Air-to-Open
(FC)
Direct-acting valve

Both PCVs are FC for Loss of IA

8/29/2014 J Windhorst 21

Incident: Controller-I/P → M/U valve FO

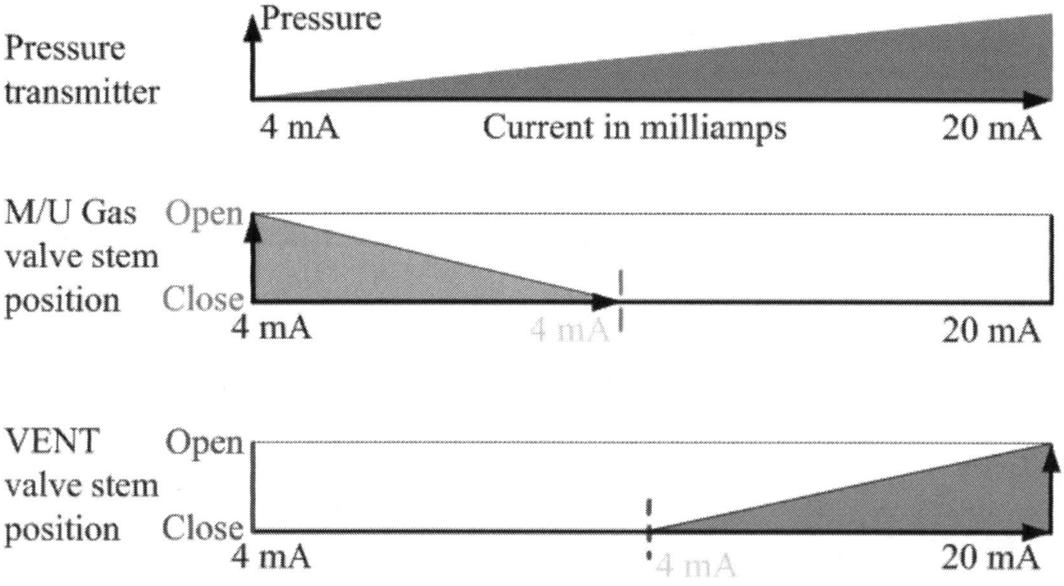

8/29/2014 J Windhorst 22

Controller-I/P → M/U valve FC (correct)

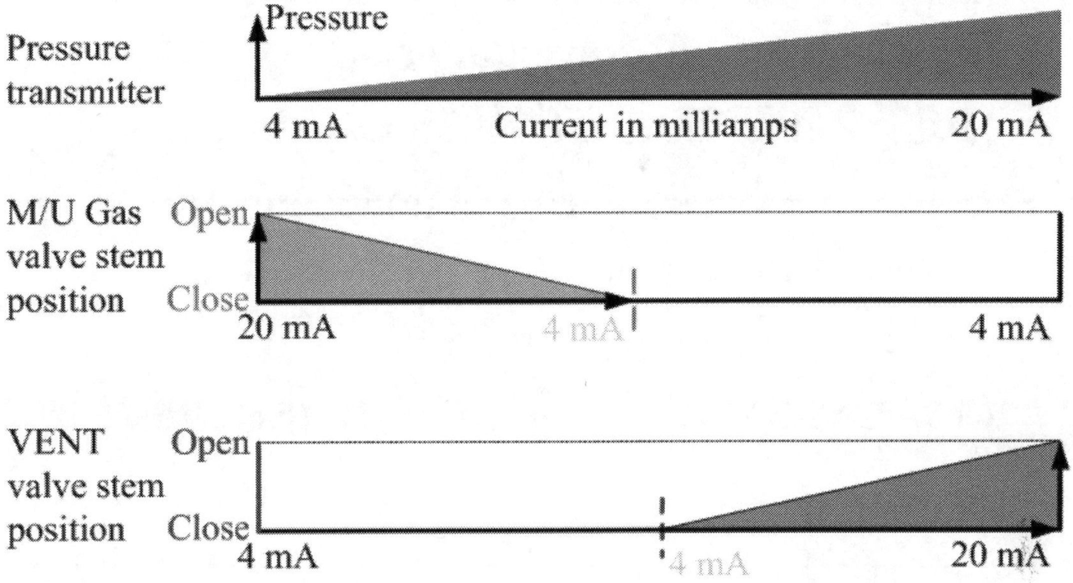

8/29/2014 J Windhorst 23

Causal Notes:

The incident was enabled by:

- S/D changes [management].
- Ignoring alarms [operators].
- Lack of understanding of control valve default positions for loss of IA as well as DCS (signal) [engineering].

Note: *This is not an usual PHA review item*!

8/29/2014 J Windhorst 24

Conclusions

- Small valves big consequences.
- Hi/Lo P interfaces are inherently dangerous.
- Ignoring process alarms is unacceptable.
- Process safety documentation has to be available and understandable.
- Enforce PSM (training, MOC, PSSR, PHAs…)

8/29/2014 J Windhorst 25

加工海洋高酸原油常减压装置设备管线的腐蚀与防护研究

朱玉明[1] 甘为民[2] 薛光亭[2]

（1.中海油气开发利用公司，北京，100029；2.中海油（青岛）重质油加工工程技术研究中心有限公司，山东 青岛，266500）

摘要：油气开发利用公司所加工的海洋高酸原油，密度大、酸值高，电脱盐困难，造成设备管线的低温盐酸腐蚀和高温环烷酸腐蚀严重，主要表现在常减压装置常减压塔顶系统设备管线腐蚀严重和减压塔高温部位填料及高温管线腐蚀损坏严重。本文通过对油气公司所属各企业的腐蚀调查和研究，归纳总结了各企业对设备管线采取的防腐蚀措施以及存在的问题，提出了改进原油电脱盐技术、原油注碱、规范和优化低温部位的“三注”工艺、加强对防腐助剂的筛选评价、合理选用设备管线材质、加强和规范腐蚀检监测工作等设备管线防腐蚀的具体措施和建议。

关键词：海洋高酸原油；常减压装置；设备管线；腐蚀与防护

Study on Corrosion and Protection of Atmospheric and Vacuum Distillation Unit Processing Marine High Acid Crude Oil

Abstract: The marine high acid crude oil processed in Oil & Gas development & Utilization Co., CNOOC has high density, high acid value and the electric desalting difficulties. Due to the characteristics of marine high acid crude oil, low temperature hydrochloric acid corrosion and high temperature naphthenic acid corrosion is the tough problem in the atmospheric and vacuum crude distillation tower system. According to the corrosion investigations and researches on subsidiaries of Oil & Gas development & Utilization Co., CNOOC, this article summarized corrosion problems and existing protection measures in subsidiaries and provided effective methods for anticorrosion. These methods include modified technology of crude oil electric desalting, alkali liquor injection technology, improved "three injection" technology at low temperature parts, standardized corrosion monitoring and so on.

Key words: Marine high acid crude oil; Atmospheric and vacuum distillation unit; Equipment and pipelines; Corrosion and protection

海洋高酸原油在加工过程中不仅高温部位设备管线环烷酸腐蚀严重，而且由于电脱盐困难，脱后含盐高，造成常减压装置初顶、常顶和减顶部位及冷凝冷却系统设备管线的严重腐蚀，有的甚至造成下游催化装置、焦化装置及加氢装置低温部位和高温部位设备管线的腐蚀。严重影响和制约了装置的长周期安全平稳运行，也影响了企业的经济效益。为了减轻设备管线腐蚀，延长装置开工周期，油气公司各企业根据各自实际情况采取了一系列设备管线防腐蚀措施，有效的减轻了装置的设备管线腐蚀，延长了开工周期，确保了装置的长周期安全平稳运行。

1. 原油的品种和存在的腐蚀问题

1.1 原油的品种及特性

油气公司各企业所加工的原油主要以 SZ 36-1、BZ25-1、CFD、LD、QHD32-6、JZ9-3 海洋高酸原油为主，这些原油大部分是高酸低硫原油，见表 1。原油的密度高、粘度大，电脱盐难度大，脱后含盐高，因此为装置的设备腐蚀留下了隐患。

表 1 各种原油与腐蚀有关的性质

原油名称	密度 g/cm^3	硫含量 %	酸值 mgKOH/g	盐含量 ug/g	Fe ug/g	Ni ug/g	V ug/g	N m%
CFD	0.9558	0.24	4.30	16.74	14.3	7.54	0.46	0.43
SZ36-1	0.9645	0.41	1.92	7.68	13	38	1.03	0.42
QHD32-6	0.9504	0.33	2.21	16.84	11.2	6.08	0.44	0.37
BZ25-1	0.9472	0.28	1.91	16.79	28.9	10.8	0.58	0.45
JZ9-3	0.9288		2.41	24.8	4.46	16.07		
LD	0.9111	0.249	1.501	5.2	8.141	13.92	0.819	0.296

2. 装置存在的设备腐蚀问题

由于所加工海洋高酸原油腐蚀性强，各企业装置普遍存在着不同程度的腐蚀问题，有的企业腐蚀相当严重，导致装置开工周期短，设备管线更换频繁，既影响了装置的经济效益，又造成检修资金的浪费。腐蚀主要发生在常减压装置的初馏塔顶、常压塔顶、减压塔顶的低温部位及塔顶冷凝冷却系统和常减压装置的高温重油部位，腐蚀的形态主要是低温部位的 $HCl-H_2S-H_2O$ 腐蚀和高温部位的高温环烷酸腐蚀。

2.1 低温部位的 $HCl-H_2S-H_2O$ 腐蚀

低温部位的腐蚀主要是由于电脱盐效果差，脱后含盐高造成的初馏塔、常压塔和减压塔顶部位以及塔顶系统冷换设备的 $HCl-H_2S-H_2O$ 腐蚀。腐蚀的形态主要表现为，初馏塔顶、常压塔顶部位碳钢塔盘腐蚀严重，使用周期短，更换频繁；初馏塔、常压塔、减压塔碳钢塔壁腐蚀减薄严重，有的甚至出现腐蚀穿孔泄漏事故；塔顶系统碳钢空冷器及冷凝冷却器腐蚀严重，见表 2。如某公司常减压装置常压塔顶部塔壁出现腐蚀穿孔泄漏，塔盘部位塔壁腐蚀减薄严重，有的减薄接近一半，常顶管线、常一线管线也腐蚀减薄严重。某公司常一线馏出线管线弯头出现腐蚀穿孔泄漏。某公司塔顶系统碳钢冷凝冷却器管箱内有大量的硫化亚铁黑色锈皮（1～2mm）和黄色腐蚀产物，管束外表面腐蚀严重，管束腐蚀减薄至为 1.9 ～2.1mm，管束表面呈腐蚀坑状分布，最大坑深约 0.5mm，见照片 1、照片 2。分析塔顶部位出现腐蚀穿孔泄漏的原因：一是原油脱后含盐高，二是塔顶挥发线注氨、注缓蚀剂、注水效果不理想。

装置加工的原油主要是 SZ36-1 和 LD 原油，从脱盐效果差，脱后含盐量比较高，脱后含盐在 5～10mg/l 之间，造成塔顶氯离子含量高，造成塔顶系统设备管线腐蚀严重。

常减顶冷凝水的 pH 值虽然基本控制在 7～9 之间，但是有波动，多数时候在 8～9 之间，偶尔会高于 9，最高时为 9.4，说明氨水注入量比较大，注氨水量一般控制冷凝排水 pH 值在 7 左右比较合适， pH 过高，不仅氨用量增大，而且生成的 NH4Cl 易溶于水，不溶于汽油，容易造成 NH4Cl 析出堵塞管线以及 NH4Cl 的垢下腐蚀[1]。

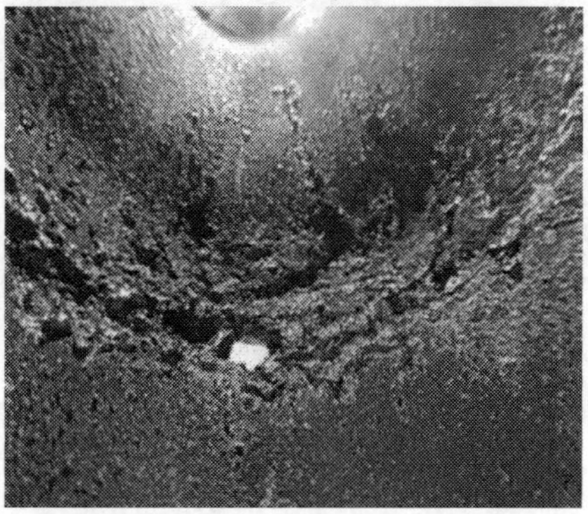

图 1 常压塔常一线大小头腐蚀穿孔外观　图 2 常压塔常一线大小头腐蚀穿孔内部

表 2 各企业常减压低温部位设备管线的腐蚀问题

序号	腐蚀问题描述
1	某公司减顶冷凝冷却器采用 Ni-P 镀管束，使用 5 个月即出现腐蚀穿孔泄漏。
2	某公司常减压装置常压塔顶部塔壁出现腐蚀穿孔泄漏。第 44、46、47、48 层塔盘部位塔壁腐蚀减薄严重，有的减薄接近一半。常顶管线、常一线管线也腐蚀减薄严重。
3	某公司常压塔顶部距离塔顶约 2.5 米处塔壁腐蚀穿孔泄漏。
4	某公司 50 万吨沥青装置常压塔上部腐蚀泄漏起火。
5	某公司塔顶系统碳钢冷凝冷却器 E-140/1、2 管箱内有大量的硫化亚铁黑色锈皮（1～2mm）和黄色腐蚀产物，管束外表面腐蚀严重，管束腐蚀减薄至为 1.9 ～ 2.1mm。
6	某公司常减压塔常一管线于2014 年7 月份共出现三次腐蚀泄露问题。见照片1、2。

2.2 高温部位的腐蚀

高温部位的腐蚀主要是指 240℃以上高温重油的环烷酸腐蚀[2]，由于海洋原油酸值高、硫含量低，环烷酸的腐蚀比其他高酸原油更为严重。其表现形式主要是减压塔减二、减三线、减四线填料及管线腐蚀严重，即使 316L 不锈钢规整填料也不能解决问题。其次，常压塔常三线、减二线、减三线、减四线管线和阀门腐蚀减薄严重，不得不提高材质，大量使用 304、321 和 316L 不锈钢材质，有的装置经常出现上述部位泵出口阀门因密封面腐蚀而关不严现象，给装置的泵的切换带来很大困难，不得不大量更换为不锈钢材质，加热炉辐射炉管、加热炉转油线、常减压高温部位塔壁也因为腐蚀严重不得不大量使用 316L 不锈钢和不锈钢复合板。

如某公司常压塔转油线因焊缝腐蚀出现泄漏着火，见照片 3、照片 4。腐蚀泄露的原因是由于内衬管（材质：00Cr17Ni14Mo2,规格尺寸：φ168×2）不锈钢层焊接未完全熔合，复合管（材质 20，规格尺寸：φ168×7）碳钢被腐蚀穿透造成。随后将高速转油线拆下进行现场剖析，发现法兰连接密封垫片内环已被腐蚀过半，利用材质分析仪进行检测，内环非 316L 材质，缠绕部分为 316L 钢带。在高速转油线 Φ168×9（7+2）内壁检查时，发现有三处局部鼓包，认为是环烷酸腐蚀所致。

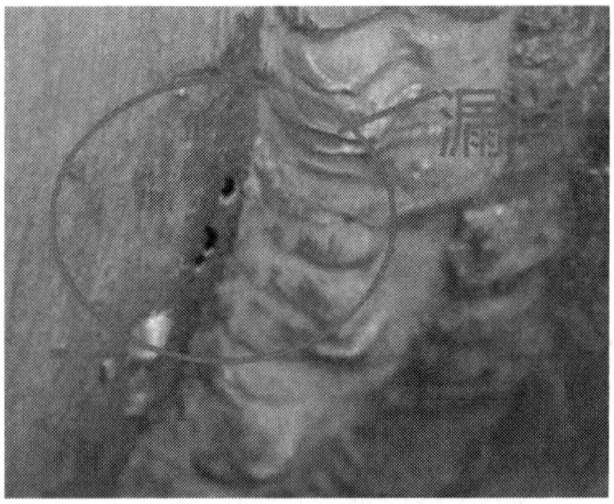

图 3 转油线漏点外观

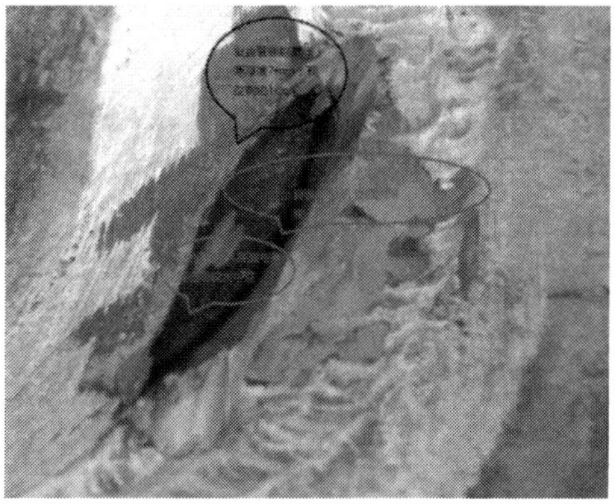

图 4 转油线漏点刨开

3. 设备管线采取的防腐措施及效果

为了应对加工海洋高酸原油带来的设备管线腐蚀问题，各企业都采取了一系列设备防腐措施，这些措施包括：加强“一脱三注”工艺防腐措施；低温部位和高温部位设备管线材质升级；加强定点测厚及腐蚀在线监测等，有的企业甚至在原油、常顶和减顶系统尝试采用注碱工艺，取得了一定的效果，但许多防腐措施还不够规范和完善，需要进一步改进。

3.1 一脱三注工艺防腐措施

3.1.1 原油电脱盐

原油电脱盐是“一脱三注”工艺防腐措施的关键和龙头措施，电脱盐效果的好坏，直接影响到常减压装置及下游装置的设备管线腐蚀。有的企业由于采取了比较先进和有效地电脱盐技术和设备，电脱盐效果较好，但也有的企业因为脱盐效果不好，脱后含盐高，能耗大，电脱盐装置长期停用。还有的企业因为常减压主装置改造扩建，但电脱盐装置并未进行相应的改造，因此电脱盐能力严重不足，原油脱后含盐高，对设备管线的腐蚀造成很大的隐患，见表 3。

分析各企业电脱盐差的原因是：随着原油深度开采及三次采油技术的广泛应用，大量聚合物驱油剂[3]、稠油降粘剂、清蜡剂应用于原油开采和储输过程中，这使得原油含水含盐高，并含有大量乳化剂、高分子聚合物等，油水乳化程度严重，乳化类型更加复杂，性质更加稳定，破乳和脱盐脱水也更加困难。对重质高酸原油电脱盐的研究和各企业的运行现状表明，电脱盐温度较常规原油电脱盐的温度高 20℃左右、注水量大 3～4%，电脱盐的电场强度稍高，停留时间长等，但脱盐脱水效果总体表现不佳。对比重大于 0.94 的重质高酸原油，脱后盐含量普遍超标，脱后原油水含量、污水油含量及能耗也较常规原油偏高。研究及调研情况表明，脱后原油含水量要达到 0.2mg/L 以下较为困难，即使在实验室条件下，要达到 0.2mg/L 以下，必须将注水（水质为新鲜水）的 PH 值调整为酸性即加酸才能达到，但相应带来设备管线的腐蚀和成本的增加，各企业为了降低成本，大都没有采取上述措施。

表 3 各企业电脱盐情况调查表

企业名称	脱盐级数	主要原油品种	原油密度 kg/m3	脱前含盐 mg/l	脱后含盐 mg/l	开工情况

湛江公司	二级	LD、CFD SZ36-1	0.9375	8～15		不开
四川公司	二级	SZ36-1	0.9589	10～15		不开
中沥公司	二级	SZ36-1、QHD32-6	0.9645	40～50~	6～9	开
泰州石化	二级	SZ36-1	0.9658	35～50	5～8	开
泰州沥青	二级	SZ36-1	0.9589	30～40		不开
营口公司	二级	SZ36-1	0.9589	8～11		不开
青岛公司	二级	SZ36-1、LD	0.9658	35	3～9	开

3.1.2 塔顶注中和剂、缓蚀剂、注水

为了减轻塔顶系统设备管线的腐蚀，各企业都采取了塔顶"三注"措施，即初顶、常顶、减顶注中和剂、缓释剂、注水措施，但注入方式、注入量、注入的种类以及注入效果不尽相同。大部分企业中和剂采用注氨，较少采用注有机胺中和剂，有的企业尝试采用注碱作为中和剂。各企业都采用了在塔顶挥发管线注缓蚀剂措施，但注水措施有的采用单独注水，有的企业采用注氨水的方式代替注水，见表 4。总体看，大部分企业在塔顶低温部位的注剂方式不够标准和规范，没有很好的发挥出"三注"对设备的保护效果。

表 4 各企业塔顶系统设备管线注中和剂、缓蚀剂、注水及 PH 值控制情况

企业名称	注中和剂位置	种类	注缓蚀剂位置	种类	注入量 ppm	注水的位置	种类	注入量	pH 控制指标
湛江公司	常减压塔顶	注碱 0.3%（3%的碱液）	常减压塔顶	中和缓蚀剂	30ppm	常减压塔顶	新鲜水循环水		7～9
四川公司	塔顶挥发线	液氨稀释成氨水	塔顶挥发线	中和缓蚀剂	30ppm	塔顶挥发线	新鲜水	塔顶馏出量 5%	6～8
中沥公司	常顶减顶	15%氨水	常顶减顶	中和缓蚀剂		常顶	新鲜水		6～9
泰州石化	常顶减顶	液氨稀释成氨水	常顶减顶	中和缓蚀剂	45-60ppm	常顶减顶	酸性水回注	2.5 吨/小时	7～9
泰州沥青			常顶减顶	中和缓蚀剂	25ppm	常顶减顶			6.5～9
营口公司	无	无	常顶减顶	中和缓蚀剂	30ppm	常顶减顶	无	连续注入	7～9

3.2 高温部位注高温缓蚀剂

为了减轻高温部位设备管线环烷酸的腐蚀，各企业都采取了注高温缓蚀剂的措施，但高温缓蚀剂的注入部位、注入方式、使用量各不相同，各企业的高温缓蚀剂使用情况见表 5。

表 5 各企业高温缓蚀剂的使用情况

企业名称	注入部位	缓释剂种类	生产厂家	注入量	注入方式	使用效果监测
湛江公司	原油	FH 高温缓蚀	济南丰华化工厂	41.2ppm	减二线减三线	分析侧线油中

	减二线 减三线	剂		34.5ppm 30ppm	返塔线注入	的 Fe^{2+}
四川公司	减二中	FH 高温缓蚀剂	济南丰华化工厂	30ppm		分析侧线油中的 Fe^{2+}
中沥公司	减二冷回流 减三冷回流	FH 高温缓蚀剂	济南丰华化工厂	30~50ppm	连续注入	分析侧线油中的 Fe^{2+}
泰州石化						
泰州沥青	减二中回流调节阀后	FH 高温缓蚀剂	济南丰华化工厂	25 ppm	连续注入	分析侧线油中的 Fe^{2+}
营口公司	常二中 常二线 减二中 减三线 减四线下部	FH 高温缓蚀剂	济南丰华化工厂	30ppm	连续注入	分析侧线油中的 Fe^{2+}
青岛公司	减一中 减二中 减三线 减四线	FH 高温缓蚀剂	济南丰华化工厂	30ppm	连续注入	分析侧线油中的酸值、Fe^{2+}

3.3 原油注碱和高温侧线油注碱的工艺防腐

油气公司各企业由于原油密度高粘度大，普遍脱盐效果不好，脱后含盐高，造成常减压塔顶系统氯离子含量很高，设备管线腐蚀严重，为此，今年在四川公司和营口公司开展了原油注碱工业试验，取得了良好的效果，减顶 Cl^- 浓度从原来的 38.8mg/l 降低到 4.0mg/l，下降了 89.56%，铁离子浓度小于在 3mg/l；常顶铁离子浓度小于 1.0mg/l，低温缓蚀剂量由 7ppm 下降到 5ppm（原油比）。大大降低了塔顶系统的氯离子含量和铁离子含量，减轻了设备管线的腐蚀。

3.4 设备管线材质升级

为了应对海洋高酸原油的环烷酸腐蚀，各企业都采取了一系列设备管线材质升级措施，见表 6。如在塔顶低温部位使用双相不锈钢复合管线，塔顶水冷器采用 AF2205 或 AF2507 双相不锈钢管束。高温部位设备管线大多采用 304L、321、和 316L 材质，如换热器管束采用 304L、321、和 316L，壳体采用 16MnR+321 或 16MnR+316L 复合板材质。高温管线大都采用 304L、321、和 316L 材质。高温阀门、高温油泵按照加工高酸原油的标准采取了升级措施。这对解决重质高酸原油的设备管线腐蚀问题起到了至关重要的作用。但有的装置在进行材质升级时，在塔顶低温部位使用了一些 304、321 奥氏体不锈钢设备管线，这些部位由于冷凝水中存在几十甚至几百 ppm 的氯离子，很容易出现氯离子引起的奥氏体不锈钢设备管线的应力腐蚀开裂。

表 6 各企业主要设备管线材质使用情况

公司名称	材质使用情况
湛江公司	常压塔：筒体碳钢+321、316 衬里，塔盘，碳钢、316L；减压塔筒体碳钢+321、316 衬里，塔内件 316L；初馏炉炉管：316L；减压炉炉管 321；初馏塔顶挥发管线碳钢；减一线碳钢；初馏塔转油线 316L；减压塔转油线 316L；常一线、常二线、减压塔顶挥发管线 321；减二线、减三线 316L。

四川公司	闪蒸塔、常压塔、常压汽提塔、减压塔、减压汽提塔采用 20R+00Cr17Ni14Mo2；常顶挥发线材质碳钢。
中沥公司	常Ⅲ常压塔下段 0Cr18Ni10Ti，上段 16MnR；常Ⅲ减压塔 16MnR+316L；常Ⅳ常压塔、常Ⅳ减压塔 16MnR；减二中油线、减三油线、减四油线、减五油线采用 321 材质；减顶馏出线水冷器管束采用 321；减二线及减二塔壁填料、采用 321、316L、317L 不锈钢；常三空冷器管束采用 AF2205 双相钢。
泰州石化	常顶空冷器管束材质 AF2205；常一、二、三空冷器管束为碳钢；闪蒸塔碳钢、常压塔、减粘分馏塔材质 16MnR+316L；常压汽提塔、减压塔、减压汽提塔材质 20R+316L；常压炉、减压炉、减粘炉炉管均为 316L；水冷器大部分为碳钢管束；高温换热器管束为 304L 和 316L 不锈钢，壳体大部分为碳钢+316L 不锈钢复合板。
泰州沥青	闪蒸塔 16MnR；常压塔、常压汽提塔、减压塔、减压汽提塔 16MnR、316L；常压炉、减压炉辐射段 316L 对流段 316L、 Cr5Mo；常压炉入口管线 316L+20＃；常炉出口管线 316L；常压转油线 316L+20＃；常顶冷油回流管线碳钢；常顶循油泵入口管线 1Cr18Ni9Ti；常顶循油泵出口管线碳钢+1Cr18Ni9Ti；常一中油泵入口管线、常一中回流管线、常一抽出管线、常二抽出管线、常二外送管线、常三外送管线碳钢；常二中回流管线、常三抽出管线、常底油泵进口管线、常底油泵出口管线、减炉入口管线、减炉出口管线、减一中回流管线、减二中回流管线、减底油进减粘炉管线、减粘炉出口管线、反应塔顶出口管线为 316L；常顶挥发线、减顶循热回流为 321。
营口公司	常顶挥发线、常一线、常二线、常三线、减顶、减二线、减四线等 220℃以下管线采用 20#碳钢；减渣油、常一中油、常一线油、减渣油、闪蒸塔退油、常压塔退油、减压塔退油、常二线油、换热后原油等 220℃以上管线采用了 20#+316L 复合管。

3.5 加强设备管线腐蚀监测

采取的腐蚀监测措施主要包括：冷凝水中 Fe^{2+}、CL^-、pH 分析监测、常减压侧线油中 Fe^{2+} 分析检测、冷凝水 PH 在线监测、腐蚀在线监测、定点测厚等。由于各企业规模和加工油种以及加工工艺不尽相同，因此所采取的腐蚀监测措施也不一样，详情见表 7。

表 7 各企业所采取的腐蚀监测措施

企业名称	所采取的腐蚀监测措施	
湛江石化	常顶、减顶冷凝水中 Fe^{2+}、CL^-、pH 分析； 电阻探针腐蚀在线监测； 定点测厚； 侧线及原油中 Fe^{2+}含量分析。	PH 试纸和便携式 PHb-4PH 计； 邻菲罗啉分光光度法； 微机盐含量测定（微库仑）； 定点测厚。
四川公司	冷凝水中 Fe^{2+}、pH 分析，2 次/月.	邻菲罗啉分光光度法； PH 试纸\ PH 计。
中沥公司	塔顶冷凝水分析 Fe^{2+} 、CL^-，不定期分析； 侧线油 Fe^{2+}分析，不定期分析； 定点测厚，每季度一次。	邻菲罗啉分光光度法； 硝酸银滴定法； PH 试纸\ PH 计。
泰州石化	常顶、减顶冷凝水 PH 值分析； 减二、减三线油 Fe^{2+}分析，1 次/月； 定点测厚，1 次/半年。	PH 试纸法； GB/T 3049-1986。
泰州沥青	常顶、减顶冷凝水 Fe^{2+}、CL^- 、pH 值分析，1 次/月； 侧线油全分析，1 次/年；	邻菲罗啉分光光度法； 硝酸银滴定法； PH 试纸\ PH 计。

	定点测厚，1次/月。	
营口公司	常顶、减顶冷凝水、PH值分析，2次/班； 常顶、减顶冷凝水Fe^{2+}、CL^{-}分析，1次/天； 侧线油Fe^{2+}分析，2次/周。	邻菲罗啉分光光度法； 硝酸银滴定法； PH试纸\ PH计； 原子吸收法。
青岛公司	塔顶冷凝水分析Fe^{2+}、CL^{-}，不定期分析； 侧线油Fe^{2+}分析，不定期分析； 定点测厚，每季度一次； 常减顶腐蚀在线监测，实时监测。	邻菲罗啉分光光度法； 硝酸银滴定法； PH试纸\ PH计。

从表7可以看出，各企业虽然对设备管线采取了一些腐蚀检（监）措施，但由于条件相差较大，所采取的措施也相差较大。有的企业由于没有专人管理，腐蚀检监测工作尚处于起步阶段，许多工作有待于健全和提高。但总的来看，各企业对设备管线腐蚀监测工作都有了一定得认识和基础，下一步需要进行一定的规范和提高。

4. 下一步应采取的防腐措施

为了减轻设备管线腐蚀，延长装置开工周期，各企业根据各自实际情况采取了一系列设备管线防腐蚀措施，油气公司针对高酸原油的设备管线腐蚀问题，专门制定了《加工高酸原油常减压装置防腐蚀实施细则》和《常减压装置检查鉴定实施细则》两个防腐和检修推荐性做法并在各企业发布实施，下一步应根据上述细则的要求，进一步指导和规范各企业的设备管线防腐蚀工作，减轻设备管线的腐蚀，延长开工周期，确保装置的长周期安全平稳运行。

4.1 改进重质高酸原油原油电脱盐技术，加强高酸原油电脱盐的管理，提高电脱盐效率，降低脱后含盐量

搞好高酸原油的电脱盐工作是常减压装置设备管线防腐蚀的关键，为此青岛研究中心联合长江电脱盐公司开展了重质高酸原油电脱盐技术集成及应用研究，目前实验室各项工作已经完成，渤海高酸原油的脱后含盐量完全达到3.0mg/l以下，下一步将在青岛研究中心工业装置上首先进行工业试验，然后推广到油气公司各企业进行推广应用。同时加强对电脱盐现有装置的管理，摸索最佳的电脱盐工艺条件和参数，适当提高天脱盐的温度和原油注水温度，筛选出适合重质高酸原油的高效破乳剂，以提高装置的电脱盐效率，降低脱后含盐量，减轻塔顶低温部位设备管线的腐蚀。

4.2 继续开展原油注碱和高温侧线油注碱的工艺防腐研究工作

原油注碱作为“一脱四注”的主要工艺防腐措施之一，曾经在炼油厂设备管线防腐蚀工作中发挥过重要作用，由于原油注碱后容易造成催化装置的催化剂中毒和换热设备结垢被废止，但原油注碱的防腐作用很难用其他方法代替。近来随着炼油装置开工周期的延长和对防腐工作的重视，原油注碱工艺又重新被研究和采用。据调查国外如日本和韩国某些炼油厂一直坚持原油注碱工艺，碱液注在加热炉之前，并且严格控制碱的注入量，一般为3～5ppm，取得了很好的防腐效果。日本有的生产沥青的炼油厂甚至在容易产生高温环烷

酸腐蚀的常减压高温重油侧线上注碱，也取得了很好的效果，较之注高温缓蚀剂更为有效。国内镇海炼油厂也在电脱盐后的原油中注碱，注碱的碱液浓度为 4%wt，注碱量为 2.5g/t 原油，注碱后常压塔顶换热设备管线的腐蚀率下降了 80%，常压塔顶冷凝水回流罐中的 Cl^{-}含量也下降了 80%，常减压塔顶中和剂的用量下降了约 30%，且对二次加工装置没有造成影响。因此建议继续开展在原油电脱盐后注碱的研究工作，找出最佳的注入位置、注入量和注入方式，然后在全系统推广。

4.3 规范和优化低温部位的“三注”工艺，提高工艺防腐的效果

“一脱三注”工艺防腐措施是减轻常减压装置低温部位设备管线腐蚀的最有效和最根本的防腐措施，油气公司各企业由于加工重质高酸原油，电脱盐效果大都不理想，因此搞好常压塔、减压塔顶的注中和剂、缓蚀剂和注水“三注”工艺就显得尤为重要。各企业的“三注”工作需要进一步优化，如有的企业只注中和缓蚀剂，有的企业只注氨水不注水，靠氨水中的水代替注水，大多数企业塔顶三注都没有设计专门的注剂喷嘴，注剂后的效果不够理想，所有这些缺陷都决定了三注的效果不会理想，因此很难保证塔顶系统设备管线的防腐蚀效果，造成塔顶部位和塔顶系统设备管线腐蚀严重，出现塔顶部位塔壁腐蚀穿孔，塔顶空冷器和水冷器管束频繁腐蚀泄漏。不仅浪费了大量的财力物力，也严重制约了装置的开工周期和经济效益，同时给装置的安全运行留下了隐患。因此应按《加工高酸原油常减压装置防腐蚀实施细则》的要求，规范和优化三顶系统的“三注”工艺，提高工艺防腐蚀效果，减轻设备管线腐蚀。

4.4 加强对防腐助剂的筛选评价，优化防腐注剂方案，提高工艺防腐效果

各企业为了减轻设备管线的腐蚀，都采用了加注电脱盐破乳剂、中和剂、低温缓蚀剂、高温缓蚀剂的工艺防腐措施，这对减轻设备管线的腐蚀，延长开工周期起到了一定的作用。但由于各企业没有对各种助剂进行评价筛选和进厂检验的措施，因此加注各种助剂的效果有待于进一步考察和验证。有的企业使用了大量的高温缓蚀剂，但这些高温缓蚀剂使用的部位、使用的浓度和选用的型号是否合适也需要进行验证和实验室试验。因此下一步应定期对各种防腐助剂进行检验和评价，对注剂的使用部位和使用量进行规范，提高防腐助剂的效率，降低防腐助剂的使用成本。

4.5 合理选用设备管线材质

油气公司各企业常减压装置原设计材质较低，加工重质高酸原油后普遍需要进行材质升级。有的企业在塔顶部位大量使用了 321 或 304L 奥氏体不锈钢管线，经过化验分析塔顶冷凝水的的氯离子高达上百 ppm，很容易出现应力腐蚀开裂。有的装置加工高酸原油后设备管线材质基本升级没有不到位，高温部位的阀门频繁出现腐蚀泄漏现象，不得不频繁更换不锈钢阀门。中海油所加工的重质高酸原油的酸值和腐蚀性远大于中石化所加工的高酸原油。因此应尽快按照《加工高酸原油常减压装置防腐蚀实施细则》的要求，对设备管线进行合理的材质升级。

4.6 加强和规范腐蚀检监测工作，预测和防范设备腐蚀事故的发生

针对装置设备管线存在的腐蚀问题，各企业采取了一些腐蚀检（监）测措施，主要包括：塔顶冷凝水 Fe^{2+}、Cl^{-}、NH_3-N、S^{2-}、pH 值分析；侧线线油中 Fe^{2+} 、酸值、硫含量分析；

管线定点测厚；腐蚀在线监测；pH 在线监测；挂片腐蚀监测；氢通量腐蚀监测等。由于各企业的分析检测方法也不尽相同，因此分析检测数据差别较大。今后应对这些检监测措施进行规范，建立一套规范的腐蚀检监侧手段和方法，包括塔顶冷凝水中 Fe^{2+}、Cl^{-}、NH3-N、S^{2-}、pH 值分析；常减压侧线油中 Fe^{2+} 、酸值、硫含量分析；管线定点测厚；腐蚀在线监测；pH 在线监测；挂片腐蚀监测等，尤其是要加强对低温部位冷凝水的分析监测和高温部位设备管线的定点测厚工作，及时发现和消除设备管线的腐蚀隐患，将腐蚀事故消灭在萌芽状态，以确保装置的安全平稳运行。

5. 参考文献：

[1] 苏亚兰.常减压蒸馏装置工艺防腐应用及进展[J].化工进展，2011,30（增刊）：1-5.

[2] 苏芳云，朱文胜，杨志强等. 减压塔填料快速腐蚀失效原因分析[J]. 石油化工腐蚀与防护，2006，23(2)：42-45.

[3] 唐晓东，杨文倩，邹雯炆等.石油化学剂对石油加工过程的影响与对策研究（Ⅱ）[J].西南石油大学学报（自然科学版），2010,32（3）：150-154.

Typical Potential Safety Hazard and Safety Measures of ChangJi Pipeline in Petrochemical Enterprises

Abstract: This paper combined with the recent domestic pipeline safety accidents, listed the screening points of potential safety hazard of petrochemical enterprise ChangJi pipeline, summarized what is typical security hidden danger, and put forward the corresponding security measures and suggestions, provided references for petrochemical enterprises to carry out the ChangJi pipeline safety hidden reviewing in order to prevent similar pipeline accidents.

Keywords: petrochemical enterprise; changJi pipeline; potential safety hazard; pipeline accident

1. 引言

近年来，随着城市油气管道的大量敷设和运行时间的延长，管道事故时有发生，造成了重大人员伤亡和经济损失，管道的安全可靠性问题日益突出。

石油化工企业厂际管道指敷设在石油化工企业、油库、油气码头等不同围墙界区之间且通过公共区域、长度小于等于30km 的输油输气管道[1]。由于厂际管道多数处于石油化工企业厂区围墙之外，涉及的危险化学品种类多，周边环境复杂，且部分厂际管道埋地敷设成为隐蔽工程设施，因此监管难度大，给企业安全检查、隐患排查、安全管理增加了难度。

2. 油气管道常见失效模式及事故案例

油气管道失效模式主要有两种[2]：一种是时间依赖的模式，如腐蚀、老化、疲劳等；另一种是与时间无关的模式，如发生频率较高的第三方损坏、施工期间造成的机械损伤等都会引发管道失效。

2.1 腐蚀、老化、疲劳等引发油气管道失效事故案例

随着时间的推移，我国许多现役管道已进入了"老龄期" 。外环境或输送介质的腐蚀影响以及管线自身的老化、疲劳等，严重影响了管道的正常运行。

2013年11月22日，山东省青岛市黄岛区发生了中石化东黄输油管道泄漏爆炸特别重大事故，事故共造成62人死亡、136人受伤，直接经济损失75172万元。发生事故的东黄输油管道于1986年投入运行，年输油能力1000万吨。该管线泄漏的原因主要是由于管线与排水暗渠交叉段的输油管道所处区域土壤盐碱和地下水氯化物含量高，同时排水暗渠内随着潮汐变化海水倒灌，输油管道长期处于干湿交替的海水及盐雾腐蚀环境，加之管道受到道路承重和振动等因素影响，导致管道加速腐蚀减薄、破裂[3]。

2014年7月31日，台湾李长荣化工公司厂际地下丙烯管道发生泄漏，泄漏的丙烯聚集在下水道内，遇到明火后发生爆炸。该管线核准建设的日期为1990年12月，泄漏的原因初步判断可能为管道老旧造成的接缝泄漏或管道腐蚀造成的管道泄漏。事故造成至少28人死亡，300人受伤。

2.2 第三方施工破坏引发油气管道失效事故案例

中国目前服役的管道中80%使用时间不足15年，管道腐蚀、老化、疲劳只是管道失效的原因之一。由于近年城市建设快速发展，管线范围内新建大量建构筑物，许多建设、施工单位存在不规范施工，在未对地下管网进行有效勘察了解的前提下就盲目施工，导致挖爆、压穿油气管道的事件屡次发生，因此第三方施工破坏成为管道失效事故高发的主要原因之一。

2014年6月30日，大连岳林建筑工程有限公司在大连为加气站配电工程实施电缆外线水平定向钻施工中，将中石油新大一线输油管线钻通，导致原油泄漏。溢出原油流入市政污水管网，在排污管网出口处出现明火，引起火灾爆炸事故，现场无人员伤亡。

2010年7月28日，扬州鸿运建设配套工程有限公司在江苏省南京市栖霞区原南京塑料四厂旧址，平整拆迁土地过程中，施工队伍盲目施工，挖掘机挖穿了地下丙烯管道，泄漏的丙烯蒸发扩散后，遇到附近餐馆明火引起大面积爆燃。事故造成13人死亡，120人入院治疗（其中重伤14人），周边近3平方公里范围的建筑物不同程度损坏。

以上事故反映出石油化工行业管道工程，特别是隐蔽的管道工程致灾隐患突出，因此石油化工企业厂际管道作为石化企业安全管理中的重要组成部分，安全形势十分严峻。

3. 石化企业厂际管道安全隐患排查要点

石化企业厂际管道安全隐患排查主要包括以下几个方面：

1、检查管道规划选址及路由周边情况。检查厂际管道的规划选址的合规性，管道与周边设施的安全防护距离的合规性，管道布置的合规性，管道与市政公用管线、民用管线等的交叉情况、管线穿跨越道路时的保护情况等。

2、检查管道的完整性。检查管线是否存在腐蚀、破损、开裂、变形、老化和跑、冒、滴、漏等现象，是否存在被违法占压现象，埋地管道上方是否存在重型车辆碾压等危害管道安全的行为。

3、检查管道安全设施设置的合规性和完整性。检查管道联锁保护、泄漏检测报警设施等的完善性和有效性，地面管线防腐保温隔热设施的完整性，管道防雷、防静电设施、阴极保护及防腐绝缘设施的设置，涵洞、管沟等场所是否采取了防止气体积聚和沿沟扩散的措施，管道低点排凝设施的设置情况，管道标识设置的合规性和完整性等。

4、检查管道安全管理及应急管理情况。检查管道日常检查、维护运行情况，管道安全技术档案的完整性，管道安全管理制度制订情况，安全管理人员配备情况，管道的安全生产应急管理情况等。

4. 石化企业厂际管道典型安全隐患

1、管线与周边设施安全间距不足。部分管线建成初期周边区域人烟稀少，但随着城镇化进程加快，工矿企业、居民区、学校、店铺等人员聚集场所相继建设，致使部分管道与周边设施过近，无法满足《输油管道工程设计规范》GB50253、《石油化工企业设计防火规范》GB50160、《石油化工金属管道布置设计规范》SH3012、《化工企业总图运输设计规范》GB50489等安全间距的要求。

2、埋地管道沿线占压情况严重。埋地管道沿线周边环境与建设之初发生变化，后期建设设施占压先期建设油气管道现象严重，这不仅给企业管道维护抢修带来不便，同时管道腐蚀穿孔产生油气的泄漏，成为人身、财产安全的巨大隐患。辽河油田公司2006年1月发布的一则通告显示，辽河油田油气管道中心线两侧各5米范围内存在建筑物 2892处，其中直接占压1506处，5米范围内非直接占压1386处。《管道保护法》实施前后，我国曾掀起清理违法占压油气管道建筑物的热潮。黄岛“11•22”事故之后，这项工作又成为各地开展油气管道隐患排查的重中之重，事故发生之后“黄岛14处重大隐患挂牌督办整改项目”中，其中有6处重大隐患为油气管道占压隐患，6处重大隐患为油气管道与市政暗渠交叉隐患，1处重大隐患为油气管道占压并与市政暗渠交叉，1处为原油管道老化腐蚀隐患。

据了解，管道被占压主要存在三种情况：

一是建构筑物占压。某些建设单位在未对地下隐蔽工程进行详细摸底的情况下，就盲目建设施工，或者对地下隐蔽工程不予理会，就将建构筑物建设在油气管线的上方。

二是树木占压。有关部门在植树绿化时容易忽视地下管网的存在，在管道附近种植乔木、灌木等根系发达的树木，随着树木生长根系不断延伸，不但可能损坏管道外防腐层，而且给管道的维护抢修也带来种种障碍。

三是建筑、生活垃圾等乱堆乱放导致的占压。乱堆乱放的垃圾、建筑材料等大大加重了埋地管道的负荷。另外，一些沿地面敷设的管墩管道，后期被周边渣土等掩埋而成为地下隐蔽工程，成为管道的安全生产运行的隐患。

3、埋地管线多处与市政管道交叉，特别是与暗沟暗渠的交叉。很多石化企业的厂际管道存在与“11•22”中石化东黄输油管道泄漏爆炸事故相同的安全隐患，即厂外埋地管线与市政暗渠存在交叉，并且管线位于暗渠盖板下方，存在泄漏的油气进入市政暗渠，从而引发大面积的火灾和爆炸事故的风险。

4、埋地管道上方重型车辆碾压情况严重。厂际埋地管线穿越三级及三级以下公路时大部分采用挖沟直接埋设的方式，未采用在套管或涵洞之内敷设的方式，但后期随着管线周边区域的发展，管线穿越道路的上方逐渐出现重型车辆通行的现象，存在重车碾压管道疲劳失效泄漏的风险。

5、管道周围施工作业安全管理不到位易导致第三方施工破坏。在管道周围实施可能危及管道安全运行的施工作业的，施工单位未将施工作业方案报管道单位，未与管道单位共同制定应急预案，未采取相应的安全防护措施就盲目施工。

6、管线建设时期不同，敷设方式不同，杂乱不方便管理。大型石化企业厂际管道数量多、输送介质品种杂，建设时期不同，敷设方式有埋地敷设也有管墩、管架敷设，存在着凌乱不方便管理的现象。特别在同一路口多条管道穿越道路时，存在地下管道杂乱密集，跟排水暗渠交叉等现象。

7、石化企业厂际管道其他典型安全问题

①未按规范要求设置管道静电跨接及防雷接地装置；

②部分埋地管线未按规范要求设置管道阴极保护及防腐绝缘设施，管线易发生腐蚀泄漏；

③在通过管涵穿越道路时，管涵内未采取防止可燃、有毒气体积聚的措施；

④管线防腐、保温、隔热等设施部分脱落，未采取措施及时修复；

⑤管线、管线转角、管线与道路交叉等处未按规范要求设置标识，地面标志缺失或不明显，易于遭到车辆碰撞和人畜破坏的管段未采取保护措施；

⑥管道安全技术档案不完善，管道管理制度不健全，未配备专人负责压力管道的安全管理工作等；

⑦其他影响管道安全运行的问题。

5. 安全措施与建议

1、建议企业对厂外管道进行腐蚀监控，定期测厚，并进行管道风险辨识，确定薄弱部位，加强监控手段，增加检测频次，以确保管线的安全运行。由于厂外埋地管道外防腐极为重要，建议企业按规范要求设置阴极保护及防腐绝缘设施并增加厂外埋地管线阴极保护的测试频次。

2、沿道路敷设的石化企业厂外管线与周边设施安全间距不足，首先应该考虑从根本上解决该问题，将管道改线或者周边设施搬迁以满足安全间距的要求。当情况特殊及受条件限制时，经过安全评估以后可以考虑将道路仅作为石油化工园区的消防道使用，严格封闭管理，禁止社会车辆通行，同时设置限速标志和安全标志，严格限制过往车辆的行驶速度。

3、管线占压问题作为油气管道隐患排查的重点，应该引起管道单位的足够重视，应及时发现、报告、处理管道占压，若等建构筑物占压形成后再做处理，往往会造成隐患的形成并长时间遗留。在违法占压没有完成处置之前，制定应急预案，并广泛宣传，进行安全告知，采取切实可行的措施，确保管道周边人生财产安全。

4、新建管道设施应尽量避免与市政管线的交叉，已建管道与市政暗渠等交叉隐患严重的应进行立即整改，例如采用管架集中敷设等，做到与市政地下管道零交叉，提高安全度。在无法避免两者交叉时，两者不能在同一个平面交叉，而应该分层交叉，暗渠和油气管道应该隔开，油气管道应该铺设在市政暗渠水泥盖板的上方，保持一定的防护间距，并采取防止油气渗漏的措施。

5、对于有埋地管道上方重型车辆碾压等危害管道安全的行为，管道单位应及时予以制止，并采取相关的保护措施，例如道路设置"禁止大车通行"标志，严格限制通行车辆，道路封闭管理，提高管道设计强度或增设套管等措施提高管道安全系数。

6、厂外管道周边若进行施工，施工企业在组织施工前应提前进行考察，全面摸清项目涉及区域地下管道的分布和走向，按相关要求将施工作业方案报管道单位，双方共同制定应急预案，采取相应的安全防范措施，同时管道单位应指派专人到现场进行管道安全防护指导工作[4]，埋地管线全线埋设管道警示带，防范第三方施工破坏引发油气管道失效事故的发生。

7、对于企业已建厂际管线建设时期不同，敷设方式不同，杂乱不方便管理且与市政管线交叉等隐患严重的情况，应将上述管道统一整合，新建管架集中敷设，提高安全度。

8、其他安全措施建议

①根据《石油与石油设施雷电安全规范》GB15599，《石油化工装置防雷设计规范》GB50650等规范的要求设置管道的静电跨接及防雷接地装置。管路两端和每隔200m~300m处，分支处，拐弯处，以及所有金属件，包括护套的金属包覆层，均应有接地装置；平行敷设的金属管道，其净间距小于100mm时，应每隔30m用金属线连接。管道交叉点净距小于100mm时，其交叉点应用金属线跨接等；

②管线穿越道路的管涵洞内应采取防止可燃、有毒气体积聚的措施，首先应避免穿越涵洞处存在阀门、法兰等易发生泄漏的管道附件，并可考虑设置可燃气体报警、视频监控设施或采用砂土填实等措施；

③管道管理方应执行严格的巡检制度，要求有专职人员、定期巡检，及时发现并处置管道沿线的异常情况。配备专门人员并配备抢维修车辆、设备和工具，合理储备管道抢修物资，对管道线路进行经常维护。

④管道沿线应设置转角桩、阴极保护测试桩、交叉和警示牌等永久性标志；对易于遭到车辆碰撞和人畜破坏的管段，应设置警示牌，并应采取保护措施，例如管道沿线、管道跨越处等设置防撞隔离墩等。

⑤根据《压力管道安全技术监察规程-工业管道》TSG D0001建立管道安全技术档案并且妥善保管，建立管道安全管理制度并且保证制度的有效实施。企业的管理层还应当配备一名人员负责压力管道安全管理工作。管道数量较多的使用单位，应当设置安全管理机构或者配备专职的安全管理人员，在使用管道的车间(分厂)、装置均应当有管道的专职或者兼职安全管理人员。

⑥建议企业针对厂际管道存在的问题和隐患制定应急预案，并定期进行演练，提升公司针对厂际管道事故的应急处置能力，减小事故风险。

6. 结束语

频发的管道事故与不断上升的伤亡数字，成为伴随着中国油气管道行业高速发展的阴影。加强石化企业厂际管道的安全管理与隐患排查治理，是保证现役管道安全可靠运行必不可少和亟待开展并需长期坚持的一项工作。

腐蚀、疲劳失效以及第三方施工破坏是引发管道事故的两个最主要原因；建构筑物距离油气管道较近，管道沿线占压，油气管道与市政管道交叉，是全国油气管道线路面临的共同问题，另外管道上方重型车辆碾压，管道周围施工作业安全管理不到位易导致第三方施工破坏，管线杂乱不方便管理，管道静电跨接、防雷接地装置、管道阴极保护及防腐绝缘设施不完善等等也是目前石化企业厂际管道的一些典型安全隐患；建议石化企业及时开展厂际管道安全专项隐患排查整治，对厂际管道进行腐蚀监控，定期测厚，并进行管道风险辨识，确定薄弱部位，加强监控手段，增加检测频次，以确保石化企业厂际管道的安全运行。

7. 参考文献

[1] 中华人民共和国住房和城乡建设部,中华人民共和国国家质量监督检验检疫总局联合发布[Z].石油化工企业设计防火规范 (GB50160-2008征求意见稿).2008-12-30 .

[2] 王婷等. 城市燃气管网的完整性管理[J].油气储运，2012,31（3）：184-187.

[3] 山东省青岛市“11•22”中石化东黄输油管道泄漏爆炸特别重大事故调查报告[R]. 安全监管总局网站：东黄输油管道泄漏爆炸特别重大事故调查组，2014.

[4] 国家安全生产监督管理总局[Z]. 危险化学品输送管道安全管理规定（第43 号）. 2011-12-31.

输油管道项目对环境的影响探析

崔海峰 段潍超 卢磊

（中国石油大学（华东），山东 青岛 266580）

摘要： 文章从输油管道建设、运行两个方面介绍了输油管道项目对环境影响。在输油管道建设期主要是对自然环境和生态环境的影响，其中生态环境受到的影响更为严重；在运行期则主要是事故状态下的环境风险。在输油管道建设期应优化工程设计，将环境影响控制在最小范围；运行期间则应加强维护，及时发现隐患并进行处理。

关键词： 输油管道；生态环境；环境风险

Analysis of Environmental Impact of Oil Pipeline Projects

Abstract: Introduced the environmental impact of pipeline projects from two aspects of pipeline construction and operation. The mainly environmental impact during pipeline construction including the impact on natural environment and ecological environment, the ecological enviroment is more serious. During pipeline operation the mainly influence is environmental risk under accdient situation. In the period of pipeline construction engineering design should be optimized to control the environmental impact at minimum scope. Runtime should strengthen the maintenance, found hidden danger and handled in a timely manner.

Keywords: oil Pipeline; ecological environment; enviroment risk

随着世界经济的高速发展，对能源的需求也越来越大，管道输送作为五大运输方式之一，已成为油品输送的重要手段之一。相比传统的陆路运输和水路运输，管道运输具有如下优点[1]：运输量大，可实现连续运输；管道埋地铺设，占地面积少；全过程密闭运输，输送过程损耗小，对环境影响小；输送过程受外界环境影响小，易于自动化管理，且运输成本低。同时管道运输也具有只能单向定点运输，不如公路、水路运输灵活的缺点。管道运输经济、环境效益显著，被誉为最经济的运输方式[2]，管道运输对社会经济的发展起到巨大的推动作用，但其对环境的影响也是不容忽视的[3]。管道工程主要分为管线勘察、管线建设、管线运营这三方面，在不同的时期所产生的环境影响也不一样。本文将从管线建设、管线运营两方面分析输油管道项目对环境的影响。

1. 输油管道建设期的环境影响

输油管道建设期对环境产生影响的环节主要在站台的建设以及管线的铺设这两部分，其对环境的影响可分为对自然环境的影响和对生态环境的影响。

1.1 对自然环境的影响

管道建设期对自然环境的影响主要是对地质环境、水环境、大气环境以及施工噪声四个部分。

1.1.1 对地质环境的影响

管道一般是采用埋地敷设的方式，当受自然条件限制时，局部地段也可采用土堤埋设或地上敷设[4]。在埋地敷设施工过程中重复的开挖、回填作业，对原地表产生扰动，损毁地表植被，造成严重的水土流失。如陕京输气管道投产后的第二年，在府谷附近的黄土冲沟就出现过管道被洪水冲毁的事故。其原因主要是由于黄土遇水湿陷的特性，黄土在受水侵蚀之后极易流失，上层覆土流失后造成管道裸露、悬空，洪水一来即将管道冲断。因此，要想预防此类事情的再次发生，关键在于减少管道沿线的水土流失，保持土体的稳定。由于管道工程其自身的特殊性，在施工期造成的水土流失的特点主要有[5]：水土流失的区域沿输油管道线路呈线状分布；输油管道线路较长，沿线地质地貌复杂，水土流失防止难度增加；沿线水土流失强度分布均匀。

1.1.2 对水环境的影响

管道建设受自然条件及经济条件的约束，不可避免要穿越一些重要河流、自然保护区和饮用水水源地等。在穿过这些水体时通常采用顶管跨越的方式，顶管跨越不会改变水流的流向和流量，但也会产生一定的环境污染，主要有以下几个方面：（1）穿越施工过程中直接在水体中挖泥或冲洗施工材料而引起水质浑浊，影响水体水质；（2）施工期间作业人员的生活污水，若不经处理而直接外排将污染水体；（3）施工机械泄漏的油料直接或经雨水冲刷之后进入水体，污染水体水质。

1.1.3 对大气和声环境的影响

管道建设施工期对大气环境的影响主要是扬尘和汽车尾气，施工扬尘主要发生在车辆运输、灰土搅合以及人员活动，通过采取洒水措施能有效的降低施工现场的扬尘。而管道建设施工期对声环境的影响则主要表现为噪声污染，这些噪声主要来自施工机械及运输车辆，会对附近的敏感点如村庄和学校产生一定影响；管道施工噪声污染是短期的可逆的，当施工结束之后，其噪声污染也立即停止。

1.2 对生态环境的影响

管道建设期对沿线生态环境的影响主要包括对植物的影响、对动物的影响以及对土壤的影响三个方面。

1.2.1 对植物的影响

管道在施工过程中，管线上层覆土中植物的地上部分以及根系均被铲除，为了保障管道运行的安全，管线两侧根系较长的植物也会被清除。施工带两侧的临时占地，例如施工车辆的临时通道、挖掘土石的堆放地、工作人员的通道等，这些临时占地的地表植被地上部分会因为车辆碾压、人员踩踏等原因被破坏甚至死亡。当施工结束，土石回填至管沟，由于土壤的理化性质发生了一定的变化，使植物的生产能力下降，植被覆盖率下降。同时，输油管道在运输过程中会具有一定的温度，这将会导致沿线管道周围土壤的水分蒸发加快，从而降低了土壤水分，不利于植物的正常生长。这种现象在站场附近尤为明显。

1.2.2 对动物的影响[6]

在施工期管线割断了部分陆生动物的栖息区域、觅食范围、活动区域、迁移途径等，对动物的生存会产生一定的影响。当施工区域内无大型野生动物时，管线工程作为人工屏障对该区域内动物的活动影响不大，否则应采取一定的措施为动物迁移提供一条通道。作业过程中人员活动、施工机械、车辆的噪声会对附近动物产生一定的惊扰，作业人员对沿线附近野生动物的狩猎将会迫使动物离开管道附近区域，这种影响随施工结束而消失。长输管道工程在施工过程中不可避免的要穿越河流水体，穿越的方式主要有大开挖、隧道、定向钻等。其中大开挖方式在施工期间会对水体产生严重的扰动，对水体中的水生动物有驱赶作用，使施工区域鱼类的密度显著降低，而另外两种方式则对水生生物的影响较小。

1.2.3 对土壤的影响

在整个输油管道工程中，重复的开挖和回填作业，土壤生态环境受到的影响最直接。管道施工对土壤环境的影响主要表现为[7]：（1）破坏土壤结构，扰乱土壤耕作层；（2）上下层土壤混合，改变土体质地；（3）土壤回填之后通过碾压或掺灰固结增加土体的紧实度，但与原有的上松下紧结构相比，不利于土壤的通气透水，影响植被生长；（4）施工和生活弃渣残留与土壤中，增加了土壤中的难降解固体废物。

2. 输油管道运行期的环境影响

输油管道运行期间对环境的影响可分为正常运行和事故状态两种情况。

2.1 正常运行状态

输油管道主要由首站、中间站、末站、管线四部分组成。在管道施工结束之后，土石回填至管沟，沿线植被逐渐恢复，生态环境也逐步向原状况恢复，在正常运行期间由于管道运输采用密闭运输的方式，沿线没有污染物的排放。对环境造成影响的主要是各个站场的排污。输油管道工程各个站场排放的大气污染物主要是储罐的大、小呼吸无组织排放的烃类气体，加热锅炉排放的燃烧烟气；水污染物则主要来自站场工作人员的生活污水，油罐清洗污水和地面冲洗水；固体污染物主要来自生活废弃物，油罐检修和清管收球作业时产生的污油渣；噪声污染主要来自各种型号的传输泵、给油泵、外输泵等中间泵站产生的噪声。

2.2 事故状态

输油管道运输的介质一般为原油或成品油，属于易燃易爆物质，且在高压条件下进行输送，一旦因事故状态造成油品泄露，将会对管线周边环境（土壤、水体、植被、人）造成严重影响，若发生火灾或爆炸将会对动植物生存环境造成毁灭性的破坏。造成事故状态的因素主要有自然因素（地震、泥石流、洪水、岩溶塌陷等）和人为因素（人类活动、人为破坏）两大类。如1976年和1984年因地震和山洪爆发分别事秦京线、铁秦线管道断裂跑油，经济损失巨大，环境污染严重；2013年2月，因不法分子在九江至樟树成品油输油管道上打孔盗油造成樵舍段发生漏油事件，40t成品油泄漏至农田和小河，油品随河流流至鄱阳湖，迫使永修县自来水厂取水口紧急关闭，6万多人饮水受到影响。输油管道泄漏之后管线周围土壤、植被受到最为直接的影响，生态环境遭到破坏。当管道处于开放空间时，油品挥发，大量烃类气体进入大气，造成严重的大气污染；处于密闭空间时，油品及可燃性气体快速聚集，遇点火源时将发生剧烈燃烧及爆炸，对人民的生命及财产安全造成严重威胁。若油品泄漏未得到及时处理，油品将会进而渗入到地下水及地表水系统，对水体造成严重破坏；油品进入水体之后由于其密度比水小，将会随着河流向下游扩散，威胁下游取水口水质的安全；油膜覆盖在水面之后将会隔绝空气与水体接触，造成水体中溶解氧减少，使水体中的动植物受到损伤，进而影响水体的生态平衡；溢油进入水体数小时之后开始乳化，不仅给清理污油的工作带来困难，而且乳化油可通过食物链富集于动物体内，最终对人造成伤害；粘度较高的原油还可附着在鸟类的羽毛及鱼类的腮上，影响鸟类和水生动物的正常生活，并导致其死亡。

3. 小结

输油管道与传统的公路运输和铁路运输相比有着不可比拟的优势，但在管道建设及运行时不可避免的会对周围环境产生一定的影响。在管道建设期主要是对生态环境的影响，管道运行期间则主要是环境风险。因此，在输油管道建设期间应优化工程设计，加强管理与监督，以降低管道工程对生态环境的影响；在管道运行期间应加强对管道的维护，完善管道的风险缓解体系，及时发现安全隐患并进行处理，将事故隐患消灭在萌芽状态。

4. 参考文献

[1] 尹士武. 西部开发中管道建设与生态环境保护的关系[J]. 石油化工环境保护，2001（4）.

[2] 高育红. 最经济的运输方式--管道运输[J]. 交通与运输，2009（4）.

[3] 鲁春霞，贾慧兰，于云江. 输油气管道工程对资源与环境损害的经济评估[J]. 资源环境，2000，22（6）.

[4] GB50253 - 2003 , 输油管道工程设计规范. 北京: 中国计划出版社,2003.

[5] 高强，刘晓璐，孙厚才.输油管道工程施工期水土流失特点及防治措施[J]. 人民长江，2009，40（6）：85-87.

[6] 程金香，马俊杰，王伯铎等． 输油管线工程生态环境的响及保护对策[J]. 水土保持研究 2003，10（4）：313-315.

[7] 程金香，马俊杰，王伯铎等． 输油管道工程环境影响的分析[J]． 水土保持学报，2003，17（6），174-177.

Design and Experimental Study of the CO_2 Pipeline Relief Equipment in Industry Scale

Abstract: In order to study CO_2 pipeline leakage diffusion characteristics, an industrial scale CO_2 pipeline (256 m long, 233 mm i.d.) discharge facility was designed. A double membrane blasting device was designed to control CO_2 release. The measurement and control system based on the CompactRIO could be used to record data from pressure transducers and thermocouples installed on the long pipe. Under certain conditions, several tests of CO_2 release was carried out. The pressures and temperatures on the pipeline, the dynamic pressures and temperatures on the near field of the discharge area, and the discharge flow state were studied. The test shows that the CO_2 pipe discharge equipment in industrial scale can meet the test requirements.
Keywords: industry scale; CO_2 pipeline; relief experiment

1. 引言

CO_2的大量排放引起了温室效应，极大影响了地球气候环境。为了应对气候变化而研发推广的CO2 捕集和封存技术被视为减少碳排放的重要方法，即CCS（Carbon Capture and Sequestration）[1]。作为CCS链的中间环节，在CO_2的运输方式中，压力管道运输最实际和经济。随着CCS的发展，大量的高压CO_2管线正在投入使用[2]。美国和加拿大已有超过 6000 km的CO_2长输管道，管道直径的范围在 0.3 m到 0.7 m之间，运输压力在 10 MPa到 20 MPa之间[3]，而我国尚未开始CO_2管道建设，只有小部分试验管道[4]。

当管道存在缺陷时，如机械损伤、腐蚀、材料缺陷等，长距离高压CO_2管道运输过程中不可避免的会发泄漏事故。一方面由于CO_2气体具有重气特征，易在地势低的区域形成高浓度区，对该区域内的人体会造成严重伤害[5]。另一方面由于在CO_2泄漏过程中造成局部温度急剧下降，管道断裂韧性随之下降，进而可能引发长距离的管道延性甚至脆性断裂[6]。因此，随着CCS技术的推广，CO_2管道的安全运行技术开发和标准的建立是急需开展的研究课题。本项目组与国际知名院校和科研单位合作，由欧盟资助（FP7，CO_2PipeHaz和CO_2QUEST）设计和建造了世界首套工业规模CO_2管道泄放实验装置，并开展试验研究。

2. 工业规模CO_2管道试验装置

2.1 主体管道的结构

工业规模CO_2管道试验装置由主体管道、双膜爆破装置、加热装置、缓冲罐、气相进口、液相进口、注气管、排空阀以及数据采集系统等组成。主体管道尺寸规格为ϕ273×20 mm，长度约为 256 m，材质为 16Mn。整条管道由二十多根混凝土柱支撑，距离地面约 1 m，并建造了可以抵抗 4×105N后座力的混凝土加固装置。该试验装置采用双膜爆破装置实现管道可控泄放，并通过在泄放口安装不同口径的孔板实现不同口径下的泄放；通过加热装置给管道升温以达到不同的初始温度，同时提高管内压力；通过数据采集系统测量记录管道上的压力和温度，以及泄放区域布置的压力、温度和扩散形态。图 1 是试验管道装置图。

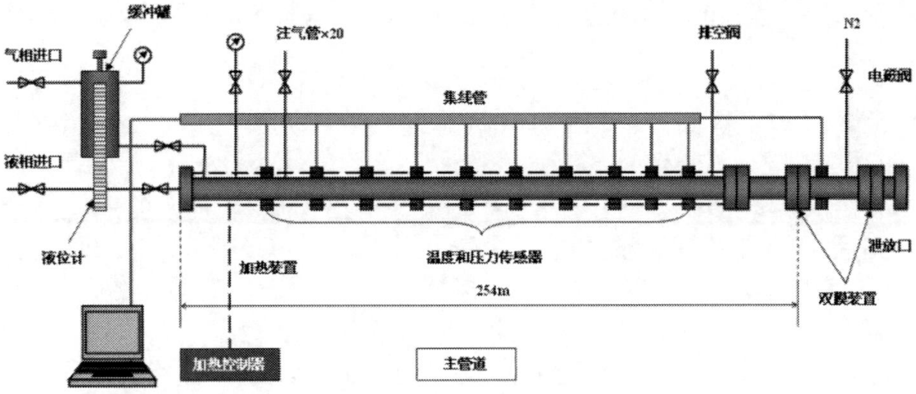

图 1 试验管道装置图

2.2 双膜爆破装置

双膜爆破装置由两个爆破片、夹持器、法兰以及管道组成，通过在两个爆破片（A和B）之间的空腔注入高压气体来控制爆破片的爆破。双膜爆破装置如图 2 所示，P_0是爆破片的爆破压力，P_1是管道内CO_2的压力，P_2是大气压，P_3是两个爆破片之间的气体压力。未爆破前，A和B之间注入一定量高压气体，当P_1-P_3<P_0以及P_3-P_2<P_0时，压力差都没有达到爆破压力P_0，这时爆破片不会爆破。当需要泄放管道中CO_2时，利用控制系统将高压气体充入空腔中，使P_3升高，当P_3-P_2>P_0时，该压力差大于爆破片B的爆破压力P_0，爆破片B立即爆破；随后，P_3急剧下降，当P_1-P_3>P_0时，爆破片A发生爆破。

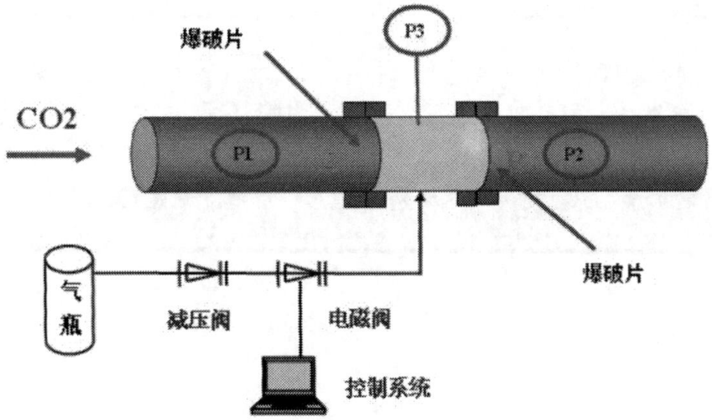

图2 双膜爆破装置示意图

2.3 数据测量方法

测量内容包括管道压力、管道顶部和底部流体温度、管壁温度、泄放区域压力和温度、泄放流量等。数据采集系统的硬件主要包括压力传感器、热电偶、差压变送器以及数据采集箱等。其中，压力传感器、热电偶和差压变送器等测量元件均满足高压低温的环境条件。数据采集主机箱采用美国 NI 公司的 CompactRIO 实时控制器，核心部件是 1 个 NI cRIO-9025 嵌入式实时控制器和 5 个 NI 9144 以太网扩展机箱，在各机箱内放置着不等数量的采集模块，总共包括 13 个 NI 9219 通用模块和 3 个 NI 9213 热电偶输入模块。数据采集系统的软件同样采用美国 NI 公司的 LabView 软件。该采集系统适合于恶劣环境条件下长距离信号输送，具备足够多的通道数量并同步测量，可保证测量过程的稳定和准确。

2.4 试验流程

CO_2管道泄放试验流程如下：（1）试验前检查装置完整性。（2）调试数据采集系统。（3）布置双膜爆破装置：包括安装爆破片和注入氮气。（4）准备就绪后，通知气体公司用槽车运输CO_2至实验现场。（5）打开槽车的气相出口阀向主管道内注入气态CO_2进行排空，一段时间后关闭排空阀，继续进气至管内压力达到 0.5～1.5 MPa时，打开槽车的液相出口阀向主管道内注入液态CO_2。（6）充装入一定量的CO_2后，停止充装；关闭所有阀门，并断开罐车的接口。（7）打开加热装置供电箱并设定温度，开始加热。（8）静置待管道内气液平衡。（9）确定实验场地内及方圆 500 m范围内没有无关人员后，实施泄放实验同时采集所有数据。（10）泄放试验完毕，关闭数据采集程序。（11）整理现场。

3. 试验条件

本项目组于 2013 年初进行了改造后的第一次工业规模CO_2管道泄放试验，试验的初始条件和环境条件如表 1。

表 1 试验初始条件和环境条件

CO_2压力 /MPa	CO_2温度 /℃	泄放口径 /mm	CO_2纯度 /%	CO_2装入量 /t	环境温度 /℃	环境压力 /MPa	环境湿度 /%
5.6	20	50	99.8%	3.4	1.5	0.10107	46.2

图 3 是CO_2相态图。如图所示R是三相点，C是临界点。RC、RB、RA分别为气液、气固、液固平衡线。在临界点C之上的物态称为超临界流体，基本上仍是气态；但密度与液体相近，有很强的溶解力，黏度小，扩散速度快[7]。本次试验的CO_2装入量为 3.4 吨，在泄放压力 5.6MPa和泄放温度 20℃下，整条管道内的液相CO_2的体积约为 2.25 m^3，气相CO_2体积约为 8.56m^3。因此本组试验的CO_2相态处于气液两相。

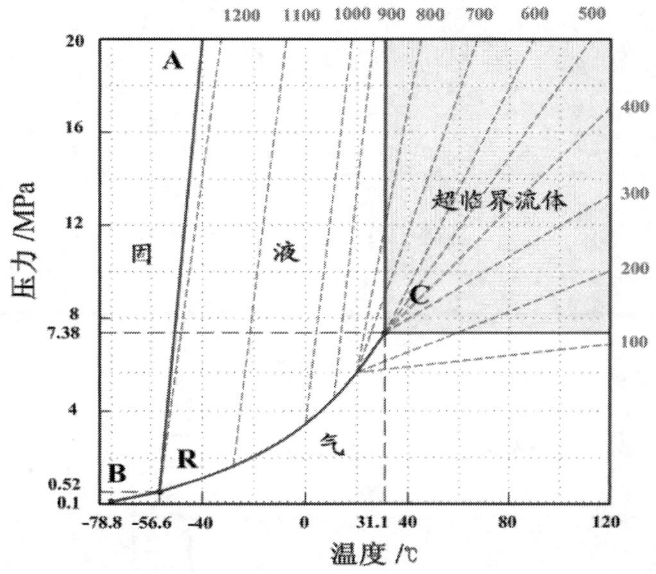

图 3 CO_2相态图

4. 结果和分析

4.1 管道上的压力和温度

图 4 是管道上压力的变化曲线。从图中可以得出，整个泄放过程中，压力从 5.6 MPa 降至大气压的时间大约为 640s，平均压降速率约为 0.01MPa/s，最大瞬时压降速率约为 1.1 MPa/s。从局部放大曲线可以看出减压波的变化。当泄放开始时，靠近泄放端的压力首先下降（如图中 13.5 m 处），减压波以该介质中的声速来传播，从距离泄放口 13.5 m 处传播至 248.6 m 处大约需要 1.2 s，其最大的减压波速度约为 198.5m/s。

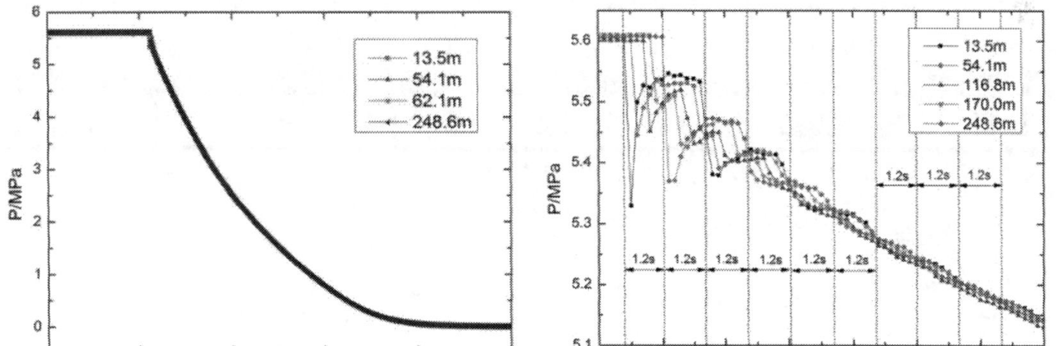

图 4 管道上压力的整体变化曲线（左图）和局部放大曲线（右图）

图 5 为管道上的温度变化曲线。从图中可以看出，当爆破片全部打开之后，管道内的流体温度开始下降，距离泄放端越远的位置的温度下降的越明显，248.6 m 处的管底流体最低温度为-50 ℃。对于相同位置（距离泄放端的距离相同），底部温度较顶部温度更低，

而管顶流体的最低温度约为-5 ℃。管壁温度主要受该处管内流体温度影响，但由于传热较慢，管壁温度下降并不明显，最大温降约为 15 ℃。

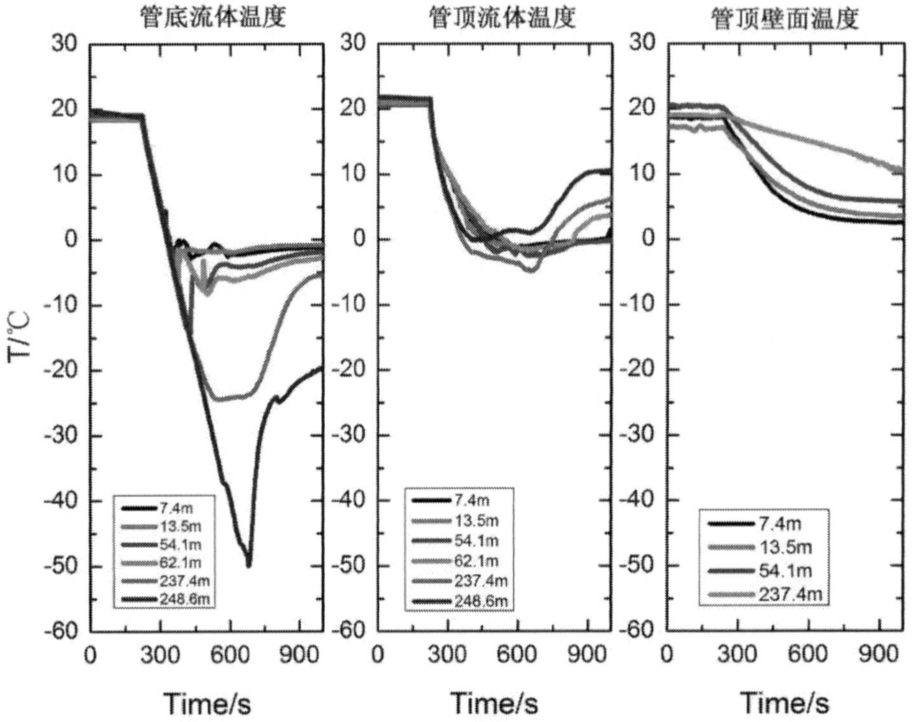

图 5 管道上温度的变化曲线

4.2 泄放近场的压力和温度

依据基于伯努利方程的皮托管原理，在距离泄放口中心 0.4 m处安装导压管并连接至差压变送器来测量该处的动压，并由此计算该处的流速和流量。图 6 为距离泄放口中心 0.4 m处的差压值和温度的变化曲线。根据下图测量到的数据和皮托管原理来计算平均流量[8]。计算得到距离泄放口中心 0.4 m处的最大流速约为 293.6 m/s，马赫数达到了 0.86 Ma，因此为跨音速流动。该处的最大流量为 17.2 kg/s。

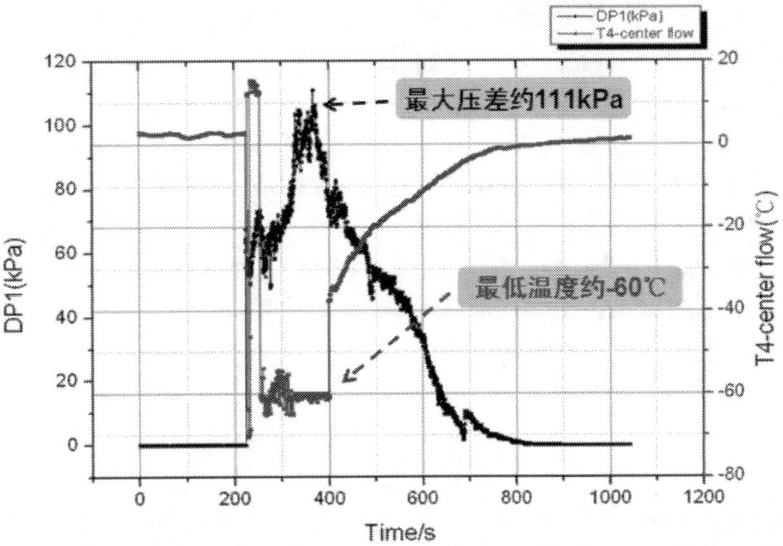

图 6 距离泄放口中心 0.4m 处的差压值和温度的变化曲线

4.3 泄放形态

图 7 为本组试验的泄放形态，白色雾状应为温度下降导致空气中的水分凝结而成的液滴，CO_2扩散的范围应较此雾区更大。管道开启后，CO_2持续泄放大约 50 s，流动形态维持不变，可定为稳定阶段。50 s之后，开始衰减且衰减速度较快；再经过大约 20 s后，扩散长度衰减为稳定阶段的一半，而后衰减开始变的缓慢，进入缓慢衰减阶段。据观察，泄放区域安全范围的长度大于 30-40 m；安全范围的宽度大于 3-4 m。

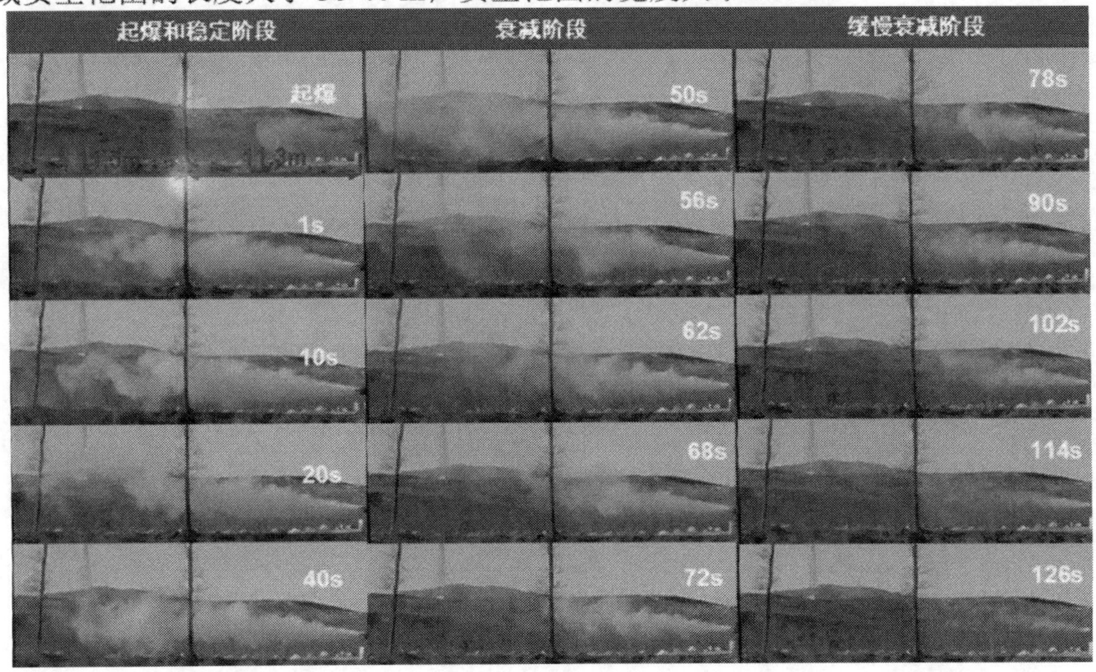

图 7 泄放形

5. 结论

本文介绍了工业规模CO_2管道泄放装置的组成，包括主体管道、双膜爆破管段、加热装置、缓冲罐、气相进口、液相进口、注气管、排空阀以及数据采集系统等；分别对双膜爆破装置和数据采集系统进行了论述。

对初始条件为泄放压力 5.6 MPa、泄放温度 20℃的气液两相CO_2进行了泄放口径 50 mm的管道泄放试验研究。试验表明：（1）整个泄放过程中泄放的时间大约为 640 s，平均压降速率约为 0.01 MPa/s，最大瞬时压降速率约为 1.1 MPa/s，减压波速度最大值约为 198.5 m/s。（2）距离泄放端越远位置的流体温度下降的越明显；对于相同位置，底部温度较顶部温度下降更低。248.6 m处的管底流体最低温度为-50 ℃。而该处管顶流体最低温度约为-5 ℃。管壁温度下降并不明显，最大温降约为 15 ℃。（3）在泄放区域近场，距离泄放口中心 0.4 m处的最大流速约为 293.6 m/s，马赫数达到了 0.86 Ma，因此为跨音速流动。该处的最大流量为 17.2 kg/s。（4）从泄放区域的CO_2形态来看，CO_2流动形态在前 50 s内维持不变，保持稳定。50 s之后，开始衰减且衰减速度较快；再过大约 20 s后，衰减开始变的缓慢，进入缓慢衰减阶段。

通过对实验装置的测试，该工业规模CO_2管道泄放装置已能用于科学研究，有望为我国工程设计标准与科学理论研究做出贡献。

6. 致谢

本文作者感谢欧盟第七框架（FP7-ENERGY-2012-1-2STAGE，协议号 309102）的资助。本文作者同样感谢本课题组成员对试验工作的支持与帮助。本文内容只代表作者的观点，欧盟对本文所持观点不负任何责任。

7. 参考文献

[1] Barker T, Bashmakov I, Bernstein L, Bogner J, Bosch P, Dave R, et al. Summary for Policymakers Drafting, Contribution of Working Group III to the Fourth Assessment Report of the Intergovernmental Panel on Climate Change. Energy, Bangkok. 2007.

[2] Mohitpour D M, Jenkins A, and Nahas G. A generalized overview of requirements for the design, construction, and operation of new pipelines for CO_2 sequestration . The Journal of Pipeline Engineering. 2008.

[3] Menso M, Corina D. Modelling of accidental from a high pressure CO_2 pipelines . Energy Procedia. 2011, 4: 2301-2307.

[4] 杜磊,湛哲,徐发龙.大规模管道长输CO_2 技术发展现状[J].油气储运. 2010, 29(2):86-89.

[5] Chow F K, Granvold P W. Modeling the effects of topography and wind on atmospheric dispersion of CO2 surface leakage at geologic carbon sequestration sites. Energy Procedia, 2009, 1(1): 1925-1931.

[6] Andrew C, David G J, Keith A. Ruptures in gas pipelines, liquid pipelines and dense phase carbon dioxide pipelines. Proceedings of the 2012 9th International Pipeline Conference. 2012.

[7] McCarty R D, Correlations for the Thermophysical Properties of Carbon Monoxide, National Institute of Standards and Technology, Boulder, CO, 1989, 0.

[8] Geankoplis C J. Transport processes and separation process principles (includes unit operations). 4th. New Jersey: Prentice Hall. 2003.

AUTHOR INDEX

AUTHOR INDEX